CARCINOGENESIS: FUNDAMENTAL MECHANISMS AND ENVIRONMENTAL EFFECTS

THE JERUSALEM SYMPOSIA ON QUANTUM CHEMISTRY AND BIOCHEMISTRY

Published by the Israel Academy of Sciences and Humanities, distributed by Academic Press (N.Y.)

1st JERUSALEM SYMPOSIUM:	*The Physicochemical Aspects of Carcinogenesis* (October 1968)
2nd JERUSALEM SYMPOSIUM:	*Quantum Aspects of Heterocyclic Compounds in Chemistry and Biochemistry* (April 1969)
3rd JERUSALEM SYMPOSIUM:	*Aromaticity, Pseudo-Aromaticity, Antiaromaticity* (April 1970)
4th JERUSALEM SYMPOSIUM:	*The Purines: Theory and Experiment* (April 1971)
5th JERUSALEM SYMPOSIUM:	*The Conformation of Biological Molecules and Polymers* (April 1972)

Published by the Israel Academy of Sciences and Humanities, distributed by D. Reidel Publishing Company (Dordrecht, Boston and London)

6th JERUSALEM SYMPOSIUM:	*Chemical and Biochemical Reactivity* (April 1973)

Published and distributed by D. Reidel Publishing Company (Dordrecht, Boston and London)

7th JERUSALEM SYMPOSIUM:	*Molecular and Quantum Pharmacology* (March/April 1974)
8th JERUSALEM SYMPOSIUM:	*Environmental Effects on Molecular Structure and Properties* (April 1975)
9th JERUSALEM SYMPOSIUM:	*Metal-Ligand Interactions in Organic Chemistry and Biochemistry* (April 1976)
10th JERUSALEM SYMPOSIUM:	*Excited States in Organic Chemistry and Biochemistry* (March 1977)
11th JERUSALEM SYMPOSIUM:	*Nuclear Magnetic Resonance Spectroscopy in Molecular Biology* (April 1978)
12th JERUSALEM SYMPOSIUM:	*Catalysis in Chemistry and Biochemistry Theory and Experiment* (April 1979)

VOLUME 13

CARCINOGENESIS: FUNDAMENTAL MECHANISMS AND ENVIRONMENTAL EFFECTS

PROCEEDINGS OF THE THIRTEENTH JERUSALEM SYMPOSIUM ON QUANTUM CHEMISTRY AND BIOCHEMISTRY HELD IN JERUSALEM, ISRAEL, APRIL 28 – MAY 2, 1980

Edited by

BERNARD PULLMAN
Université Pierre et Marie Curie (Paris VI)
Institut de Biologie Physico-Chimique
(Fondation Edmond de Rothschild) Paris, France

PAUL O. P. TS'O
The Johns Hopkins University
Baltimore, Maryland, U.S.A.

and

HARRY GELBOIN
National Cancer Institute
Bethesda, Maryland, U.S.A.

D. REIDEL PUBLISHING COMPANY

DORDRECHT : HOLLAND / BOSTON : U.S.A.

LONDON : ENGLAND

Library of Congress Cataloging in Publication Data

CIP

Jerusalem Symposium on Quantum Chemistry and Biochemistry, 13th, 1980.
Carcinogenesis, fundamental mechanisms and environmental effects.

(The Jerusalem symposia on quantum chemistry and biochemistry ; v. 13)
Includes indexes.
1. Carcinogenesis–Congresses. 2. Environmentally induced diseases–Congresses. I. Pullman, Bernard, 1919- II. Ts'o, Paul On Pong, 1928- III. Gelboin, Harry Victor, 1929- IV. Title. V. Series: Jerusalem symposia on quantum chemistry and biochemistry ; v. 13. [DNLM: 1. Carcinogens–Pharma-codynamics–Congresses. W3 JE56 v. 13 1980 / QZ 202 J56 1980c]
RC268.5.J47 1980 616.99'4071 80-20725
ISBN-13: 978-94-009-9106-4 e-ISBN-13: 978-94-009-9104-0
DOI: 10.1007/978-94-009-9104-0

Published by D. Reidel Publishing Company,
P.O. Box 17, 3300 AA Dordrecht, Holland.

Sold and distributed in the U.S.A. and Canada
by Kluwer Boston Inc.,
190 Old Derby Street, Hingham, MA 02043, U.S.A.

In all other countries, sold and distributed
by Kluwer Academic Publishers Group,
P.O. Box 322, 3300 AH Dordrecht, Holland.

D. Reidel Publishing Company is a member of the Kluwer Group.

TABLE OF CONTENTS

PREFACE

The 13th Jerusalem Symposium on Quantum Chemistry and Biochemistry was dedicated to the memory of Ernst David Bergmann, one of the founders of this series of Symposia, at the occasion of the 5th anniversary of his death. The opening session was honored by the presence of H. E. Yzhak Navon, President of the State of Israel and of Baron Edmond de Rothschild whose generous and constant support makes these Symposia possible. To both these distinguished guests we address the expression of our deep gratitude.

Our thanks are also due to the Israel Academy of Sciences and Humanities, in particular to its President Professor Aryeh Dvoretzky and to the Hebrew University of Jerusalem for their hospitality and helpful association.

I had the privilege this time to join efforts in the organization of this Symposium with two distinguished American colleagues, Prof. P. O. P. Ts'o from Johns Hopkins University and Dr. Harry Gelboin from the National Institute for Cancer Research. I wish to thank them for their invaluable help and the two Institutions which they represent for important financial support.

The subject of the 13th Jerusalem Symposium brings us back to that of the 1st Symposium held in 1967. The comparison of the two Proceedings enables us to evaluate the important developments which occurred in this field of research during the last 13 years.

Bernard PULLMAN

CARCINOGENICITY OF POLYCYCLIC AROMATIC HYDROCARBONS: THE BAY-REGION THEORY

D. M. Jerina, J. M. Sayer, D. R. Thakker, and H. Yagi
NIH, NIAMDD, Bethesda, MD 20205

W. Levin, A. W. Wood and A. H. Conney
Dept. of Biochemistry and Drug Metabolism
Hoffmann-La Roche Inc., Nutley, N.J. 07110

ABSTRACT

The bay-region theory proposes that diol epoxides will have high biological activity when they are on angular benzo-rings and the epoxide group forms part of a bay-region of a carcinogenic polycyclic aromatic hydrocarbon. In general, predicted chemical reactivity for positional isomers of benzo-ring diol epoxides of individual hydrocarbons has been paralleled by their respective mutagenicity and/or tumorigenicity. Although relative reactivity of bay-region diol epoxides from a series of different hydrocarbons provides a fairly respectable index of tumorigenicity for the series, metabolic and stereochemical factors are also critically important determinants of biological activity. Several hydrocarbons have now been tested, either directly or indirectly, and have been found to meet the predictions of the theory.

INTRODUCTION

By the end of 1975 several lines of evidence had begun to point to the conclusion that a benzo[a]pyrene 7,8-diol-9,10-epoxide would be the first ultimate carcinogen to be identified for any member of the polycyclic aromatic hydrocarbon class of carcinogens (reviewed in ref. 1 and 2). Metabolically, two diastereomers of the diol epoxide are possible in which the benzylic hydroxyl group is either cis (isomer-1) or trans (isomer-2) to the epoxide oxygen (Figure 1). In an attempt to formulate a general theory to explain and predict carcinogenicity or the lack of it for a given hydrocarbon based on the concept of diol epoxides, we questioned whether the high chemical reactivity of the benzo[a]pyrene 7,8-diol-9,10-epoxides might be responsible for their high biological activity. This reasoning led to the conclusion that, for a given polycyclic hydrocarbon, benzylic positions on saturated, benzo-rings would be most reactive when such positions formed part of a "bay-region" of the hydrocarbon. The simplest example of a bay-region is the area between the 4- and 5-positions of phenanthrene. Such sterically hindered regions result when a benzo-ring is angularly fused to a polycyclic aromatic hydrocarbon.

B. Pullman, P. O. P. Ts'o and H. Gelboin (eds.), Carcinogenesis: Fundamental Mechanisms and Environmental Effects, 1–12.

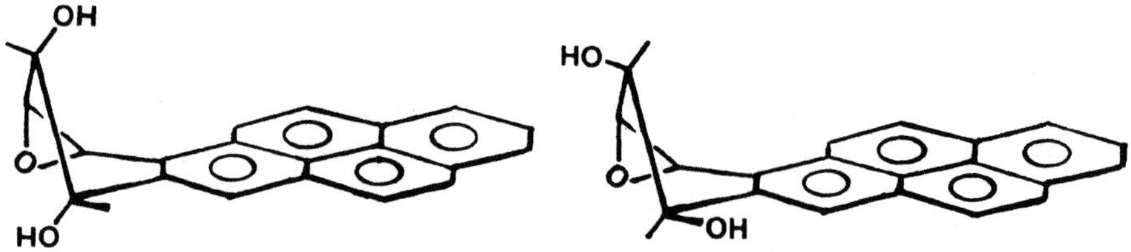

(+)-Diol Epoxide-1 (+)-Diol Epoxide-2

Figure 1. Benzo[a]pyrene 7,8-diol-9,10-epoxides. Isomer-1 prefers the conformation in which the hydroxyl groups are quasi-diaxial whereas isomer-2 prefers the conformation in which the hydroxyl groups are quasi-diequatorial[25]. Absolute stereochemistry for the enantiomers of the principal diastereomers formed metabolically is as shown[24].

Through application of Dewar perturbational molecular orbital calculations[3] to the opening of benzo-ring diol epoxides to triol carbonium ions (Figure 2), it was possible to predict that the positional isomer of a diol epoxide in which the epoxide group formed part of a bay-region of the hydrocarbon would be more reactive than other isomeric diol epoxides possible for the hydrocarbon. Thus we proposed that bay-region diol epoxides would be important ultimate carcinogens *when they were metabolically formed* from tumorigenic polycyclic aromatic hydrocarbons[4-8].

Figure 2. Application of the Dewar perturbational model to predict the ease with which a diol epoxide opens to a triol carbonium ion. Larger values of ΔE_{deloc} in units of β imply greater reactivity.

As an additional use of the molecular orbital calculations, we sought to determine whether or not predicted reactivity of the bay-region diol epoxide of a given hydrocarbon could be used to rank the parent hydrocarbon for its carcinogenic activity. In view of the complexity of chemical carcinogenesis, such a ranking could not be expected to be perfect, and numerous exceptions were noted[5]. Nonetheless, a respectable ranking (Table 1) of the parent hydrocarbons for tumorigenic activity, based on the predicted reactivity of benzylic bay-region positions

TABLE 1

Comparison of Predicted Ease of Carbonium-ion Formation from Bay-region Diol Epoxides with Approximate Tumorigenicity of the Parent Hydrocarbons.

Diol Epoxide	$\Delta E_{deloc}/\beta$	Approximate Relative Carcinogenicity (Parent Hydrocarbon)
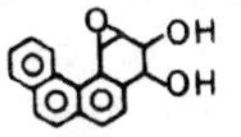	0.600	+
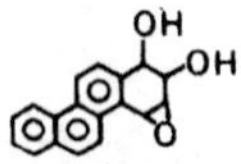	0.640	+
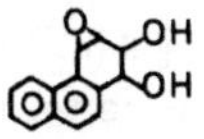	0.658	–
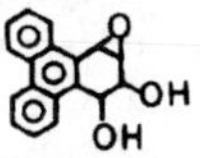	0.664	–
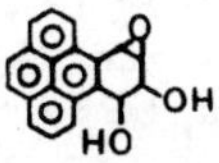	0.714	–
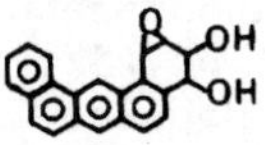	0.722	+
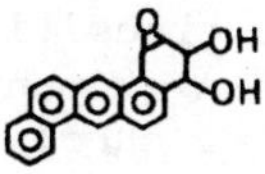	0.738	+++
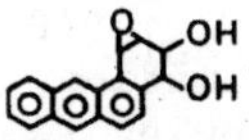	0.766	+
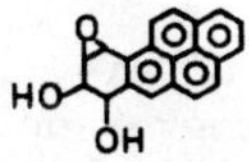	0.794	++++
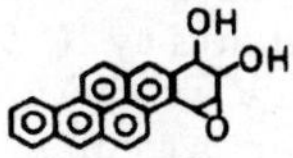	0.845	++++
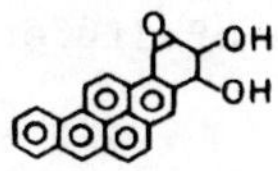	0.870	++++

Data are from ref. 6 and citations therein.

on saturated angular rings, was generated. Prior to our correlations (cf. ref. 6), the important K-region calculations of the Pullmans[9] had been the dominant theory in the field for over twenty years.

The past four years have proven the bay-region theory to be exceptionally useful in predicting the structures of ultimate carcinogens from several different polycyclic aromatic hydrocarbons. As much of this work has recently been reviewed[10], the purpose of the present chapter will be to give some perspective on the overall area and to discuss the role of stereochemical factors, an aspect of hydrocarbon-induced carcinogenicity which the bay-region theory made no attempt to take into account.

COMPARISON OF PREDICTED VERSUS OBSERVED CHEMICAL REACTIVITY

Since it is likely that the biological activities of the diol epoxides of polycyclic aromatic hydrocarbons are a result of covalent binding to cellular constituents, understanding of simple chemical models for the electrophilic reactivity of the bay-region diol epoxides provides necessary background for the elucidation of their biochemical mechanism of action. One such chemical model is solvolysis in aqueous solution. Solvolyses of the bay-region diol epoxides and other similar model compounds[11-13] all exhibit both a hydrogen-ion catalyzed (k_{H^+}) and a spontaneous (k_o) reaction pathway, with rate = ($k_{H^+}a_{H^+}$ + k_o)[epoxide]. To test the applicability of the Dewar calculations to observed reactivity, free-energy plots of log k_o for the spontaneous reaction, in 10% dioxane-water, ionic strength 0.1M ($NaClO_4$) at 25°, versus $\Delta E_{deloc}/\beta$ were constructed and were found to produce two separate lines for the isomer-1 and isomer-2 series (Figure 3). For the four sets of compounds shown, an acceptable correlation exists between transition-state (k_o) and carbonium ion stabilization ($\Delta E_{deloc}/\beta$), in agreement with the idea that this calculated value can be used to predict chemical reactivity when partial positive charge develops at carbon in the transition state[3]. Although the linear correlation found between $\Delta E_{deloc}/\beta$ and log k_o for these compounds is satisfying as a first approximation, it represents somewhat of an over-simplification for the following reasons:

(1) The products of the spontaneous "solvolysis" reactions are different for the different compounds: for isomer-2 only tetraols are formed, whereas for the three least reactive members of the isomer-1 series, tetraol products along with a large quantity (>50%) of the dihydroxy ketone are observed (ref. 12 and this study). With the more reactive benzo[a]pyrene diol epoxide-1, smaller amounts of the ketone are formed, and the predominant products are the tetraols[11]. Such changes in product distrubution are suggestive of differences in their mechanisms of reaction.

(2) Isomer-1 of benzo[c]phenanthrene diol epoxide has a much greater preference for the conformation (see Figure 6) with the hydroxyl

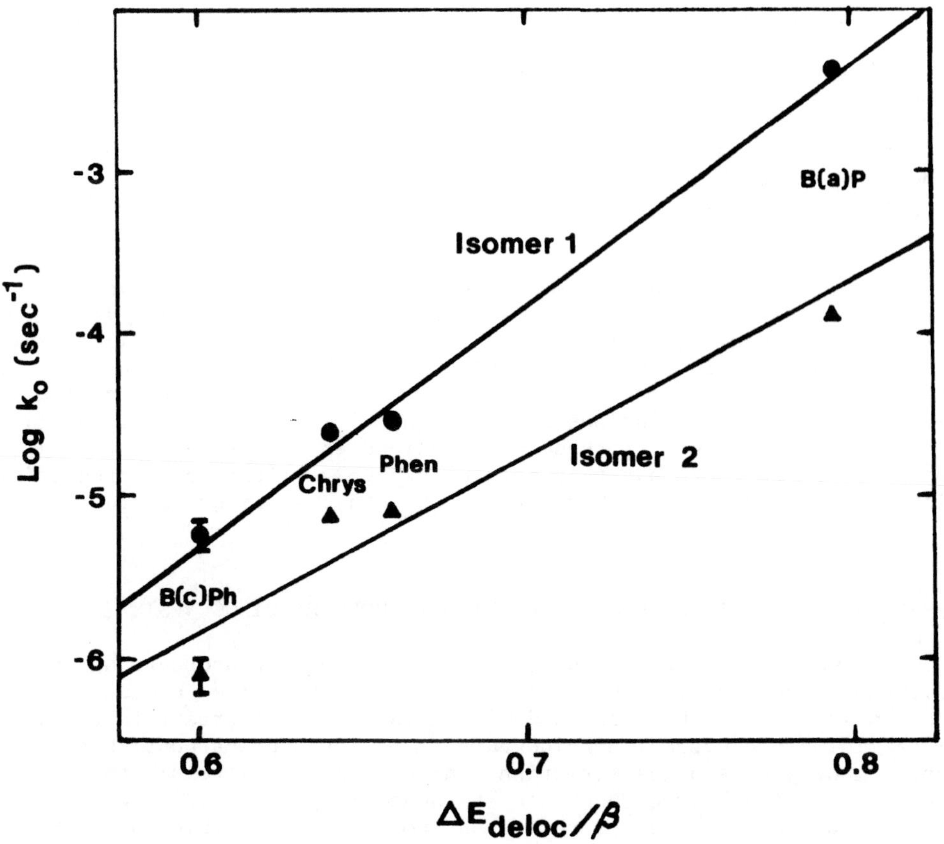

Figure 3. Correlation of log k_o for spontaneous solvolysis of bay-region diol epoxides of the isomer-1 (•) and isomer-2 (▲) series, in 10% dioxane-water, ionic strength 0.1M ($NaClO_4$) at 25°C, with $\Delta E_{deloc}/\beta$ for carbonium ion formation from the parent hydrocarbons. Values of k_o for benzo[a]pyrene diol epoxides are from ref. 11. Values of k_o for phenanthrene diol epoxides in water (ref. 12) have been corrected by a factor of 4.2 for the solvent effect of 10% dioxane.

groups quasi-diequatorial when compared with other members of the isomer-1 series, a factor which might well have been expected to affect its reactivity relative to that of the other diol epoxides[12]. Assessment of the contributions of such conformational effects to chemical reactivity, as well as of the correlation between rates of solvolysis and product distributions will require a more extensive series of compounds than that available to date.

Correlations of log k_{H^+} with $\Delta E_{deloc}/\beta$ show a similar trend of increasing reactivity with increasing carbonium ion stabilization for chrysene, phenanthrene and benzo[a]pyrene diol epoxides. In contrast to the data for the spontaneous pathway (Figure 3), both isomers of benzo[c]phenanthrene diol epoxide (expected to be the least reactive of the compounds) show equal or greater reactivity toward hydrogen-ion catalyzed solvolysis (unpublished observations) than the bay-region diol epoxides of phenanthrene and chrysene. Apparently factors other than the stabilization of a positive charge at carbon are able to assume major importance in the acid catalyzed as opposed to the uncatalyzed reaction.

In summary, based on the limited data presently available, the extent of stabilization of the bay-region benzylic carbonium ion provides a reasonable empirical means of predicting the chemical reactivity of diol epoxides in spontaneous solvolysis reactions. The influence of factors, such as conformation, that may lead to deviations from the linear free energy relationships of Figure 3 remains to be explored. An additional feature of these calculations which deserves particular mention is the fact that they have been highly effective in predicting inherent mutagenicity of benzo-ring tetrahydro epoxides toward bacterial cells[29].

EFFECTS OF SUBSTITUENTS ON METABOLISM INDUCED CARCINOGENESIS

In many tumor experiments, the dose of applied hydrocarbon is completely metabolized. Thus, the extent to which a given hydrocarbon is metabolized to its bay-region diol epoxide is an important factor in determining the overall carcinogenicity of the hydrocarbon. Metabolic control of diol epoxide formation can occur at two levels: the percent conversion of a hydrocarbon into its benzo-ring dihydrodiol with a bay-region double bond and the extent to which this dihydrodiol is converted into its bay-region diol epoxides.

The regiospecificity of the cytochromes P-450 and the efficiency of epoxide hydrolase determine the extent to which a particular dihydrodiol of a given hydrocarbon is formed. Thus, although the bay-region diol epoxides of benzo[a]anthracene are predicted to be more reactive than those of dibenzo[a,h]anthracene (Table 1), the latter hydrocarbon is known to be more carcinogenic. At least part of the reduced activity for benzo[a]anthracene is thought to stem from the fact that only 1-3% of its total metabolism results in the formation of its 3,4-dihydrodiol with a bay-region 1,2-double bond, whereas >25% of the total metabolism of dibenzo[a,h]anthracene is represented by such a metabolite[14]. Similarly, very low conversion of benzo[e]pyrene to its 9,10-dihydrodiol[16] is a factor which contributes to the inactivity of this hydrocarbon as a carcinogen. We had previously drawn attention to the fact that substituents on the critical angular benzo-ring of a carcinogenic hydrocarbon result in a marked decrease in activity[4-8]. Since it is widely recognized[22] that substituents on formal aromatic double bonds either block or greatly reduce the rate of epoxidation by mono-oxygenases at such positions, reduced rates of dihydrodiol formation are easily explained when such substituents are present. Thus both the regiospecificity of the cytochrome P-450 system as well as the presence of substituents control the extent to which a particular hydrocarbon is converted into a specific dihydrodiol.

Although the double bond of a benzo-ring dihydrodiol is clearly the easiest place to form an epoxide from a thermodynamic standpoint, this is not necessarily the primary site at which the cytochromes P-450 metabolize dihydrodiols. Studies of the metabolism of benzo[a]pyrene 9,10-dihydrodiol[17] and benzo[e]pyrene 9,10-dihydrodiol[18] have shown that very little benzo-ring diol epoxide is formed (Figure 4). In each case,

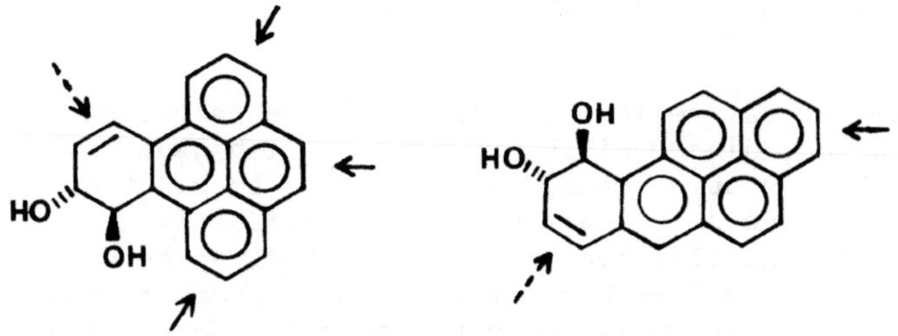

Figure 4. The quasi-diaxial dihydrodiols of benzo[e]pyrene (left) and benzo[a]pyrene (right) are primarily metabolized (dark arrows) at sites other than the benzo-ring double bond (dashed arrows).

steric factors cause these bay-region dihydrodiols to adopt the unusual quasi-diaxial conformation[19]. In this conformation, dihydrodiols are more polar than their quasi-diequatorial counterparts[20]. These observations aided Sims and coworkers[21] in their identification of metabolically formed dihydrodiols based in part on their chromatographic elution order. These workers also confirmed that a methyl group adjacent to the diol group of a dihydrodiol would act sterically like a bay-region and cause the dihydrodiol to exist mainly in the quasi-diaxial conformation. In formulating the bay-region theory[4], we had drawn attention to the fact that alkyl and halogen substituents adjacent to the critical angular, benzo-ring, but not in the bay-region (e.g., the 5-position in benzo[a]anthracene), decreased tumorigenic activity relative to the parent hydrocarbon. Although the fundamental basis of this "peri-effect" was not entirely clear, Hecht and coworkers[23] have suggested that a peri fluoro substituent reduces the extent to which the adjacent dihydrodiol is formed. With alkyl substituents adjacent to the dihydrodiol such as the 2- or 5-position (Figure 5) of the benzo[a]anthracene 3,4-dihydrodiol, the basis for the peri-effect is more apparent since any dihydrodiol which does form will be mainly quasi-diaxial. Based on our previous studies[17,18], poor conversion of such quasi-diaxial

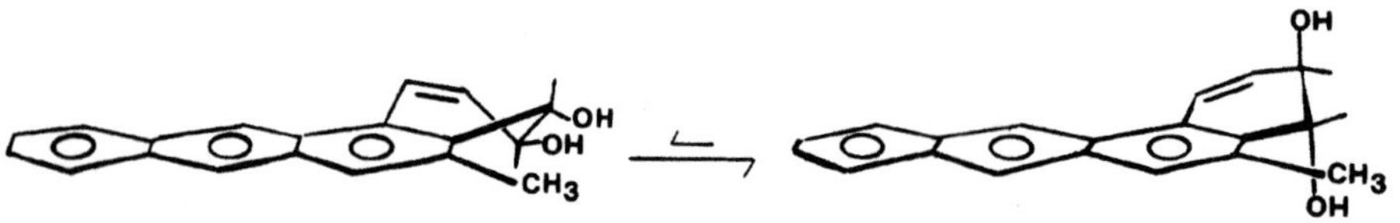

Figure 5. The presence of a bulky 5-methyl substituent on the 3,4-dihydrodiol of benzo[a]anthracene causes the substituted dihydrodiol to adopt the quasi-diaxial conformation rather than maintain the predominantly quasi-diequatorial conformation of the unsubstituted dihydrodiol.

dihydrodiols to bay-region diol epoxides can be expected. In cases where benzo-ring dihydrodiols are formed which have bay-region substituents on their double bonds, the effect on epoxidation by cytochrome P-450 is yet unknown. Even when such substituents are absent and when quasi-diequatorial dihydrodiols are compared, extents of conversion to bay-region diol epoxides vary greatly. For example, benzo[a]pyrene 7,8-dihydrodiol is extensively converted into its bay-region diol epoxides[24] by liver microsomes from 3-methylcholanthrene-treated rats while the 3,4-dihydrodiol of benzo[a]anthracene is only poorly converted into such metabolites[15].

ABSOLUTE STEREOCHEMISTRY

When enantiomers are taken into consideration, four different benzo[a]pyrene 7,8-diol-9,10-epoxides are metabolically possible. Remarkably, only (+)-diol epoxide-2 (Figure 1) shows high tumorigenicity when injected into newborn mice. In addition, (+)-diol epoxide-2 is the predominant isomer formed from benzo[a]pyrene by liver microsomal enzymes. The role of the absolute stereochemistry of benzo[a]pyrene metabolites in the tumorigenicity and mutagenicity of benzo[a]pyrene has been reviewed[24]. Recent results[24] (and unpublished observations) indicate that an exactly parallel situation exists for the hydrocarbon benzo[a]anthracene. Stereoselective metabolism occurs to form predominantly (+)-diol epoxide-2 which is far more tumorigenic than the other three metabolically possible diol epoxides. Notably when the bay-regions and the angular benzo-rings of the two hydrocarbons are aligned, their strongly tumorigenic diol epoxides are completely superimposible. One possible explanation is that there is a "receptor site" for ultimate carcinogens from the hydrocarbons which has extremely high stereochemical requirements for covalent binding. Studies of other hydrocarbons are in progress.

CONFORMATION AND CARCINOGENICITY

In the absence of unusual steric or electronic factors, bay-region diol epoxides in the isomer-1 series tend to prefer the conformation in which the hydroxyl groups are quasi-diaxial while those in the isomer-2 series prefer the conformation in which the hydroxyl groups are quasi-diequatorial (cf. Figure 1). Benzo[e]pyrene 9,10-diol-11,12-epoxide isomer-2, whose benzylic hydroxyl group is located in a sterically hindered bay-region, is forced by this hindrance to adopt the conformation, unusual in the isomer-2 series, with quasi-diaxial hydroxyl groups[25]. This compound is a very weak mutagen and tumorigen[26] despite its relatively high value of $\Delta E_{deloc}/\beta$. Since we were unaware of any examples of diol epoxide-1 diastereomers with strong tumorigenic activity, we had wondered whether this might be due to the preferred quasi-diaxial nature of their hydroxyl groups. To probe this possibility, we sought to prepare a diol epoxide-1 diastereomer for which the diequatorial conformation was preferred. Inspection of molecular models suggested that the high steric hindrance in the bay-region of benzo[c]phenanthrene might well result in such a conformation. When these two bay-region diol epoxides were synthesized, their nuclear magnetic resonance spectra indicated that both the diol epoxides-1 and 2 are mainly in the desired conformation with quasi-diequatorial hydroxyl groups (Figure 6). Evaluation of their mutagenicity toward bacterial and mammalian cells shows both diastereomers to be highly active[27]. Furthermore, when examined as tumor-initiators on mouse skin[28], both diastereomers are of comparable activity (Table 2). Even more striking, however, is the fact that both diastereomers are more active than any other diol epoxide yet tested on skin. Comparative data for benzo[a]anthracene, whose diol epoxides were the most active heretofore, are shown in Table 2. The three- to four-fold higher tumorigenic activity of the benzo[c]phenanthrene diol epoxides, relative to the more active benzo[a]anthracene diol epoxide-2, may actually underestimate their activity because the doses used were either at or above the top of the linear dose response range for the benzo[c]phenanthrene compounds. Both the low values of $\Delta E_{deloc}/\beta$ and the low rates of solvolysis (pointed out earlier) for the benzo[c]-phenanthrene diol epoxides gave us little reason to anticipate such a marked tumorigenic response.

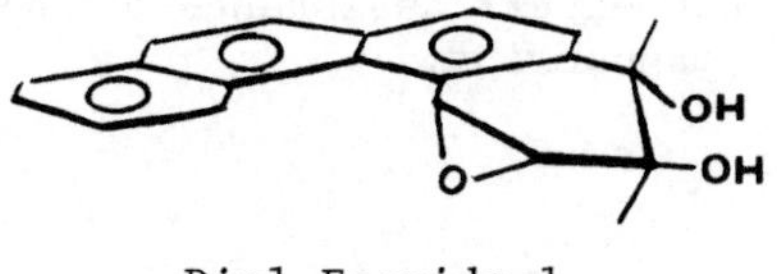

Diol Epoxide-1

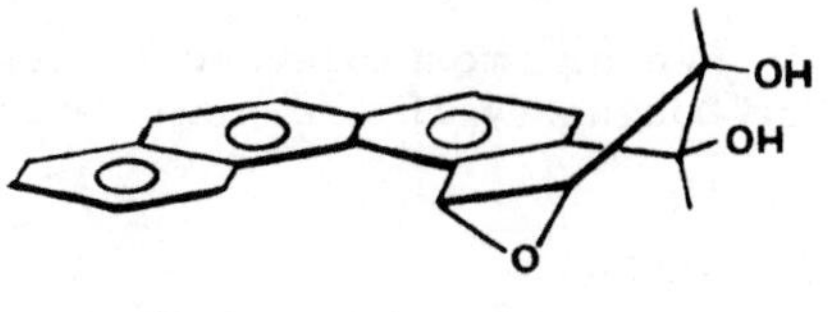

Diol Epoxide-2

Figure 6. Preferred conformation of bay-region diol epoxides of the highly hindered hydrocarbon benzo[c]phenanthrene. Part of the high tumorigenicity of the diol epoxide-1 isomer may be related to its unusual quasi-diequatorial conformation for the hydroxyl groups.

TABLE 2

Comparative Tumorigenicity of Bay-Region Diol Epoxides from Benzo[a]anthracene and Benzo[c]phenanthrene by Initiation-Promotion on Mouse Skin.

Compound (0.4 μmoles)	Percent of Mice with Tumors	Papillomas per Mouse
Control	3	0.03
Benzo[a]anthracene	7	0.07
3,4-diol-1,2-epoxide-1	43	0.60
3,4-diol-1,2-epoxide-2	70	1.90
Benzo[c]phenanthrene	17	0.17
3,4-diol-1,2-epoxide-1	87	6.47
3,4-diol-1,2-epoxide-2	80	7.13

Compounds were applied once on the backs of CD-1 mice followed, starting one week later, by twice weekly applications of 12-O-tetradecanoylphorbol-13-acetate for twenty weeks.

CONCLUSION

In every case examined to date, the bay-region theory has correctly predicted which positional isomer of a diol epoxide from a given hydrocarbon would have the highest tumorigenicity. However, quantitative predictions of relative activity for a series of hydrocarbons, although useful, can be misleading. We have now found that stereochemical factors, such as relative configuration, absolute configuration and conformation, play extremely important roles in the tumorigenesis of the polycyclic aromatic hydrocarbon class of carcinogens. Refinements of the bay-region theory will have to take these factors into consideration.

ACKNOWLEDGEMENT

We are most pleased to acknowledge the expert assistance of Mrs. Dot Dougherty in the preparation of this manuscript.

REFERENCES

1. Conney, A. H., Levin, W., Wood, A. W., Yagi, H., Lehr, R. E., and Jerina, D. M. in "Advances in Pharmacology and Therapeutics (Proceedings of the Seventh International Congress of Pharmacology, Paris, 1978), Vol. 9, Toxicology" (Y. Cohen, Ed.) Pergamon Press, Oxford, 1979, pp. 41-52.

2. Levin, W., Wood, A. W., Wislocki, P. G., Chang, R. L., Kapitulnik, J., Mah, H. D., Yagi, H., Jerina, D. M. and Conney, A. H. in "Polycyclic Hydrocarbons and Cancer: Vol. 1, Environment, Chemistry and Metabolism" (H. V. Gelboin and P. O. P. Ts'o, Eds.) Academic Press, New York, 1978, pp. 189-202.
3. Dewar, M. J. S.: "The Molecular Orbital Theory of Organic Chemistry", McGraw-Hill, New York, 1969, pp. 214-217, 304-306.
4. Jerina, D. M. and Daly, J. W. in "Drug Metabolism-from Microbe to Man" (D. V. Parke and R. L. Smith, Eds.) Taylor and Francis Ltd., London, 1976, pp. 13-32.
5. Jerina, D. M., Lehr, R. E., Yagi, H., Hernandez, O., Dansette, P. M., Wislocki, P. G., Wood, A. W., Chang, R. L., Levin, W., and Conney, A. H. in "In Vitro Metabolic Activation in Mutagenesis Testing" (F. J. de Serres, J. R. Fouts, J. R. Bend, and R. M. Philpot, Eds.) Elsevier/North-Holland Biomedical Press, Amsterdam, 1976, pp. 159-177.
6. Jerina, D. M. and Lehr, R. E. in "Microsomes and Drug Oxidations" (V. Ullrich, I. Roots, A. Hildebrandt, R. W. Estabrook, and A. H. Conney, Eds.) Pergamon Press, Oxford, 1977, pp. 709-720.
7. Jerina, D. M., Lehr, R., Schaefer-Ridder, M., Yagi, H., Karle, J. M., Thakker, D. R., Wood, A. W., Lu, A. Y. H., Ryan, D., West, S., Levin, W. and Conney, A. H. in "Origins of Human Cancer" (H. H. Hiatt, J. D. Watson and J. A. Winsten, Eds.) Cold Spring Harbor Laboratory, Cold Spring Harbor, N. Y., 1977, pp. 639-658.
8. Jerina, D. M., Yagi, H., Lehr, R. E., Thakker, D. R., Schaefer-Ridder, M., Karle, J. M., Levin, W., Wood, A. W., Chang, R. L. and Conney, A. H. in "Polycyclic Hydrocarbons and Cancer: Vol. 1, Environment, Chemistry and Metabolism" (H. V. Gelboin and P. O. P. Ts'o, Eds.) Academic Press, New York, 1978, pp. 173-188.
9. Pullman, A. and Pullman, B.: 1955, Adv. Cancer Res. 3, pp. 117-169.
10. Nordqvist, M., Thakker, D. R., Yagi, H., Lehr, R. E., Wood, A. W., Levin, W., Conney, A. H. and Jerina, D. M. in "Molecular Basis of Environmental Toxicity" (R. S. Bhatnagar, Ed.) Ann Arbor Science Publishers, Inc., Ann Arbor, Mich., 1980, pp. 329-357.
11. Whalen, D. L., Montemarano, J. A., Thakker, D. R., Yagi, H., and Jerina, D. M.: 1977, J. Am. Chem. Soc. 99, pp. 5522-5524.
12. Whalen, D. L., Ross, A. M., Yagi, H., Karle, J. M. and Jerina, D. M.: 1978, J. Am. Chem. Soc. 100, pp. 5218-5221.
13. Rogers, D. Z. and Bruice, T. C.: 1979, J. Am. Chem. Soc. 101, pp. 4713-4719. Becker, A. R., Janusz, J. M., and Bruice, T. C.: 1979, J. Am. Chem. Soc. 101, pp. 5679-5687.
14. Nordqvist, M., Thakker, D. R., Levin, W., Yagi, H., Ryan, D. E., Thomas, P. E., Conney, A. H. and Jerina, D. M.: 1979, Mol. Pharmacol. 16, pp. 643-655.
15. Thakker, D. R., Levin, W., Yagi, H., Tada, M., Conney, A. H., and Jerina, D. M. in "Fourth International Symposium on Polynuclear Aromatic Hydrocarbons" (A. Bjorseth and A. J. Dennis, Eds.) Battelle Press, Columbus, Ohio, in press.
16. MacLeod, M. C., Levin, W., Conney, A. H., Lehr, R. E., Mansfield, B. K., Jerina, D. M. and Selkirk, J. K.: 1980, Carcinogenesis 1, pp. 165-173.

17. Thakker, D. R., Yagi, H., Lehr, R. E., Levin, W., Buening, M., Lu, A. Y. H., Chang, R. L., Wood, A. W., Conney, A. H. and Jerina, D. M.: 1978, Mol. Pharmacol. 14, pp. 502-513.
18. Wood, A. W., Levin, W., Thakker, D. R., Yagi, H., Chang, R. L., Ryan, D. E., Thomas, P. E., Dansette, P. M., Whittaker, N., Turujman, S., Lehr, R. E., Kumar, S., Jerina, D. M. and Conney, A. H.: 1979, J. Biol. Chem. 254, pp. 4408-4415.
19. Jerina, D. M., Selander, H., Yagi, H., Wells, M. C., Davey, J. F., Mahadevan, V. and Gibson, D. T.: 1976, J. Am. Chem. Soc. 98, pp. 5988-5996. Lehr, R. E., Taylor, C. W., Kumar, S., Mah, H. D. and Jerina, D. M: 1978, J. Org. Chem. 43, pp. 3462-3466.
20. Thakker, D. R., Yagi, H. and Jerina, D. M. in "Methods in Enzymology, Vol. 52, Biomembranes (Part C)" (S. Fleischer and L. Packer, Eds.) Academic Press, New York, 1978, pp. 279-296.
21. Tierney, B., Burden, P., Hewer, A., Ribeiro, O., Walsh, C., Rattle, H., Grover, P. L. and Sims, B.: 1979, J. Chromatogr. 176, pp. 329-335.
22. Daly, J. W., Jerina, D. M. and Witkop, B.: 1972, Experientia 28, pp. 1129-1149.
23. Hecht, S. S., Mazzarese, R., Amin, S., LaVoie, E. and Hoffmann, D. in "Polynuclear Aromatic Hydrocargons: Third International Symposium on Chemistry and Biology-Carcinogenesis and Mutagenesis", (P. W. Jones and P. Leber, Eds.) Ann Arbor Science Publishers, Ann Arbor, Mich., 1979, pp. 733-752.
24. Jerina, D. M., Yagi, H., Thakker, D. R., Karle, J. M., Mah, H. D., Boyd, D. R., Gadaginamath, G., Wood, A. W., Buening, M., Chang, R. L., Levin, W. and Conney, A. H. in "Advances in Pharmacology and Therapeutics (Proceedings of the Seventh International Congress of Pharmacology, Paris, 1978) Vol. 9, Toxicology" (Y. Cohen, Ed.) Pergamon Press, Oxford, 1979, pp. 53-62.
25. Yagi, H., Thakker, D. R., Lehr, R. E. and Jerina, D. M.: 1979, J. Org. Chem. 44, pp. 3439-3442.
26. Lehr, R. E., Kumar, S., Levin, W., Wood, A. W., Chang, R. L., Buening, M. K., Conney, A. H., Whalen, D. L., Thakker, D. R., Yagi, H., and Jerina, D. M. in "Fourth International Symposium on Polynuclear Aromatic Hydrocarbons (A. Bjorseth and A. J. Dennis, Eds.) Battelle Press, Columbus, Ohio, in press.
27. Wood, A. W., Chang, R. L., Levin, W., Ryan, D. E., Thomas, P. E., Croisy-Delcey, M., Ittah, Y., Yagi, H., Jerina, D. M. and Conney, A. H.: Cancer Res., in press.
28. Levin, W., Wood, A. W., Chang, R. L., Ittah, Y., Croisy-Delcey, M., Yagi, H., Jerina, D. M. and Conney, A. H., submitted for publication in Cancer Res.
29. Wood, A. W., Levin, W., Chang, R. L., Yagi, H., Thakker, D. R., Lehr, R. E., Jerina, D. M. and Conney, A. H. in "Polynuclear Aromatic Hydrocarbons: Third International Symposium on Chemistry and Biology-Carcinogenesis and Mutagenesis", (P. W. Jones and P. Leber, Eds.) Ann Arbor Science Publishers, Ann Arbor, Mich., 1979, pp. 531-551.

INACTIVATION OF ØX174 AND SV40 VIRAL DNA REPLICATION BY DIOL EPOXIDE DERIVATIVES OF CARCINOGENIC POLYCYCLIC HYDROCARBONS

Ronald G. Harvey, Wen-Tah Hsu, George T. Chang, and Samuel B. Weiss
Ben May Laboratory for Cancer Research
The University of Chicago
Chicago, Illinois 60637

ABSTRACT

Interaction of reactive diol epoxide metabolites of polycyclic hydrocarbons with nucleic acids has recently been implicated as a critical event in the mechanism of hydrocarbon carcinogenesis. To study such interactions we have devised two convenient assays involving inhibition of infectivity of ØX174 and SV40 viral DNA in E. coli spheroplasts and monkey kidney cells, respectively. Anti-BPDE is shown to efficiently alkylate both ØX174 and SV40 viral DNA, inhibiting their infectivities 100% and 75-80%, respectively. Comparison of the ØX174 DNA viral replication inhibitory activities of the diol epoxide derivatives of a series of hydrocarbons varying in their carcinogenic activities reveals no correlation with theoretically predicted reactivity. In particular, bay region diol epoxides do not exhibit exceptional reactivity. Substantial recovery of SV40 DNA synthesis takes place following anti-BPDE treatment, although infectivity remains depressed. These results suggest that the DNA synthesized following BPDE treatment possibly is defective. The significance of these findings with respect to the mechanism of carcinogenesis is discussed.

INTRODUCTION

Reactive diol epoxide metabolites are implicated by an increasing body of evidence as the ultimate carcinogenic forms of polycyclic aromatic hydrocarbons (1), including the potent carcinogens benzo[a]pyrene, 7,12-dimethylbenz[a]anthracene (2-4), and 3-methylcholanthrene (5-7). The isomeric diol epoxide derivatives in which the epoxide function is located in a bay molecular region are found to be the principal metabolites of carcinogenic hydrocarbons bound covalently to nucleic acids in cells (8). It is proposed by Jerina and associates (9,10) on the basis of molecular orbital calculations that diol epoxide metabolites in bay regions are distinguished by exceptional reactivity which favors their reaction with nucleic acids or other cellular targets, leading to tumor induction. However, direct chemical evidence concerning the comparative chemical reactivities of the various diol epoxide and epoxide derivatives of polycyclic hydrocarbons is currently inadequate to evaluate the validity of this hypothesis (11).

B. Pullman, P. O. P. Ts'o and H. Gelboin (eds.), Carcinogenesis: Fundamental Mechanisms and Environmental Effects, 13–31.

Since interaction of the reactive diolepoxides with nucleic acids is apparently a critical event in the multistage process of tumorigenesis, we have undertaken studies directed towards elucidation of the detailed mechanistic pathways of hydrocarbon nucleic acid interactions. Prime objectives are to obtain basic information concerning the roles of steric, electronic, and other structural factors in determining the extent of reaction and the regiospecificity of attack on DNA and RNA of the diol epoxides and related hydrocarbon derivatives, and to determine the consequences of hydrocarbon adduct formation on the structure and stability of the nucleic acids products.

INACTIVATION OF ØX174 DNA BY ANTI-BPDE[1]

The ØX174 viral DNA assay procedure devised by Weiss and his associates at the University of Chicago (12-14) was utilized in our initial studies. When intact ØX174 DNA is mixed with E. coli spheroplasts, the nucleic acid is adsorbed by the host and initiates the production of new virus. The latter may be assayed by plating the mixture with cells of a sensitive strain of E. coli on agar plates, and counting viral plaque formation. When the infectious virus is alkylated with reactive diol epoxides prior to incubation with spheroplasts, viral replication is inhibited, and the extent of inhibition can be assayed using only μg quantities of the relatively precious hydrocarbon derivatives.

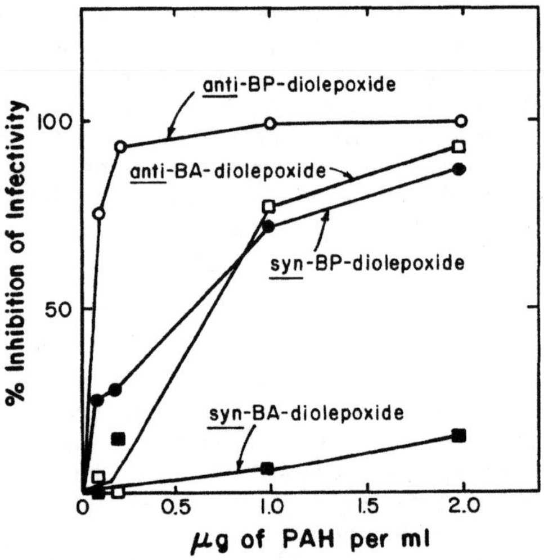

Fig. 1. Effect of the diol epoxides of BP and BA on the infectivity of ØX174 DNA. Details of the experimental methodology are described by Hsu et al. (12).

The effects of the isomeric anti and syn diol epoxides of BP and BA on the infectivity of ØX174 DNA are shown in Fig. 1. Anti-BPDE, implicated as the ultimate carcinogenic form of BP (1), showed the greatest response, inhibiting viral replication virtually quantitatively at low concentrations, while the syn isomer was markedly less effective even at much higher dosages. Similarly, the anti-7,8-diol-9,10-epoxide of BA exhibited considerably higher activity than the corresponding syn isomer.

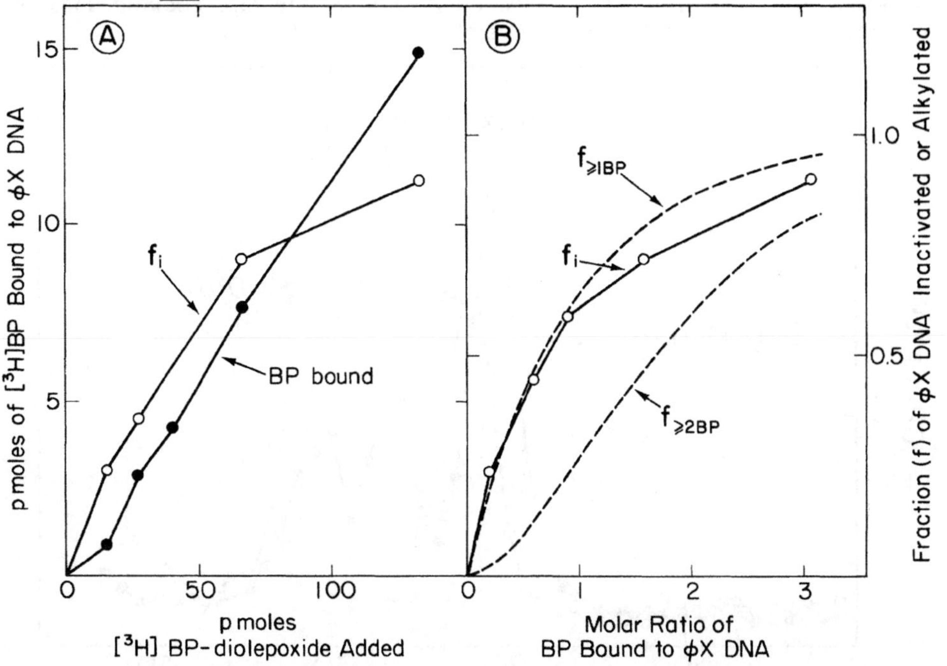

Fig. 2. Relationship between the molar quantity of ØX DNA alkylated by anti-BPDE and the amount of infectivity lost. (A) The amounts of [^{3}H] BPDE indicated were incubated individually with 4.8 mmol of ØX DNA, then the [^{3}H] BP-DNA complex was isolated and its DNA, radioactivity, and infectivity content were determined as described by Hsu et al. (13). The term "f_i" represents the fraction of infectious ØX DNA which lost infectivity as a consequence of the diol epoxide treatment. (B) The fraction of ØX DNA inactivated (f_i) is plotted as a function of the molar ratio (MR) of the average number of BP molecules bound per molecule of ØX DNA, calculated from the radioactivity and DNA content of the isolated [^{3}H] BP-DNA complex. The terms $f_{\geq 1BP}$ and $f_{\geq 2BP}$ represent the calculated theoretical Poisson distributions for one and two hit inactivation, respectively. Reprinted from (13).

The relationship between the extent of binding of tritium labelled anti-BPDE to ØX174 DNA and its effect upon the infectivity of the alkylated DNA is shown in Fig. 2A. With increasing concentration of anti-BPDE, the amount of the anti-BPDE-DNA adduct formed increases linearly. The fraction (f_i) of infectious viral DNA inactivated at equivalent dosages follows a parallel relationship through most of the range. When the molar ratio of hydrocarbon bound to DNA is plotted against f_i, the experimental curve (Fig. 2B) matches rather closely the calculated theoretical Poisson distribution ($f \geqslant 1$) for single hit inactivation of viral replication by anti-BPDE.

When the ØX174 DNA complexed with anti-BPDE (molar ratio = 8) is subjected to neutral sucrose gradient analysis, the bound radioactive hydrocarbon sediments in a homogeneous manner at a slightly slower rate than the untreated viral DNA (Fig. 3). Thus, alkylation by the diol epoxide does not appear to result in significant breakage of the DNA molecule.

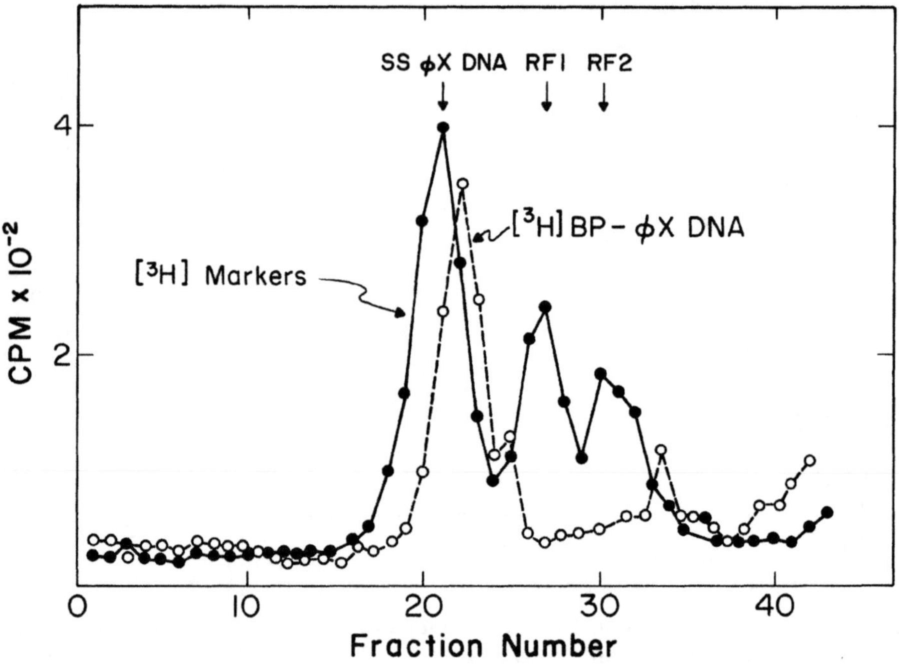

Fig. 3. Sucrose gradient analysis of ØX174 DNA complexed with [^{3}H]-BPDE. ØX DNA (8 μg) was treated with [^{3}H]BPDE, and the [^{3}H]BP-bound DNA was isolated (13). An average of eight [^{3}H]BP molecules bond per molecule of viral DNA was estimated. The [^{3}H]BP-ØX DNA (o) was subjected to neutral sucrose gradient analysis at 39,000 rpm for 5 hr. A second gradient contained the radioactive markers (•), ØX DNA, RF1, and RF2. Reprinted from (13).

In order to determine the effect of the bound hydrocarbon on replication, DNA synthesis was examined using the alkylated ØX DNA as a template. When analyzed by centrifugation in a neutral sucrose gradient (Fig. 4), the product DNA labelled with [^{3}H] thymidine sediments at a rate similar to that of the ØX DNA replicative form 2 (RF2). The latter sediments in higher fractions than either the single-stranded form or the double stranded replicative form 1. The kinetics of the polymerization reaction on unmodified and anti-BPDE modified ØX DNA templates were examined by determination of the incorpora-

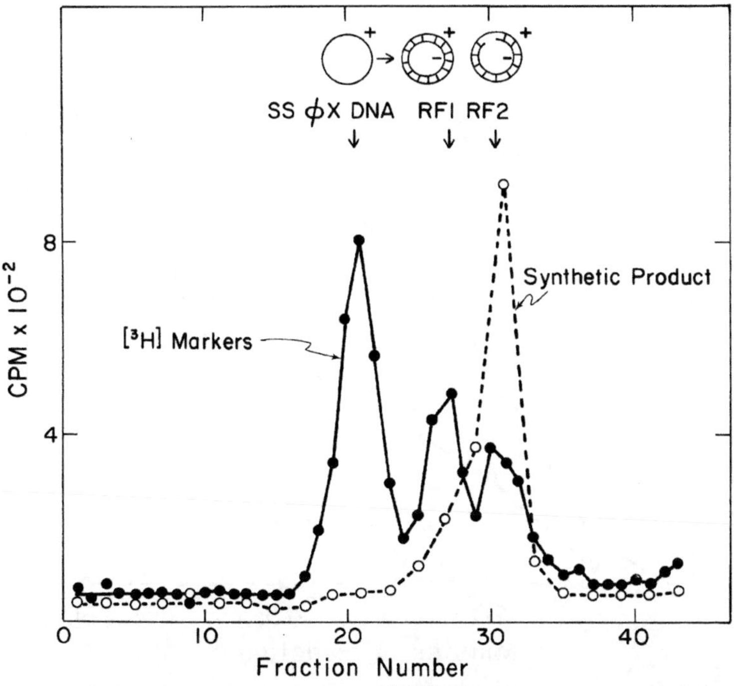

Fig. 4. Sucrose gradient analysis of the enzymatic reaction product directed by ØX DNA. In vitro synthesis directed by ØX DNA was carried out for 20 min as described by Hsu et al. (13). After Pronase treatment and ethanol precipitation, the resuspended product was subjected to sedimentation in neutral sucrose as described for Fig. 3. ●, tritium labelled markers; O, synthetic product. Reprinted from (13).

tion of ^{3}H-dTMP into DNA (Fig. 5). A significantly reduced rate of DNA synthesis was observed for the alkylated product at a molar ratio of anti-BPDE/DNA = 3. At higher ratios, the rate is reduced even further. The DNA products of these reactions were shown by sucrose gradient centrifugation (Fig. 6) to sediment like replicative intermediate RF2, regardless of the extent of alkylation. However, less RF2 product was formed for those templates containing higher ratios of bound anti-BPDE.

When the RF2 products were examined by sedimentation under alkaline conditions, marked differences were observed in the profiles of the dissociated complementary strands polymerized on the alkylated and nonalkylated templates (Fig. 7). Under alkaline conditions the RF2 product made with ØX DNA alkylated with anti-BPDE serving as template sediments more slowly and less uniformly than the normal RF2 product made with the unmodified ØX DNA template. This suggests that alkylation with BPDE blocks nucleotide polymerization at the alkylated site and leads to the synthesis of complementary DNA strands that are fragmentary. With higher molar ratios of hydrocarbon bound to DNA or longer reaction time, the products become increasingly heterogeneous.

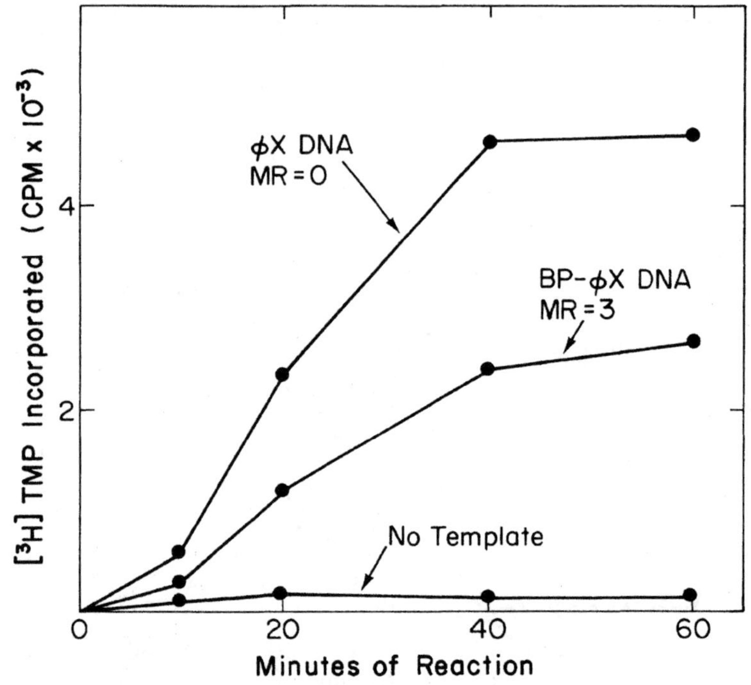

Fig. 5. Kinetics of the BP-ØX DNA-directed synthesis of DNA. Details of the experimental methodology are described by Hsu et al. (13). Both unmodified and BPDE-modified ØX DNA were used as templates in separate reaction mixtures. At the time periods indicated, samples were withdrawn, and the amount of [^{3}H] dTMP incorporated into DNA was determined by acid precipitation. The BP-ØDNA used in this experiment had an average molar ratio (MR) (BP/ØX DNA) of 3, as estimated from Fig. 2, based on the fraction of infectivity inactivated, f_i. Reprinted from (13).

COMPARATIVE REACTIVITY OF DIOL EPOXIDE DERIVATIVES WITH ØX174 DNA

Since alkylation of nucleic acids is generally assumed to be a critical event in carcinogenesis, it was of interest to use the ØX174 viral DNA assay to compare the effectiveness of a series of diol epoxide derivatives in the alkylation of DNA. This system has the advantages of being simple, rapid, and requiring only tiny quantities of hydrocarbon metabolites. From our synthetic studies, we had on hand in our laboratory the diol epoxide derivatives, both bay region and nonbay region, of a wide range of both carcinogenic and noncarcinogenic hydrocarbons (1,15-21).

Utilizing this assay, we investigated the comparative viral replication inhibitory activities of the series of hydrocarbon derivatives shown in Fig. 8 (22). These compounds are arranged in order of decreasing calculated ΔE_{deloc}.

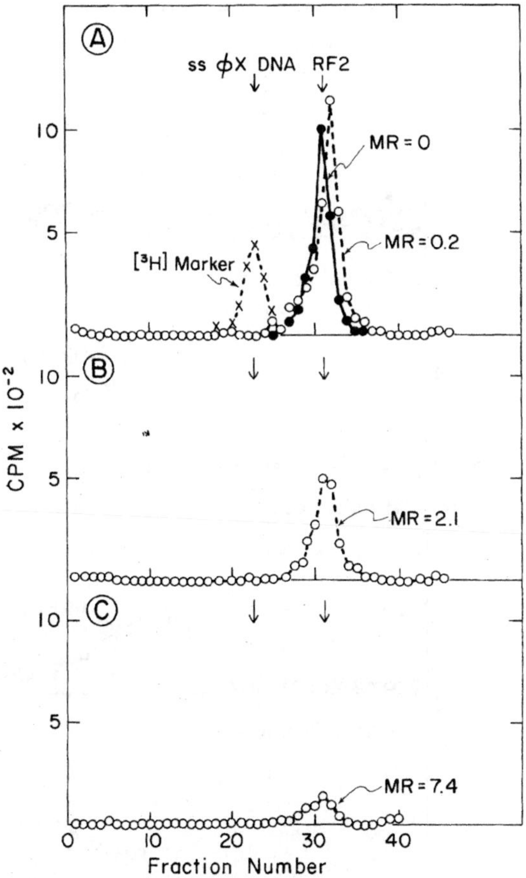

Fig. 6. Sedimentation analysis of the synthetic DNA products under neutral conditions. Separate reaction mixtures contained unmodified and BP-modified ØX DNA as templates at different molar ratios (MR) of BP/ØX DNA. After 20 min of incubation at 25°, the individual reaction products were isolated as described by Hsu et al. (13) and subjected to centrifugation in neutral sucrose gradients for 6 hr at 39,000 rpm. The gradient region corresponding to RF2 was collected and precipitated with alcohol, and the precipitate was resuspended in 0.3 ml of Tris/EDTA. A sample of 0.1 ml of this suspension was again subjected to centrifugation in neutral sucrose; the remaining material was used for alkali gradient analysis (Fig. 7). Single-stranded ØX[^{3}H] DNA was used as marker in a separate gradient and is shown only in A. Reprinted from (13).

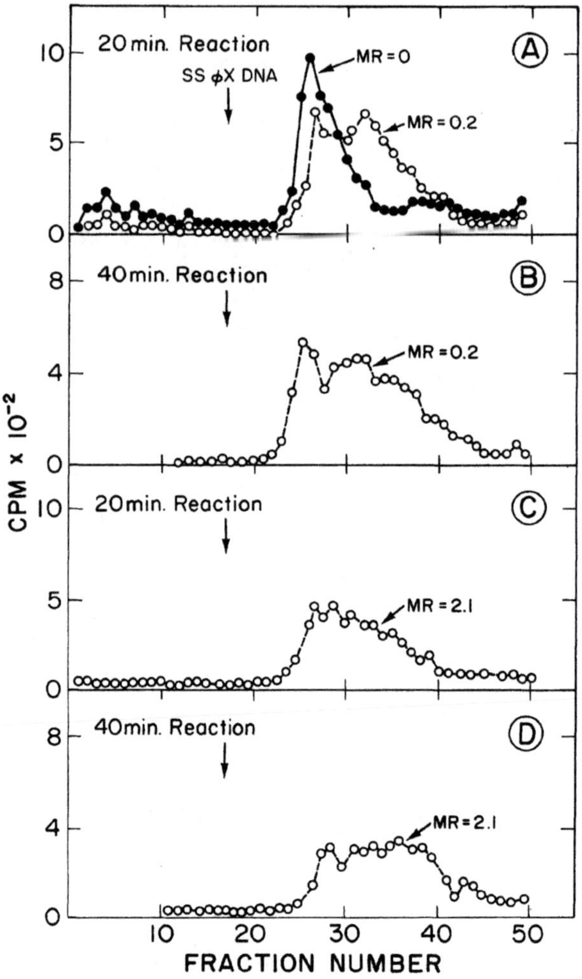

Fig. 7. Alkali sucrose-gradient analysis of the synthetic DNA products. The synthetic RF2 reaction products isolated by sedimentation in neutral sucrose in the experiment described in Fig. 6 were subjected to alkaline sucrose gradient centrifugation for 10 hr at 39,000 rpm. Experimental details are described by Hsu et al. (13).

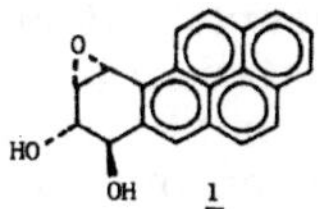

1
(>99;0.794)

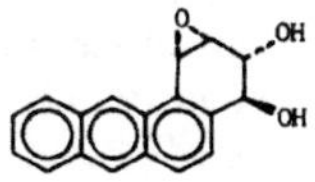

6
(46;0.766)

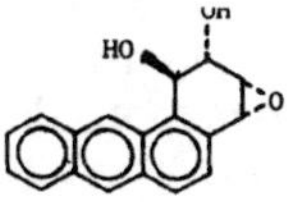

11
(55;0.628)

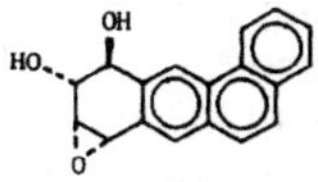

16
(32;0.526)

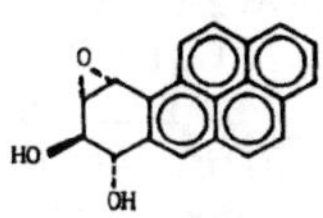

2
(46;0.794)

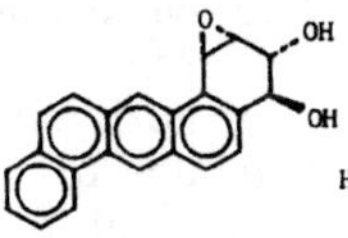

7
(66;0.738)

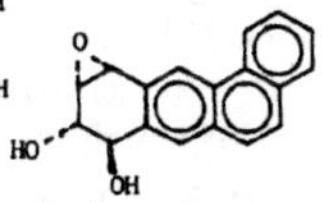

12
(78;0.572)

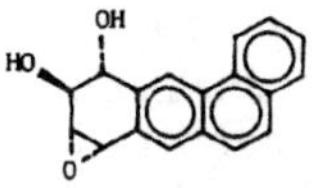

17
(18;0.526)

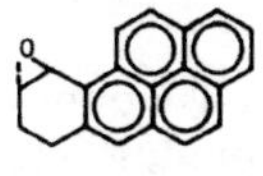

3
(>99;0.794)

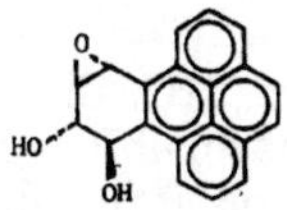

8
(0;0.714)

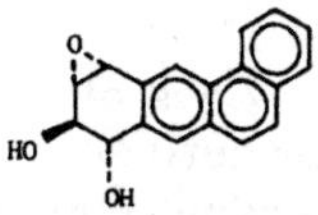

13
(7;0.572)

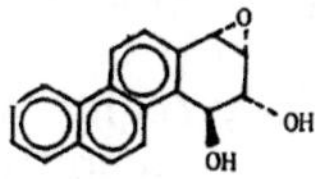

18
(0;0.526)

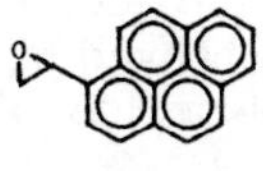

4
(>99;0.794)

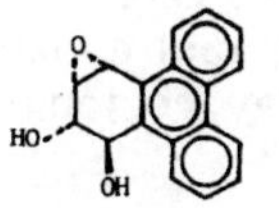

9
(12;0.644)

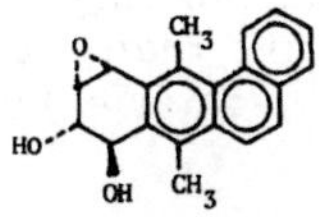

14
(64;0.572)

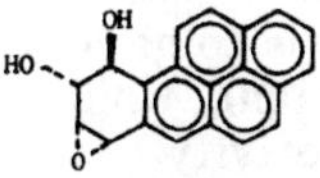

19
(72;0.488)

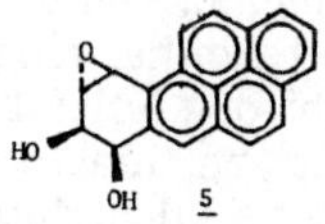

5
(90;0.794)

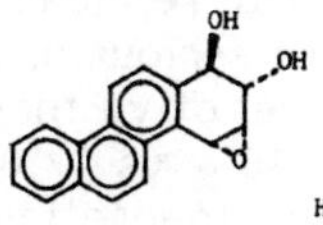

10
(91;0.640)

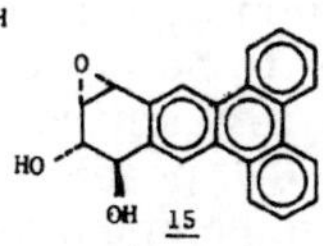

15
(38;0.544)

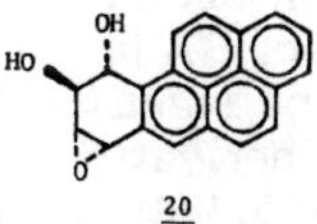

20
(46;0.488)

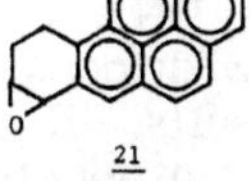

21
(78;0.488)

Fig. 8. Inhibition of ∅X174 DNA infectivity in E. coli spheroplasts by diol epoxides and related compounds. Numbers in parenthesis are % inhibition and $\Delta E_{deloc.}$ in β units. Reprinted from (25).

energies (in β units) of the corresponding benzylic carbonium ions expected to be formed on opening the epoxide ring. In these experiments, 0.5 μg of the polycyclic aromatic hydrocarbon derivatives were incubated with 10 μg of ØX174 DNA in 0.1 ml of aqueous solution at pH 7.5 for 10 min at 25°. The DNA was then reisolated (free from unreacted BPDE) and assayed in triplicate for infectivity in E. coli spheroplasts. Many of the compounds exhibited potent viral inhibitory activity. Anti-BPDE inhibited viral replication essentially quantitatively at the dosage employed. The syn isomer showed lower activity (46%), in line with its reported lower mutagenic activity. Both 9,10-epoxy-H_4BP and 1-oxiranylpyrene were equally as active as anti-BPDE. Since the former compound lacks the two hydroxyl groups of anti-BPDE and the latter lacks also the 7,8-carbon atoms, these structural features apparently do not contribute significantly to the viral inhibitory activity of anti-BPDE. The cis isomer of anti-BPDE in which the 7,8-hydroxy groups are cis rather than trans also showed high activity (90%).

The bay region diol epoxides of BA and DBA displayed moderate activity (46% and 66% inhibition, respectively) in reasonable agreement with expectation. The bay region anti-diol epoxides of BeP and triphenylene showed essentially no activity, in agreement with their inactivity as complete carcinogens, but inconsistent with theoretical prediction of reactivity based on $\Delta E_{deloc.}$. On the other hand, the chrysene bay region diol epoxide, predicted on the same basis to be less active than the BeP diol epoxide, proved highly active (91% inhibition), while the isomeric non bay region diolepoxide of chrysene showed no activity.

In general, there is no satisfactory correlation evident in the data in Fig. 8 between inhibition of ØX174 DNA viral replication and the calculated β-delocalization energies. Among the more serious discrepancies, the bay region diol epoxide of BA predicted to be more reactive than the terminal ring isomer is found to be significantly less active (46% vs 78% inhibitory activity). The 7,8-epoxide derivatives of BP predicted theoretically to be least reactive ($\Delta E_{deloc}/\beta = 0.488$) of all the compounds tested, proved to be all highly active. We conclude that the electronic index of reactivity, ΔE_{deloc}, does not reliably predict the inhibition of ØX174 viral DNA infectivity. It should be kept in mind that ØX174 DNA is single stranded viral DNA. Interaction of the same compounds with double stranded DNA of mammalian origin might yield quite different results. It is also conceivable that some of the compounds are less active than anticipated, due to their hydrolytic instability. However, the exceptionally high activities of some of the compounds predicted to be least reactive, particularly the 7,8-oxide of 7,8,9,10-tetrahydro-BP, is difficult to rationalize as due to anything other than efficient alkylation of DNA. Unpublished experiments confirm that the 7,8-oxide of 7,8,9,10-tetrahydro-BP also efficiently alkylates intact double stranded calf DNA.

If it is assumed that the compounds in Fig. 8 which exhibit significant viral inhibitory activity act through a common mechanism involving alkylation of DNA, the results suggest that a wide range of epoxide and diol epoxide derivatives react readily with nucleic acids. Therefore, this property alone would not appear to uniquely distinguish the bay region diol epoxide derivatives of carcinogenic hydrocarbons as proposed in the "bay region theory" (9,10).

INACTIVATION OF SV40 REPLICATION BY ANTI-BPDE

Since it was of obvious interest to determine whether viruses containing double stranded genomes would respond in a similar manner, we chose to examine the effect of anti-BPDE and other reactive hydrocarbon derivatives on the replication of simian virus 40 (23). The SV40 virus contains infectious duplex DNA of known sequence, is of a size comparable to that of the small bacterial viruses, and replicates in animal rather than bacterial cell lines. Continuous African green monkey kidney cell lines (CV-1) were employed for these studies.

The infectivity of SV40 DNA isolated from infected CV-1 cells 24 hours after treatment with anti-BPDE was assayed by a procedure analogous to that used in the ØX DNA studies. Alkylation of SV40 DNA with anti-BPDE is shown in Table 1 to significantly lower the specific infectivity of the virus. However, complete inhibition of virus production was not observed even at molar binding ratios as high as 10. In these and other experiments, inhibition attained a maximum of only 75-80% (23). These data suggest that inactivation does not follow a single-hit mechanism as observed with ØX DNA.

Table 1. Effect of Alkylation of SV40 DNA with BPDE on Infectivity

Moles of BPDE bound per mole of SV40 DNA	SV40 titer	
	PFU/ng	% Inhibition
0	230	0
4.6	90	59
9.1	60	74
10.1	57	75

To determine the effect of anti-BPDE on SV40 DNA synthesis, we pulsed infected cells with tritiated thymidine in the presence or absence of the diol epoxide. In Fig. 9 are shown the profiles of the product DNA obtained on neutral sucrose gradient centrifugation analysis. The levels of viral DNA synthesis in infected cells exposed to 0.1 and 0.2 μg/ml of BPDE for 26 hours were 75% and 25%, respectively, of the control level. In contrast, when infected cells were pulsed with tritium labelled thymidine 2 hours after BPDE treatment, the inhibition of DNA synthesis was reduced to 15% of the level for untreated cells. Evidently, substantial recovery of the capacity for DNA synthesis takes place between 2 and 26 hours following treatment.

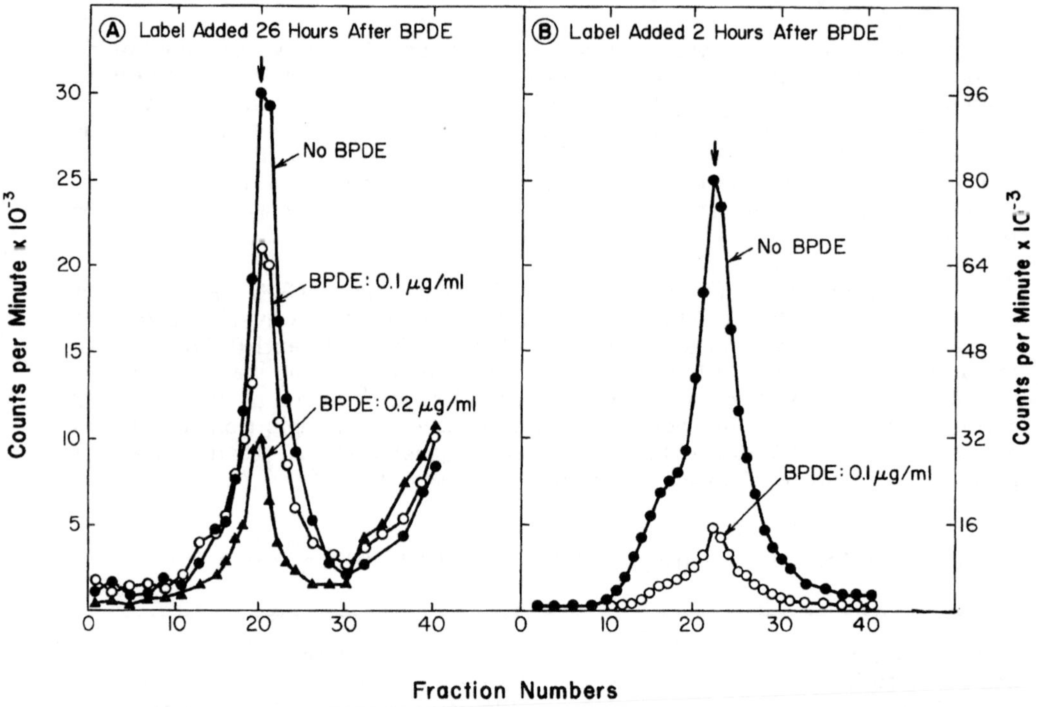

Fig. 9. Effect of BPDE on SV40 DNA synthesis in infected cells. Details of the experimental methodology are described by Chang et al. (23). (A) Confluent CV-1 cells were infected with SV40 (MOI=16), and 2 hours later were treated with a given concentration of anti-BPDE. Twenty-six hours after BPDE exposure, [^{3}H] thymidine (1 mCi/ml at 6.7 Ci/mmol) was supplied to each infected monolayer, and the cells were harvested 2 hours later. Hirt supernatants were prepared and subjected to neutral sucrose gradient centrifugation. (B) The experimental conditions were essentially the same as in (A) except that infected cells were treated with anti-BPDE 26 hours after infection, and 2 hours later were pulsed with [^{3}H] thymidine for 40 minutes. Reprinted from (23).

The time course of SV40 DNA synthesis in infected CV-1 cells following anti-BPDE treatment is shown in Fig. 10. The data indicate that incorporation of thymidine into viral DNA is suppressed about 70% for the first 12 hours after exposure to BPDE, but within 36 hours after treatment, DNA synthesis has largely recovered to control levels. If the inhibition of DNA synthesis is only temporary, how does one explain the persistent reduction in viral infectivity following anti-BPDE treatment? An attractive hypothesis is that the viral DNA synthesized on BPDE-alkylated DNA is partially defective and therefore less infectious.

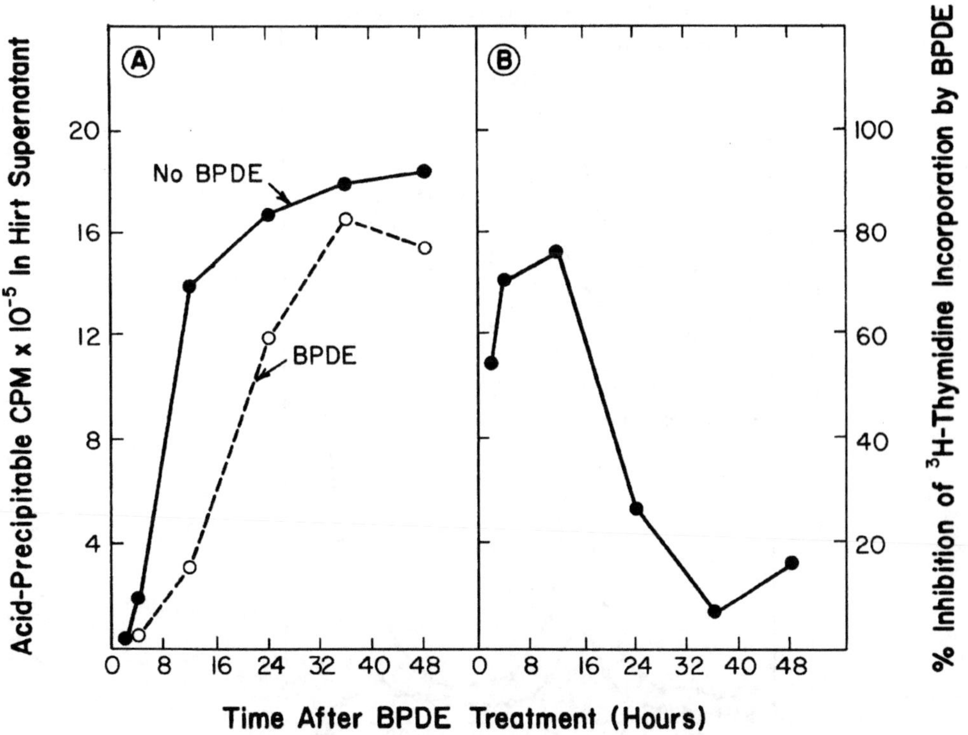

Fig. 10. Time course of SV40 DNA synthesis following BPDE treatment. (A) Confluent CV-1 cells were infected with SV40 (MOI=15); 26 hours later, BPDE was added, and 60 minutes after BPDE [^{3}H] thymidine was supplied to each culture as described in Figure 9A. At the times after BPDE treatment indicated, cells were washed, Hirt supernatants were prepared as described in Figure 9B, and acid-insoluble counts were determined by precipitation with trichloroacetic acid. (B) The percent inhibition shown for [^{3}H] thymidine incorporation was derived from the difference between the acid-insoluble counts found for the untreated and BPDE-treated infected cells at the times shown in (A). Reprinted from (23).

Some additional experiments are relevant to this point. In order to verify that alkylation of viral DNA takes place in SV40 infected CV-1 cells, Hirt supernatants from infected cells treated with [^{3}H] -anti-BPDE 2 hours prior to viral DNA isolation were analyzed by sucrose gradient centrifugation analysis. The bound radioactive hydrocarbon was found to sediment under neutral or alkaline conditions virtually identically with normal SV40 DNA tagged with [^{14}C] thymidine (Fig. 11). Thus, alkylation of SV40 DNA by BPDE does indeed take place in viral infected CV-1 cells, and alkylation does not result in significant intracellular breakage of the viral DNA, at least not after relatively short periods of BPDE exposure.

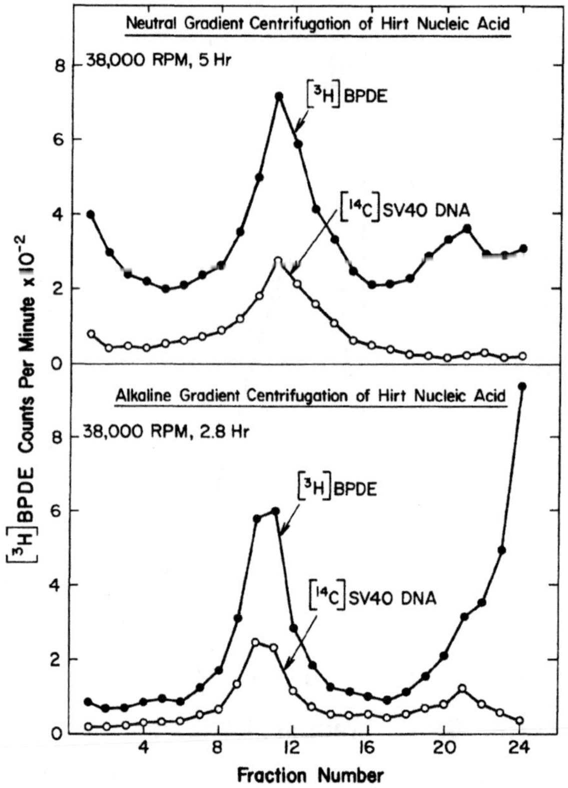

Fig. 11. In vivo alkylation of SV40 DNA. Monkey kidney cells (CV-1) were infected with SV40 virus (strain 776) and exposed to [^{3}H] anti-BPDE (0.1 mg per ml of medium; specific activity, 160 cpm per ng) 49 to 52 hours following infection. At the end of this time, the medium was removed and the cells washed with fresh MEM medium. Viral DNA was extracted from these cells and subjected to neutral and alkaline sucrose gradient centrifugation as previously reported (23). In each run, marker [^{14}C] SV40 DNA was also included.

The stability of SV40 DNA in infected cells was investigated by alkaline sucrose gradient analysis of pre-labelled viral DNA at several time intervals following BPDE treatment. At intervals of 5 and 14 hours, the levels of form I SV40 DNA declined to 60% and 30%, respectively, in relation to infected cells that were not exposed to BPDE (Fig. 12). It seemed reasonable to assume that the disappearance of labeled viral DNA in infected cells was the result of selective breakdown of BPDE alkylated SV40 DNA molecules. Viral infected cells were treated with tritium labelled BPDE and the alkylated viral DNA isolated at several time intervals following BPDE treatment and analyzed by agarose gel electrophoresis (Fig. 13A). Two hours after treatment with labelled BPDE, a significant amount of diol epoxide was found associated with form I viral DNA (as expected from the experiment of Fig. 11). However, at longer intervals after BPDE treatment only 40-50% of the labelled BP derivative was

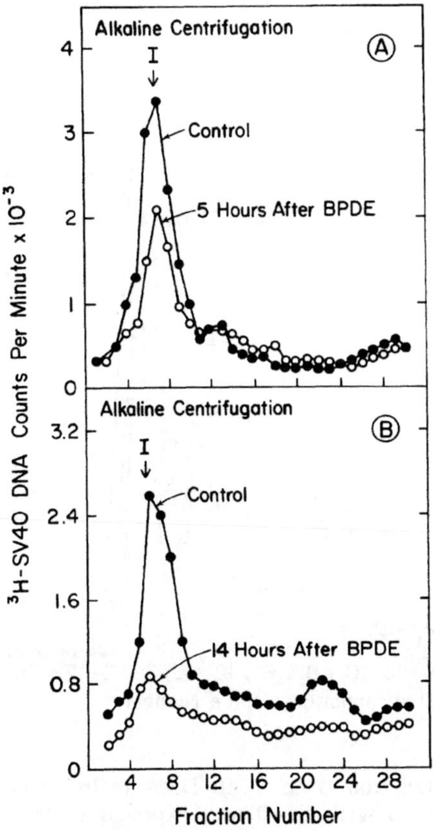

Fig. 12. The effect of BPDE on prelabelled SV40 DNA. SV40 infected CV-1 cell monolayers were exposed to [^{3}H] thymidine (20 mCi/ml of medium; specific activity, 5 Ci/mM) 26 to 27 hours after infection. The labelled media was removed, cells were washed with fresh medium, and then chased for 6 hours with non-radioactive thymidine (5 x 10^{-4}M) contained in MEM medium. Immediately after chasing, BPDE was added to the infected cultures (0.1 mg per ml of medium) and viral DNA isolated at the time intervals shown above. Centrifugation under alkaline sucrose conditions and the isolation of viral DNA were as described for Fig. 11.

still bound to form I DNA. The reduction of alkylated SV40 DNA seems not to be a consequence of viral nucleic acid turnover, since viral DNA pulse-labelled with [^{3}H] thymidine is quite stable even after a 24 hour chase period (Fig. 13B). Thus, there appears to be a selective degradation of alkylated viral DNA in these cells. The fact that even after 24 hours following [^{3}H] BPDE treatment a significant amount of the viral DNA still contains labelled adducts may be important but remains unexplained.

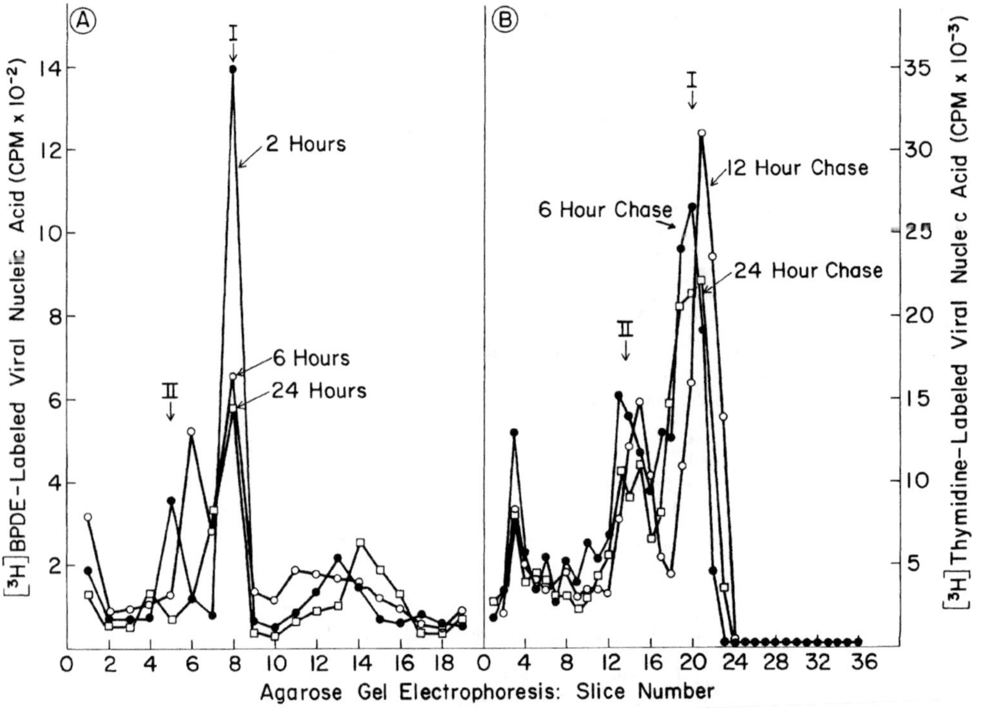

Fig. 13. The fate of [^{3}H]BPDE bound to SV40 DNA in infected cells. (A) CV-1 cell monolayers were infected with SV40 and exposed to tritiated BPDE under conditions similar to those described in Fig. 11. Viral DNA was isolated from the infected BPDE-treated cells at the time intervals shown above and subjected to electrophoresis on 0.7% agarose gels formed in a buffer containing 0.04M Tris base, 0.02M sodium acetate, 0.15% glacial acetic acid and 0.001M EDTA, adjusted to pH 8. After electrophoresis, the gels were cut into 3 mm slices, dissolved by heating in 1 ml of H_2O, and their isotope content determined by scintillation counting. Forms I and II SV40 [^{14}C] DNA applied to the gels served as markers. (B) CV-1 cells, infected with SV40 under the same conditions as in (A), but not exposed to hydrocarbon, were treated with [^{3}H] thymidine for one hour and then chased with cold thymidine as described in Fig. 12. At various time periods after chasing, viral DNA was isolated and subjected to gel electrophoresis as described in (A).

The apparent breakdown of the SV40 DNA alkylated with BPDE takes place within the same time frame as the observed inhibition of DNA synthesis (Fig. 10). It is conceivable that DNA repair is occuring in this time interval. The fact that DNA synthesis is restored to essentially normal levels while infectivity remains depressed suggests that repair, if it is taking place, is partially defective. Alternative explanations are also conceivable. For example, alkylation on the N-7 of guanosine, for which there is precedent (24), could result in depurination leading to mutation.

SUMMARY AND CONCLUSIONS

The results of the foregoing studies are significant, we believe, with respect to the mechanism of hydrocarbon tumorigenesis. The experiments with ØX174 DNA show that anti-BPDE efficiently alkylates single-stranded viral DNA, rendering it noninfectious. One molecule of bound diol epoxide appears sufficient to inhibit the replication of a single molecule of ØX DNA. The propagation of synthetic DNA strands is blocked so that incomplete complementary strands are assembled. The ØX174 DNA assay provides a convenient method for the rapid determination on small scale of the comparative reactivities of the diol epoxide derivatives of polycyclic hydrocarbons varying in their carcinogenic activities. The results indicate that a wide range of epoxide and diol epoxide derivatives undergo facile reaction with nucleic acids. The bay region diol epoxide derivatives do not appear to be uniquely distinguished in this respect contrary to prediction of the "bay region theory" (9,10).

The experiments with SV40 DNA demonstrate that anti-BPDE also efficiently alkylates double-stranded viral DNA, inhibiting infectivity a maximum of 75-80%, even at molar binding ratios as high as 10. Although substantial recovery of DNA synthesis takes place within 24-36 hours following treatment, infectivity remains depressed. It is proposed as a working hypothesis that the new viral DNA synthesized on BPDE-alkylated DNA is altered, possibly as a consequence of defective enzymatic repair. Alternative explanations are also admissable. In any case, these results appear significant with respect to the mechanism of carcinogenesis, since they implicate a hereditable genetic change consequent upon alkylation of DNA with a carcinogenic hydrocarbon metabolite.

ACKNOWLEDGEMENT. This research was supported by grants from the National Cancer Institute, DHEW (CA 11968) and the American Cancer Society (BC-132) and by funds from the Dept. of Energy, Contract No. DE-AC0280EV. The following figures are reprinted with permission from the sources indicated: Fig. 1-7, Hsu et al., Proc. Natl. Acad. Sci. USA, 1977 (12,13); Fig. 8, Harvey, "Safe Handling of Chemical Carcinogens, Mutagens, Teratogens, and Highly Toxic Substances", Ann Arbor Science Publishers, 1980 (25); and Fig. 9 and 10 and Tables 1 and 2, Chang, et al., Biochem. Biophys. Res. Commun., 1979 (23).

[1] ABBREVIATIONS: BP, benzo[a]pyrene; anti(syn)-BPDE, trans-7,8-dihydroxy-anti(syn)-9,10-epoxy-7,8,9,10-tetrahydro-BP; BA, benz[a]anthracene; DBA, dibenz[a,h]anthracene; BeP, benzo[e]pyrene.

REFERENCES

1. "Polycyclic Hydrocarbons and Cancer", Gelboin, H. V. and Ts'o, P.O.P., Eds., Academic Press, New York, N.Y., 1978.

2. Slaga, T. J., Gleason, G. L., DiGiovanni, J., Sukumaran, K. B., and Harvey, R. G.: 1979, Cancer Res. 39, pp. 1934-1936.

3. Malaveille, C., Bartsch, H., Tierney, B., Grover, P. L., and Sims, P.: 1978, Biochem. Biophys. Res. Commun., 83, pp. 1468-1473.

4. Dipple, A. and Nebzydoski, J. A.: 1978, Chem.-Biol. Interact. 20, pp. 17-26.

5. Malaveille, C., Bartsch, H., Marquardt, H., Baker, S., Tierney, B., Hewer, A., Grover, P. L., and Sims, P.: 1978, Biochem. Biophys. Res. Commun. 85, pp. 1568-1574.

6. Thakker, D. R., Levin, W., Wood, A. W., Conney, A. H., Stoming, T. A., and Jerina, D. M.: 1978, J. Amer. Chem. Soc. 100, pp. 645-647.

7. King, H.W.S., Osborne, M. R., and Brookes, P.: 1977, Int. J. Cancer 20, pp. 564-571.

8. Weinstein, I. B., Jeffrey, A. M., Leffler, S., Pulkrabek, P., Yamasaki, H., and Grunberger, D.: 1978, "Polycyclic Hydrocarbons and Cancer", Gelboin, H. V. and Ts'o, P.O.P., Eds., Academic Press, New York, N. Y., Vol. 2, Chap. 1, pp. 3-36.

9. Jerina, D. M., and Lehr, R. E.: 1977, "Microsomes and Drug Oxidations", Ullrich, V. et al., Eds., Pergamon, Oxford, pp. 709-720.

10. Fu, P. P., Harvey, R. G., and Beland, F. A.: 1978, Tetrahedron 34, pp. 857-866.

11. Harvey, R. G. and Fu, P. P.: 1978, "Polycyclic Hydrocarbons and Cancer," Gelboin, H. V. and Ts'o, P.O.P., Eds., Academic Press, New York, N.Y., Vol. I, Chap. 6, pp. 133-165.

12. Hsu, W. T., Harvey, R. G., Lin, E. J., and Weiss, S. B.: 1977, Proc. Natl. Acad. Sci. USA 74, pp. 1378-1382.

13. Hsu, W. T., Lin, E. J., Harvey, R. G., and Weiss, S. B.: 1977, Proc. Natl. Acad. Sci. USA 74, pp. 3335-3339.

14. Sagher, D., Harvey, R. G., Hsu, W., T., and Weiss, S. B.: 1979, Proc. Natl. Acad. Sci. USA 76, pp. 620-624.

15. Cortez, C. and Harvey R. G.: 1978, Org. Syn. 58, pp. 12-17.

16. Harvey, R. G., Lee, H. M., and Shyamasundar, N.: 1979, J. Org. Chem. 44, pp. 78-83.

17. Sukumaran, K. B. and Harvey, R. G.: 1979, J. Amer. Chem. Soc. 101, pp. 1353-1354.

18. Fu, P. P. and Harvey, R. G.: 1979, J. Org. Chem. 44, pp. 3778-3784.

19. Lee, H. M. and Harvey, R. G.: 1979, J. Org. Chem. 44, pp. 4948-4953.

20. Harvey, R. G. and Fu, P. P.: 1980, J. Org. Chem. 45, pp. 169-171.

21. Lee, H. M. and Harvey, R. G.: 1980, J. Org. Chem. 45, pp. 588-592.

22. Hsu, W. T., Lin, E. J., Fu, P. P., Harvey, R. G., and Weiss, S. B.: 1979, Biochem. Biophys. Res. Commun. 88, pp. 251-257.

23. Chang, G. T., Harvey, R. G., Hsu, W. T., and Weiss, S. B.: 1979, Biochem. Biophys. Res. Commun. 88, pp. 688-695.

24. Osborne, M. R., Harvey, R. G., and Brookes, P.: 1978, Chem.-Biol. Interact. 20, pp. 123-130.

25. Harvey, R. G.: 1980, "Safe Handling of Chemical Carcinogens, Mutagens, Teratogens, and Highly Toxic Substances", Walters, D. B., Ed., Ann Arbor Science, Ann Arbor, Mich., Vol. 2, Chap. 23, pp. 439-468.

DIHYDRODIOLS AND DIOL-EPOXIDES IN THE ACTIVATION AND DETOXIFICATION OF POLYCYCLIC HYDROCARBONS

Peter Sims
Chester Beatty Research Institute, Institute of Cancer Research:Royal Cancer Hospital, Fulham Road, London SW3 6JB, England.

ABSTRACT

The weak carcinogen, benz(a)anthracene, is metabolized to a number of dihydrodiols and diol-epoxides. Hydrolysates of the DNA of mouse skin or of hamster embryo cells treated with the hydrocarbon contain adducts arising from the reactions with the nucleic acid of both the 'bay-region' anti-3,4-diol-1,2-epoxide and the non-'bay-region' anti-8,9-diol-10,11-epoxide. However, only the 8,9-diol-10,11-epoxide appears to be involved in the reaction of the hydrocarbon with the RNA of hamster embryo cells. In contrast, the activation of the moderately active carcinogen, 7-methylbenz(a)anthracene and the potent carcinogen, 7,12-dimethylbenz(a)anthracene on mouse skin appears to involve mainly the 'bay-region' 3,4-diol-1,2-epoxides.

1. INTRODUCTION

It is now widely accepted that most carcinogens, including the polycyclic aromatic hydrocarbons, react covalently with cellular macromolecules such as nucleic acids and proteins and that these reactions may be concerned in the initiation of malignancy. Since the hydrocarbons are chemically inert, it is also believed that their metabolic activation within cells to electrophilic species is a necessary prerequisite for these covalent interactions. It was suggested many years ago by Boyland (1950) that the primary metabolic products were simple epoxides and it is now known that the enzymes concerned in their formation are the microsomal NADPH-dependent mono-oxygenases. Although the simple epoxides, and, in particular, those formed at the K-regions that are present in most hydrocarbons, appeared to be suitable candidates for the electrophilic species (Sims and Grover, 1974), it was found (Baird et al., 1973) that the hydrocarbon-nucleoside adducts formed when cells or tissues were treated with a polycyclic hydrocarbon did not arise through the interactions of their K-region epoxides.

The simple epoxides are further metabolized by the microsomal enzyme epoxide hydratase (Oesch, 1972) to trans-dihydrodiols and these can be further metabolized to diol-epoxides (Booth and Sims, 1974). These diol-

B. Pullman, P. O. P. Ts'o and H. Gelboin (eds.), Carcinogenesis: Fundamental Mechanisms and Environmental Effects, 33–42.

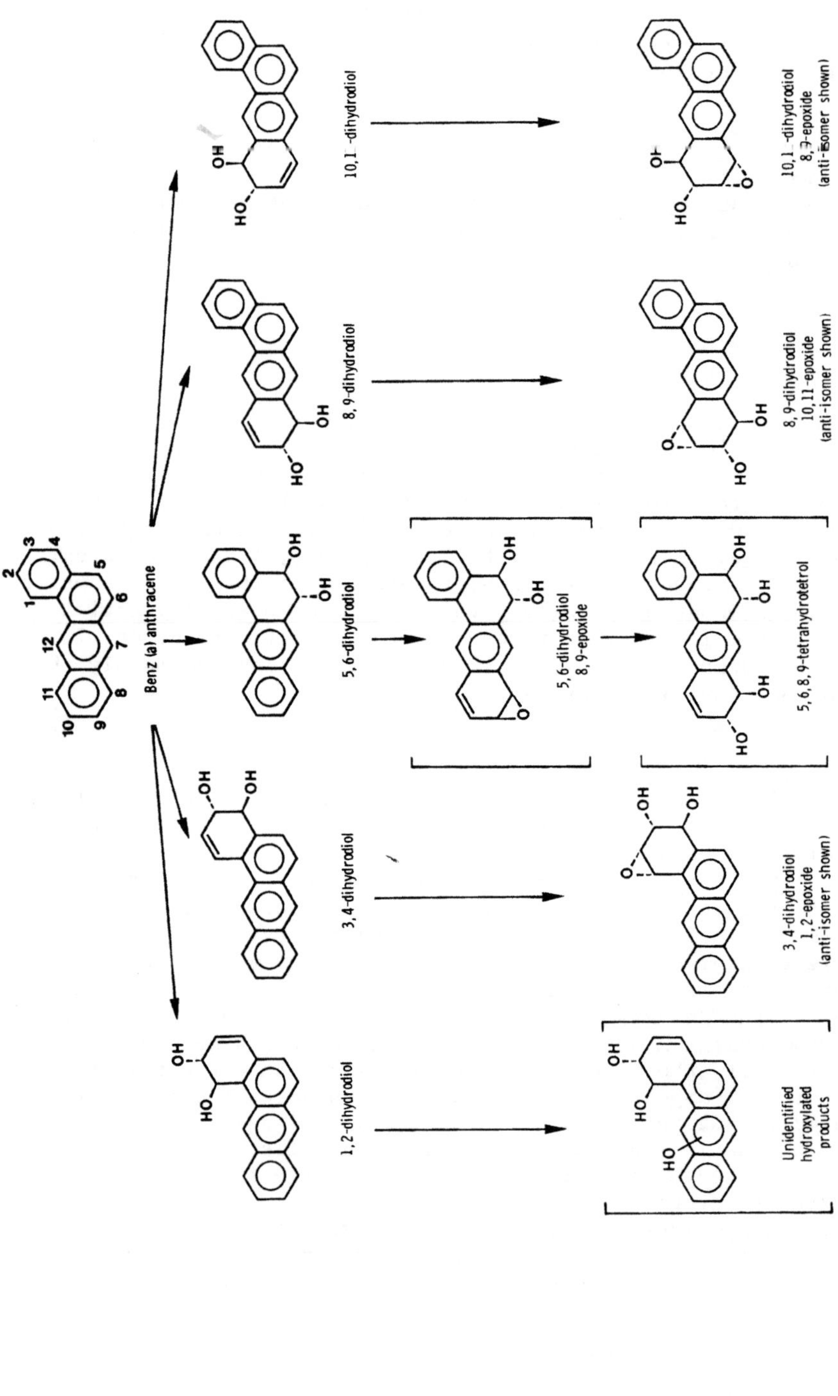

Figure 1. Metabolic formation of dihydrodiols and diol-epoxides derived from benz(a)anthracene

epoxides can exist in stereoisomeric forms according to whether the epoxide group is on the same side (the syn-isomers) or the opposite side (the anti-isomers) of the ring to the benzylic hydroxyl groups. It is now apparent that diol-epoxides formed from certain non-K-region dihydrodiols by metabolism on the adjacent olefinic double bonds are involved in the reactions with nucleic acids that occur when a tissue is treated with a carcinogenic hydrocarbon and thus compounds of this type appear to be the ultimate carcinogenic species. It has been pointed out by Jerina and his colleagues (Jerina and Daly, 1977; Jerina and Lehr, 1977) that epoxides formed adjacent to the 'bay-region' (see formulae I, II and VI) of polycyclic hydrocarbons would, on theoretical grounds, be more reactive chemically and thus possibly more biologically active than epoxides formed elsewhere on a hydrocarbon molecule. The purpose of this article is to consider the involvement of diol-epoxides, both 'bay-region' and non-'bay-region', in the reactions with nucleic acids that occur in cells or tissues treated with either benz(a)anthracene, 7-methylbenz(a) anthracene or 7,12-dimethylbenz(a)anthracene.

2. METABOLIC ACTIVATION OF BENZ(a)ANTHRACENE

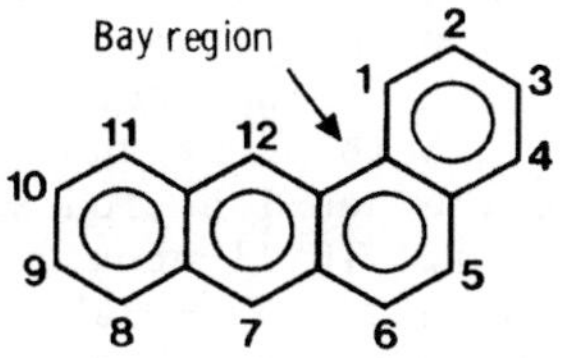

I

Benz(a)anthracene (I) is generally regarded as a weak carcinogen and tumour-initiator. The dihydrodiols and diol-epoxides formed in the metabolism of the hydrocarbon are shown in Figure 1. The 1,2-dihydrodiol is only a minor product of metabolism and, since the hydroxyl groups are adjacent to the 'bay-region', it exists in a quasi-diaxial conformation (Lehr et al., 1977). Although the further metabolism of this dihydrodiol has not been examined, other non-'bay-region' dihydrodiols, such as 9,10-dihydrodiol of benzo(e)pyrene, that have diaxial conformations appear to be metabolized to products other than vicinal diol-epoxides (Wood et al., 1979). It seems likely therefore, that the 1,2-dihydrodiol is a detoxification product since its further metabolism probably does not lead to a vicinal diol-epoxide. The 3,4-dihydrodiol can give rise on further metabolism to a 'bay-region' 3,4-diol-1,2-epoxide and evidence presented below suggests that it is mainly the anti-isomer. The K-region 5,6-dihydrodiol does not possess an olefinic double bond and therefore cannot give rise directly to vicinal diol-epoxides. However, studies (Keysell et al., 1975) on its further metabolism, using glutathione as a trapping agent, indicated that metabolism probably occurred on the 8,9-bond, presumably through the formation of the non-vicinal 5,6-diol-8,9-epoxide, which could then yield the 5,6,8,9-tetrahydrotetrol by hydration of the epoxide ring. The 8,9-dihydrodiol is metabolized mainly to the anti-isomer of the 8,9-diol-10,11-epoxide (Sims and Grover, 1976), although there is evidence that some metabolism on the 5,6-bond also occurs (Keysell *et al.*, 1975). The 10,11-dihydrodiol appears to be further metabolized mainly to the *anti*-isomer of the 10,11-diol-8,9-epoxide, although a little of the *syn*-isomer is also formed (C.S. Cooper, unpublished observations).

Early studies by Swaisland *et al.* (1974), using Sephadex LH20 chromatography, of the nucleoside adducts obtained from the hydrolysis of DNA of hamster embryo cells treated with benz(a)anthracene indicated that an 8,9-diol-10,11-epoxide might be involved in the metabolic activation of the hydrocarbon. However, later studies showed that the 3,4-dihydrodiol, which could give rise to the 'bay-region' 3,4-diol-1,2-epoxide on further metabolism, was the most active of the five dihydrodiols, includin the 8,9-dihydrodiol, of benz(a)anthracene, both as a mutagen (Wood *et al.*, 1976) and as a tumour initiator (Wood *et al.*, 1977), and this was interpreted as indicating that the 'bay-region' diol-epoxides might be involved in the metabolic activation of the hydrocarbon.

More recently, detailed chromatographic examinations of the nucleoside adducts obtained from the DNA of both mouse skin and hamster embryo cells treated with ^{3}H-labelled benz(a)anthracene have been carried out (Cooper *et al.*, 1980a). The hydrocarbon-nucleoside adducts obtained from both samples of DNA were identical when chromatographed on Sephadex LH20 and they both yielded one major peak of radioactivity. Comparisons betwee the chromatographic properties of these peaks and those of the nucleoside adducts obtained chemically from the non-labelled *anti*-isomers of the 3,4-diol-1,2-epoxide and the 8,9-diol-10,11-epoxide and the *syn*-isomer of the 10,11-diol-8,9-epoxide showed that the ^{3}H-labelled and the non-labelle adducts eluted from Sephadex LH20 in the same positions. The position of elution of the adducts obtained from the *syn*-isomer of the 8,9-diol-10,11-epoxide differed from those of the ^{3}H-labelled adducts. By eluting the columns with borate buffer-methanol mixtures in place of water-methanol mixtures as described by King *et al.* (1976), it was also shown that the ^{3}H-labelled adducts probably arose from *anti*- and not from *syn*-diol-epoxides since their positions of elution changed when borate buffer was used.

By the use of HPLC columns eluted with methanol-water gradients, it was possible to show that, although the nucleoside adducts obtained from the *anti*-isomers of the 3,4-diol-1,2-epoxide and the 8,9-diol-10,11-epoxide could be resolved into two peaks (probably derived from the diasteriomeric forms of the adducts since the diol-epoxides were prepared from the racemic dihydrodiols), only one peak was obtained from each of the ^{3}H-labelled adducts, indicating that the metabolic reactions leadin to the formation of these adducts from the hydrocarbon were stereospecific One of the pair of unlabelled adducts obtained from each of the diol-epoxides could not be separated by HPLC from the ^{3}H-labelled adducts. However, when the adducts were acetylated and the products examined by HPLC, the ^{3}H-labelled adducts were separated into two components, one of which was identical to one of the adducts obtained from the *anti*-3,4-diol-1,2-epoxide and the other to one of the adducts obtained from the *anti*-8,9-diol-10,11-epoxide. Thus both of these diol-epoxides appear to be involve in the reactions with DNA that occur in mouse skin and hamster embryo cell treated with benz(a)anthracene.

In a second series of experiments, Cooper et al.(1980b) examined the interactions of benz(a)anthracene with the RNA of hamster embryo cells using methods similar to those outlined above. The major product involved in the reactions of benz(a)anthracene with the nucleic acid appears to be the anti-isomer of the 8,9-diol-10,11-epoxide and little or no reactions involving reactions of the anti-3,4-diol-1,2-epoxides were detected.

Confirmation of the importance of the non-'bay-region' 8,9-diol-10,11-epoxide in the in vivo reactions of benz(a)anthracene with DNA has been provided by studies on the fluorescence spectral properties of the nucleoside adducts obtained from DNA that was isolated from mouse skin or hamster embryo cells treated with the hydrocarbon (Vigny et al., 1980). Using an excitation wavelength of 250 nm, fluorescence spectra were obtained that were similar to that of the major nucleoside adduct derived from RNA that was treated in vitro with syn-8,9-diol-10,11-epoxide. Thus one of the moieties that react with the DNA of mouse skin and hamster embryo cells treated with benz(a)anthracene appears to have retained an intact phenanthrene nucleus.

3. METABOLIC ACTIVATION OF 7-METHYLBENZ(a)ANTHRACENE

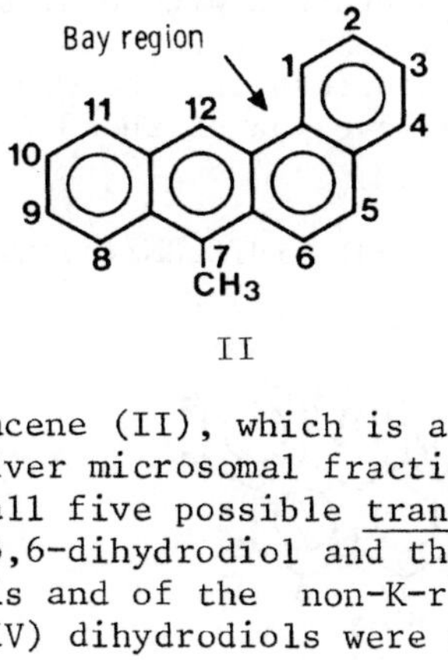

II

7-Methylbenz(a)anthracene (II), which is a moderately active carcinogen, is metabolized by rat-liver microsomal fractions and by mouse-skin maintained in organ culture to all five possible trans-dihydrodiols (Tierney et al., 1977), the K-region 5,6-dihydrodiol and the non-K-region 1,2-, 3,4-, 8,9- and 10,11-dihydrodiols and of the non-K-region dihydrodiols, the 3,4- (III) and the 8,9- (IV) dihydrodiols were the major products.

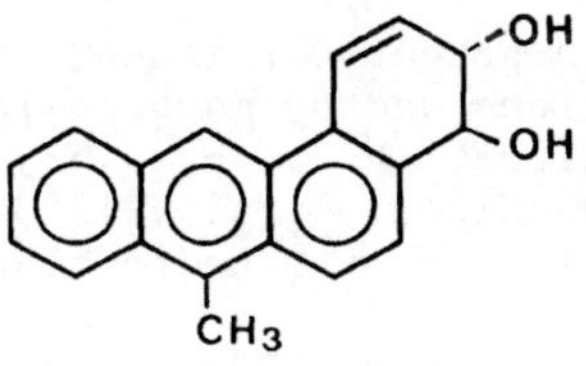

III

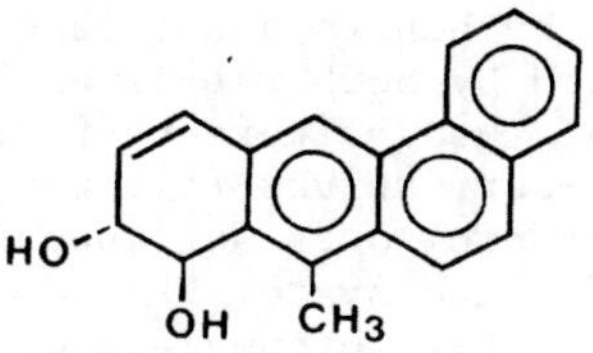

IV

The nucleoside adducts obtained from the DNA of mouse skin or mouse embryo cells treated with 7-methylbenz(a)anthracene were separable into two products by Sephadex LH20 chromatography (Baird and Brookes, 1973; Tierney et al., 1977). Comparisons of the elution profiles of the nucleoside adducts obtained from the DNA of treated mouse skin with those obtained when the nucleoside adducts derived from DNA that was allowed to react chemically with the anti-isomers of either the 'bay-region' 3,4-diol-1,2-epoxide (V) or the non-'bay-region' 8,9-diol-10,11-epoxide showed

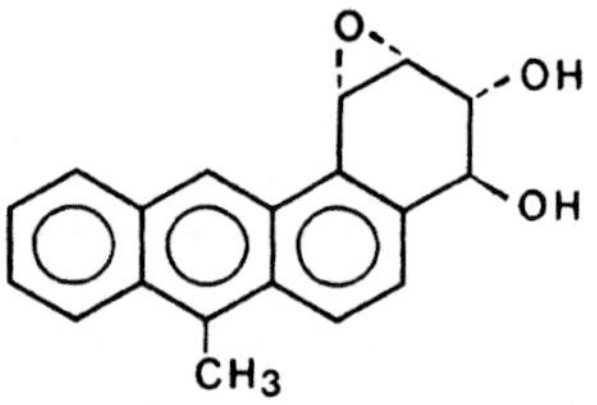

V

that the adducts arising from the anti-3,4-diol-1,2-epoxide were chromatographically similar to the adducts obtained from mouse skin whereas those derived from the anti-8,9-diol-10,11-epoxide were not (Tierney et al., 1977).

Fluorescence spectral evidence (P. Vigny, unpublished observations) also implicates the involvement of the 3,4-diol-1,2-epoxide in the in vivo reactions of 7-methylbenz(a)anthracene with DNA, since the fluorescence spectrum of the major nucleoside adduct derived from mouse skin was anthracene- rather than phenanthrene-like and was similar to that of the major adduct derived from the reaction of the 3,4-diol-1,2-epoxide with DNA.

4. METABOLIC ACTIVATION OF 7,12-DIMETHYLBENZ(a)ANTHRACENE

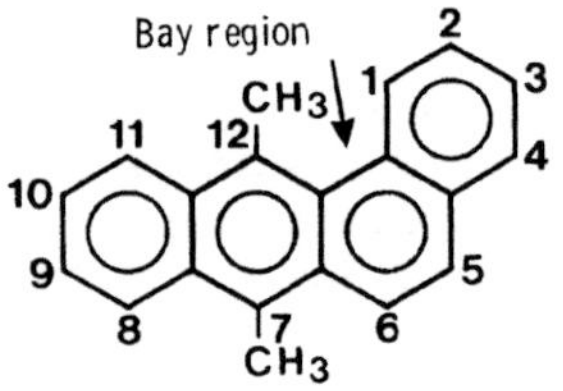

VI

7,12-Dimethylbenz(a)anthracene (VI), which is a potent carcinogen, is metabolized by both rat-liver microsomal fractions and by mouse skin maintained in organ culture to four trans-dihydrodiols, the 3,4-, 5,6-, 8,9- and 10,11-derivatives (Tierney et al., 1978; MacNicoll et al., 1980). The 1,2-dihydrodiol does not appear to be a metabolite of the hydrocarbon in either of these systems.

The mixture of nucleoside adducts obtained from the DNA of mouse skin treated with the hydrocarbon can be resolved by Sephadex chromatography into at least three products (Cooper et al., 1980c) but attempts to identify these by chromatographic comparisons with the nucleoside adducts obtained from DNA that was treated with either the 'bay-region' 3,4-diol-1,2-epoxide (VII) or the non-'bay-region' 8,9-diol-10,11-epoxide and 10,11-diol-1,2-epoxide have so far been only partially successful. The adducts arising from the 3,4-diol-1,2-epoxide eluted in similar

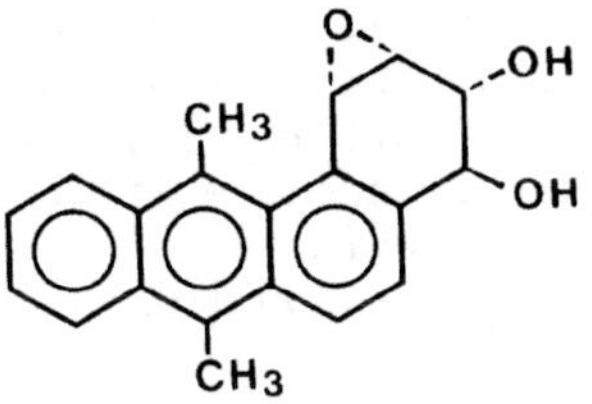

VII

fractions to two of the products obtained in the in vivo experiments, but the possibility that the other diol-epoxides were involved in the reactions could not be eliminated.

Studies on the fluorescence spectra of the three products showed that all possessed anthracene- rather than phenanthrene-like spectra (P. Vigny, unpublished observations). These results suggest that the 'bay-region' 3,4-diol-1,2-epoxide is the important metabolic intermediate in the in vivo reactions of 7,12-dimethylbenz(a)anthracene with the DNA of mouse skin

5. DISCUSSION

The presence of adducts arising from the reaction of the non-'bay-region' anti-8,9-diol-10,11-epoxide with the nucleic acids of mouse skin and hamster embryo cells treated with benz(a)anthracene is perhaps unexpected since the 8,9-dihydrodiol, which yields this diol-epoxide on further metabolism shows only low biological activity. Thus, for example, when the five dihydrodiols and the parent hydrocarbon were tested for mutagenicity towards Salmonella typhimurium in the presence of a metabolizing system, only the 3,4-dihydrodiol was highly active (Wood et al., 1976). However, the 8,9-dihydrodiol was more active than the parent hydrocarbon, confirming similar observations by Malaveille et al. (1975). Of the dihydrodiols tested in Chinese hamster V79 cells, the 3,4-dihydrodiol was also the most active mutagen (Slaga et al., 1978).

In tests on the abilities of the dihydrodiols to act as tumour initiators on mouse skin both Wood et al. (1977) and Slaga et al. (1978) found that the 3,4-dihydrodiol was much more active than either benz(a) anthracene itself or any of the other dihydrodiols tested. Similarly, the 3,4-dihydrodiol was active in inducing pulmonary tumours in newborn mice whereas the other dihydrodiols showed little or no activity (Wislocki et al., 1978).

The anti-3,4-diol-1,2-epoxide was also the most active of a number of diol-epoxides derived from benz(a)anthracene that were tested by Slaga et al. (1978) as tumour initiators on mouse skin.

Thus, although the biological data suggest that the 3,4-dihydrodiol and its related 'bay-region' anti-3,4-diol-1,2-epoxide are biologically active, the biochemical data indicate that the 8,9-dihydrodiol and its related non-'bay-region' anti-8,9-diol-10,11-epoxide are involved in the metabolic activation of the hydrocarbon. The reasons for the relatively low activities shown by the latter dihydrodiol and diol-epoxide when tested in biological systems have yet to be explained. When tested in the ØX174 DNA system of Hsu et al. (1979), the anti-8,9-diol-10,11-epoxide was more active than the anti-3,4-diol-1,2-epoxide, indicating either that the 8,9-diol-10,11-epoxide reacts more readily than the 3,4-diol-1,2-epoxide with DNA or that the latter epoxide is less stable than the former. The anti-10,11-diol-8,9-epoxide is also positive in this system but appears to play no part in the metabolic activation of benz(a)anthracene.

The evidence so far available on the metabolic activation of 7-methylbenz(a)anthracene and 7,12-dimethylbenz(a)anthracene indicates that it is the 'bay-region' 3,4-diol-1,2-epoxides that are involved in the metabolic activation of the hydrocarbons. With both these hydrocarbons, the biological evidence supports the biochemical evidence so that, for example, the 3,4-dihydrodiols are highly active as mutagens (Malaveille et al., 1977, 1978) and as tumour initiators on mouse skin (Chouroulinkov et al., 1977; Slaga et al., 1979).

The reasons why the 8,9-dihydrodiol of benz(a)anthracene is involved in the metabolic activation of this hydrocarbon whereas the 8,9-dihydrodiols are not involved in the metabolic activation of 7-methylbenz(a) anthracene and 7,12-dimethylbenz(a)anthracene are not yet clear. Approximately equal amounts of the 3,4- and 8,9-dihydrodiols are formed from benz(a)anthracene by mouse skin maintained in organ culture, whereas more of the 8,9- than the 3,4-dihydrodiols are formed from both 7-methylbenz(a)anthracene and 7,12-dimethylbenz(a)anthracene in this system (MacNicoll et al., 1980). However, although the 8,9-dihydrodiol of benz(a)anthracene exists in a preferred quasi-diequatorial conformation, those of 7-methylbenz(a)anthracene and 7,12-dimethylbenz(a)anthracene adopt quasi-diaxial conformations. It has been shown that 'bay-region' dihydrodiols, which also adopt diaxial conformations, are metabolized to products other than vicinal diol-epoxides (Wood et al., 1979). It is thus possible that the 8,9-dihydrodiol of 7-methylbenz(a)anthracene and 7,12-dimethylbenz(a)anthracene are not metabolized to any large extent to the corresponding vicinal diol-epoxides. In any event, the overall levels of reaction of benz(a)anthracene to the DNA of mouse skin are much lower than those of the carcinogens, 7-methylbenz(a)anthracene and 7,12-dimethylbenz(a) anthracene (Phillips et al., 1979).

ACKNOWLEDGEMENTS

I wish to acknowledge the contributions made to this work by my colleagues, Philip Grover, Colin Cooper, Alan Hewer, Alan MacNicoll, Kalyani Pal, David Phillips, Odartey Ribeiro, Brian Tierney and Christine Walsh at the Institute of Cancer Research, and to Paul Vigny and his colleagues at the Institut Curie in Paris. The work was supported by grant from the Medical Research Council and the Cancer Research Campaign and by grant No. CA21959 from the National Cancer Institute, USPHS.

REFERENCES

Baird, W.M. and Brookes, P.: 1973, Cancer Res. 33, pp 2378-2385.

Baird, W.M., Dipple, A., Grover, P.L., Sims, P. and Brookes, P.: 1973, Cancer Res. 33, pp 2386-2392.

Booth, J. and Sims, P.: 1974, FEBS Lett. 47, pp 30-33.

Boyland, E. : 1950, Biochem.Soc.Symp. 5, pp 40-54.

Chouroulinkov, I., Gentil, A., Tierney, B., Grover, P.L. and Sims, P.: 1977, Cancer Lett. 3, pp 247-253.

Cooper, C.S., Ribeiro, O., Hewer, A., Walsh, C., Pal, K., Grover, P.L. and Sims, P.: 1980a, Carcinogenesis, in press.

Cooper, C.S., Ribeiro, O., Farmer, P.B., Hewer, A., Walsh, C., Pal, K., Grover, P.L. and Sims, P.: 1980b, Chem.-Biol.Interactions, submitted

Cooper, C.S., Ribeiro, O., Hewer, A., Walsh, C., Grover, P.L. and Sims, P.: 1980c, Chem.-Biol.Interactions, 29, pp 357-367.

Hsu, W.T., Lin, E.J., Fu, P.P., Harvey, R.G. and Weiss, S.B.: 1979, Biochem.Biophys.Res.Commun. 88, pp 251-257.

Jerina, D.M. and Daly, J.W.: 1977, In Parke, D.V. and Smith, R.L. (eds.) Drug Metabolism - from Microbe to Man, Taylor & Francis. London, pp 13-32.

Jerina, D.M. and Lehr, R.E.: 1977, In Ullrich, V., Roots, I., Hildebrandt, A.G., Esterbrook, R.W. and Conney, A.H. (eds.) Microsomes and Drug Oxidation, Pergamon , Oxford, U.K. pp 709-720.

Keysell, G.R., Booth, J., and Sims, P.: 1975, Xenobiotica, 5, pp 439-448.

King, H.W.S., Osborne, M.R., Beland, F.A., Harvey, R.G. and Brookes, P.: 1976, Proc.Natl.Acad.Sci.USA, 73, pp 2679-2681.

Lehr, R.E., Schaefer-Ridder, M., and Jerina, D.M.: 1976, J.Org.Chem. 42, pp 736-743.

MacNicoll, A.D., Grover, P.L. and Sims, P.: 1980, Chem.-Biol.Interactions, 29, pp 169-188.

Malaveille, C., Bartsch, H., Grover, P.L. and Sims, P.: Biochem.Biophys. Res.Commun. 66, pp 693-700.

Malaveille, C., Tierney, B., Grover, P.L., Sims, P. and Bartsch, H.: 1977, Biochem.Biophys.Res.Commun. 75, pp 427-433.

Malaveille, C., Bartsch, H., Tierney, B., Grover, P.L. and Sims, P.: 1978, Biochem.Biophys.Res.Commun. 83, pp 1468-1473.

Oesch, F. : 1972, Xenobiotica, 3, pp 305-340.

Phillips, D.H., Grover, P.L. and Sims, P.: 1979, Int.J.Cancer 23, pp 201-208.

Sims, P. and Grover, P.L.: 1974, Adv.Cancer Res. 20, pp 165 274.

Sims, P. and Grover, P.L.: 1976, Med.Biol.Environ. 4, pp 315-329.

Slaga, T.J., Huberman, E., Selkirk, J.K., Harvey, R.G., and Bracken, W.M.: 1978, Cancer Res. 38, pp1699-1704.

Slaga, T.J., Gleason, G.L., DiGiovanni, J., Sukumaran, K.B. and Harvey, R.G.: 1979, Cancer Res. 39, pp 1934-1936.

Swaisland, A.J., Hewer, A., Pal, K., Keysell, G.R., Booth, J., Grover, P.L and Sims, P.: 1974, FEBS Lett. 47, pp 34-38.

Tierney, B., Hewer, A., Walsh, C., Grover, P.L. and Sims, P.: 1977, Chem.-Biol.Interactions 18, pp 179-193.

Tierney, B., Hewer, A., MacNicoll, A.D., Gervasi, P.G., Rattle, H., Walsh, C., Grover, P.L. and Sims, P.: 1978, Chem.-Biol.Interactions 23, pp 243-257.

Vigny, P., Kindts, M., Duquesne, M., Cooper, C.S., Grover, P.L. and Sims, P: 1980, Carcinogenesis 1, pp 33-36.

Wislocki, P.G., Kapitulnik, J., Levin, W., Lehr, R., Schaefer-Ridder, M., Karle, J.M., Jerina, D.M. and Conney, A.H.: 1978, Cancer Res. 38, pp 693-696.

Wood, A.W., Levin, W., Lu, A.Y.H., Ryan, D., West, S.B., Lehr, R.E., Schaefer-Ridder, M., Jerina, D.M. and Conney, A.H.: 1976, Biochem.Biophys. Res.Commun. 72, pp 680-686.

Wood, A.W., Levin, W., Chang, R.L., Lehr, R.E., Schaefer-Ridder, M., Karle, J.M., Jerina, D.M. and Conney, A.H.: 1977, Proc.Natl.Acad.Sci.USA, 74, pp 3176-3179.

Wood, A.W., Levin, W., Thakker, D.R., Yagi, H., Chang, R.L., Ryan, D.E., Thomas, P.E., Dansette, P.M., Whittaker, N., Turujman, S., Lehr, R.E., Kumar, S., Jerina, D.M. and Conney, A.H.: 1979, J.Biol.Chem. 254, pp 4408-4415.

METABOLICALLY GENERATED FREE RADICALS FROM MANY TYPES OF CHEMICAL CARCINOGENS AND BINDING OF THE RADICALS WITH NUCLEIC ACID BASES

Chikayoshi Nagata, Masahiko Kodama, Teruyuki Kimura
and Misako Aida
Biophysics Division, National Cancer Center Res. Institute

Potent chemical carcinogens such as benzo[a]pyrene, 2-hydroxybenzo[a]-pyrene, 7,12-dimethylbenz[a]anthracene, N,N-dimethyl-4-aminoazobenzene and its derivatives, acetylaminofluorene and naphthylamines have been found to be easily converted to free radicals after being activated to the proximate forms. These free radicals were sufficiently reactive to bind covalently with nucleic acid bases. Anthanthrene which lacks bay region and 10-aza-benzo[a]pyrene in which no diol-epoxide formation is expected were found to be activated to reactive free radicals. These findings suggest that the metabolically formed free radicals are important in chemical carcinogenesis.

1. INTRODUCTION

It is generally recognized that most chemical carcinogens are required to be metabolized to proximate and ultimate forms for induction of tumor. Abundant data have been accumulated supporting the proposition that active electrophiles metabolically formed bind to nucleophilic sites in nucleic acid bases (1,2). Such heterolytic reaction is the most probable candidate for the initiation reaction in chemical carcinogenesis. However, this does not necessarily exclude other route of binding reaction of chemical carcinogens with nucleic acid bases. Thus, we have found that almost chemical carcinogens are metabolically converted to reactive free radicals, which suggests the possible significance of homolytic reaction in the initial step of carcinogenesis.

2. ENZYMATIC FORMATION OF FREE RADICALS FROM AROMATIC HYDROCARBONS

2.1. Hydroxylated Benzo[a]pyrene (B[a]P)

Previously we have found that B[a]P is converted to 6-oxy-B[a]P radical after being metabolized to 6-hydroxy-B[a]P in liver- as well as skin- and lung-microsome systems (3-5). In this connection, it is of interest to know whether or not other hydroxylated B[a]P are active in

B. Pullman, P. O. P. Ts'o and H. Gelboin (eds.), Carcinogenesis: Fundamental Mechanisms and Environmental Effects, 43–54.

forming the free radical in microsome system (Table i). Twelve hydroxy B[a]P can be classified into three groups. Group 1: NADPH is required for the free radical formation and 2-, 4-, 9-, and 11-OH derivatives belong to this group. Group 2: No ESR signal was detected in the presence and absence of NADPH and 3- and 12-OH derivatives belong to this group. Group 3: NADPH is dispensable for the formation of the free radical and 1-, 5-, 6-, 7-, 8-, and 10-OH derivatives belong to this group. It is likely that oxy radical is formed from 8-OH derivative by comparing the linewidth of the radical formed from 8-OH derivative with that of 6-oxy-B[a]P radical. Linewidths of other OH derivatives are smaller than that of oxy radical and rather close to the values of B[a]P-semiquinone radicals (5.2G). By standing the ESR tubes at room

Table 1. Free radicals produced by incubating hydroxylated B[a]P with rat liver microsomes (MC-induced) in the presence and absence of NADPH

	Incubation with rat liver microsomes			Incubation with rat liver microsomes	
	-NADPH	+NADPH		-NADPH	+NADPH
2-OH-B[a]P	0	+ (7.7G)	1-OH-	+ (11.0G)	+ (8.7G)
4-OH-	0	+ (5.4G)	5-OH-	+ (4.4G)	+ (7.1G)
9-OH-	0	+ (4.3G)	6-OH-	+++ (13.7G)	++ (13.7G)
11-OH-	0	+ (5.4G)	7-OH-	+ (5.5G)	+ (5.5G)
3-OH-	0	0	8-OH-	+ (13.0G)	+ (13.0G)
12-OH-	0	0	10-OH-	+ (4.4G)	+ (4.4G)

Sign + designates that the ESR signal is detected and the amount of the radical is proportional to increasing number of sign +. No ESR signal is detected in the case of 0. The values in parentheses are the linewidths.

temperature exposing to the air, enzymatically formed free radical from 1-, 2-, 4-, 5-, 9-, and 11-OH-B[a]P increased gradually. This shows

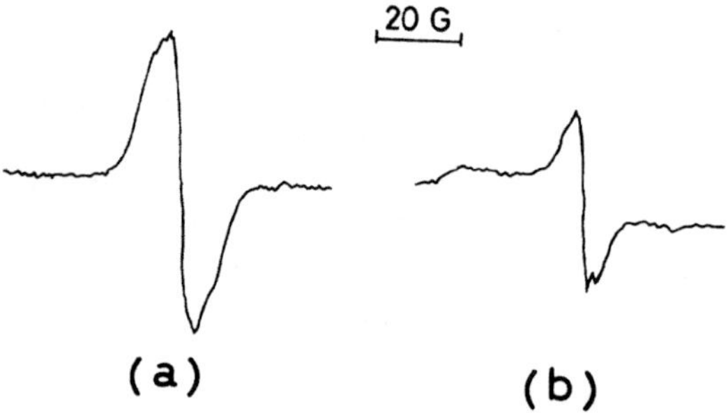

Figure 1. (a) ESR signal of the free radical obtained by incubating 2-OH-B[a]P with rat liver microsomes (MC-induced) and (b) covalently bound adduct of the radical with poly(G). The experimental condition is given in ref. 7.

that metabolites formed from these hydroxy derivatives are easily converted oxidatively to the free radicals.

Recently, potent carcinogenicity of 2-OH-B[a]P has been found (6). In relevance to the importance of bay region in carcinogenesis of aromatic hydrocarbons, it is of interest to know whether this compound give rise to diol or diol-epoxide in microsome system. However, to our knowledge, no metabolic formation of diol or diol-epoxide has been reported. Interestingly we observed that a free radical is produced enzymatically from 2-OH-B[a]P and this radical binds covalently with poly(G) (Fig. 1 b).

2.2. Anthanthrene and 10-Aza-B[a]P

No diol-epoxide formation is expected for anthanthrene and 10-aza-B[a]P, because of lack of bay region in the former compound and aza-replacement at the 10 position in the latter compound. As these two compounds have been reported to be moderately carcinogenic, an activation pathway other than epoxidation at the bay region must be involved in the metabolic activation of these compounds. As a plausible candidate for this activation pathway, the formation of free radicals was considered. Incubation of these compounds with liver microsomes fortified with NADPH resulted in the appearance of ESR signals; g=2.005 and linewidth was 9.8G for the free radical produced from anthanthrene and g=2.004 and linewidth was 14.3G for 10-aza-B[a]P. From the calculated spin densities and coupling constants, the former radical was assigned to 6,12-anthanthrene-semiquinone radical and the latter one to 6-oxy-10-aza-B[a]P radical (7,8). These free radicals were fairly reactive to bind covalently with poly(G) (Fig. 2).

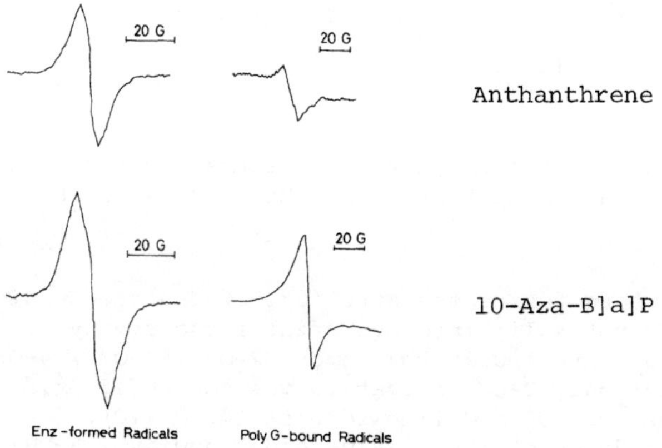

Figure 2. The ESR signal of the free radical formed by incubating anthanthrene or 10-aza-B[a]P with rat liver microsomes (MC-induced) and the signal of the covalently bound adduct of the radical with poly(G). Reaction condition and sample preparation are given in ref. 7.

2.3. 7,12-Dimethylbenz[a]anthracene (DMBA)

Carcinogenicity of DMBA is known to be the most potent among aromatic hydrocarbon carcinogens. Nonetheless, the metabolic pathways of this compound have not been extensively studied. Recently Cooper et al. (9) showed that some of the hydrocarbon-DNA products formed in mouse skin treated DMBA arose from the reaction of DNA with diol-epoxide at the bay region. They found, however, that other unidentified adducts were also formed suggesting the involvement of other mechanism than the diol-epoxide formation. In this sense, it is of importance to test the enzymatic formation of free radical from DMBA.

After incubation of DMBA with rat liver microsomes (MC-induced) plus supernatant fortified with NADPH, fairly intense ESR signal was observed with g=2.005 and linewidth=14.9G (Fig. 3. a). Microsomes alone was less effective in yielding free radical and no free radical was formed when supernatant alone was used. No signal was detected when microsomes plus supernatant from non-induced rats were used.

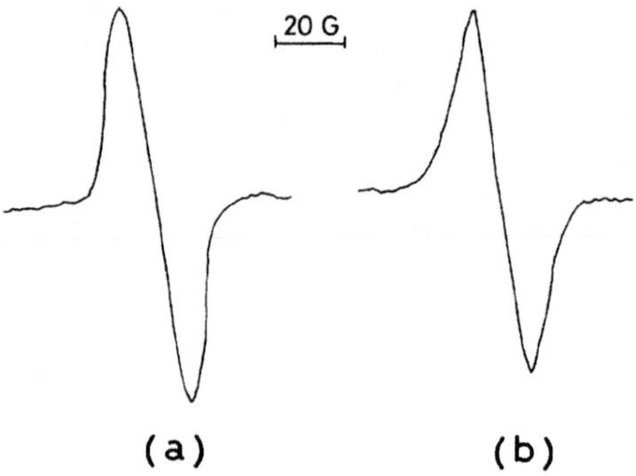

Figure 3. (a) The ESR signal of the enzymatically formed free radical from DMBA. (b) Covalently bound adduct of 2-oxy-DMBA radical with poly(G).

It was difficult to characterize the structure of the free radical, because the ESR signal did not split into hyperfine structure by degassing. However, we found previously that among 2-OH-, 3-OH-, 4-OH-, 5-OH-, 9-OH- and 10-OH-DMBA, only 2-OH derivative was converted oxidatively to oxy radical with g=2.005 and linewidth of 14.9G (10). Considering from the coincidence of the g value and linewidth, 2-OH-DMBA radical is thought to be the most probable candidate for the enzymatically formed free radical from DMBA. 2-OH-DMBA radical was reactive to bind covalently with poly(G) (Fig. 3, b).

3. ENZYMATIC FORMATION OF FREE RADICALS FROM DIMETHYL AMINOAZOBENZENE AND ITS DERIVATIVES

3.1. Free Radical Formation in Enzymatic System in vitro

Incubation of potent hepatocarcinogen, 3'-methyl-N,N-dimethyl-4-aminoazobenzene (3'-Me-DAB) with rat liver microsomes fortified with NADPH or NADH resulted in the formation of free radical composed of six hyperfine structure (Fig. 4) (11). Based on the extensive studies, Miller and Miller have established two main metabolic pathways of aminoazo dyes; the activation via N-hydroxy-N-methyl-4-aminoazobenzene (N-OH-MAB) and detoxification pathway via N-hydroxy-4-aminoazobenzene (N-OH-AB) (1,2). It is of importance, therefore, to clarify the relation between the free radical formation and these two metabolic pathways. Since the enzymatically formed free radical gave hyperfine structure, identification of the chemical structure was undertaken by synthesis of authentic samples of N-OH-MAB and N-OH-AB and by calculation of the coupling constants using the INDO (Intermediate Neglect of Differential Overlap) molecular orbital method.

Figure 4. The ESR signal obtained by incubating 3'-Me-DAB with liver microsomes from MC-treated rats. The reation mixture, in a total volume of 50 ml contained 15 mg of microsomal protein, 2.5 mmole Tris-HCl buffer (pH 7.5), 7.75 mmole KCl, 40 μmole NADPH, 0.15 mmole $MgCl_2$, and 5 μmole of 3'-Me-DAB (added in 0.5 ml of methanol just prior to incubation). After incubation (37°,10 min), the reaction mixture was extracted twice with 50 ml of benzene. Following evaporation of the extract to dryness, the residue was dissolved in 0.3 ml of benzene for the ESR measurement.

Synthesized N-OH-MAB and N-OH-AB, especially the former compound, were extremely labile and easily converted to free radical by dissolving in benzene (Fig. 5). By comparing these signals with the ESR signal in Fig. 4 and also with calculated coupling constants, the enzymatically formed free radical was identified as nitroxide radical produced from N-OH-MAB (11).

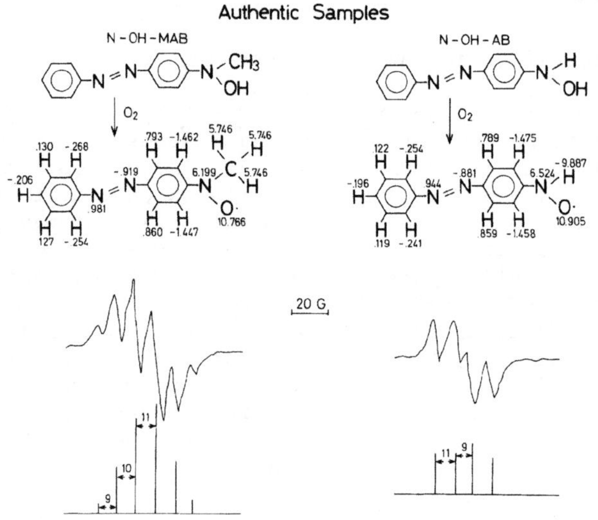

Figure 5. ESR signals of authentic samples of N-OH-MAB and N-OH-AB in benzene and coupling constants calculated by means of INDO MO method.

Very intense ESR signal with the same hyperfine structure as that of 3'-Me-DAB was observed when 3'-Me-MAB was used. Inhibitory effect of CO was observed for 3'-Me-DAB but not for 3'-Me-MAB (Table 2), indicating that the enzyme other than the cytochrome P-450 is involved in the N-hydroxylation of 3'-Me-MAB. Thus, we propose the metabolic pathway as shown in Fig. 6.

N-Hydroxylation has been considered to be an obligatory step in the metabolic conversion of carcinogenic aminoazo dyes to ultimate carcinogen (1,2). However, N-OH-MAB has never been detected directly by incubating MAB or DAB with rat liver microsomes, although it was

Table 2. Effect of CO on the formation of free radicals from 3'-Me-DAB and 3'-Me-MAB

	air	N_2:air 1 : 1	N_2:air 19 : 1	CO:air 1 : 1	CO:air 19 : 1
3'-Me-DAB	100	115	105	61	20
3'-Me-MAB	100	93	135	108	113

detected in purified enzyme system (12). We have detected nitroxide radical which was easily produced from N-OH-MAB and this is the reason why N-OH-MAB has never been detected in microsome system.

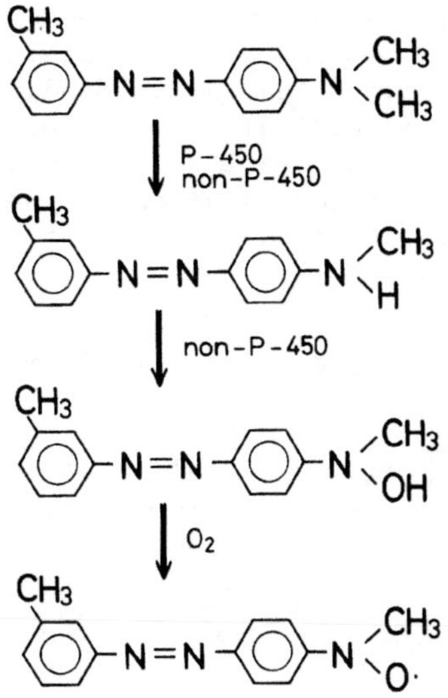

Figure 6. Proposed metabolic pathways of 3'-Me-DAB.

Free radical produced from N-OH-MAB is fairly reactive. Thus, after incubation of N-OH-MAB with poly(G), the covalently bound adduct showed a strong ESR signal (Fig. 7, a). On the other hand, only a very weak signal was observed when N-OH-AB which is slightly or non-carcinogenic (13) was subjected to incubation with poly(G) (Fig. 7 b).

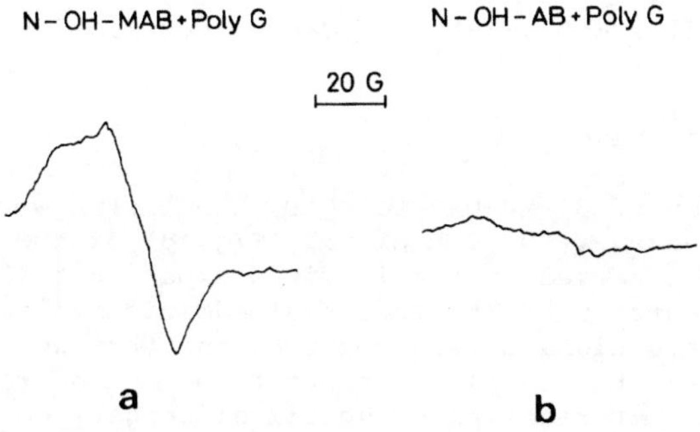

Figure 7. ESR signals of the covalently bound adducts of (a) N-OH-MAB and (b) N-OH-AB with poy(G). The reaction condition and sample preparation are the same as the case of Fig. 2., except that ethanol/H_2O (1:1) was replaced by DMSO/H_2O (1:1).

It is generally believed that N-OH-MAB is further activated to the sulfate compound in order to bind with tissue components (1,2). However, the above result indicates that without further activation, N-OH-MAB is sufficiently reactive to bind with nucleic acid bases. Thus, it is not unreasonable to consider two activation pathways leading to binding with nucleic acid bases; one is via nitroxide radical and another is via sulfate compound. Additional support to the idea that nitroxide radical is involved in the carcinogenesis proceses is given by the fairly good correlation between the amounts of produced free radicals and carcinogenicity of DAB derivatives (Table 3).

Table 3. Comparison between the amounts of the free radicals formed enzymatically from aminoazo dyes and their hepatocarcinogenicities

Compound	Amount of free radical formed enzymatically*	Carcinogenicity**
3'-Me-DAB	0.80 ± 0.15	10-12
3'-NO_2-DAB	0.74 ± 0.21	5(9)
3'-Cl-DAB	0.60 ± 0.07	5-6
DAB	0.52 ± 0.11	6
2'-Cl-DAB	0.45 ± 0.10	2
4'-Cl-DAB	0.43 ± 0.05	1-2
2'-Me-DAB	0.32 ± 0.08	2-3
2'-NO_2-DAB	0.29 ± 0.04	3
4'-Me-DAB	0.26	< 1
4'-NO_2-DAB	0.15 ± 0.04	0
2-Me-DAB	~0	0
4'-OH-DAB	~0	0

* The ratio of ESR signal of the sample to that of standard sample of Mn^{++} is normalized by the protein concentration. Each value is the average of five experiments.

** Taken from Miller,J.A.,Miller,E.C.,:1953,Adv.Cancer Res.1,pp339-396.

3.2. Free Radical Formation in vivo

After oral administration of 3'-Me-DAB to rats, the livers were removed and extracted with benzene. The ESR signal (Fig. 8) of the extract was the same as that observed in the in vitro experiment (Fig. 4). The clear signal was detected at 6 hr after oral administration. After 18 hr and 24 hr, the same signals were observed but they were a little smaller than that at 6 hr. Sample prepared from control rats which were pretreated with MC but received olive oil alone gave no ESR signal.

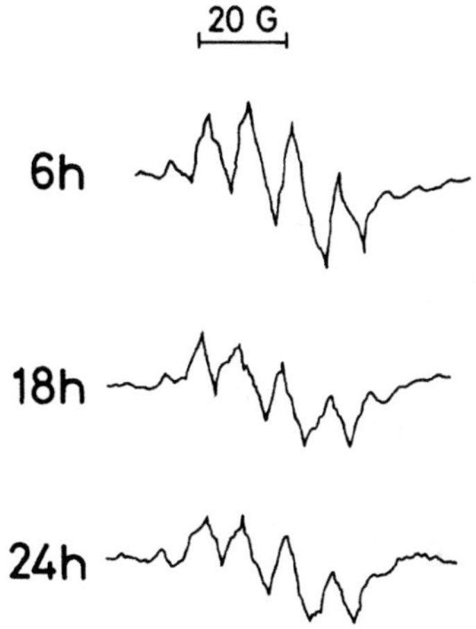

Figure 8. The ESR signal of the free radical produced from 3'-Me-DAB in vivo. 3'-Me-DAB (40 mg/2 ml olive oil) was administered through a stomach tube to male Sprague-Dawley rats (5 weeks old and MC-induced) and after intervals (6, 18 and 24 hr), five rats were killed and livers were homogenized with 8 vol. 0.15M KCl-0.05M Tris-HCl buffer (pH 7.5) in polytron for 1 min. After extraction with an equal volume of benzene, it was evaporated to dryness and dissolved in 0.3 ml of benzene for the ESR measurement.

4. ENZYMATIC FORMATION OF FREE RADICAL FROM AROMATIC AMINES

4.1. Acetylaminofluorene (AAF)

Incubation of AAF with rat liver microsomes (MC-induced) fortified with NADPH resulted in the formation of large amount of free radical, ESR signal of which is shown in Fig. 9. c. From the proposed activation scheme of AAF (1,2), nitroxide radical and nitroso radical are considered as possible candidate. By comparing the calculated coupling constants (Fig. 9, a and b) with the observed value, it was readily concluded that the produced free radical has a nitroso structure. Although it is not clear whether the nitroso radical is formed via nitroxide radical or formed directly from AAF, it may be reasonable to consider that the nitroso radical is formed via nitroxide radical in view of the proposed metabolic pathways of AAF (1,2).

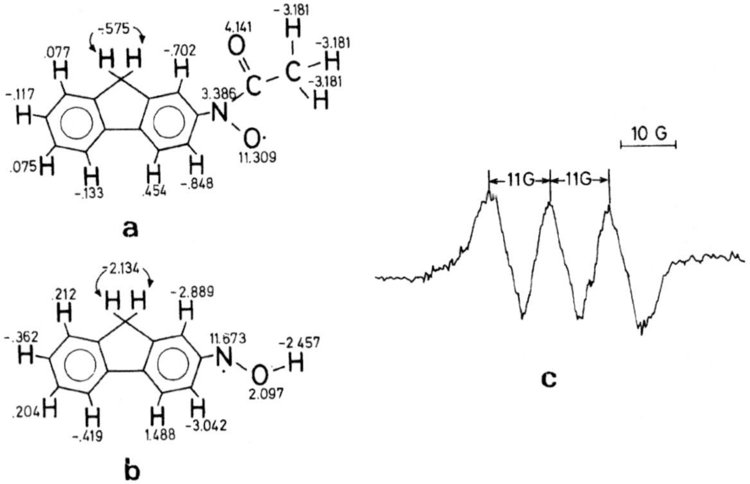

Figure 9. Coupling constants for nitroxide radical (a) and nitroso radical (b). (c) is the ESR signal of the free radical produced by incubating AAF with rat liver microsomes (MC-induced).

4.2. Naphthylamines

A large amount of free radical was formed by incubating bladder carcinogen 2-naphthylamine with rat liver microsomes (MC-induced)

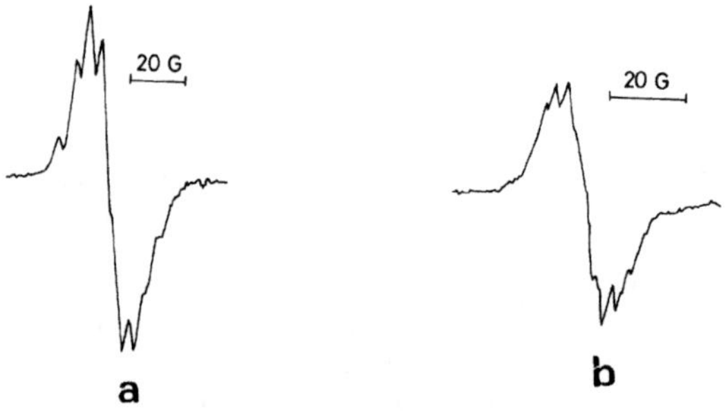

Figure 10. (a) The ESR signal of the free radical produced by incubating 2-naphthylamine with rat liver microsomes (MC-induced) and (b) the ESR signal of the free radical produced from the authentic sample of 2-aminó-1-naphthol in benzene.

fortified with NADPH, signal of which is indicated in Fig. 10, a. When 1-naphthylamine was used, the signal was far smaller than that for 2-naphthylamine. Since 2-amino-1-naphthol and 2-naphthylhydroxylamine have been regarded as proximate form of 2-naphthylamine, ESR signals of these compounds in benzene were compared with that of the enzymatically formed radical. As seen in Fig. 10, the enzymatically formed free radical can be identified as 2-amino-1-oxide radical, and this result is consistent with the proposed metabolic pathways of 2-naphthylamine, in which 2-amino-1-naphthol is regarded as a more proximate form than 2-naphthylhydroxylamine (14).

5. CONCLUSION

Activation pathways leading to the formation of free radicals from many types of potent carcinogens are indicated in Fig. 11. In addition

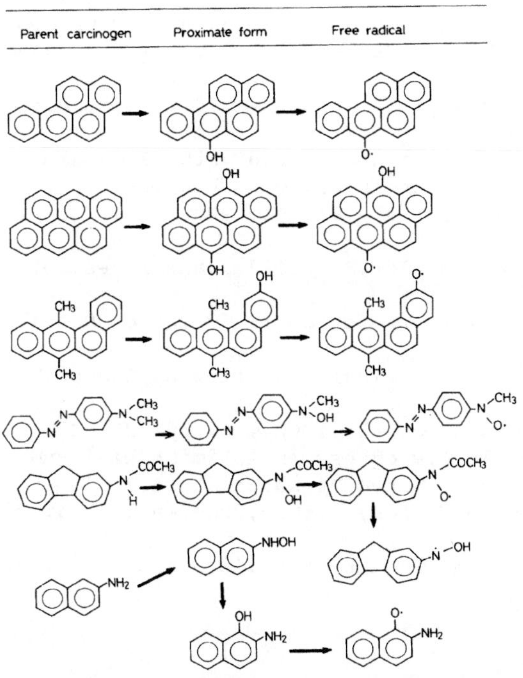

Figure 11. Enzymatic formation of free radicals from many types of potent chemical carcinogens.

to them, 4-nitroquinoline 1-oxide (15), mitomycin (16) and other quinone antibiotics (17) have been found to give rise to free radicals by reductive activation. Direct carcinogen, N-methyl-N'-nitro-N-nitrosoguanidine was found to be easily converted to free radicals without participation of enzymes (18). These findings strongly suggest that free radicals formed from chemical carcinogens are involved in the initiation of tumor production.

References

1. Miller,J.A.:1970,Cancer Res.30,pp.559-573.
2. Miller,J.A.,and Miller,E.C.:1977,Origin of Human Cancer,Book B, Mechanism of Carcinogenesis(ed.H.H.Hiatt,J.D.Watson and J.A.Winstein)Cold Spring Harbor Lab.,pp.605-627.
3. Nagata,C.,Inomata,M.,Kodama,M.,and Tagashira,Y.:1968,GANN,59, pp.289-298.
4. Nagata,C.,Tagashira,Y.,and Kodama,M.:1974, The Biochemistry of Diseases 4,Chemical Carcinogenesis(ed.P.O.P.Ts'o and J.A.DiPaolo) Marcel Dekker Inc., New York,pp.87-111.
5. Kimura,T.,Kodama,M.,and Nagata,C.:1977,Biochem.Pharmacol.26, pp.671-674.
6. Wislocki,P.G.,Chang,R.L.,et al.:1977,Cancer Res.pp.2608-2611.
7. Kodama,M.,Kimura,T.,Nagata,C.,and Shudo,K.:1978,GANN,69,pp.865-866.
8. Nagata,C.,Kodama,M.,Kimura,T.,and Yamaguchi,T.:1979,Int.J.Quantum. Chem.16,pp.917-930.
9. Cooper,C.S.,Ribeiro,O.,Hewer,A.,Walsh,C.,Grover,P.L.,Sims,P.: 1980,Chem.-Biol.Interactions,29,pp.357-367.
10. Nagata,C.,Kodama,M.,Ioki,Y.,and Kimura,T.:1980,Free Radicals and Cancer (ed.R.A.Floyd)Marcel Dekker Inc.,New York,pp.1-74.
11. Kimura,T.,Kodama,M.,and Nagata,C:1979,Biochem.Pharmacol.28,pp.557-560.
12. Kadlubar,F.F.,Miller,J.A.,and Miller,E.C.:1976,Cancer Res.36, pp.1196-1206.
13. Miller,E.C.,Kadlubar,F.F.Miller,J.A. et al.:1979,Cancer Res.39, pp.3411-3418.
14. Arcos,J.C.,and Argus,M.F.:1974,Chemical Induction of Cancer,Vol II B.Acad.Press Inc.
15. Nagata,C.,Kataoka,N.,Imamura,A.,et al.:1966,GANN,pp.323-335.
16. Nagata,C.,and Matsuyama,A.:1970,Progress in Antimicrobial and Anticancer Chemotherapy,Univ.of Tokyo Press,pp.423-427.
17. Bachur,N.R.,Gordon,S.L.,Gee,M.V.,Kon,H.:1979,Proc.Natl.Acad.Sci., USA,76,pp.954-957.
18. Nagata,C.,Ioki,Y.,Kodama,M.,and Tagashira,Y.:1973,Ann.N.Y.Acad. Sci.222,pp.1031-1047.

NUCLEOPHILICITY OF DNA. RELATION TO CHEMICAL CARCINOGENESIS.

BERNARD PULLMAN and ALBERTE PULLMAN
Institut de Biologie Physico-Chimique, Laboratoire de Biochimie Théorique, associé au C.N.R.S., 13, rue Pierre et Marie Curie, PARIS 75005, France.

It seems well established today that the active forms of a large number of carcinogenic substances (the so-called proximate or ultimate carcinogens) are electrophilic agents : positive ions or uncharged molecules with reactive centers at their electron deficient atoms (see e.g. 1,2). This is true both for carcinogens considered as being active per se (a number of relatively simple alkylating agents) and those which necessitate a metabolic activation (aromatic hydrocarbons, aromatic amines, aflatoxin B_1 etc.) (Fig. 1).

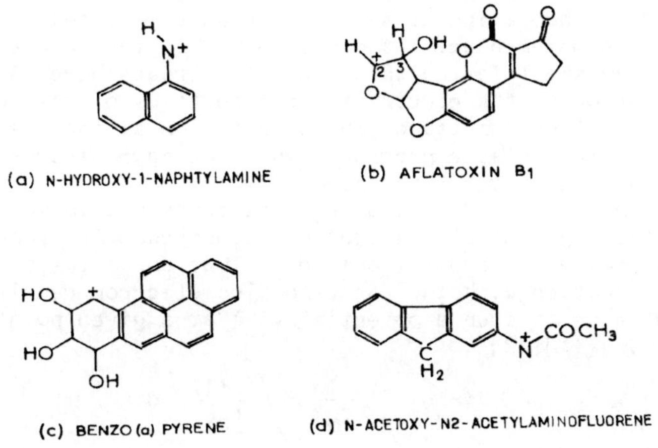

Figure 1. Proximate or ultimate metabolic electrophilic forms of carcinogens.

B. Pullman, P. O. P. Ts'o and H. Gelboin (eds.), Carcinogenesis: Fundamental Mechanisms and Environmental Effects, 55–66.

Among the known targets of these substances, presumed to be important for their physiopathological activity, are the nucleic acids in which the attacked centers are primarily the electron rich atoms of the purine and pyrimidine bases, generally their lone-pair carrying heteroatoms nitrogens and oxygens (but in some cases also their carbons) and the negatively charged phosphates.

This situation draws attention to the problem of the nucleophilicity of the nucleic acids and their constituents and because of the apparent electronic complementarity of the interacting entities suggests a possible important role of electrostatic factors in this mutual affinity. I wish to underline immediately that important does not mean exclusive, does not even necessarily mean most important. The total interaction energy between two species, as analysable by appropriate quantum-mechanical procedures, is known to be composed of at least four essential components : the electrostatic, the polarization, the charge transfer and the exchange repulsion contributions (see e.g. 3-8). There is also the role of the solvent and counterions to be considered. A precise account of the detailed reaction specificities and preferences undoubtedly needs to take into consideration all the factors involved (see e.g. [3-8]).

Nevertheless the fundamental importance of the electrostatic factor can hardly be questioned in particular for the early stages of the interactions and it seems thus useful to investigate to what an extent is this factor able to account for some of the characteristic features of the interaction of carcinogens with the nucleic acids and their constituents. Following the indications of such a preliminary screening an attempt may then be carried out to explicit the unaccounted features, operating at shorter distances, by taking into consideration the complementary factors involved in the interaction.

We shall follow this line of approach here. As an appropriate representation of the electrostatic properties of the substrates we shall utilize the results of the computation of their electrostatic molecular potentials, a procedure which we have discussed already at two previous Jerusalem Symposia [9,10].

May I remind that under this denomination we consider the electrostatic potential created in the neighbouring space by the nuclear charges and the electronic distribution of the system. For a given wave function with the corresponding electron density distribution $\rho(i)$ the value of such a potential $V(P)$ at a given point P in space is given by [11][12] :

$$V(P) = \sum_{\alpha} \frac{Z_{\alpha}}{r_{\alpha i}} - \int \frac{\rho(i)}{r_{Pi}} \, d\tau_i$$

where Z_{α} is the nuclear charge of nucleus α. This quantity has the double advantage of being directly obtainable from the wave fucntion (and avoiding thus any arbitrary partitioning into charges or populations) and of being an expression of the global molecular reality, clearly related to what a reactant "feels" upon approaching the substrate.

We shall concentrate our attention here essentially on the purine and pyrimidine bases of the nucleic acids and shall consider the evolution of the electrostatic molecular potentials associated with these moieties in the series : free bases, nucleosides, nucleotides, single helices and finally double helices.

TABLE I
Electrostatic molecular potential minima (in kcal/mole) at the nucleophilic sites of the four nucleic acid bases, either isolated or incorporated into different compounds.

	Single bases	Nucleosides	Nucleotides	Single helices	Double helices
N7(G)	-88.2	-88.0	-146.8	-420.6	-682.6
N3(G)	-55.6	-58.9	-100.9	-373.3	-670.0
N3(A)	-62.6	-65.8	-107.1	-390.5	-668.8
O2(T)	-56.2	-75.4	-110.5	-367.9	-662.8
O6(G)	-75.0	-75.4	-115.2	-387.7	-654.4
N7(A)	-67.1	-66.9	-127.2	-385.2	-650.2
O2(C)	-82.3	-84.7	-122.6	-410.4	-645.1
C8(G)5'	-8.0	-9.6	-73.3	-350.7	-630.3
C8(G)3'	-8.0	-5.5	-68.2	-366.5	-623.4
N2(G)3'	3.3	2.7	-29.5	-302.1	-623.1
N2(G)5'	3.3	0.8	-31.6	-260.1	-622.6
O4(T)	-55.0	-55.3	-97.2	-353.2	-611.7
C8(A)5'	0.9	-5.1	-69.2	-341.9	-610.5
N4(C)3'	-8.3	-8.2	-51.4	-299.6	-601.9
N6(A)5'	-12.9	-14.0	-52.2	-313.3	-600.4
N6(A)3'	-12.9	-13.0	-50.9	-302.7	-597.7
C8(A)3'	0.9	1.3	-60.0	-337.3	-597.2
N4(C)5'	-8.3	-9.3	-52.9	-322.2	-593.1
C5-C6(T)3'	1.0	-1.6	-69.3	-325.7	-591.6
C5-C6(C)5'	-3.2	-5.6	-70.8	-349.0	-584.1
C5-C6(T)5'	1.0	2.8	-61.1	-325.7	-583.6
C5-C6(C)3'	-3.2	-1.7	-64.2	-311.0	-569.0
N3(C)	-85.7	-85.2	-126.7	-404.4	-
N1(A)	-70.4	-71.3	-104.0	-367.4	-

Table I presents the results of a series of computations on the most prominent electrostatic potential minima associated with these different structures. Details of the computations may be found in ref. [13]. Here we would simply like to underline that the results for the double helices were obtained for a helical turn of a B-DNA type model of poly(dG.dC) and poly(dA.dT). The results for the single helices were derived simply by separating the two strands, without alteration of their internal geometry. They should thus not be confused with the usual representations of homopolynucleotides which correspond to different geometrical and spatial configurations and in some of their forms involve inter-base interactions. It seems to us that they may nevertheless be considered as a first approximation to the representation of partially denatured DNA or of a "breathing" double helix and to some extent translate certain aspects of the differences between essentially single stranded and essentially double stranded polynucleotides.

The results indicate that the most prominent electrostatic potential minima around the bases occur in their planes in the vicinity of their ring nitrogens and carbonyl oxygens, the deepest such minimum being generally associated with N7 of guanine. Out of plane minima are also present and they are associated with the amino groups of the bases, carbons C8 of the purines and the C5=C6 bonds of the pyrimidines.

We have paid special attention to the evolution of the values of such minima at related sites in the series : single bases, nucleosides, nucleotides, single helices, double helices. What appears as the most striking result in this respect is the general constant, progressive increase in the absolute value (depth) of the potential minima when we follow this series of substrates in order of increasing complexity [13-15]. Particularly striking are the strong increases in these values in the last three columns of Table I. They are obviously due to the penetration of the strong potential generated by the phosphate(s) [16,17] into the vicinity of the bases and its (their) superposition upon the potential inherent to the bases themselves. (The potentials of the phosphates extend far in space so that their effect is a long distance one [18]).

Noticeable exceptions to this rule occur upon the transition from a single helix to a double helix for the ring nitrogens such as N1 of adenine and N3 of cytosine, which are exposed in the former and involved in hydrogen bonding in the latter. Such a hydrogen bonding produces a very strong depletion of the associated molecular potential minimum [19,20]. On the other hand, the minima associated with hydrogen bonded carbonyl oxygens and amino groups follow the general trend.

Another significant feature, also evident from Table I is the possible reversal of the relative ordering of related nucleophilic centers when we go over from one type of substrate (say, single helices) to another (say, double helices).

These essential results correlate with and provide a possible explanation for some essential aspects of the interaction of the electrophilic carcinogens with nucleic acids and their constituents. Thus the result about the increase of the electrostatic molecular potential at the minima associated with the ring N and O atoms of the bases, their nucleosides or nucleotides, when those are incorporated into double stranded nucleic acids (with possibly an intermediate situation in single stranded polynucleostides or denatured DNA) can be correlated with numerous observations on the general parallel increase of reactivity towards carcinogenic agents. Among striking examples we may quote in the first place in the group of simple alkylating agents :

a) The alkylation at N7 of guanine by the diethylethyleneimonium ion $(C_2H_5)_2 N^+ \langle CH_2, CH_2 \rangle$, the cyclized form of the diethylmustard N,N-diethyl-2-chlorethylamine, $CH_3N(CH_2CH_2Cl)_2$, which proceeds at a rate 50-fold greater in native DNA and 7-fold greater in denatured DNA than in monomeric guanine ribosides and ribotides [21] .

b) A similar greater reactivity towards the same reagent of denatured DNA and of poly(A) with respect to 5'-adenylic acid [21] . On the other hand adenine units in native DNA are rather unreactive, due to the involvement in hydrogen bonding with thymine of the N1 position of adenine, the major alkylation site in the unhindered base.

c) The absence of a reaction between N-methyl-N'-nitro-N-nitrosoguanidine $O = N - N(CH_3) - C(=NH) - NH - NO_2$ and guanosine and the ability of this agent to methylate guanine residues at N7 and N3 in poly(G), poly(G.C) [22,23] and in DNA [24] .

c) Similarly, N-methyl-N-nitrosourea $O = N - N(CH_3) - C(=O) - NH_2$ and dimethyl sulfate $(CH_3)_2 SO_4$ methylate the N3 position of guanine in polymers, but do not seem to carry out this reaction at the nucleoside level [25] .

e) Diethylsulfate, $(C_2H_5)_2 SO_4$, does not react with O6 of guanosine at neutrality [26] but reacts with that site in RNA [23] .

f) Ethylation by diethylsulfate and ethylmethanesulfonate, $C_2H_5CH_3SO_3$, of poly (C), at N3 and N4, is greater than that of cytidine [27]. The restricted alkylation by these agents of the cytidine unit in poly(dG).poly(dC) may obviously be attributed to the screening of N3 of cytidine by hydrogen bonding and the displacement of the reactive sites to the guanine unit.

g) As concerns the possible inversion of the relative ordering of the reactivity of related nucleophilic centers when we go over from one type of substrate to another it may be illustrated by the reaction of ethylnitrosourea, $O = N - N(C_2H_5) - C(O) - NH_2$, an electrophile showing a strong preference for attacking oxygen atoms, with the oxygens of the bases in double stranded DNA and single stranded RNA. Thus this order is O2(T), O6(G) > O2(C), O4(T) in double stranded DNA and O2(C) > O6(G), O2(U) > O4(U) in single stranded RNA [28] . In as much as we may consider our results for the single helix representative of the situation in single stranded RNA and thymine as representative of uracil, this experimental situation parallels the results of the computations. Those account also for the observation, which appeared surprising to its authors [28] , that considerable alkylation occurs on the oxygens of the bases involved in hydrogen bonding to the complementary base in double-stranded DNA : O6(G), O2(C) and O4(T). The computations show that an appreciable potential remains associated with these atoms.

Remarkably, the general rule on the increase of the affinity of the electrophilic carcinogens towards the substrates as a function of their complexity in Table I holds also for the more voluminous carcinogen for which metabolic activation is a necessary condition for activity. Thus :

h) The arylamination and arylation of O6 of guanine by the carcinogen N-hydroxy-1-naphtylamine is greater in DNA than in poly(G) or denatured DNA and does not occur with nucleosides or nucleotides [29]. The active agent in this case probably is the naphtylnitrenium ion (Fig. 1a).

i) The reaction of the carcinogen N-hydroxy-N-2-aminofluorene, which occurs mostly at C8 of the guanine ring, is greater with native than with denatured DNA [30].

j) The adduct formation between the probable carbonium ion reactive metabolic form of aflatoxin B_1 (Fig. 1b) and, principally, N7 of the guanine moiety occurs, as investigated with the model compound of aflatoxin B_1-2,3-dichloride, to a larger extent with DNA than with RNA or poly G (the other homopolynucleotides react only to a small extent). The reaction with mononucleotides is much less than with polynucleotides (greater with GMP, AMP, CMP than with UMP or dTMP) [31][32][33].

k) A more spectacular example of a possible significance of the computations concerns the interactions involving the metabolites of carcinogenic polycyclic hydrocarbons, such as the 7,8-dihydrodiol-9,10-oxide of benzo[a]pyrene and the triol carbonium cation derived from it by the opening of the epoxide ring (Fig. 1c), considered by many as the probable ultimate carcinogenic form of this hydrocarbon. These metabolites interact in vitro and in vivo with the NH_2 groups of DNA, in the order guanine > adenine > cytosine [34-37]. Now, the NH_2 groups of the bases (N2 in guanine, N6 in adenine, N4 in cytosine) are generally considered as secondary sites for electrophilic attacks on these structures and in fact, as shown in Table I, the energy minima associated with the isolated bases are either very small, as in the case of cytosine, or may even correspond to a repulsion, as in the case of guanine. The order of nucleophilicity of these groups in the free bases is thus A > C > G. It can be seen from Table I that the potentials associated with these groups become negative in the nucleotides of the three bases concerned and that in the double helical DNA the order of their dephts becomes G > A > C and correlates with the relative affinity of the bases for the bay-region metabolites of PAH. It may be interesting to add that while the reaction of the benz[a]pyrene diol epoxide with deoxyadenosine apparently yields the same product as with DNA (substitution at the NH_2 group of A) although in lower yield, its reaction with deoxyguanosine gives, in a poor yield, a different type of product, probably an adduct at N7 of G. This situation may be correlated with the different signs of the potential minima at the NH_2 groups of these nucleosides indicated in Table I. On the other hand, the relative abundance of the different nucleoside adducts formed when the anti isomer of the metabolite reacts with DNA is only slightly influenced by the type and state of the DNA [39].

l) Similarly, the carcinogenic 7-bromomethylbenz[a]anthracene reacts with DNA also preferentially at the NH_2 groups of the bases, in aqueous solutions, in the order G > A > C [19]. In this case Michelson and Pochon [40] have shown that native DNA is more effective in that reaction than denatured DNA and that the reactivity is about 10 fold less in guanylic or adenylic acid with respect to DNA. (These authors were considering C8 as the probable site of interaction).

m) A related although apparently somewhat more complicated situation seems to occur in the interaction with DNA of a different type of carcinogen, the 2-acetylaminofluorene, in fact its metabolite N-acetoxy-N2-acetylaminofluorene (A-Ac-AAF) (active form Fig. 1d). It concerns the competition between C8 and the amino sites of purines for the binding of this carcinogen [41-43]. When this reaction takes place with denatured DNA, the binding of N-Ac-AAF occurs nearly exclusively at the C8 site of guanine. However, when the reaction takes place with native DNA only 80-85% of the binding is to that site, while 15-20% of the reactant is bound at the guanine amino group. This trend seems to be accountable for by our theoretical results on the relative values of the electrostatic molecular potentials at these sites in the two states of DNA. Thus, our results strongly suggest that for denatured DNA, the guanine C8 potential minimum is very considerably deeper (by about 60 kcal/mole) than that of its NH_2 and the observed practically exclusive binding of N-Ac-AAF at this site is therefore explicable. For the double helical DNA, the guanine C8 potential still remains the deepest one but is, however, followed closely (7 kcal/mole) by that of its amino group. The occurence of a fraction of the binding at this latter site seems therefore reasonably accounted for. The absence of reactivity of adenine, whether in the denatured or native DNA is also satisfactorily accounted for by the computations. It may also be noted that the distribution of the potentials in the free bases is quite different from that in either denatured or double-stranded DNA models and could not be used to explain the observed binding data of N-Ac-AAF.

Altogether, it appears thus from this review that the electrostatic factor seems to be able to account satisfactorily for the evolution of the reactivity of a large number of different types of carcinogenic electrophiles towards a given type of site in a series of nucleotide structures. This is to some extent a remarkable result because there is no doubt that the different types of substrate considered impose different steric and environmental constraints on the interactions established and that the increased nucleophilicity of DNA with respect to its components, as indicated by the calculations on the electrostatic molecular potentials, has to be weighted against the other factors complementary in the interaction with these electrophiles and which may play either in the same or in a reverse direction. Thus e.g. the possible occurence of intercalation or of partial stacking with the bases of some of the large carcinogenic species considered above and the associated deformations of the attacked substrate are expected to influence the ease of the reaction. Nevertheless the correlations presented above involve compounds for which intercalating covalent binding is strongly advocated such as N-acetoxy-N2-acetylaminofluorene [43] (see however, [44]), doubtful such as aflatoxin B_1 [32,33] and those for which it is rather refuted such as the dihydrodiol epoxide of benz[a]pyrene [45-46]. Apparently these factors are of somewhat lesser importance for the type of comparison carried out here than is the electrostatic one, the consideration of which seems therefore to introduce a unifying principle for an important aspect of the interaction of carcinogens with one of their probably major receptor.

We do not expect, of course, that such a single principle should be of general validity. Moreover, there is obviously one fundamental aspect of the experimental data which cannot be accounted for by the electrostatic factor alone. This concerns the preferential affinity of the different electrophilic reactants for different nucleophilic centers on the nucleic acids and their constituents. This differential affinity is obvious from the experimental data quoted and some of its essential aspects in relation with the carcinogens mentioned are illustrated in Fig. 2. It is obvious that the other factors involved in the interaction namely polarization, charge transfer and exchange repulsion must play a fundamental role in this selection.

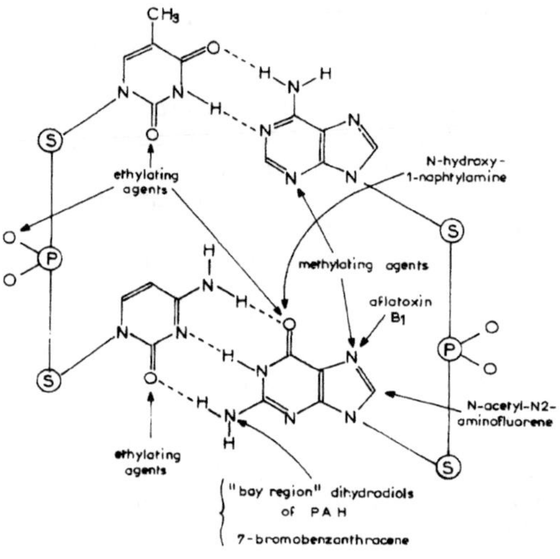

Figure 2. Principal site of interaction of principal types of carcinogens with DNA.

The problem is difficult to study theoretically with huge, complex substrates. It was, however, investigated on some simple systems. An illustrative study was concerned with the relative reactivity of the ring nitrogens and the carbonyl oxygens of the nucleic acid bases towards protonation and different alkylation reactions. It seems established that protonation occurs always preferentially on the nitrogens [8][47]. On the other hand different classes of alkylating agents show different preferences : thus for instance, while both in RNA and in DNA, N7 of guanine is the preferred site of base methylation under neutral conditions, its 06 becomes a competitive site for ethylation with certain ethylating agents such as ethylnitrosourea (although not with others, such as diethylsulfate) [48]. Under similar conditions, N3 of cytidine is the preferred site of base methylation by a large number of methylating agents including methylnitrosourea but its 02 becomes the preferred site of base ethylation with ethylnitrosourea (but not by diethylsulfate) [48][49].

Without entering here into details about the mechanism of action specific for the different alkylating agents, it is interesting, in view of these differences, to explore from the theoretical point of view what is the intrinsic preference of a methyl and of an ethyl cation with respect to attachement on a nitrogen or an oxygen atom of a nucleic base. Such a study was performed on the example of cytosine by computations implying all the electronic factors involved [5]. The result presented in Table II show an interesting evolution of the situation when going from H^+ to CH_3^+ and $C_2H_5^+$. Thus protonation is found to favor the nitrogen over the oxygen position, essentially owing to the coulomb attractive term. Going over to CH_3^+, one observes that the binding energies at equilibrium (with respect to the reactants infinitely separated) become practically identical for the methylation of N3 and 02. The components of the binding energy for CH_3^+ as for H^+ reveal that the pure electrostatic attraction is larger towards the nitrogen atom than towards the oxygen, that the delocalization term is very similar for the two positions, but that the exchange repulsion is appreciably larger for an approach towards N3 than towards 02 so that a practical cancellation of tendancies occurs, making the binding energy indifferent to the position of attack. Finally for the ethyl cation the preference observed in the final equilibrium values is now in favor of the oxygen. The factors involved appear again in the calculation of the components of the binding energy : at equilibrium, the pure coulomb attraction would favor the nitrogen, the delocalization adding somewhat to this trend, but the repulsion strongly counteracts this tendency in such a way as to make the total energy more favorable for fixation on oxygen.

Hence, it appears that the intrinsic preference of H^+, CH_3^+ and $C_2H_5^+$ towards the two essential possible binding sites in cytosine, N3 and 02, undergoes a continuous modification with the increase in size and complexity of the cation, the phenomenon being due to the increasing effect of the exchange repulsion counteracting the intrinsically larger attraction for the nitrogen position. The exchange repulsion is thus the crucial element in the evolution of the reaction from H^+ to $C_2H_5^+$ fixation. It may also be added that altogether the rate and extent of base modifications is generally lower with ethylation than with methylation [39][50][51]. This is in agreement with the relative values of the corresponding interaction energies, as indicated in Table II.

TABLE II
The binding energy ΔE and its components[(a)] (in kcal/mole) for protonation and alkylation of nucleophilic sites on cytosine.

Site	ΔE	E_C	E_{Del}	E_{Ex}
		Protonation		
N3	-293.3	-105.4	-187.9	
O2	-291.5	-85.6	-205.9	
		Methylation		
N3	-201.7	-136.6	-199.3	134.3
O2	-202.0	-105.5	-201.9	105.4
		Ethylation		
N3	-151.0	-136.6	-193.2	178.8
O2	-156.4	-104.7	-187.8	136

(a) E_C = electrostatic, E_{Del} = delocalization = polarization + charge transfer, E_{Ex} = exchange repulsion (absent in the case of a proton, because of absence of electronic cloud).

This example shows the complexity and diversity of the distribution of the different components of interaction energies, even in the case of apparently closely related reactions and contains an implicit explanation for the multiplicity and selectivity of the affinities. We are extending in our laboratory this exploration to other systems (see e.g. [7]) for the extension to the phosphate group) with the hope that these investigations will contribute to the unravelling of the mechanism of action of carcinogens.

Acknowledgement. This work was supported by the contract n°14 of the Institut National de la Santé et de la Recherche Médicale of France with its ATP 77-79-109 on Chemical Carcinogenesis.

BIBLIOGRAPHY

[1] Miller,E.C. : Cancer Research, 38, 1479 (1978).

[2] Miller, J.A. and Miller, E.C. : in Origins of Human Cancer, Cold Spring Harbor Laboratory, 1977, p. 605.

[3] Perahia, D., Pullman, A. and Pullman, B. : Theoret. Chim. Acta, 42, 23 (1976).

[4] Perahia, D., Pullman, A. and Pullman, B. : Theoret. Chim. Acta, 43, 207 (1977).

[5] Pullman, A. and Armbruster, A-M. : Theoret. Chim. Acta, 45, 249 (1977).

[6] Pullman, A. and Perahia, D. : Theoret. Chim. Acta, 48, 29 (1978).

[7] Pullman, A. and Armbruster, A-M. : Theoret. Chim. Acta, 50, 359 (1979).

[8] Pullman, A. in :Mecanismes d'Altération et de Réparation du DNA, Relations avec la Mutagénèse et la Cancérogénèse Chimique. Colloque du C.N.R.S. n°256, 1976, p. 103.

[9] Pullman, A. in : Chemical and Biochemical Reactivity, Proceedings of the 6th Jerusalem Symposium on Quantum Chemistry and Biochemistry (E.D. Bergmann and B. Pullman Eds.) Reidel Publishing Co. Dordrecht, Holland, 1974, p.1.

[10] Pullman, B. in : Catalysis in Chemistry and Biochemistry, Proceedings of the 12th Jerusalem Symposium on Quantum Chemistry and Biochemistry (B. Pullman Ed.) Reidel Publishing Co., Dordrecht, Holland, 1979, p.1.

[11] Scrocco, E. and Tomasi, J. : Topics in Current Chemistry, 42, 95 (1973).

[12] Scrocco, E. and Tomasi, J. : Adv. Quantum Chemistry, 11, 116 (1978).

[13] Pullman, B., Perahia, D. and Cauchy, D. : Nucleic Acid Research, 6, 3821 (1979).

[14] Pullman, A., Zakrzewska, Ch. and Perahia, D. : Intern. J. Quant. Chem. 16, 395 (1979).

[15] Perahia, D. and Pullman, A. : Theoret. Chim. Acta, 50, 351 (1979).

[16] Pullman, A. and Berthod, H. : Chem. Phys. Lett. 32, 233 (1975).

[17] Pullman, A. and Berthod, H. : Chem. Phys. Lett. 41, 205 (1976).

[18] Perahia, D., Pullman, A. and Pullman, B. : Theoret. Chim. Acta, 51, 349 (1979).

[19] Perahia, D. and Pullman A. : Theoret. Chim. Acta, 48, 263 (1978).

[20] Pullman, A. and Berthod, H. : Theoret. Chim. Acta, 48, 269 (1978).

[21] Price, C.C., Gaucher, G.M., Koneru, P., Shibakawa, R., Socoa, J.R. and Yamaguchi, M. : Biochim. Biophys. Acta, 166, 327 (1968).

[22] Singer, B. and Fraenkel-Conrat, H. : Biochemistry, 8, 3260 (1969)

[23] Singer, B. : Progress in Nucleic Acid Research and Molecular Biology, 15, 219 (1975).

[24] Craddock, V.M. : Chem. Biol. Interactions, 1, 234 (1969).

[25] Lawley, P.D. and Shah, S.A. : Biochemistry, J., 128, 117 (1972).

[26] Singer, B. : Biochemistry, 22, 3939 (1972).

[27] Sun, L. and Singer, B. : Biochemistry, 13, 1903 (1974).

[28] Singer, B., Bodell, W.J., Cleaver, J.E., Thomas, G.H., Rajewsky, M.F. and Than, W. : Nature, 276, 85 (1978).

29 Kadlubar, F.F., Miller, J.A. and Miller, E.C. : Cancer Research, 38, 3628 (1978).
30 Spodheim-Maurizot, M., Saint-Ruf, G. and Leng, M. : Nucleic Acid Research, 6, 1683 (1979).
31 Swenson, D.H., Miller, E.C. and Miller, J.A. : Biochem. Biophys. Res. Comm. 60, 1036 (1974).
32 Lin, J-K., Miller, J.A. and Miller, E.C. : Cancer Res., 37, 4430 (1977).
33 Swenson, D.H., Miller, J.A. and Miller, E.C. : Cancer Res. 35, 3811 (1975).
34 Jeffrey, A.M., Jennette, K.W., Blobstein, S.H., Weinstein, I.B., Beland, E.A., Harvey, R.G., Kasai, H., Miura, I. and Nakanishi, K.: J. Am. Chem. Soc. 97, 5714 (1976).
35 Weinstein, I.B., Jeffrey, A.M., Jennette, K.W., Blobstein, S.H., Harvey, R.G., Harris, C., Autrup, H., Kasai, H. and Nakanishi, K. : Science, 193, 592 (1976).
36 Jennette, K.W., Jeffrey, A.M., Blobstein, S.H., Belard, F.A., Harvey, R.G., and Weinstein, I.B. : Biochemistry, 16, 932 (1977).
37 Nakanishi, K, Kasai, H., Cho, H., Harvey, R.G., Jeffrey, A.M., Jennette, K.W. and Weinstein, I.B. : J. Am. Chem. Soc. 99, 258 (1977).
38 Osborne, M.R., Beland, F.A., Harvey, R.G. and Brookes, P. : Int. J. Cancer, 18, 362 (1976).
39 Pulkrabek, P., Leffler, S., Grunberger, D. and Weinstein, I. : Biochemistry, 18, 5128 (1979).
40 Michelson, M. and Pochon, F. : Biochimie, 54, 17 (1972).
41 Poirier, M.C., Yuspa, S.H., Weinstein, I.B. and Blobstein, S.H. : Nature, 270, 186 (1977).
42 Yamasaki, H., Pulkrabek, P., Grunberger, D. and Weinstein, I.B. : Cancer Res. 37, 3756 (1977).
43 Drinkwater, N.R., Miller, J.A., Miller, E.C. and Yang, N.C. : Cancer Res. 38, 3246 (1978).
44 Nordden, B. : Biophysical Chemistry, 8, 385 (1978).
45 Geacintov, N.E., Gagliano, A., Ivanovic, V. and Weinstein, I.B. : Biochemistry, 17, 5256 (1978).
46 Pulkrabek, P., Leffler, S., Weinstein, I.B. and Grunberger, D. : Biochemistry, 16, 3127 (1977).
47 Singer, B. : J. of Toxicology and Environmental Helath, 2, 1279 (1977).
48 Singer, B. : FEBS Letters, 63, 85 (1976).
49 Jensen, D.E. and Reed, D.J. : Biochemistry, 17, 5098 (1978).
50 Jensen, D.E. : Biochemistry, 17, 5108 (1978).

AN ANALYSIS OF THE REACTIVITIES OF EPOXIDE RINGS IN SOME CYCLIC HYDROCARBONS

Peter Politzer and Peter Trefonas III
Department of Chemistry
University of New Orleans
New Orleans, LA, 70122, U.S.A.

ABSTRACT

The reactivities of the C-O bonds in the epoxide derivatives of several cyclic hydrocarbons are analyzed by means of ab initio self-consistent-field molecular orbital calculations. Of particular interest is the degree to which epoxide ring opening may be facilitated by either or both of two special factors: (1) The proximity, in the molecule, of a double bond, and (2) intramolecular hydrogen bonding between an appropriately oriented hydroxyl group and the epoxide oxygen. It is found that while the presence of a double bond can very markedly enhance the ease of opening both epoxide C-O bonds, no such effect can be attributed to a hydroxyl group, even when it does hydrogen bond to the epoxide. The mechanism by which a double bond can weaken the C-O bonds is not understood; it appears, however, that a hydroxyl group can interfere with it. The results of this investigation are of particular importance because of their relevance to the diol epoxides that are implicated as ultimate carcinogenic forms of several polycyclic aromatic hydrocarbons.

I. INTRODUCTION

In recent years, a great deal of attention has focused upon diol epoxide metabolites of polycyclic aromatic hydrocarbons. A considerable body of evidence indicates that certain of these diol epoxides are ultimate carcinogenic forms of the parent hydrocarbons [1-6]. The prime examples are the anti and syn isomers of 7,8-dihydroxy-9,10-epoxy-7,8,9,10-tetrahydrobenzo[a]pyrene, I and II, which have been studied the most extensively; however a diol epoxide of ben[a]anthracene has now been shown to be an ultimate carcinogen [7], and there are strong indications that the same conclusion will soon be reached for diol epoxides of other polycyclic aromatic hydrocarbons [6,8].

B. Pullman, P. O. P. Ts'o and H. Gelboin (eds.), Carcinogenesis: Fundamental Mechanisms and Environmental Effects, 67–79.

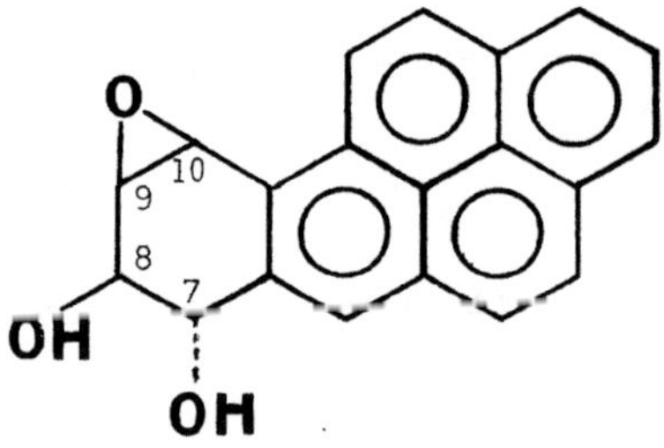

I: anti isomer

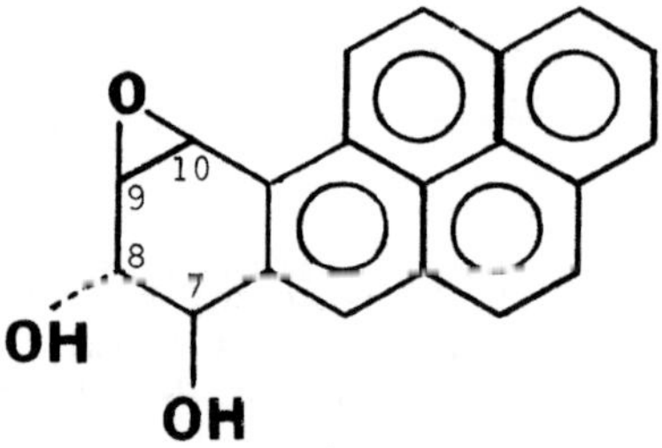

II: syn isomer

Both I and II react with nucleophilic centers on nucleic acids, DNA, and RNA; the epoxide ring opens and the interaction occurs through C(10) [2,4,9-11]. An intriguing aspect of the reactive behaviors of I and II concerns the role of the hydroxyl groups. Thus, while both of these diol epoxides exhibit high reactivity toward nucleophiles, the syn isomer, II, appears to be the more reactive [4,12-14]. For instance, its second-order rate constant with sodium 4-nitrothiophenolate is 163 times as great as that of the anti isomer, I -- and about 500 times as great as that of 9,10-epoxy-7,8,9,10-tetrahydrobenzo[a]pyrene, III, which lacks the two hydroxyl groups [12,14].

The greater reactivity of diol epoxide II, compared to I, has been attributed to intramolecular hydrogen bonding in II, between the 7-OH and the epoxide oxygen, which supposedly facilitates the opening of the epoxide ring [15-17]. In diol epoxide I, according to this reasoning, the 8-OH is not situated appropriately for such intramolecular hydrogen bonding, and so there is no corresponding enhancement of the reactivity of the epoxide ring.

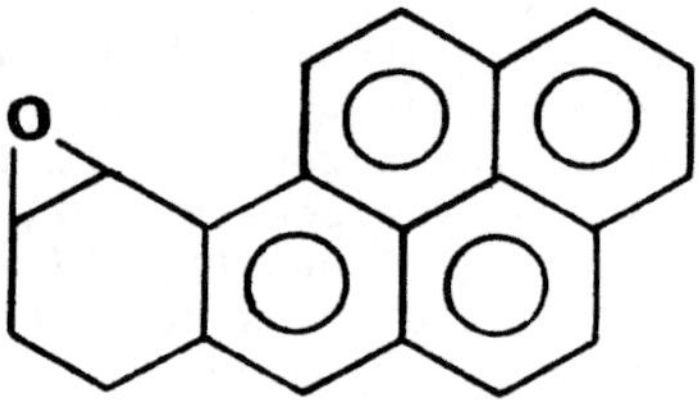

III

This concept, that hydroxyl groups can provide anchimeric assistance in the opening of nearby epoxide rings through intramolecular hydrogen bonding, has previously been invoked for several other systems, including some hydroxy-epoxy-cholestanes [18,19], the analeptic drug picrotoxinin [20], and the triptolide (IVa) and tripdiolide (IVb) molecules [21,22]. The latter two are naturally-occurring anti-leukemic agents in which, according to their NMR spectra, there is strong hydrogen bonding between the hydroxyl group on C(14) and the C(9)-C(11) epoxide. It has been found that the anti-leukemic activity of IVa is not present in the isomer in which the 14-OH is trans to the epoxide ring, nor is it present in the derivative in which the 14-OH has been oxidized to a ketone [22]. For these and other reasons, it has been proposed that anchimeric assistance by the 14-OH in the opening of the epoxide ring is a key factor in the biological mode of action of these molecules.

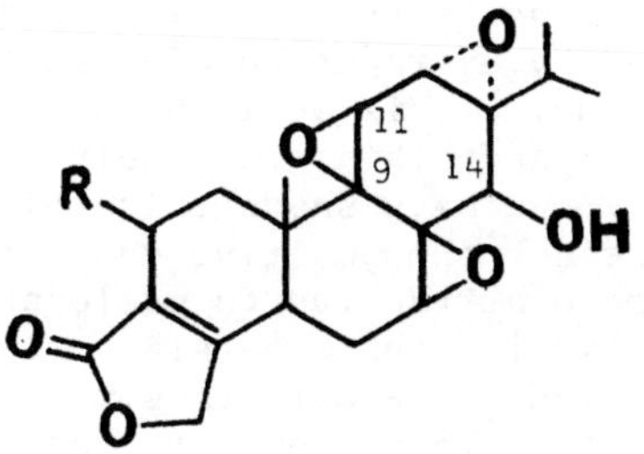

IVa: R = H

IVb: R = OH

Another interesting aspect of the role of the hydroxyl groups in the anti and syn diol epoxides of benzo[a]pyrene, I and II, relates to mutagenic activity. 9,10-epoxy-9,10-dihydrobenzo[a]pyrene, V, which

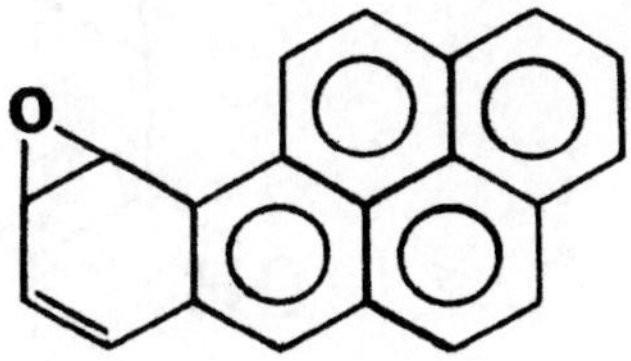

V

contains no hydroxyls and in which the C(7)-C(8) bond is unsaturated, is only weakly mutagenic, in contrast to the diol epoxides I and II, both of which are very highly mutagenic [12,23,24]. On the other hand, 9,10-epoxy-7,8,9,10-tetrahydrobenzo[a]pyrene, III, which still contains no hydroxyls but in which the C(7)-C(8) bond is now saturated, is an extremely potent mutagen, similar to the diol epoxides I and II [12,23,24]. These observations have led to the suggestion that perhaps it is not the hydroxyl groups themselves, in I and II, but rather saturation of the C(7)-C(8) bond that may be the key to mutagenicity in molecules I, II, III and V [23].

II. OUR PREVIOUS WORK

As part of our continuing study of the reactive properties of epoxides [25-28], and especially their relationship to chemical carcinogenesis, we have been interested in developing a better understanding of the effect upon these properties that is produced by the hydroxyl groups in the diol epoxides of benzo[a]pyrene. The idea that epoxide ring opening may be facilitated by hydrogen bonding to the epoxide oxygen is not an unreasonable one. It is well known that epoxide ring-opening reactions are acid catalyzed; this presumably involves the protonation of the oxygen [29]. We have shown that this protonation weakens the C-O bonds and increases the positive character of the carbons (thus rendering them more attractive to nucleophiles) [26,27,30,31]. It seemed, initially, quite possible that hydrogen bonding to the epoxide oxygen might have effects similar to those of protonation, although probably to a lesser extent, and therefore might also catalyze the ring opening.

We have investigated this matter in two earlier computational studies, using two diol epoxides of benzene, VI and VII, as models for the *anti* and *syn* diol epoxides of benzo[a]pyrene, I and II [27,28]. The sizes of the latter make it impractical to carry out extensive *ab initio* calculations, while on the other hand, VI and VII do contain the key features that are of primary importance for present purposes.

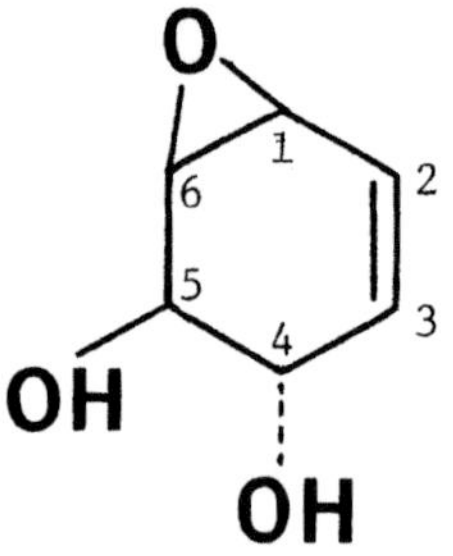

VI: *anti* isomer

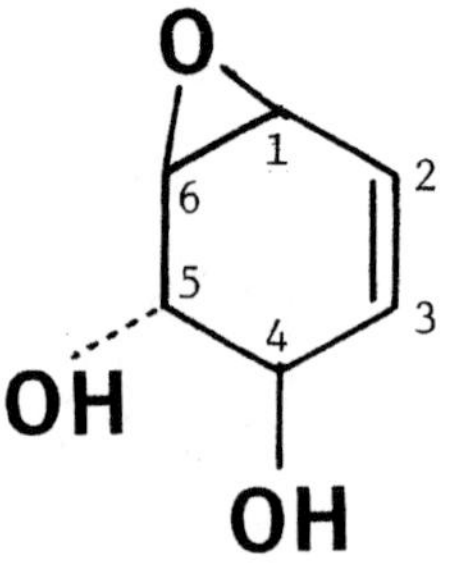

VII: *syn* isomer

We used an ab initio self-consistent-field (SCF) molecular orbital procedure (GAUSSIAN 70, with an STO-5G basis set [32]) to compute optimized structures for VI and VII [28]. The final structure obtained for VII, the syn isomer, revealed the 4-OH group to be oriented toward the epoxide oxygen, and was in every way indicative of the presence of an intramolecular hydrogen bond. However the anti isomer, VI, showed no tendency whatsoever for intramolecular hydrogen bonding; the 5-OH group, which would have to be involved, was preferentially oriented away from the epoxide oxygen. (Lavery and Pullman have arrived at similar conclusions regarding hydrogen bonding in VI and VII [33], as as have Yeh et al, by a semi-empirical technique, for I and II [34].) Furthermore, we found, in the case of the syn isomer, VII, that both the C(1)-O and also the C(6)-O bonds were weaker and easier to break in the presence of the intramolecular hydrogen bond than when the latter was disrupted by forcing the 4-OH hydrogen to rotate by 90° away from the epoxide oxygen [27,28]. Indeed the C(1)-O bond appeared to be particularly weakened by the hydrogen bond, more so than the C(6)-O.

All of these results fully support the anchimeric assistance hypothesis; they are even consistent with the experimental observation, mentioned above, that the diol epoxides of benzo[a]pyrene react with nucleophiles preferentially at C(10) (which corresponds, in VII, to C(1)). A problem was encountered, however, when we examined another system, that of ethylene oxide with a hydrogen-bonded water molecule, (VIII) [27]. Despite a very careful and thorough analysis, we found that the hydrogen bond (which was unquestionably present) had absolutely no weakening effect upon the epoxide C-O bonds. This appeared to be a direct contradiction of what had been determined in the case of the syn diol epoxide of benzene.

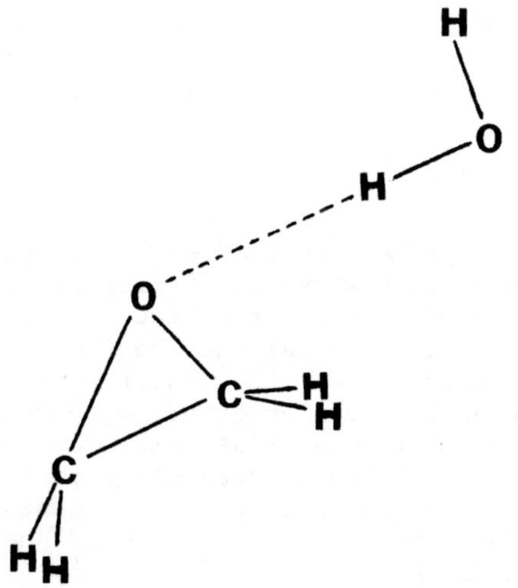

VIII

Another complication is the fact that the preferential opening of the epoxide ring at C(1) for VI and VII and at C(10) for I and II can be predicted on the basis of another factor that is completely unrelated to the hydroxyl groups. That other factor is the double bond that is present in VI and VII, and in the corresponding rings in I and II. If the C-O bond opening is accompanied by the development of a positive charge at the carbon (as is widely assumed), then a simple resonance argument suggests that the epoxide ring would tend to open at C(1) (or C(10)) because of the possibility of charge delocalization in the transition state (see structures IX and X). Of course this does not explain why the syn isomer is more reactive; furthermore, it should be mentioned that our calculations have revealed no tendency for the C-O bond stretching to be accompanied by a charge separation such as is shown in IX (unless there was prior protonation of the oxygen) [27]. The same observation has been made by Kaufman et al in computational studies of various benzo[a]pyrene diol epoxides [35].

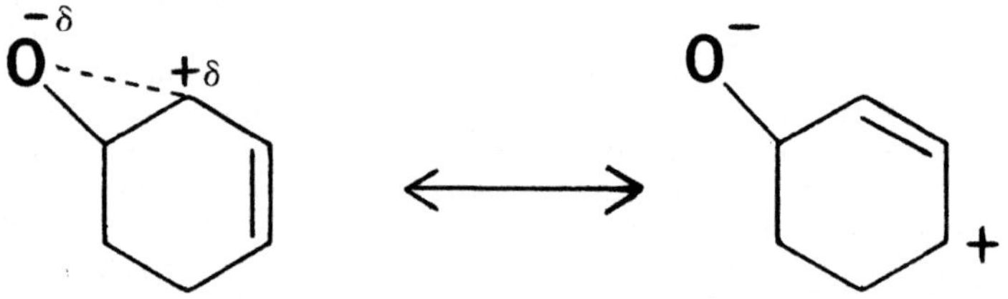

IX X

It is clear that the role of the hydroxyl groups in these diol epoxides is not well understood, and warrants further investigation. The results of such an investigation will be reported in this paper.

III. PROCEDURE

All of the studies to be described were carried out computationally, using the ab initio SCF GAUSSIAN 70 program with an STO-5G basis set [32]. For each system investigated, the structure was optimized fairly thoroughly; this included varying the C-C and C-O bond lengths and many bond and dihedral angles. The major exception to this was that the length of the C-C double bond in the benzene diol epoxides, VI and VII, and in cyclohexadiene oxide, XII, was constrained to a standard aromatic value of 1.400 A, very close to the actual separation, 1.410 A, of the corresponding carbons in benzo[a]pyrene [36]. (The optimization of the structures of VI and VII has been described in detail in reference 28.)

The ease of opening of each epoxide C-O bond in the various systems was assessed in terms of the same two criteria as have been used in the past [27,28]. The first of these involves computing the amount of energy needed to stretch the bond by an arbitrary amount (taken to be 0.050 A) while keeping all other bond lengths unchanged. (All of the bond angles of the epoxide ring are of course distorted in this process.) The magnitude of this energy should be a direct measure of the difficulty of stretching and opening the bond.

Second, an "effective ring-distortion force constant" was calculated for each epoxide C-O bond. While one of the C-O and all of the C-C distances were held constant, the total energy, E, of the system was computed for two non-equilibrium values of the other C-O bond length; these were greater and smaller by 0.050 A than its optimum value, R_e. These two energies plus that for the equilibrium structure were used to determine a parabolic E vs. R_{C-O} relationship. A force constant was then obtained by evaluating the quantity $(d^2E/dR^2_{C-O})_{R_e}$. (This process again involves distorting all of the bond angles of the epoxide ring, so that the result should not be regarded as simply a stretching force constant.) The larger the magnitude of this force constant for a given C-O bond, the more difficult it should be to open that bond.

IV. RESULTS AND DISCUSSION

Since our aim is to elucidate the possible effects of both hydroxyl groups and a C-C double bond upon epoxide reactivity, a reasonable starting point is cyclohexene oxide, XI, which contains neither. The results for this system will be a useful point of reference, since they correspond to a case in which both of the factors that may enhance epoxide ring opening are absent. For our optimized cyclohexene oxide structure, the energy required to stretch either C-O bond by 0.050 A is 7.3 kcal/mole, and the epoxide ring distortion force constant is 11.3 atomic units. (All calculated stretching energies and force constants are listed in Table 1.)

XI

Table 1. Calculated Stretching Energies and Force Constants.[a]

Molecule	Structure and Bond	Stretching Energy	Force Constant
Cyclohexene Oxide	XI	7.3	11.3
Cyclohexadiene Oxide	XII		
	C(1)-O	1.3	2.0
	C(6)-O	1.4	1.9
Benzene Oxide	XIII	1.6	1.8
Benzene Diol Epoxide (anti)	VI		
	C(1)-O	1.5	2.0
	C(6)-O	1.3	1.9
Benzene Diol Epoxide (syn)	VII		
	C(1)-O	1.3	1.8
	C(6)-O	1.5	2.0
Benzene Diol Epoxide (syn)	VII		
(4-OH hydrogen rotated	C(1)-O	1.5	1.9
110° counterclockwise)	C(6)-O	1.5	2.0
Benzene Diol Epoxide (syn)	VII		
(4-OH hydrogen rotated	C(1)-O	5.8	6.8
90° clockwise)	C(6)-O	5.8	6.9

[a] The stretching energies are given in kcal/mole; the force constants are in atomic units.

In order to see the effects of double bonds, the next two systems considered will be cyclohexadiene oxide, XII, and benzene oxide, XIII. The presence of the double bond in cyclohexadiene oxide, in the same position relative to the epoxide group as in the diol epoxides, very markedly enhances the ease of epoxide ring opening (Table 1). For each C-O bond, both the bond-stretching energy and also the force constant decrease by about 80%, to approximately 1.4 kcal/mole and 1.9 atomic units, respectively. It is important to note, however, that the double bond has this very significant effect on both C-O bonds, since the resonance/charge delocalization interpretation discussed in section II can only account for enhanced opening of the C(1)-O bond.

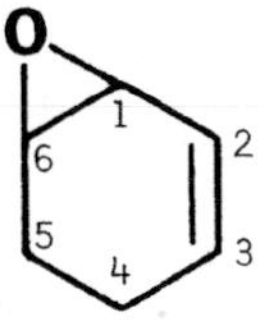

XII

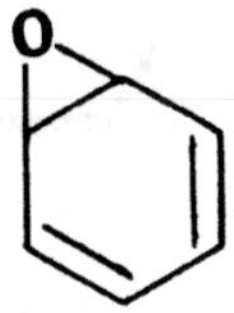

XIII

As was mentioned in section III, the C(2)-C(3) bond in XII was assigned a length of 1.400 A, considerably greater than the usual C-C double bond value of approximately 1.34 A. In order to see whether a greater degree of double bond character would increase the effect upon the epoxide ring, no constraints were imposed upon the C-C distances in benzene oxide, XIII. The optimum structure was found to be a diene, with C(2)-C(3) and C(4)-C(5) distances of 1.340 A. However the presence of two very definite double bonds did not further weaken the epoxide C-O bonds to any significant extent; the bond stretching energy (1.6 kcal/mole) and the force constant (1.8 atomic units) are very much the same as for cyclohexadiene oxide. (Our computed benzene oxide structure is given in Figure 1. There have been several other recent studies of this molecule, both semi-empirical and *ab initio* [25,31,37]; the present energy, -304.44503 hartrees, is the lowest among the *ab initio* treatments.)

Proceeding now to the diol epoxides, and considering first the *anti* isomer, VI, the epoxide C-O bond stretching energies and force constants are seen in Table 1 to again be very nearly the same as in cyclohexadiene oxide, XII. The presence of the hydroxyl groups has no obvious effect. None was really anticipated, of course, since it has been shown that there is no hydrogen bonding to the epoxide oxygen in this system [28,33,34].

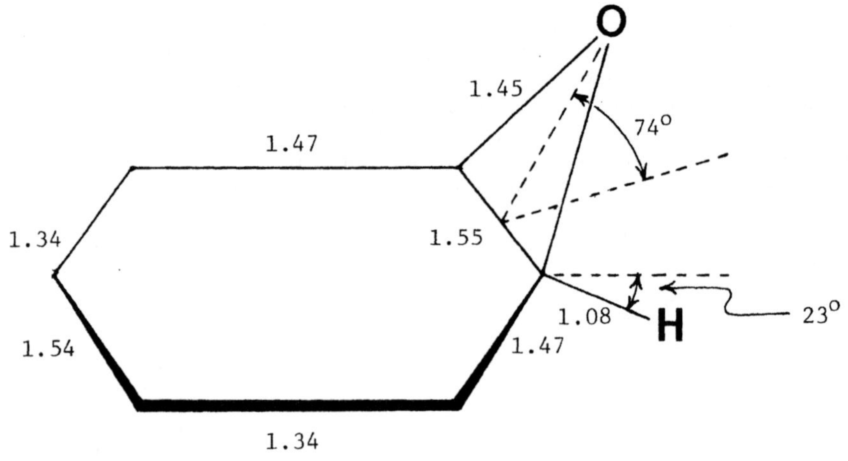

Figure 1. Calculated structure for benzene oxide. All of the carbons and hydrogens are coplanar except for the hydrogens on the two carbons that form part of the epoxide ring. All distances are in Angstroms.

The final system to be discussed is the syn diol epoxide, VII. It has been established quite clearly that there is a hydrogen bond in this molecule, between the 4-OH and the epoxide oxygen [28,33,34]. Nevertheless, the stretching energies and force constants for both the C(1)-O and the C(6)-O bonds are very similar to the corresponding values for the anti isomer, VI, in which there is no hydrogen bonding, and for cyclohexadiene oxide, XII, which does not even possess hydroxyl groups. Thus, the intramolecular hydrogen bond in VII does not weaken the epoxide C-O bonds beyond the degree that can be attributed to the presence of the C-C double bond. This was confirmed by disrupting the hydrogen bonding; the 4-OH hydrogen was forced to rotate away from the epoxide oxygen, counterclockwise, through an angle of about 110°. (This put it in the general neighborhood of C(5).) This rotation, and the disruption of the hydrogen bonding, produced essentially no change in the C-O stretching energies and force constants (Table 1).

What is particularly interesting about this result is that a 90° rotation of the 4-OH hydrogen in the clockwise direction does very definitely affect the C-O bonds, as was found earlier [27,28] and was discussed in section II. The energies needed to stretch the epoxide C-O bonds increase to 5.8 kcal/mole, and the force constants jump to approximately 6.9 atomic units. Thus, this rotation has had the effect of considerably strengthening the C-O bonds, not to the extent found in cyclohexene oxide (Table 1) but certainly making them much more difficult to open than in the various other systems (containing double bonds) that have been examined in this work.

In trying to understand this phenomenon, it seems important to note that a clockwise rotation through an angle of 90° puts the 4-OH hydrogen in the vicinity of C(3), the separation being 2.23 A. Since the sum of the carbon and hydrogen van der Waals radii is roughly 3.0 A [38], it can be assumed that there is some interaction between the 4-OH hydrogen and C(3). Some evidence of this is provided by the calculated atomic charges; as a result of the 90° clockwise rotation, C(3) becomes more negative by a small but significant amount, going from -0.055 in the equilibrium (hydrogen-bonded) structure to -0.074, while C(2) and the epoxide oxygen become very slightly less negative. If the mechanism by which the double bond facilitates epoxide ring opening is related in any way to the resonance/charge delocalization argument given in section II, then it would not be surprising that something which causes C(3) to become more negative would counteract, to some extent, the effect of the double bond upon the epoxide ring.

To summarize, it seems quite clear that the C-C double bonds in the systems studied facilitate the opening of the epoxide C-O bonds. The manner in which they do this is not understood. The resonance/charge delocalization concept may provide a partial explanation, but it fails to account for our finding that <u>both</u> C-O bonds are affected to approximately the same extent. Perhaps a second factor to consider is that the presence of the double bond in addition to the epoxide ring produces a high degree of strain in the system, which is relieved when one of the epoxide C-O bonds opens.

On the other hand, no evidence has been found to indicate that the ease of opening the epoxide ring is enhanced by the hydroxyl groups in VI and VII, whether or not they engage in intramolecular hydrogen bonding to the epoxide oxygen. It is very interesting to observe, however, that a hydroxyl group can be made to quite strongly interfere with the effect that the C-C double bond exerts upon the epoxide ring. This interference appears to involve an interaction between the hydrogen of the hydroxyl group and one of the carbons of the double bond.

This study has clarified some points, but it has also raised some intriguing questions. For instance, what is the mechanism by which the C-C double bond affects the epoxide C-O bonds? Some hitherto attractive hypotheses regarding diol epoxide behavior, dealing with the effects of hydrogen bonding and resonance/charge delocalization on epoxide C-O bond strengths, must be re-examined; this is true for diol epoxides of polycyclic aromatic hydrocarbons as well as for the simpler molecules that have been treated here. Once again, we see the need for extreme caution in attempting to interpret the behavior of complex systems.

ACKNOWLEDGMENT

We greatly appreciate the extremely helpful discussions with Dr. Ieva Ruks Politzer. We are also very grateful for the financial support of the University of New Orleans Computer Research Center.

REFERENCES

[1] Sims, P. and Grover, P. L.: 1974, Adv. Cancer Res. 20, 165.
[2] Weinstein, I. B., Jeffrey, A. M., Jennette, K. W., Blobstein, S. H., Harvey, R. G., Harris, C., Autrup, H., Kasai, H., and Nakanishi, K.: 1976, Science 193, 592.
[3] Jerina, D. M., Yagi, H., Lehr, R. E., Thakker, D. R., Schaefer-Ridder, M., Karle, J. M., Levin, W., Wood, A. W., Chang, R. L., and Conney, A. H.: 1978, Polycyclic Hydrocarbons and Cancer, Vol. 1, H. V. Gelboin and P. O. P. Ts'o, eds., Academic Press, New York, p. 173.
[4] Yang, S. K., Roller, P. R., and Gelboin, H. V.: 1978, Carcinogenesis, Vol. 3: Polynuclear Aromatic Hydrocarbons, P. W. Jones and R. I. Freudenthal, eds., Raven Press, New York, p. 285.
[5] Tsang, W.-S. and Griffin, G. W.: 1979, Metabolic Activation of Polynuclear Aromatic Hydrocarbons, Pergamon Press, London.
[6] Jerina, D. M., Thakker, D. R., Yagi, H., Levin, W., Wood, A. W., and Conney, A. H.: 1978, Pure & Appl. Chem. 50, 1033.
[7] Levin, W., Thakker, D. R., Wood, A. W., Chang, R. L., Lehr, R. E., Jerina, D. M., and Conney, A. H.: 1978, Cancer Res. 38, 1705.
[8] Jeffrey, A. M., Weinstein, K. B., and Harvey, R. G.: 1979, Proc. Am. Assoc. Cancer Res. & Amer. Soc. Clinical Oncology 20, 131; Wood, A. W., Chang, R. L., Levin, W., Yagi, H., Jerina, D. M., and Conney, A. H.: ibid., 222.
[9] Jeffrey, A. M., Jennette, K. W., Blobstein, S. H., Weinstein, I. B., Beland, F. A., Harvey, R. G., Kasai, H., Miura, I., and Nakanishi, K.: 1976, J. Amer. Chem. Soc. 98, 5714.
[10] Koreeda, M., Moore, P. D., Wislocki, P. G., Levin, W., Conney, A. H., Yagi, H., and Jerina, D. M.: 1978, Science 199, 778.
[11] Brookes, P., King, H. W. S., and Osborne, M. R.: 1978, Polycyclic Hydrocarbons and Cancer, Vol. 2, H. V. Gelboin and P. O. P. Ts'o, eds., Academic Press, New York, 43.
[12] Jerina, D. M., Yagi, H., Hernandez, O., Dansette, P. M., Wood, A. W., Levin, W., Chang, R. L., Wislocki, P. G., and Conney, A. H.: 1976, Carcinogenesis, Vol. 1: Polynuclear Aromatic Hydrocarbons, R. I. Freudenthal and P. W. Jones., eds., Raven Press, New York, 91.
[13] Yagi, H., Hernandez, O., and Jerina, D. M.: 1975, J. Amer. Chem. Soc. 97, 6881.
[14] Whalen, D. L., Ross, A. M., Yagi, H., Karle, J. M., and Jerina, D. M.: 1978, J. Amer. Chem. Soc. 100, 5218.
[15] Yagi, H., Hernandez, O., and Jerina, D. M.: 1975, J. Amer. Chem. Soc. 97, 6881.
[16] Hulbert, P.: 1975, Nature 256, 146.

[17] Keller, J. W., Heidelberger, C., Beland, F. A., and Harvey, R. G.: 1976, J. Amer. Chem. Soc. 98, 8276.
[18] Barton, D. H. R. and Houminer, Y.: 1973, J. C. S. Chem. Commun., 839.
[19] Houminer, Y.: 1975, J. C. S. Perkin Trans. 1, 1663.
[20] Dalzell, H. C., Razdan, R. K., and Sawdaye, R.: 1976, J. Org. Chem. 41, 1650.
[21] Kupchan, S. M., Court, W. A., Dailey, Jr., R. G., Gilmore, C. J., and Bryan, R. F.: 1972, J. Amer. Chem. Soc. 94, 7194.
[22] Kupchan, S. M. and Schubert, R. M.: 1974, Science 185, 791.
[23] Wood, A. W., Levin, W., Lu, A. Y. H., Ryan, D., West, S. B., Lehr, R. E., Schaefer-Ridder, M., Jerina, D. M., and Conney, A. H.: 1976, Biochem. Biophys. Res. Commun. 72, 680.
[24] Levin, W., Wood, A. W., Wislocki, P. G., Chang, R. L., Kapitulnik, J., Mah, H. D., Yagi, H., Jerina, D. M., and Conney, A. H.: 1978, Polycyclic Hydrocarbons and Cancer, Vol. 1, H. V. Gelboin and P. O. P. Ts'o, eds., Academic Press, New York, 189.
[25] Politzer, P. and Daiker, K. C.: 1977, Excited States in Organic Chemistry and Biochemistry, B. Pullman and N. Goldblum, eds., D. Reidel Publishing Co., Dordrecht, Holland, 331.
[26] Politzer, P., Daiker, K. C., Estes, V. M., and Baughman, M.: 1978, Internat. J. Quantum Chem., Quantum Biol. Symp. No. 5, 291.
[27] Politzer, P. and Estes, V. M.: 1979, Catalysis in Chemistry and Biochemistry. Theory and Experiment, B. Pullman, ed., D. Reidel Publishing Co., Dordrecht, Holland, 305.
[28] Politzer, P., Daiker, K. C., and Estes, V. M.: 1979, Internat. J. Quantum Chem., Quantum Biol. Symp. No. 6, 47.
[29] Morrison, R. T. and Boyd, R. N.: 1973, Organic Chemistry, 3rd ed., Allyn & Bacon, Inc., Boston, pp. 562-571.
[30] Hopkinson, A. C., Lieu, M. H., Csizmadia, I. G., and Yates, K.: 1978, Theoret. Chim. Acta 47, 97.
[31] Ferrell, Jr., J. E. and Loew, G. H.: 1979, J. Amer. Chem. Soc. 101, 1385.
[32] Hehre, W. J., Lathan, W. A., Ditchfield, R., Newton, M. E., and Pople, J. A.: Program available from Quantum Chemistry Program Exchange, Indiana University, Bloomington, IN, 47401, program #236.
[33] Lavery, R. and Pullman, B.: 1979, Internat. J. Quantum Chem. 15, 271.
[34] Yeh, C. Y., Fu, P. P., Beland, F. A., and Harvey, R. G.: 1978, Bioorg. Chem. 7, 497.
[35] Kaufman, J. J., Popkie, H. E., Palalikit, S., and Hariharan, P. C.: 1978, Internat. J. Quantum Chem. 14, 793; Hariharan, P. C., Kaufman, J. J., and Petrongolo, C.: 1979, Internat. J. Quantum Chem. Quantum Biol. Symp. No. 6, 223.
[36] Iball, J., Scrimgeour, S. N., and Young, D. W.: 1976, Acta Cryst. B32, 328.
[37] Hayes, D. M., Nelson, S. D., Garland, W. A., and Kollman, P. A.: 1980, J. Amer. Chem. Soc. 102, 1255.
[38] Pauling, L.: 1960, The Nature of the Chemical Bond, 3rd ed., Cornell University Press, Ithaca, N.Y., pp. 257-264.

BIOLOGICAL EFFECTS OF SPECIFIC HYDROCARBON-DNA REACTIONS

Peter Brookes and Robert F. Newbold

Institute of Cancer Research, Pollards Wood Research Station, Nightingales Lane, Chalfont St. Giles, Bucks, U.K.

ABSTRACT.

Encouraged by earlier work on hydrocarbon-DNA reaction in relation to carcinogenesis, a study was made of the induction of $HPRT^-$ mutants in V79 Chinese hamster cells. Within a series of hydrocarbon epoxides, differences in mutagenic effectiveness and efficiency were related to carcinogenic potency. The cell-mediated mutation system was used to compare the mutagenic efficiency of a series of parent hydrocarbons. Quantitative studies of the mutagenicity of the *syn*- and *anti*-diolepoxides of benzo(a)pyrene revealed differences in absolute efficiency reminiscent of data obtained with methylating agents which gave different DNA alkylation products. The significance of these data to the role of mutation in cancer initiation is discussed

INTRODUCTION

Before discussing current data on the relationship between the carcinogenicity and mutagenicity of polycyclic aromatic hydrocarbons and their products of reaction with DNA, it is worth recalling briefly the stages which have led to the present state of knowledge.

It was the correlation between the extent of a hydrocarbon's covalent binding to the DNA of mouse skin and its carcinogenic potency which led to the hypothesis that such reactions were an essential step in tumour induction (Brookes and Lawley, 1964). There followed a period of a few years during which this concept became increasingly accepted and extended to many other classes of chemical carcinogens.

The requirement for metabolism prior to DNA reaction was implicit from the beginning, and led to much work on the mechanism of metabolic activation of hydrocarbons and the identification of the ultimate carcinogen. Following the suggestion by Boyland (1950), epoxides were likely candidates, and many studies favoured the K-region epoxides as being responsible for DNA reaction. However, this was disputed on the grounds that the DNA reaction products obtained with

B. Pullman, P. O. P. Ts'o and H. Gelboin (eds.), Carcinogenesis: Fundamental Mechanisms and Environmental Effects, 81–90.

these derivatives were not those found by an *in vivo* binding of the parent hydrocarbons (Baird and Brookes, 1973; Baird et al., 1973, 1975). The problem was resolved by the identification by Sims et al., (1974) of the diolepoxide of benzo(a)pyrene (BP) and the demonstration that the *anti*-isomer of this metabolite reacted with DNA to give the same products as obtained *in vivo* (King et al., 1976).

Attention was now directed to the nature of the diolepoxide-DNA reaction products. It was established that the major site of reaction was the 2-amino group of guanine, although minor reactions at other sites in both guanine and adenine moieties were detected (Osborne et al., 1976; Jeffrey et al., 1976; Straub et al., 1977).

Despite the very considerable progress of the past 15 years we are still a long way from understanding the molecular mechanism of hydrocarbon carcinogenesis. For example it is clear that not all hydrocarbon derivatives which react with the 2-amino group of guanine are equally carcinogenic e.g. 7-bromomethylbenz(a)anthracene and the K-region epoxides. Furthermore even with the diolepoxide of BP considerable differences in carcinogenic activity have been reported (Slaga et al., 1977; Kapitulnik et al., 1978) for the *anti*- and *syn*-isomers (Table I; formulae I and II respectively) despite the fact that they react similarly with DNA.

Our initial hypothesis of the role of covalent DNA reaction in carcinogenesis encouraged us to believe that a correlation should exist between mammalian cell mutation and carcinogenicity. The system we decided to study was the induction of 8-azaguanine resistant mutations in V79 Chinese hamster cells. We have described previously the factors which enable this system to be used to yield reliable, quantitative data for both reactive hydrocarbon derivatives (Duncan and Brookes, 1973; Newbold et al., 1975) and those hydrocarbons requiring metabolic activation (Newbold et al., 1977).

The present paper will consider how effectively mutation at the HPRT-locus of V79 cells correlates with carcinogenicity and if the system can be used to learn more about the mechanism of carcinogenesis.

MUTAGENIC EFFECTIVENESS AND EFFICIENCY

When comparing the mutagenicity of various compounds it is important to consider on what basis the comparison is being made. Ehrenberg (1971) clearly recognised this problem and introduced the terms mutagenic effectiveness and mutagenic efficiency. In the former case mutagenicity is related to the dose of the chemical used and is particularly important, for example, when considering the likely hazard from environmental pollution or industrial exposure. Mutagenic efficiency, defined as mutagenicity per lethal event, is more relevant when considering the molecular mechanism of the mutational event. On the assumption that both mutation and cytotoxicity are a result of DNA damage (which is very often but not always the case), then mutagenic efficiency becomes a particularly valuable concept. Of even greater relevance is absolute mutagenic efficiency, defined as mutation per unit extent of DNA reactions, but this can only be assessed when a method of measuring DNA reaction (such as radioactive chemical) is available.

As mentioned above, epoxides have long been considered as likely metabolites of hydrocarbons which would be expected to react covalently with DNA and to be mutagenic. In fact, when a series of such epoxides (supplied by Professor R.G. Harvey, University of Chicago) were tested in V79 cell system, all were found to be mutagenic. However, considerable differences were found in their mutagenic effectiveness (see Table I) which, as discussed previously, we believe can be understood in terms of the stability of the derived carbonium ions (Newbold et al., 1979). For the present discussion it is sufficient to note that those compounds having low values of mutagenic effectiveness (Table I, compounds V-X) are either non- or very weak carcinogens.

TABLE I

THE MUTAGENIC EFFECTIVENESS OF A SERIES OF HYDROCARBON DERIVATIVES

Compound	Dose, µg/ml	Survival % Survival	Mutagenic effectiveness [1]	Compound	Dose, µg/ml	Survival % Survival	Mutagenic effectiveness [1]
I 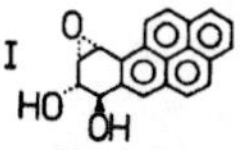	0.1	96	1050	V 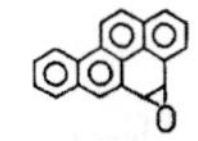	0.5	88	30
	0.2	92	920		1.0	75	30
	0.3	63	1290		2.0	40	40
II 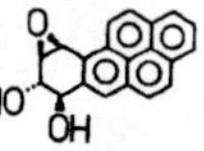	0.1	93	270	VI 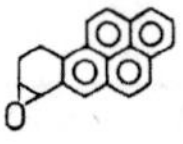	0.5	83	54
	0.5	53	180		1.0	71	88
	0.75	20	260		2.0	14	72
III 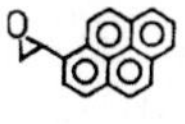	0.1	86	260	VII 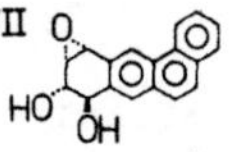	1.0	91	18
	0.15	68	270		2.5	56	50
IV 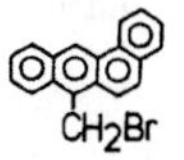	0.05	78	1500	VIII 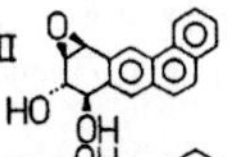	1.0	80	5
	0.1	30	1730		2.5	63	6
				IX 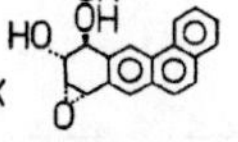	0.2	100	53
					1.5	98	21
					4.0	37	37
				X 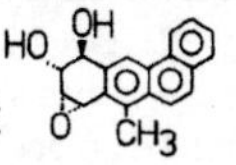	0.1	74	52
					1.0	51	52

1. Mutagenic effectiveness at each dose level used is expressed as mutation frequency per 10^5 survivors per µg/ml of treatment.

This suggests that potent carcinogens must be effective mutagens. However the data obtained with compounds I-IV (Table I) indicate that this is not a sufficient criterion since, as mentioned above, *anti*-BP-diolepoxide (I) is a more potent carcinogen than compounds II-IV, which are also very effective mutagens.

A much better correlation with carcinogenicity emerges when the mutagenic efficiency of the same series of compounds is considered (Figure 1).

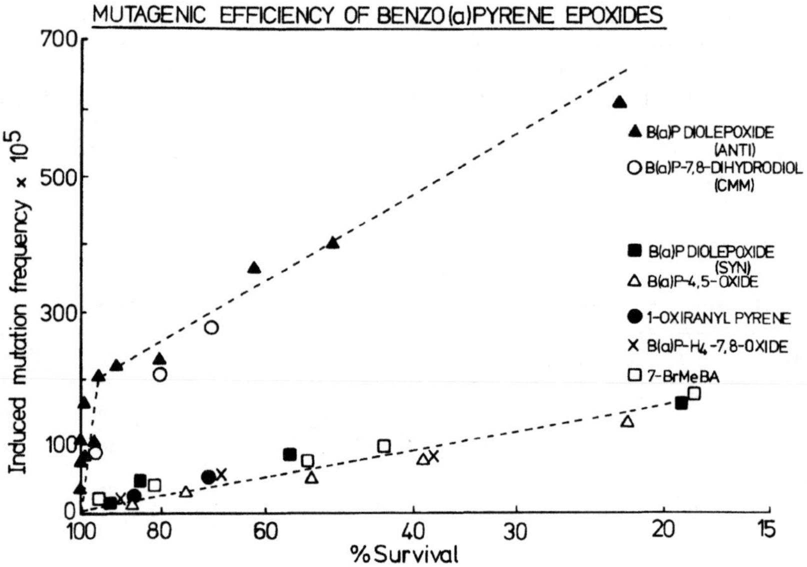

Figure 1. The mutagenic efficiency of a number of hydrocarbon derivatives assessed in the 8-azaguanine V79 Chinese hamster cell system. For full details see Newbold et al. (1979).

It is now apparent that only *anti*-BP-diolepoxide and its precursor BP-7,8-dihydrodiol (assayed in the cell mediated system) show any deviation from the response found with all the compounds tested; compounds II-IV now resemble compound V in keeping with their similar weak carcinogenicity. Furthermore, this difference persisted when the absolute mutagenic efficiency was obtained using tritium-labelled hydrocarbon derivatives (Figure 2).

Although we find these differences in mutagenic efficiency of interest, we recognise that any discussion of relative carcinogenicity of chemicals is limited unless data is available on the extent of reaction of the compounds within the target cell. Such data are available in only a very limited number of cases (e.g. Lawley, 1980).

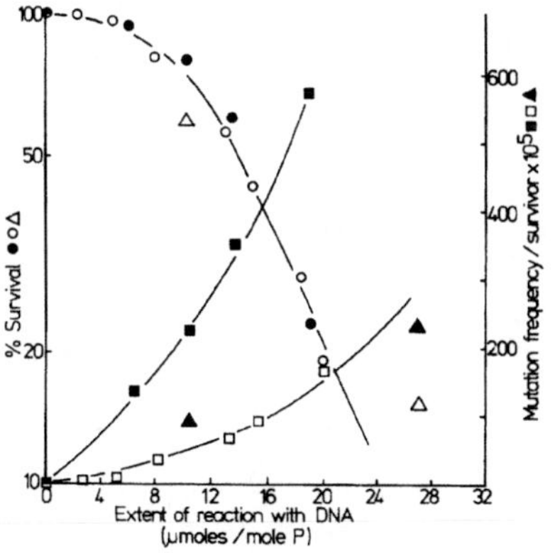

Figure 2. Survival and absolute mutagenic efficiency data for *anti*- (● and ■ respectively) and *syn*- (△ and ▲ respectively)-BP-diol-epoxides, and 7-bromomethylbenz(a)-anthracene (○ and □ respectively). The data was obtained using V79 cells as described in Newbold et al.(1979).

MUTATION STUDIES WITH A SERIES OF HYDROCARBONS

The bay-region diolepoxide hypothesis of hydrocarbon activation rests heavily on data obtained with BP although it seems likely that a similar mechanism of activation occurs in the case of 3-methyl-cholanthrene (King et al., 1977, 1978; Thakker et al., 1978), 7-methyl-benz(a)anthracene (Tierney et al., 1977), 7,12-dimethylbenz(a)-anthracene (Dipple and Nebzydoski, 1978) and dibenz(a,h)anthracene (Wood et al., 1978).

In the absence of the tritium-labelled diolepoxides of any hydro-carbons other than BP, we assessed the absolute mutagenic efficiency of a series of parent hydrocarbons by using the labelled compounds in the cell-mediated mutation system (Wigley et al., 1979). We had previously shown that in this system the DNA lesion responsible for the mutation of the V79 cells is the same as that resulting from metabolism of the hydrocarbon by the feeder-layer of hamster fibroblasts (Newbold et al., 1977) i.e. presumably the bay-region diolepoxide derivatives. Considering the technical problems involved in this approach we would suggest that the data obtained (Table II) indicate that there is no significant difference in mutagenic efficiency between the hydrocarbons tested or between BP and its ultimate carcinogenic metabolite (*anti*-BP-diolepoxide) although the value for the *syn*-isomer is significantly lower.

The difference in carcinogenic activity of the various hydro-carbons was, as suggested previously (Duncan et al., 1969), reflected by the value of binding index for each compound i.e. the extent of DNA reaction which results per unit of hydrocarbon metabolised.

TABLE II

ABSOLUTE MUTAGENIC EFFICIENCY OF HYDROCARBONS AND DERIVATIVES

	anti-diol-epoxide	*syn*-diol epoxide	7BrMe-BA	BP	7MBA	3MC	DMBA	BA
Extent of hydrocarbon DNA reaction, μmole/mole P.	13	11	15	15	6.5	10.5	9	NIL
Induced mutation (8-azaG^r) frequency per 10^5 survivors	360	90	90	227	73	169	208	1.2
Absolute mutagenic efficiency i.e. induced mutations per μmole/mole P reaction with DNA	28	8	6	15	11	16	23	NIL

Abbreviations: 7BrMeBa, 7-bromomethylbenz(a)anthracene; 7MBA, 7-methylbenz(a)anthracene; 3MC, 3-methylcholanthrene; DMBA, 7,12-dimethylbenz(a)anthracene; BA, benz(a)anthracene.

MUTAGENICITY OF *SYN*- AND *ANTI*-BP-DIOLEPOXIDES

The difference in mutagenic efficiency of *syn*- and *anti*-BP-diolepoxides, as illustrated in Figure 1 and 2, was reminiscent of a similar difference found by Loveless (1969) for the alkylating agents methyl methanesulphonate (or dimethyl sulphate, DMS), and methyl nitrosourea (MNU) in studies of the induction of mutation in bacteriophage. In this case the explanation lay in the greater yield of the promutagenic O^6-methylguanine formed on alkylation of DNA with MNU compared to DMS. When these two compounds were compared as mutagens in the V79 system a corresponding difference was found, which could again be related to the amount of O^6-methylguanine formed (Newbold et al., 1980). Presentation of the data from this publication in the same manner as used for the diolepoxides in Figure 2, revealed a very similar pattern of response (Figure 3). For each pair of compounds equal extents of overall reaction lead to equal cytotoxicity but MNU and *anti*-BP-diolepoxide have the greater absolute mutagenic efficiency.

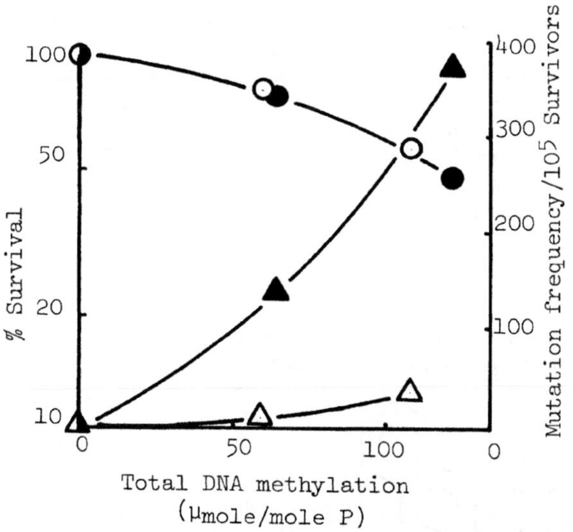

Figure 3. Survival and absolute mutagenic efficiency for MNU (● and ▲ respectively) and DMS (○ and △ respectively). The data was obtained using V79 cells as described in Newbold et al. (1980).

In view of the above it was tempting to speculate that *anti*-BP-diolepoxide might also be inducing a promutagenic lesion in DNA not produced by the *syn*-isomer. In fact we have previously reported (Osborne et al., 1978; King et al., 1979) that only the *anti*-isomer gave an N^7-guanine derivative on reaction with DNA *in vitro*. However, we were unable to identify this product from *in vivo* reaction of BP or its diolepoxide, and it is not readily apparent why N^7-guanine reaction, which leads to depurination, should be an effective promutagenic lesion.

From the data of Figure 3 and the determined percentage of alkylation at the O^6-position of guanine it can be calculated that for MNU the induced mutation frequency per μmole O^6-methylguanine moiety in DNA of V79 cells is about 20, which is of the same order as found per μmole total DNA reaction for the *anti*-diolepoxide and other hydrocarbons for which data is available (Table II). Could this mean that O^6-guanine methylation and *anti*-diolepoxide lesions in DNA are inducing mutations by similar mechanisms. We are currently attempting to answer this question by a study of the $HPRT^-$ mutants induced by these two agents.

The relevance of the above data to the mechanism of carcinogenesis is supported by the work of Frei et al. (1978) who found a quantitative relationship between the extent of O^6-alkylation of the

DNA of thymus and bone marrow, and the yield of thymic lymphomas in mice following a single injection of MNU of ENU.

MUTAGENICITY AND CARCINOGENICITY

Much has been written on the correlation between carcinogenicity and bacterial mutagenicity and the use of tests based on this correlation as pre-screens for environmental chemicals. In the case of the hydrocarbons such tests are of little value possibly due to the inappropriate metabolic activation by the rat liver microsome preparations used. Since we know that the cell-mediated mutation system does not suffer from this problem (Newbold et al., 1977; Wigley et al., 1979) we decided to test the 12 possible monomethylbenz(a)anthracenes (kindly supplied by Professor M.S. Newman, University of Columbus, Ohio).

TABLE III

CORRELATION BETWEEN CARCINOGENICITY AND MUTAGENICITY OF MONOMETHYLBENZ(a)ANTHRACENES

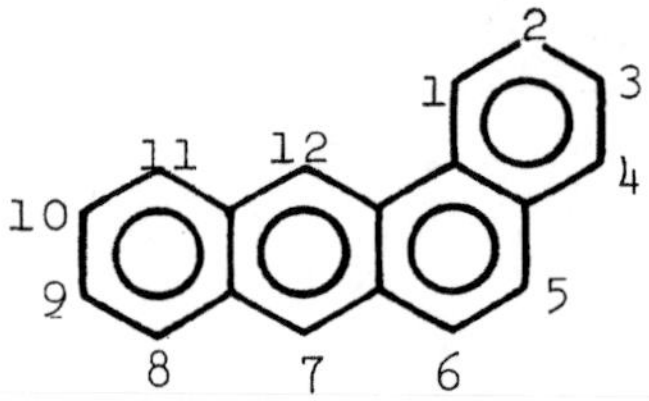

Compound	Carcinogenicity[1]	Mutagenicity[2]
Unsubstituted	0	0
1-methyl	0	0
2-methyl	0	1
3-methyl	0	0
4-methyl	0	0
5-methyl	0	0
6-methyl	++	9
7-methyl	+++	40
8-methyl	++	7
9-methyl	0	0
10-methyl	0	0
11-methyl	0	0
12-methyl	++(+)	12

[1]Carcinogenicity data taken from Badger (1948).

[2]Induced mutant frequency/10^5 surviving cells following treatment with 1 µg/ml for 48 hours in the cell mediated mutation system as described by Wigley et al. (1979).

This group of compounds was particularly suitable for this study since the carcinogenicity of all 12 hydrocarbons has been investigated by several authors and the series contains inactive, weakly active and potent carcinogens. As shown in Table III, the mutagenic activity in V79 cells with cell-mediated activation, relates well with the carcinogenic potency.

Further evidence of the value of the V79 -HPRT locus is provided by the study of Wood et al. (1977) who compared the mutagenicity of the (+) and (-) enantiomers of both *syn* and *anti*-BP-diolepoxide in this system and in *S. typhimurium* TA98 and TA100. All four derivatives were found to be potent mutagens in the bacterial systems - but in the V79 cells only (+) *anti*-BP-diolepoxide was an efficient mutagen consistent with its carcinogenicity for both skin and lung of mice.

THE NEXT PHASE

We are encouraged by the above results to believe that a detailed understanding of the molecular mechanism of mutation induction by hydrocarbons in V79 cells will be relevant to the mechanism of cancer initiation. However, we also believe that work must now be directed towards an understanding of the cellular events leading to "transformation" of cells in culture. With this objective we are studying hamster embryo fibroblasts and taking as our initial criterion of transformation the growth of the cells beyond the period of normal senescence. The ability of various chemicals of known mutagenic efficiency to induce this "transformation" is being assessed in relation to the known nature and extent of DNA damage.

ACKNOWLEDGEMENTS

The authors are grateful to Ms Jacqueline Amos for expert technical assistance with the work reported in this paper. The investigation was initially supported by NCI (USA) Contract No. NO1-CP-33367 and more recently by Grant Number 1-RO1-CA25807 awarded by the N.C.I., DHEW, U.S.A., also by grants to the Institute of Cancer Research from the Medical Research Council and the Cancer Research Campaign.

REFERENCES

Badger, G.M.: 1948, Brit. J. Cancer, 2, pp. 309-350.
Baird, W.M., and Brookes, P. : 1973, Cancer Res., 33, pp. 2378-2385.
Baird, W.M., Dipple, A., Grover, P.L., Sims, P., and Brookes, P.: 1973, Cancer Res., 33, pp. 2386-2392.
Baird, W.M., Harvey, R.G., and Brookes, P.: 1975, Cancer Res., 35, pp. 54-57.
Boyland, E.: 1950, Biochem. Soc. Symp., 5, pp. 40-54.
Brookes, P., and Lawley, P.D.: 1964, Nature, 202, pp. 781-784 and J. Cell Comp. Physiol., Suppl.1, 64, pp. 111-127.
Dipple, A., and Nebzydoski, J.A.: 1978, Chem.-Biol. Interactions, 20, pp. 17-24.

Duncan, M., Brookes, P., and Dipple, A.: 1969, Int. J. Cancer, 4, pp. 813-819.
Duncan, M.E., and Brookes, P.: 1973, Mutation Res., 21, pp. 107-118
Ehrenberg, L.: 1971, In: A Hollaender (ed.) Chemical Mutagens. Principles and Methods for their Detection, Plenum Press, New York, London, Vol. 2, pp. 365-386.
Frei, J.V., Swenson, D.H., Warren, W., and Lawley, P.D.: 1978, Biochem. J., 174, pp. 1031-1044.
Jeffrey, A.M., Jennette, K.W., Blobstein, S.H., Weinstein, I.B., Beland, F.A., Harvey, R.G., Kasai, Miura, I., and Nakanishi, K.: 1976, J. Amer. Chem. Soc., 98, pp. 5714-5715.
Kapitulnik, J., Wislocki, P.G., Levin, W., Yagi, H, Jerina, D.M., and Conney. A.H.: 1978, Cancer Res., 38, pp. 354-358.
King, H.W.S., Osborne, M.R., Beland, F.A., Harvey, R.G., and Brookes,P: 1976, Proc. Natl. Acad. Sci. (Wash.), 73, pp. 2679-2681.
King, H.W.S., Osborne, M.R. and Brookes, P.: 1977, Int. J. Cancer, 20, pp. 564-571.
King, H.W.S., Osborne, M.R., and Brookes, P.: 1978, Chem.-Biol. Interactions, 20, pp. 367-371.
King, H.W.S., Osborne, M.R., and Brookes, P.: 1979, Chem.-Biol. Interactions, 24, pp. 345-353.
Lawley, P.D.: 1980, Brit. Med. Bull., 36, pp. 19-24.
Loveless, A.: 1969, Nature, 223, pp. 206-208.
Newbold, R.F., Brookes, P., Arlett, C.F., Bridges, B.A., and Dean, B.: 1975, Mutation Res., 30, pp. 143-148.
Newbold, R.F., Wigley, C.B., Thompson, M.H. and Brookes, P.: 1977, Mutation Res., 43, pp. 101-116.
Newbold, R.F., Brookes, P., and Harvey, R.G.: 1979, Int. J. Cancer, 24, pp. 203-209.
Newbold, R.F., Warren, W., Medcalf, A.S.C., and Amos, J.: 1980, Nature, 283, pp. 596-599.
Osborne, M.R., Beland, F.A., Harvey, R.G., and Brookes, P.: 1976, Int. J. Cancer, 18, pp. 362-368.
Sims, P., Grover, P.L., Swaisland, A., Pal, K., and Hewer, A.: 1979, Nature, 252, pp. 326-327.
Slaga, T.J., Bracken, W.M., Viaje, A., Levin, W., Yagi, H., Jerina, D.M., and Conney, A.H.: 1977, Cancer Res., 37, pp. 4130-4133.
Straub, K.M., Meehan, T., Burlingame, A.L., and Calvin, M.: 1977, Proc. Natl. Acad. Sci. (Wash.), 74, pp. 5285-5289.
Thakker, D.R., Levin, W., Wood, A.W., Conney, A.H., Stoming, T.A., and Jerina, D.M.: 1978, J. Amer. Chem. Soc., 100, pp. 645-647.
Tierney, B., Hewer, A., Walsh, C., Grover, P.L. and Sims, P.: 1977, Chem.-Biol. Interactions, 18, pp. 179-193.
Wigley, C.B., Newbold, R.F., Amos, J. and Brookes, P.: 1979, Int. J. Cancer, 23, pp. 691-696.
Wood, A.W., Levin, W., Thomas, P.E., Ryan, D., Karle, J.M., Yagi, H., Jerina, D.M., and Conney, A.H. : 1978, Cancer Res., 38, pp.1967-1973
Wood, A.W., Chang, R.L., Levin, W., Yagi, H., Thakker, D.R., Jerina, D.M., and Conney, A.H.: 1977, Biochem. Res. Comm. 77, pp.1389-1396.

THE EFFECT OF BASE MODIFICATION ON FIDELITY IN TRANSCRIPTION

B. Singer
Department of Molecular Biology and Virus Laboratory
University of California, Berkeley, California 94720

ABSTRACT

Ribopolynucleotides containing a small number of various modified nucleotides were examined for fidelity in transcription using DNA-dependent RNA polymerase in the presence of Mn^{2+}. When normal hydrogen bonding is blocked by a substituent, total ambiguity results with transcription continuing beyond the modified nucleotide. Examples are 3-MeC, 3-MeU, 1-MeA and O^4-MeU. In the case of modifications of the exocyclic amino groups, in addition to the possible tautomeric shift with C, the orientation of the substituent plays a major role in determining fidelity. Thus N^4-methoxy C acts only as U; N^4-hydroxy C, N^4-MeC, and N^2-MeG act ambiguously; and N^4-acetyl C base-pairs only with G. Three different substituents on the N^6 of A did not affect A•U pairing, indicating an anti conformation.

INTRODUCTION

In the classical concept of Watson and Crick, no changes are permitted on certain positions necessary for hydrogen bonding. Thus A and U (or T) have two hydrogen bonds, with the N^6 of A donating a proton to the O^4 of U (T) and the N-3 of U (T) donating a proton to the N-1 of A. Any disruption of this bonding would necessarily lead to less than the two required hydrogen bonds. Similarly, G and C are paired in a prescribed manner. If positions are modified on nonhydrogen bonded sites, such as halogenation of the 5 positions of pyrimidines, base-pairing with the normal complementary base occurs although there are detectable changes in thermal stability.

When complex formation of modified polynucleotides is examined (reviewed by Singer (1)), we find that, in general, the rule is obeyed although the stability of the complex may be low. For example, poly (2-aminoadenosine) complexes with poly (U), poly (7-methylguanosine) complexes with poly (C) and poly (8-oxyadenosine) complexes with poly (U). All this is expected. What is unexpected is that

B. Pullman, P. O. P. Ts'o and H. Gelboin (eds.), Carcinogenesis: Fundamental Mechanisms and Environmental Effects, 91–102.

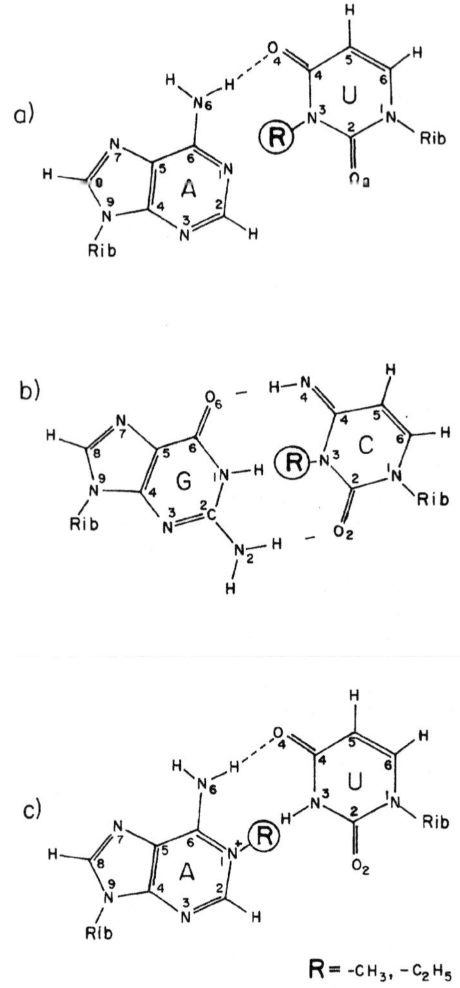

Figure 1. Effect on Watson Crick-hydrogen bonding of substituents blocking the N-3 of U (a), the N-3 of C (b) and the N-1 of A (c). The result of alkylation of these positions is total ambiguity in transcription (4; Table 1, Section I).

poly (N^6-methyladenosine) does not complex with poly (U). This brings up the important point that steric factors play a significant role in determining whether polymers can form Watson-Crick complexes. The exocyclic amino groups of C, A and G can rotate and if a substituent blocks or shields the hydrogen-bonding site, then normal base-pairing cannot occur. This point will be expanded in a later section.

Our research has centered around the practical question of how a small number of modified bases in a polynucleotide can affect transcription. This is related to the general question of what is a mutagenic modification. Most of the data presented are on the effect of methylation and ethylation at the sites shown to be modified by the N-nitroso carcinogens (2,3). The technique used was to copolymerize, using polynucleotide phosphorylase, a modified nucleoside diphosphate with an unmodified one and to transcribe the resulting ribopolynucleotide using DNA-dependent RNA polymerase in the presence of Mn^{2+}. Nearest neighbor analysis of the transcript was found to be a sensitive method for determining whether a modified base caused misincorporation (4).

The data are presented in a series of Figures and Tables. Each Figure[1] illustrates the effect on transcription of a type of base modification with or without steric hindrance.

EFFECT OF SUBSTITUENTS BLOCKING THE N-1 OF PURINES OR THE N-3 OF PYRIMIDINES.

These are the hydrogen-bonding positions which are considered to be essential for base-pairing. Figure 1 illustrates that any substituent should block transcription. However, this does not happen. Rather, the polymerase continues transcription by almost randomly putting any nucleotide into the transcript ((4), Table 1, Section I). This we term "total ambiguity" and it is reminiscent of the ambiguity reported when partially depurinated alternating deoxypolynucleotides are transcribed using the natural conditions for DNA transcription (5). We did not postulate that any hydrogen bonds are formed but that the misincorporated nucleotide is probably held in the helix by stacking forces. Nearest neighbor analysis, of course, proved that the mismatched nucleotide is actually incorporated into the transcript and is not an artifact due to the polymerase.

TABLE 1. EFFECT OF MODIFIED NUCLEOSIDES ON TRANSCRIPTION: DERIVATIVES SHOWING TOTAL AMBIGUITY

		Simulates the Presence Of[a]			
Modified Nucleoside		A	G	U	C
I	3-Methyl Cyd[b]	++	++	++	+
	3-Methyl Urd	++	±	+	+
	1-Methyl Ado	++	+	++	+
II	O^4-Methyl Urd	±	±	+	+
III	N^2-Methyl Guo	+	+	++	±
	N^4-Hydroxy Cyd	+	++	++	±

a Data were obtained using nearest neighbor analysis. Copolymers contained 3-16% modified nucleoside. The qualitative data presented are the summary of many quantitative experiments. The frequency of misincorporation ranges from 50-100%, based on analysis of the modified nucleoside in the polymer. There was no detectable uridine in any polymer, as determined by HPLC analysis after enzyme digestion.

b 3-Ethyl Cyd incorporates A, U or C in transcription. Nearest neighbor analysis was not performed and neither was the normal G incorporation tested.

EFFECT OF METHYLATION OF O^4 OF U.

O^4- methyl U in polynucleotides also acts with ambiguity but in this case it is possible to form a G•O^4-alkyl U wobble pair (Figure 2 (a)). All other misincorporations observed ((6) and Table 1,

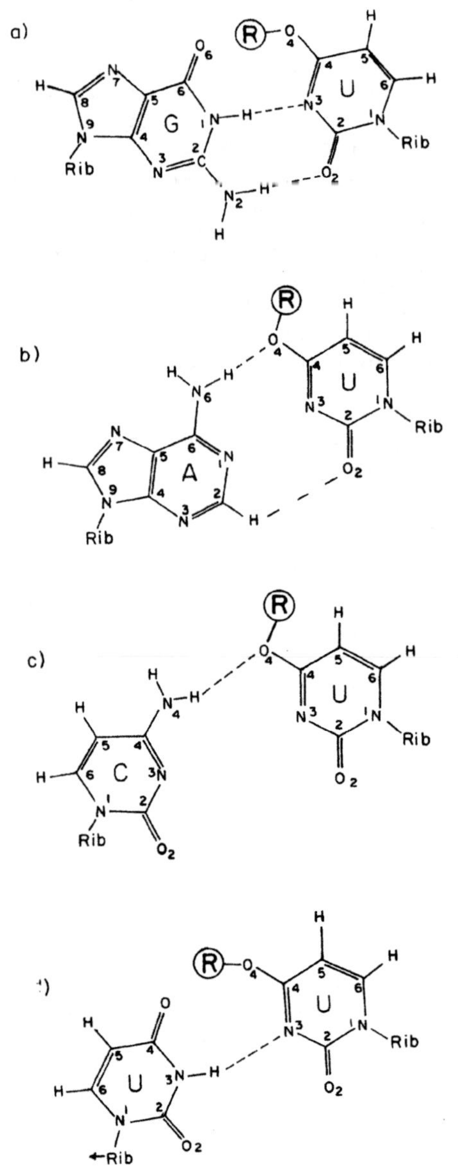

Figure 2. Effect of methylation of the O^4 of U on base-pairing. R = $-CH_3$. Two hydrogen bonds can be formed with G (a), possibly two bonds can be formed with A (b), but only a single one with C (c), and a single one with U, with displacement of ribose (d). O^4-MeU in transcription shows total ambiguity but with a preference for simulating C (a) and U (b) (Table 1, Section II).

Section II) must occur without more than a single hydrogen bond being formed. This is illustrated in Figure 2. The greatest distortion is that of O^4-alkyl U•U. However this mispair, as well as O^4-alkyl U•C, was observed also in a different system, namely doublet-stimulated tRNA binding to ribosomes (7).

EFFECT OF ALKYLATION OF O^6 OF G.

Much attention has been paid in recent years to the formation and repair of O^6-methyl G and O^6-ethyl G in vivo since the formation of O^6-alkyl G appears to correlate with the carcinogenicity of alkylating agents (8). Gerchman and Ludlum (9) reported that, in polynucleotides, O^6-methylguanosine could be transcribed, but misincorporation of A and U resulted. There exist as yet no definitive data on other misincorporations or on misincorporation resulting from O^6-ethyl G. In Figure 3(a), O^6-ethyl G is shown making two hydrogen bonds with C. This normal base pair (G•C) could result if the ethyl group were oriented so that there was no steric hindrance. Similarly, the reported misincorporations, shown in Figure 3(b) and (c) can result, on paper at least, in two hydrogen bonds with either U or A. Naturally, as was the case for O^4-alkyl U,

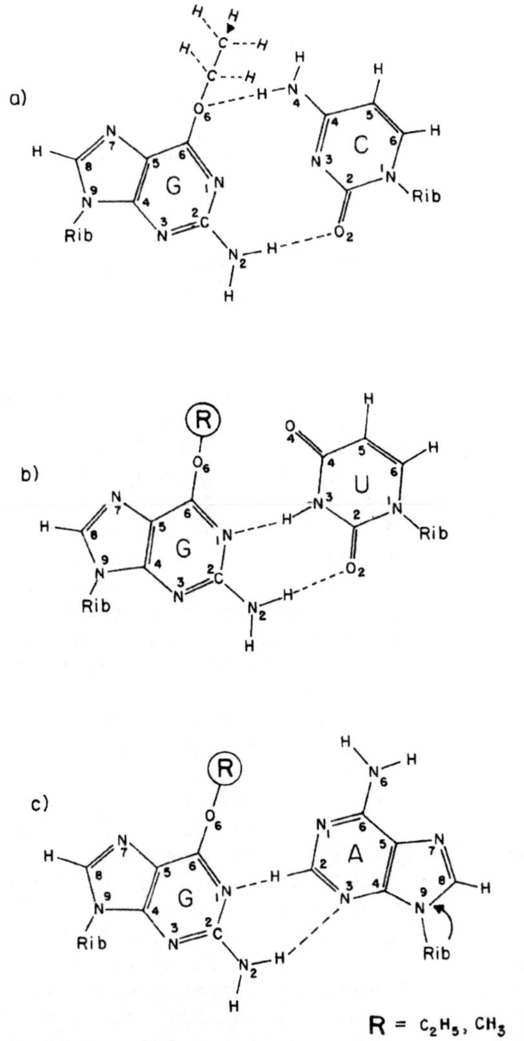

Figure 3. Effect of ethylation of the O^6 of G on base-pairing. The ethyl group is assumed to be proximal and thus unable to shield Watson-Crick base-pairing. Two hydrogen bonds can be formed with C and no mispairing results (a), two hydrogen bonds can be formed with U (b) and two bonds with A although there is strain (c). The mispairings shown in (b) and (c) were reported for O^6-MeG by Gerchman and Ludlum (9).

if the alkyl group is oriented so as to cause distortion, no hydrogen bonds could be drawn.

EFFECT OF ALKYLATION OF THE O^2 OF U.

The O^2 of U is not normally needed to form hydrogen bonds with A. However when it is substituted, the N-3 can no longer be a hydrogen bond donor (Figure 4). O^2-ethyl U in polynucleotides causes misincorporation of G (6). This differs from the total ambiguity found when polymers containing O^4-methyl U are transcribed. However O^4-methyl U has the hydrogen-bonding capability to form two bonds with either G or A and directs these into the transcript with preference while O^2-alkyl U can only form a single hydrogen bond with G if the alkyl group shields the potential second hydrogen bond (Figure 4b). The fact that O^2-methyl U and O^2-ethyl U preferentially direct incorporation of GMP indicates that there is no steric hindrance. If there were, as shown in Figure 4, then either total ambiguity or termination of transcription would be likely.

O^2-ethylthymidine, like O^6-ethyldeoxyguanosine, is a major product

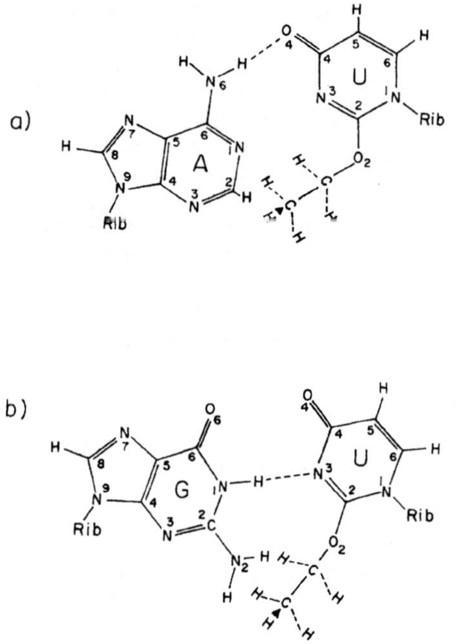

Figure 4. Effect of ethylation of the O^2 of U on base-pairing. The ethyl group is drawn so as to illustrate steric hindrance which may not occur if the smaller methyl group is the substituent. If the ethyl group is rotated away from the Watson-Crick sites, then O^2-alkyl U can pair with A (a), and with G (b). Since these are the effects observed in transcription of O^2-ethyl U, it is likely that there is no steric hindrance.

of reaction of DNA in vivo with ethylnitrosourea (3). This derivative is likely to have a mutagenic effect in transcription and its mispairing may be one additional factor in the observed high carcinogenicity of ethylnitrosourea.

EFFECT OF SUBSTITUENTS ON THE EXOCYCLIC AMINO GROUPS.

We have recently completed a study of transcription of polymers containing substituents on the N^4 of C, the N^2 of G and the N^6 of A. All of the exocyclic amino groups are theoretically free to rotate and also it is assumed that the two hydrogens do not differ in their availability to substituents. Given these caveats, the results obtained can be interpreted in terms of the orientation of the substituent. In the case of N^4-hydroxy C and N^4-methoxy C, the tautomeric equilibrium also plays a role.

Modification of the N^4 of C gave unexpected results. The well-known tautomeric shifts of N^4-hydroxy C and N^4-methoxy C were confirmed in the transcription system. However, while N^4-methoxy C simulated U, it did not even partially act as C. The percent of total transcription found to be due to the tautomer pairing like U exactly correlated with the percent N^4-methoxy C in the polymer. We conclude that N^4-methoxy C exists only as the tautomer (Figure 5(b)).

N^4-Hydroxy C differs in that it not only acts as U and C but also simulates G and A to a lesser extent. We conclude that N^4-hydroxy C exists partly as the tautomer (Figure 5(b)) and partly with the hydroxyamino group becoming syn to the ring N-3, rotated so that no specific base is recognized (10).

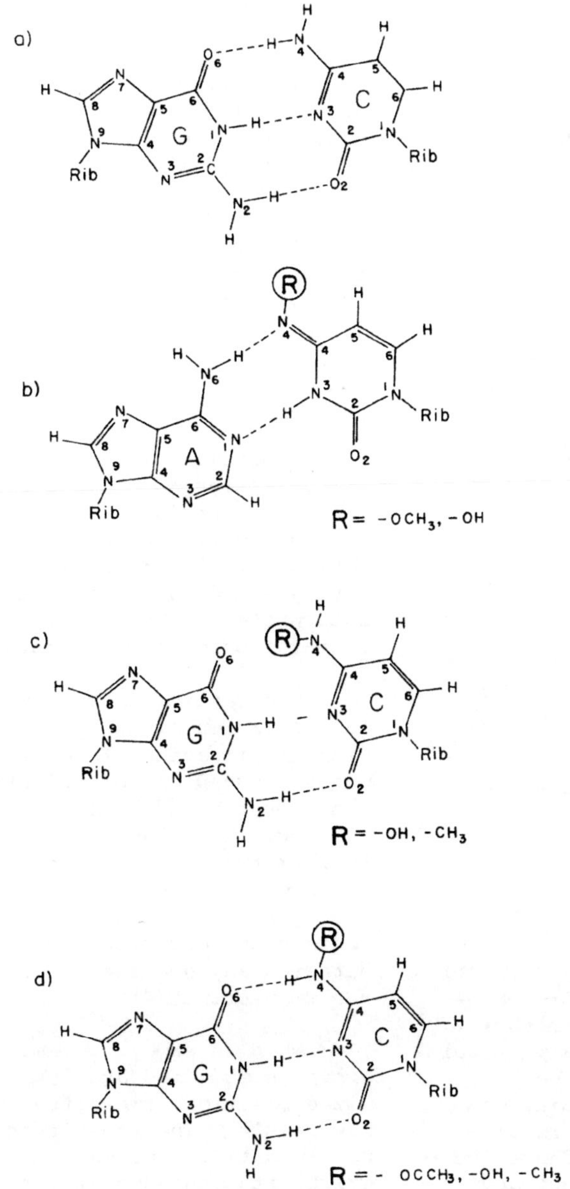

Figure 5. Effect of substituents on the N^4 of C. The normal G•C base-pair is shown in (a). When the substituent if $-OCH_3$, only the base-pairing attributed to the tautomeric form of N^4-methoxy C is observed (b). When the substituent is $-OCCH_3$, no changed base-pairing occurs (d). However both $-OH$ and $-CH_3$ substituents lead to total ambiguity (b), (c), (d) and Table 1, Section III.

Another and different result is obtained when N^4-methyl C is transcribed. In this case, the predominant orientation (Figure 5(d)) allows the modified nucleoside to base-pair normally. However, rotation probably occurs occasionally causing random mispairing.

The fourth modified cytidine was N^4-acetyl C. The acetyl group is known from crystallographic data to be proximal to the C(5), anti to the ring N-3 and would not shield the Watson-Crick base-pairing (11). Our data are in complete agreement since even poly (C, 70% N^4-acetyl C) does not direct any nucleotide into the transcript, other than GMP.

The one amino substituted guanosine derivative used, N^2-methyl G directed all four nucleotides in transcription although there was a preference for AMP (Table 1, Section III). Figure 6 shows that a methyl group on the N^2 does not necessarily change base-pairing but, if rotated, there can be a weak N^2-MeG•A pair. If the substituent were larger, e.g. ethyl, then only one hydrogen bond with A may be possible and the ambiguity may be more random.

N^6 substituted adenosines are in striking contrast to N^4 substituted cytidines or N^2 substituted guanosine. Neither methyl nor hydroxyethyl nor isopentenyl substituents caused any mispairing (Figure 7) and we conclude that the substituents are all and always anti to the ring N-1.

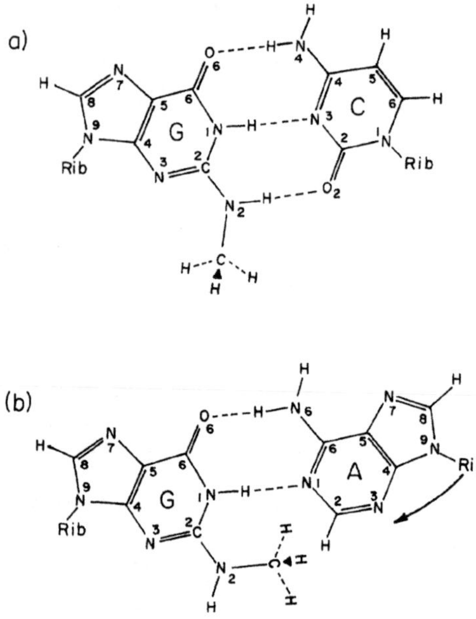

Figure 6. Effect of methylation of the N^2 of G. The normal G•C base-pair is unaffected if the methyl group is oriented as in (a). Regardless of the orientation, the strained base-pair, N^2-MeG•A, can occur although the ribose is displaced. Experimental evidence indicates that N^2-MeG can direct any nucleotide into a transcript although there is a preference for C and A (Table 1, Section III).

The results of all transcription data for amino substituted nucleosides is shown in Table 2.

DISCUSSION

Experimental results using transcription as an indirect tool to determine whether, and how, a modified base can participate in transcription have yielded data which can be interpreted in terms of orientation of substituents.

The earlier work from this laboratory on ambiguity dealt with modification of positions which, regardless of orientation, must block normal base-pairing. This lack of ability to form two or three hydrogen bonds with any base did not block transcription, but instead the modified base directed any nucleotide into the transcript (4).

When an exocyclic amino group is modified and is in a fixed position, two situations may result. The substituent may be anti to the Watson-Crick sites and not affect base-pairing or it may be syn and therefore act as a block or shield. In addition rotation can occur from syn to anti and vice-versa.

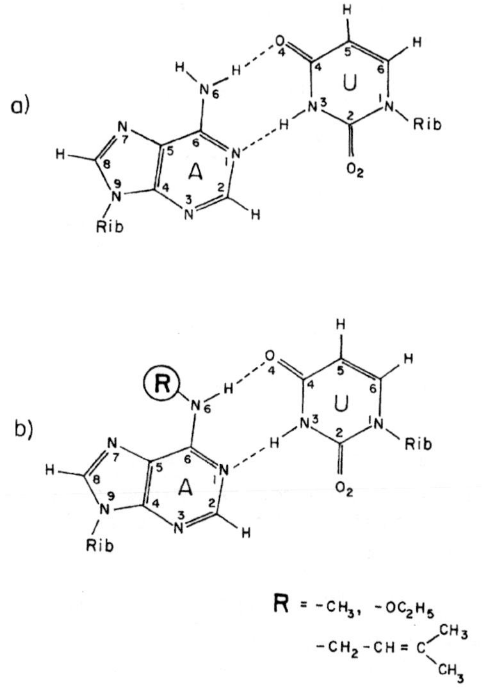

Figure 7. Effect of substituents on the N^6 of A. The normal A•U base-pair is shown in (a). None of the three substituents changed the base-pairing, and therefore must be anti to the N-1 (b).

X-ray crystallographic data indicate that the acetyl group in N^4-acetylcytidine is fixed in the anti position. Our data confirm this since only GMP is incorporated when poly (C, N^4-acetyl C) is transcribed. Crystal structures of N^2-MeG, N^4-MeC or N^6-MeA have not yet been reported. However nmr data on these modified structures are available although not in complete agreement with our results, possibly a consequence of using free bases or different experimental conditions (12, 13). Shoup et al. (12) conclude, on the basis of nmr at -19° in dimethylformamide, that 1-methyl, N^4-methylcytosine is predominately (95%) in the syn configuration. When protonated, as in aqueous solution, the N^4-methyl substituent is anti to the N-3 (14). Our data indicate that the methyl group is anti since the normal complementary base is directed into the transcript. Studies of the secondary structure and base-pairing of poly (C, N^4-MeC) also support the anti conformation (15).

Engel and von Hippel (13) believe that the methylated amino group of G freely rotates and our data can be interpreted as agreeing with this structure. On the other hand, they report that in N^6-methyladenine the methyl group is syn to the ring N-1 which would block or shield any pairing. If this conformation existed in polymers, then N^6-isopentenyl A could not act as A, which it does in transcription. The other N^6 substituted derivatives, N^6-MeA and N^6-hydroxyethyl A also act only as A. We conclude, on the basis of transcription results, that all three substituents are fixed in the anti configuration in a polynucleotide.

Two modified derivatives of special interest are N^4-hydroxy C and N^4-methoxy C. Both have been found to exhibit a tautomeric shift to the imino form (16) and, in various types of biochemical experiments, act like U to a significant extent (17, 18). The crystal structure of

TABLE 2. EFFECT OF MODIFIED NUCLEOSIDES ON TRANSCRIPTION: FIDELITY OF TRANSCRIPTION WHEN EXOCYCLIC AMINO GROUPS ARE MODIFIED

	Simulates the Presence of[b]			
Modified Nucleoside[a]	A	G	U	C
N^4-Methyl Cyd	±	±	±	++
N^4-Acetyl Cyd	−	−	−	++++
N^4-Hydroxy Cyd	+	++	++	±
N^4-Methoxy Cyd	−	−	++++	−
N^2Methyl Guo	+	+	++	±
N^6-Methyl Ado	++++	−	−	−
N^6-Hydroxyethyl Ado	++++	−	−	−
N^6-Isopentenyl Ado	++++	−	−	−

a For several derivatives, varying percents of modified nucleoside were studied. The misincorporation observed increased with increasing amounts of modified nucleoside.

b See Table 1, footnote a. Poly (C, N^4-acetyl C) contained 70% N^4-acetyl Cyd.

1-methyl-N^4-hydroxycytosine hydrochloride has recently been reported by Birnbaum et al. (10) to be only in the imino form, but it is unclear from these results whether the amino form (normal C configuration) exists to any extent in the crystal and whether rotation of the hydroxyamino group could occur. We find that, in polynucleotides, N^4-hydroxy C acts primarily as U (imino) but can direct GMP incorporation. This would indicate that the amino form exists, although there appears to be about 10:1 preference for the imino tautomer. In addition there is nonspecific incorporation of UMP and CMP. This ambiguity suggests that rotation occurs with a reasonable frequency, even for the imino form. Birnbaum et al. (10) do not consider that rotation from syn to anti occurs.

In the case of N^4-methoxy C, the transcription results are striking since this derivative acts only as U and only AMP is incorporated. No misincorporation or GMP incorporation was observed. On this basis we conclude that N^4-methoxy C exists only in the imino form and that the substituent is anti to the ring N-3.

A general hypothesis may be derived from these studies on fidelity in transcription. Ambiguity, rather than termination, will result if the appropriate number of hydrogen bonds cannot be formed. This may be due to steric hindrance or shielding as well as substitution of a hydrogen bonding site.

FOOTNOTES

[1]The Figures are based on the complementary base-pair structures used by Professor B. Pullman in Catalysis in Chemistry and Biochemistry, Theory and Experiment, 1979, pp. 1-10, B. Pullman (Ed.), D. Reidel Publishing Com.

ACKNOWLEDGMENT

This work was supported by Grant CA12316 from the National Cancer Institute, National Institues of Health, Bethesda, Md., USA. I wish to express my appreciation to Dr. S. Spengler who participated in some of this work and who contributed significantly in the conception of the Figures.

REFERENCES

1. Singer, B. and Kröger, M.: 1979 Prog. in Nucleic Acids Res. and Mol. Biol. 23, pp. 151-194

2. Singer, B.: 1976 Nature (London) 264, pp. 333-339

3. Singer, B., Bodell, W. J., Cleaver, J. E., Thomas, G. H., Rajewsky, M. G. and Thon, W.: 1978 Nature (London) 276, pp. 85-88.

4. Kröger, M. and Singer, B.: 1979, Biochemistry 18, pp. 3493-3500

5. Shearman, C. W. and Loeb, L. A.: 1977 Nature (London) 270, pp. 537-538

6. Singer, B., Fraenkel-Conrat, H. and Kuśmierek, J. T.: 1978 Proc. Natl Acad. Sci (USA) 75, pp. 1722-1726

7. Singer, B., Pergolizzi, R. G. and Grunberger, D.: 1979 Nucleic Acids Res. 6, pp. 1709-1719

8. Singer, B.: 1979 J. Natl. Cancer Inst. 62, pp. 1329-1339.

9. Gerchman, L. L. and Ludlum, D. B.: 1973 Biochim. Biophys. Acta 308, pp. 310-316

10. Birnbaum, G. I., Kulikowski, T. and Shugar, D.: 1979 Can. J. Biochem 57, pp. 308-313

11. Parthasarathy, R., Ginell, S. L., De, N. C., and Chheda, G. B.: 1978 Biochem. Biophys. Res. Comm. 83, pp. 657-663

12. Shoup, R. R., Miles, H. T., and Becker, E. D.: 1972 J. Phys. Chem. 76, pp. 64-70

13. Engel, J. D. and von Hippel, P. H.: 1974 Biochemistry 13, pp. 4143-4158

14. Becker, E. D., Miles, H. T. and Bradley, R. B.: 1965 J. Amer. Chem. Soc. 87, pp. 5575-5582

15. Brimacombe, R. L. C. and Reese, C. B.: 1966 J. Mol. Biol. 18, pp. 529-540

16. Brown, D. M., Hewlins, M. J. E. and Schell, P.: 1968 J. Chem. Soc. (C), pp. 1925-1929

17. Budowsky, E. I., Sverdlov, E. D., and Spasokukotskaya, T. N.: 1972 Biochim. Biophys. Acta 287, pp. 195-210

18. Müller, W., Weber, H., Meyer, F. and Weissmann, C.: 1978 J. Mol. Biol. 124, pp. 343-358

MODIFICATION OF DNA BY CARCINOGENIC AROMATIC AMINES IN VIVO AND IN VITRO WITH POSSIBLE PROMUTAGENIC CONSEQUENCES

E. Kriek
Chemical Carcinogenesis Division, Antoni Van Leeuwenhoek-Huis, The Netherlands Cancer Institute, 121 Plesmanlaan, 1066 CX Amsterdam, The Netherlands

Carcinogenic aromatic amines form a group of compounds which can induce tumors in a variety of tissues in various species. In a number of situations certain arylamines may also pose a health hazard to humans. Several properties are responsible for the hazardous nature of this group of compounds, which may be summarized as follows:
(1) they can undergo a variety of metabolic activation reactions;
(2) the interaction with DNA leads to the formation of several reaction products, some of which may persist for prolonged periods of time. In particular these persistent products may lead to possible cumulative effects of biological significance;
(3) certain reaction products, particularly those of guanine, are potentially pro-mutagenic. Some typical examples of specific base modifications will be discussed.

INTRODUCTION

Aromatic amines represent a large and still expanding group of chemicals which are extensively used in a variety of industrial processes. These include the dyeing of textiles, wood, pharmaceuticals and miscellaneous household articles. A number of aromatic amines has importance as dye intermediates and in the manufacture of drugs, pesticides and plastics. Several aromatic amines constitute a group of well-known cancer producing agents, both in experimental animals as well as in humans (1,2). Among these, 2-naphthylamine, 4-biphenylamine and benzidine have now definitely been established as being responsible for almost all of the known industrial bladder cancer. The carcinogenic aromatic amines exhibit a great diversity in species and in tissue specificity, depending partly on their chemical structure. Although actually very little is known of the reasons for this difference in specificity, some common properties seem to be important:
(a) the type of aromatic ring system;
(b) the position of the amino group on the aromatic ring, and
(c) the presence of other substitutents. For practical reasons it is convenient to subdivide the large group of carcinogenic aromatic

B. Pullman, P. O. P. Ts'o and H. Gelboin (eds.), Carcinogenesis: Fundamental Mechanisms and Environmental Effects, 103–111.

amines according to their chemical structure. Some typical examples are given in Table 1.

Table 1. Carcinogenic aromatic amines subdivided according to their chemical structure

Group	Examples
A. Substituted analines	Phenacetin, o-Toluidine
B. Extended anilines	Benzidines, Biphenylamines
C. Fused ring amines	Fluorenylamines, Naphthylamines
D. Azo dyes	N,N-Dimethyl-4-aminoazobenzene
E. Heterocyclic amines	4-Nitroquinoline-1-oxide

The group of substituted anilines has been studied less extensively as compared to the other groups. Consequently much less information is available with regard to their carcinogenic properties and metabolism. Generally, polynuclear amines are more potent carcinogens than mononuclear amines. This is reflected in the required dosage of the compounds for the induction of tumors. Upon dietary administration polynuclear amines require 50-800 ppm, whereas mononuclear amines need 10,000 ppm. However, since many of the older tests of substituted anilines have been considered as inadequate (3), the carcinogenic activity of this group of compounds should be interpreted with caution. Several substituted anilines are being re-examined at present for carcinogenic activity.

Aromatic amines with a methyl group ortho to the amine function are often more carcinogenic than the corresponding unsubstituted amines or than the meta- or para-substituted isomers. For example o-toluidine is carcinogenic inducing predominantly subcutaneous tumors in rats and blood-vessel tumors in mice, whereas no significant carcinogenic activity was observed for m- and p-toluidine in rats (4). Other interesting examples are found in the naphthalene and biphenyl series. Whereas 2-naphthylamine and 4-biphenylamine usually lead to liver and bladder cancer in rodents, the corresponding o-methylarylamines induce colon cancer in male animals, and breast and colon cancer in females (5,6). The reasons for this change in tissue specificity are not known. At present the group of o-methylarylamines is of particular interest, because of their relation to environmental chemicals. Heterocyclic amines have been detected recently by Japanese investigators (7,8) in tryptophan and protein pyrolysates. One of the products from tryptophan has been identified as an o-methylarylamine type of compound. Heterocyclic amines are also characteristic of many of the constituents of tobacco and tobacco smoke. The o-methylarylamines o-toluidine and 2-methyl-1-naphthylamine have been recently detected in cigarette smoke (9).

METABOLIC FORMATION OF REACTIVE INTERMEDIATES

The metabolism of aromatic amines is complex because of the large variety of reactions these compounds can undergo. Since a number of reviews has appeared in recent years (see ref. 10 for pertinent literature), only some of the most important pathways leading to the formation of chemically reactive intermediates will be outlined briefly in this section. In most species, with the exception of dogs, a rapid equilibrium is being established between acetylation of the arylamine and deacetylation of the N-acetyl compound, which under normal conditions is in favor of acetylation. Further studies in experimental animals, originally with the carcinogen N-acetyl-2-aminofluorene, have shown that N-oxidation is required as a first activation reaction for carcinogenic activity. In this way aryl acetamides are converted to aryl acethydroxamic acids, which are then further activated by a variety of enzymatic reactions, including esterification, deacetylation, N,O-transacetylation and one-electron oxidation.

Esterification of Aryl Acethydroxamic Acids

O-Sulfates. Indirect evidence for the formation of N,O-sulfates as ultimate carcinogenic metabolites in rat liver has been obtained with in vitro assays dependent on the presence of an activated sulfate ester, 3'-phosphoadenosine-5'-phosphosulfate (11,12). This reaction is catalyzed by the enzyme sulfotransferase present in the cytosol of rat liver. The presence and level of hepatic sulfotransferase correlated well with the susceptibility to carcinogenesis by N-hydroxy-AAF under a variety of conditions. The sulfate ester of N-hydroxy-AAF has been prepared synthetically and is a highly unstable compound which reacts rapidly at neutral pH with nucleic acids and proteins. Unlike AAF-N-sulfate, which is unstable to obtain in pure form, the sulfuric acid esters (potassium salts) of N-hydroxy-4-acetylaminobiphenyl, N-hydroxy-4-acetylamino-4'-fluorobiphenyl and N-hydroxy-2-acetylaminophenanthrene have been prepared by direct chemical synthesis in analytically pure form. All compounds have been shown to react readily with nucleophiles (methionine, guanosine, adenosine, RNA and DNA) in aqueous solution. These compounds have been very valuable in model studies of their reactivity towards nucleic acids and proteins and elucidation of various reaction products with nucleic acid bases (see next section, Table 2).

O-Acetates. Despite of numerous examples of biological N-acetylation of aromatic amines, the enzymatic O-acetylation of aryl acethydroxamic acids has never been demonstrated. However, the synthetic compound N-acetoxy-AAF is easily prepared in pure form and is also highly reactive towards nucleophilic groups in nucleic acids and proteins. It is being used extensively in model studies with DNA and RNA, particularly in DNA repair studies.

Formation of Arylhydroxylamines

Arylhydroxylamines may be formed metabolically by various pathways, including direct oxidation of arylamines and enzymatic deacetylation of aryl acethydroxamic acids. For example, N-hydroxy-AAF is readily deacetylated by enzymes present in the microsomal fraction of livers from rats, rabbits, hamsters and Guinea pigs to form N-hydroxy-AF, but the rates of deacetylation differ appreciably among the species. Arylhydroxylamines may be further activated to arylnitrenium or carbocation electrophiles under mildly acidic conditions (13-15), which react rapidly with nucleic acids (see next section).

An interesting reaction, in which N,O-acetyltransferase catalyzes the transfer of N-acetyl groups from aryl acethydroxamic acids to the oxygen to form O-acetyl derivatives, was first reported by Bartsch et al. (16). In later years this reaction has been studied extensively by King et al. (17). Although the O-acetates of arylhydroxylamines are too unstable to be isolated, their formation can be detected by trapping them with various nucleophiles, such as methionine, guanosine or tRNA. N,O-transacetylase activity has been found in a variety of tissues, including liver, stomach, kidney, small intestine, colon and ear-duct sebaceous gland. Enzymatic deacetylation and N,O-transacetylation are important metabolic pathways of activation of aryl acethydroxamic acids, and probably involved in the covalent binding of arylamine residues to cellular macromolecules in the tissues cited.

INTERACTION WITH DNA

The covalent binding of a number of aromatic amines to DNA has been studied extensively by several investigators (see ref. 10), leading to the structural identification of a number of products in vitro as well as in vivo. The identified sites of reaction are presented in Table 2.

Table 2. Identified reaction sites of aryl acethydroxamic acid esters with deoxyribonucleic acid in vitro

Reactive ester	Site of reaction C	A	G	Ratio C-8/NH_2	Refs.
N-acetoxy-AAF	-	-	C-8; NH_2	7.8	18-20
N-OSO_3K-AABP	-	-	C-8; NH_2	1.3	20
N-OSO_3^--AABP4'F	-	-	C-8; NH_2	1.3	21
N-OSO_3K-AAP	-	NH_2	C-8	1.0	22
N-Bzo-MAB	-	-	C-8; NH_2	2.3	23
N-acetoxy-AAS	N-3	N-1	N-1; O-6	-	24

Most of these products have been obtained synthetically by reacting esters of aryl acethydroxamic acids as indicated with nucleosides or nucleotides. The ratios C-8/NH_2 have been calculated from the reaction with DNA and are dependent on salt concentration and pH (20). From the results it can be concluded that only AAF and MAB react predominantly with ring carbon-8 of guanine. Although the biphenyl derivatives also react specifically with guanine, the react about equally well with C-8 and exocyclic nitrogen. AAP reacts to the same extent with adenine and guanine. Only N-acetoxy-AAS is an exception in this series, because it behaves more like an alkylating agent. Similar products have been identified in rat liver in vivo, with the exception of AAP and AAS. Scribner and Koponen (25) recently reported the failure of the hydroxamic acid ester model for in vivo activation of AAP. At least six adducts were found in the DNA hydrolysates, but all products found lacked the N-acetyl group. Gaugler et al. (26) report the identification of all four products in rat liver rRNA after administration of trans-4-dimethylaminostilbene (AAS) to female rats, but these products represented only a small fraction of the total RNA-bound carcinogen. The authors conclude that also in the activation of AAS other mechanisms must be operative in addition to the hydroxamic acid esterification.

Under in vivo conditions the aryl acetamide products represent only a minor proportion of the DNA-bound carcinogen. In the case of AAF only 30% of the DNA-bound material is present as AAF derivatives; the other products are deacetylated. Kriek (15) showed that the N-hydroxy derivative of 2-aminofluorene, 4-aminobiphenyl and 2-naphthylamine react with DNA at slightly acidic pH values (4.0-6.0). In this reaction N-hydroxy-AF has been shown to form almost exclusively N-(deoxyguanosin-8-yl)-2-aminofluorene (27,28). The interaction of arylhydroxylamines with DNA has been studied less extensively as compared with reactive esters of arylhydroxamic acids. A limited number of data has recently become available and is listed in Table 3.

Table 3. Identified reaction sites of arylhydroxylamines with deoxyribonucleic acid in vitro

Arylhydroxylamine	Site of reaction C	A	G	Ratio C-8/NH_2	Refs.
N-hydroxy-AF	-	-	C-8	-	27,28
N-hydroxy-1-NA	-	-	O-6	-	29
N-hydroxy-2-NA	-	NH_2	C-8;NH_2	1.1	30
N-hydroxy-4-AQO	-	NH_2	O-6	-	31

Since N-hydroxy derivatives are known urinary metabolites of arylamine bladder carcinogens, their conversion to electrophiles in the normally acidic bladder lumen of humans and dogs, and their subsequent binding to epithelial cell DNA have been proposed by Kadlubar et al. (32) and

Radomski et al. (33) as critical steps in arylamine induced bladder cancer. The adducts formed by reaction of DNA with N-hydroxy-1-NA in vitro were identified as N-(deoxyguanosin-O^6-yl)-1-naphthylamine and 2-(deoxyguanosin-O^6-yl)-1-naphthylamine (29). In contrast, N-hydroxy-2-NA gave a completely different substitution pattern, and reacted at the exocyclic amino groups of guanine and adenine and at the C-8 position of guanine (30). The latter product was isolated from DNA hydrolysates as a purine ring-opened derivative and identified as 1-[5-(2,6-diamino-4-oxopyrimidinyl-N^6-deoxyriboside)]-3-(2-naphthyl)urea. The opening of the imidazole ring of guanine-8-arylamines seems to be a general reaction, because Kriek and Westra (34) recently found that upon treatment of dGuo-AF with 0.1 N NaOH it is hydrolyzed across the 7-8 guanine bond to form two pyrimidine derivatives. These products were isolated by Sephadex LH-20 chromatography and identified as stereoisomers of 1-[6-(2,5-diamino-4-oxopyrimidinyl-N^6-deoxyriboside)]-3-(2-fluorenyl)urea (Figure 1, structure III). Similar products were obtained by Spodheim-Maurizot et al. (35) and found to differ only in their circular dichroism spectra, which supports the stereoisomerism of the compounds. The ring opening of dGuo-AF occurs already at a measurable rate at pH 9.5 and 37°C, and is catalyzed by metal ions and also by alkaline phosphatase from E. coli. These observations might explain the mutagenicity which has been found for N-hydroxy-AF in bacterial test systems (36). The purine ring-opened products were also isolated when DNA modified with N-hydroxy-AF was hydrolyzed with venom phosphodiesterase and alkaline phosphatase at pH 9 (37). However, good yields of dGuo-AF were obtained when the hydrolysis was performed at pH 8 as described by Beland et al. (28). These results led us to search for the presence of purine ring-opened products in rat liver DNA following administration of (G-^{3}H)AAF. For this purpose a rapid isolation procedure of DNA had to be developed (38), because during the isolation method according to Kirby and Cook (39) appreciable hydrolysis of dGuo-AF into the pyrimidine derivatives was found (27). The relative proportion of products in rat liver DNA following a single dose of (G-^{3}H)-AAF was determined by a combination of two procedures; first hydrolysis in trifluoroacetic acid (27) giving the relative amounts of products I and II (see Figure 1), and secondly enzymatic hydrolysis with nuclease S_1, spleen phosphodiesterase and acid phosphatase followed by Sephadex LH-20 chromatography (34). The latter procedure provided the relative amounts of products III and IV. The proportion of the products is dependent on dose, strain, and whether AAF or its N-hydroxy metabolite is being administered. As yet it is not known whether the hydrolytic ring opening of dGuo-AF occurs already in vivo. The metabolism and persistence of dGuo-AF and its hydrolysis products are being investigated at present.

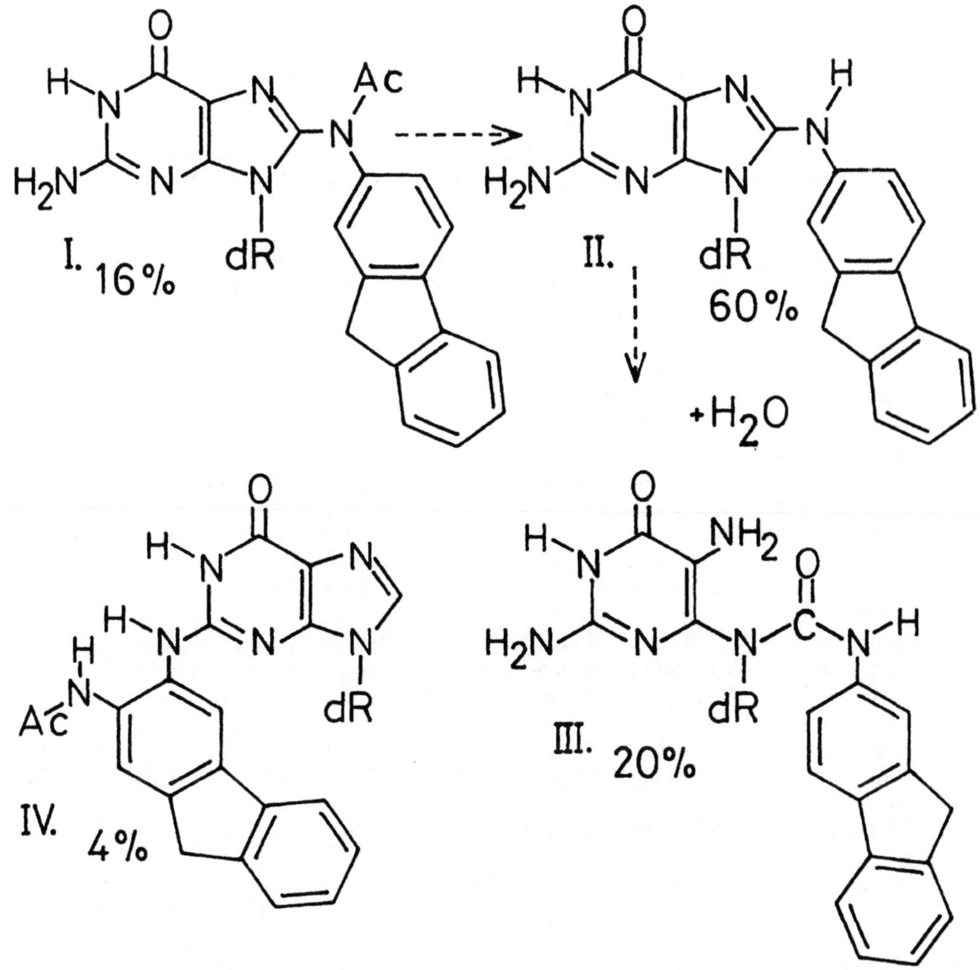

Figure 1. Reaction products in rat liver DNA detected 24 hours following a single dose (0.022 mmol/kg body weight) of (G-^{3}H)AAF (specific radioactivity: 400 Ci/mol).

Approximate percentages of total substitution are given.
Broken arrows indicate possible metabolic pathways.
Ac, acetyl; dR, deoxyribose.

REFERENCES

1. Clayson, D.B. and Garner, R.C.: 1976, In: Searle, C.E., ed., Chemical Carcinogens, ACS Monograph 173, American Chemical Society, Washington D.C., pp.366-461.
2. Parkes, H.G.: 1976, In: Searle, C.E., ed., Chemical Carcinogens, ACS Monograph 173, American Chemical Society, Washington D.C., pp.462-480.
3. IARC Monographs on the Evaluation of the Carcinogenic Risk of Chemicals to Man: Some Aromatic Amines and Related Nitro Compounds - Hair Dyes, Colouring Agents and Miscellaneous Industrial Chemicals, 1978, Vol.16, IARC, Lyon, France.
4. Weisburger, E.K., Russfield, A.B., Homburger, F., Weisburger,J.H., Boger, E., Van Dongen, C.G., and Chu, K.C.: 1978, J. Environ. Pathol. Toxicol., 2, pp.325-356.
5. Walpole, A.L., Williams, M.H.C., and Roberts, D.C.: 1952, Br.Ind. Med. 9, p.255.
6. Miller, E.C., Sandin, R.B., Miller, J.A., and Rusch, H.P., 1956, Cancer Res., 16, pp.525-534.
7. Kosuge, T., Tsuji, K., Wakabayashi, T., Okamoto, K., Shudo, Y., Itai, T., Iitaka, A., Sugimura, T., Kawachi, T., Nagao, M., Yahagi, T., and Seino, Y.: 1978, Chem. Pharm. Bull. (Japan) 26, pp.611-619.
8. Hashimoto, Y., Takeda, K., Shudo, K., Okamoto, T., Sugimura, T., and Kosuge, T.: 1978, Chem.-Biol. Interact. 23, pp.137-140.
9. Patrianakos, C., and Hoffmann, D.: 1979, J. Anal. Toxicol., 3, pp.150-154.
10. Kriek, E., and Westra, J.G.: 1979, In: Grover, P.L., ed., Chemical Carcinogens and DNA, Vol. 2, CRC Press Inc., Boca Raton, FA., pp. 1-28.
11. DeBaun, J.R., Miller, E.C., and Miller, J.A.: 1970, Cancer Res., 30, pp.577-595.
12. King, C.M., and Phillips, B.: 1968, Science 169, pp.1351-1352.
13. Heller, H.E., Hughes, E.D., and Ingold, C.K.: 1951, Nature 168. pp.909-910.
14. Boyland, E., Manson, D., and Nery, R.: 1962, J. Chem. Soc. (London) pp.606-611.
15. Kriek, E.: 1965, Biochem.Biophys.Res.Commun., 20, pp.793-799.
16. Bartsch, H., Dworkin, M., Miller, J.A., and Miller, E.C.: 1972, Biochim.Biophys.Acta 286, pp.272-298.
17. Weeks, C.E., Allaben, W.R., Louie, S.C., Lazear, E.J., and King, C.M.: 1978, Cancer Res., 38, pp.613-618.
18. Kriek, E., Miller, J.A., Juhl, U., and Miller, E.C.: 1967, Biochemistry, 6, pp.177-182.
19. Westra, J.G., Kriek, E., and Hittenhausen, H.: 1976, Chem.-Biol. Interact., 15, pp.149-164.
20. Kriek, E.: 1979, Cancer Lett., 7, pp.141-146.
21. Kriek, E., and Hengeveld, G.M.: 1978, Chem.-Biol. Interact., 21, pp.179-201.

22. Scribner, J.D., and Naimy, N.K.: 1975, Cancer Res., 35, pp.1416-1421.
23. Beland, F.A., Tullis, D.L., Kadlubar, F.F., Straub, K.M., and Evans, F.E.: 1980, Cancer Res. 40, in press.
24. Scribner N.K., and Scribner, J.D.: 1979, Chem.-Biol.Interact., 26, pp.47-55.
25. Scribner, J.D., and Koponen, G.: 1979, Chem.-Biol. Interact., 26, pp.201-209.
26. Gaugler, B.J.M., Neumann, H.-G., Scribner, N.K., and Scribner, J.D.: 1979, Chem.-Biol. Interact., 27, pp.335-342.
27. Westra, J.G., and Visser, A.: 1979, Cancer Lett., 8, pp.155-162.
28. Beland, F.A., Allaben, W.T., and Evans, F.E.: 1980, Cancer Res., 40, pp.751-757.
29. Kadlubar, F.F., Miller, J.A., and Miller, E.C.: 1978, Cancer Res., 38, pp.3628-3638.
30. Kadlubar, F.F., Unruh, L.E., Beland, F.A., Straub, K.M., and Evans, F.E.: 1980, Carcinogenesis 1, pp.139-150.
31. Kawazoe, Y.: 1980, J. Natl. Cancer Inst., Monograph, in press.
32. Kadlubar, F.F., Miller, J.A., and Miller, E.C.: 1977, Cancer Res., 37, pp.805-814.
33. Radomski, J.L., Hearn, W.L., Radomski, T., Moreno, H., and Scott, W.: 1977, Cancer Res. 37, 1757-1762.
34. Kriek, E., and Westra, J.G.: 1980, Carcinogenesis, 1, submitted.
35. Spodheim-Maurizot, M., Dreux, M., Saint-Ruf, G., and Leng, M.: 1979, Nucl. Acids Res., 7, pp.2347-2356.
36. Sakai, S., Reinhold, C.E., Wirth, P.H., and Thorgeirsson, S.S.: 1978, Cancer Res., 38, pp.2058-2067.
37. Kriek, E., and Westra, J.G.: 1980, J. Natl. Cancer Inst., Monograph, in press.
38. Westra, J.G. and Visser, A., in preparation.
39. Kirby, K.S., and Cook, A.E.: Biochem. J., 104, pp.254-257, 1967.

ABBREVIATIONS

N-hydroxy-AF,	N-hydroxy-2-aminofluorene
N-hydroxy-AAF,	N-hydroxy-N-acetyl-2-aminofluorene
N-acetoxy-AAF,	N-acetoxy-N-acetyl-2-aminofluorene
N-OSO_3K-AABP,	Sulfate ester (potassium salt) of N-hydroxy-N-acetyl-4-aminobiphenyl
N-OSO_3K-AABP4'F,	Sulfate ester (potassium salt) of N-hydroxy-N-acetyl-4-amino-4'fluorobiphenyl
N-OSO_3K-AAP,	Sulfate ester (potassium salt) of N-hydroxy-N-acetyl-2-aminophenanthrene
N-BzO-MAB,	N-benzoyloxy-N-methyl-4-aminoazobenzene
N-acetoxy-AAS,	N-acetocy-N-acetyl-4-aminostilbene
N-hydroxy-1-NA,	N-hydroxy-1-naphthylamine
N-hydroxy-2-NA,	N-hydroxy-2-naphthylamine
N-hydroxy-4-AQO,	N-hydroxy-4-aminoquinoline-1-oxide
d Guo-AF,	N-(deoxyguanosin-8-yl)-2-aminofluorene

BIOCHEMICAL CONSIDERATIONS OF THE ENZYMOLOGY ASSOCIATED WITH QUINONE AND TETROL FORMATION DURING BENZO(A)PYRENE METABOLISM

J. Capdevila, Y. Saeki, R.A. Prough, and R.W. Estabrook
Department of Biochemistry, Southwestern Medical School,
The University of Texas Health Science Center at Dallas,
Dallas, Texas, 75235, U.S.A.

A. Abstract

While considerable effort has been directed toward understanding the molecular events involved in chemical carcinogenesis, a detailed description of the enzymatic machinery involved and the precursor-product relationships which exist have only begun during the last five years. The efforts of our research groups have focused on the enzymatic formation of benzo(a)pyrene quinones and on the spatial relationship between the requisite enzymes sequestered in the microsomal membrane. The mechanism of quinone formation has been studied using ^{18}O-incorporation experiments coupled with high performance liquid chromatographic and mass spectrometric techniques. The results indicate that *in vitro* a significant proportion of the quinones appears to be formed as a result of a cytochrome P-450-dependent monooxygenase process. However, a measurable amount of the quinone product contains oxygen incorporated from water suggesting at least two mechanisms of quinone formation.

The spatial relationship between the microsomal enzymes involved in the production of 7,8-dihydrodiol-9,10-epoxide of benzo(a)pyrene (measured as tetrols) has been tested by evaluating the existence of a free 7,8-dihydrodiol pool formed by release of the dihydrodiol from the enzyme, epoxide hydrase. Using dual isotope experiments, it was noted that under certain conditions there is limited exchange between exogenously added 7,8-dihydrodiol and the 7,8-dihydrodiol generated enzymatically from benzo(a)pyrene. This result can be explained by the existence of an organized complex of membrane-bound proteins which facilitates 7,8-dihydrodiol-9,10-epoxide formation from benzo(a)pyrene.

B. Introduction

The metabolic transformation of benzo(a)pyrene leads to a diversity of products, notably phenols, dihydrodiols, epoxides, quinones and more polar compounds (1). The ratio of products formed

B. Pullman, P. O. P. Ts'o and H. Gelboin (eds.), Carcinogenesis: Fundamental Mechanisms and Environmental Effects, 113–124.

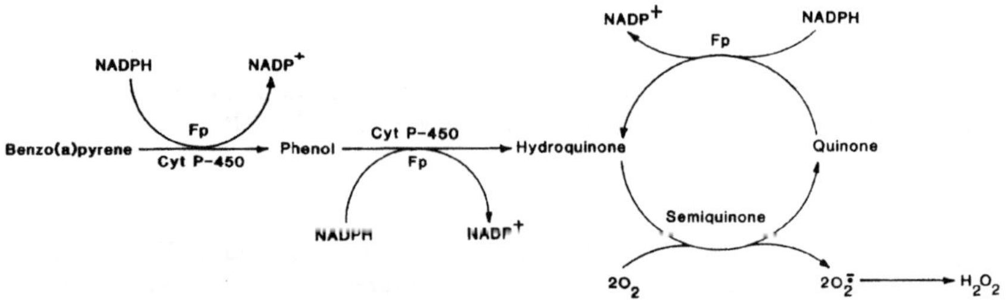

Figure 1. The postulated mechanism of benzo(a)pyrene quinone formation and the redox cycle involving the quinone/hydroquinone pair.

can vary depending on the source of the enzyme system (obtained from different species, sex, and organs), and the history of the animals exposure to diets, pollutants, and various agents capable of stimulating the synthesis of different forms of the hemoprotein, cytochrome P-450 (2-5). It has been established that some of the products formed during benzo(a)pyrene metabolism are primary metabolites that can be further oxidatively converted to more polar products, although it is difficult to demonstrate a precursor-product relationship *in vitro* unless limiting concentrations of benzo(a)pyrene are employed.

Recently, we have described (6) studies directed toward understanding the mechanism of quinone formation during benzo(a)pyrene metabolism. Quinones can serve either as redox active compounds which when catalytically reduced can react with oxygen to generate hydrogen peroxide or the superoxide anion (7) or when oxidized serve to accept reducing equivalents derived from reduced pyridine nucleotides (8). These properties provide quinones with unique characteristics not shared by other metabolites of benzo(a)pyrene (Figure 1).

The catalyst for the activation of oxygen required for the formation of many products of benzo(a)pyrene metabolism is a family of membrane-bound hemoproteins called cytochrome P-450. This hemoprotein functions together with a flavoprotein, termed NADPH-cytochrome P-450 reductase, to form an electron transport system generally localized in the endoplasmic reticulum. Not only does cytochrome P-450 serve as a monooxygenase in the presence of oxygen and NADPH, but it can also function as a peroxidase in the presence of organic hydroperoxides or hydrogen peroxide. It is well established that a diversity of products are formed when liver microsomes are incubated with benzo(a)pyrene, NADPH, and oxygen. In contrast, quinones are the major products generated when liver microsomes are incubated with benzo(a)pyrene and cumene hydroperoxide (10). In this case, various benzo(a)pyrene phenols and their oxo radicals are presumed to be intermediates in the

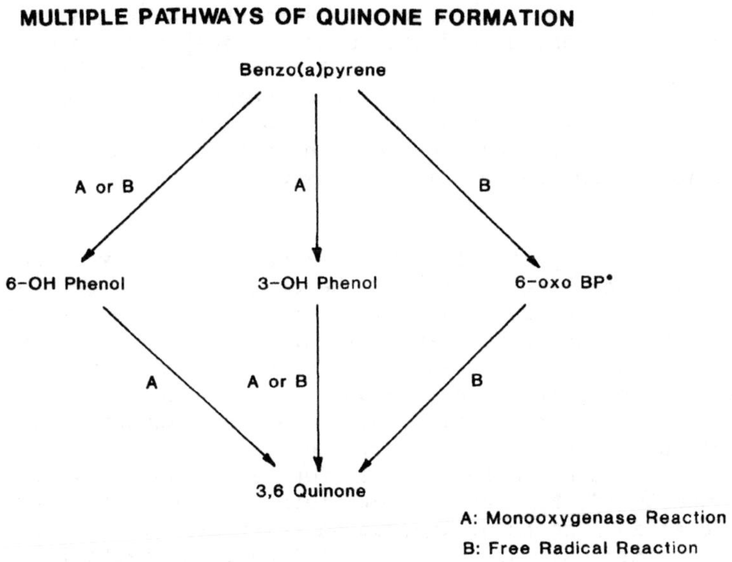

Figure 2. Possible metabolic pathways for the generation of the benzo(a)pyrene-3,6-quinone.

formation of quinones. However, several pathways may be operative, as illustrated in Figure 2, for the formation (for example) of the 3,6-quinone of benzo(a)pyrene. As shown in Figure 2, one initial step in the metabolism of benzo(a)pyrene which leads to quinone formation may be either the generation of the 3- or 6-phenols, via a monooxygenase reaction, or the formation of a 6-oxo radical formed by a one-electron oxidation mechanism. Comparable reactions may occur in the second step, thereby leading to quinone formation.

C. ^{18}O Studies During Quinone Formation.

Recently we have undertaken experiments to evaluate the proportionate role of cytochrome P-450-dependent monooxygenation reactions and of free radical reactions in the formation of quinones. The mechanism of quinone formation was studied using stable isotope incorporation experiments with either $^{18}O_2$ or $H_2{}^{18}O$ and subsequent analysis by hplc and mass spectrometry. This technique allows us to elucidate the role of molecular oxygen and oxygen derived from water in the formation of the respective benzo(a)pyrene quinones and dihydroxy metabolites. It is assumed that a classic monooxygenase reaction would incorporate stoichiometric amounts of an oxygen atom from atmospheric oxygen into the substrate. Conversely, those reactions which do not incorporate stoichiometric amounts of atmospheric oxygen may involve a free radical mechanism and will most probably contain oxygen derived from water. We are aware that molecular oxygen may be incorporated into substrate during certain free radical reactions, but have restricted the arguments presented in this paper since we have no data to substantiate the function of an alternate mechanism of oxygen incorporation under the conditions used in the present series of experiments.

As a preliminary step for these experiments, it was necessary to determine the exchange rate of the oxygens of the quinones with the oxygen of water, particularly during the process of extraction and isolation by high performance liquid chromatography (hplc). In addition, the exchange of the oxygen in the quinone molecule may be potentiated during its reduction to the hydroquinone by the microsomal NADPH-dependent quinone reductase and subsequent oxidation by atmospheric oxygen (8). Therefore, the 3,6-quinone was incubated under acid conditions in $H_2{}^{18}O$ in order to form a compound which could be readily monitored by mass spectrometry to evaluate exchange rates during the various steps of incubation, isolation, and purification. It was of interest to note the difficulty in replacing both oxygen atoms of the purified quinone by the ^{18}O of water even under these stringent conditions. It was observed that approximately 10-15 percent of the quinones oxygens exchanged when pure quinone was incubated in $H_2{}^{18}O$ with liver microsomes and NADPH for 5 minutes followed by extraction and hplc analysis. This data suggests that the exchange of quinone oxygens with the oxygen of water in the reaction media is a slow process and should not pose a serious obstacle to the analysis of results obtained during the metabolic conversion of benzo(a)pyrene to quinones.

With this important control defined, liver microsomes from phenobarbital treated male rats were incubated with 0.1 mM benzo(a)pyrene for 5 minutes in the presence of an excess of NADPH using an atmosphere of 20 percent oxygen which was 90 percent enriched in $^{18}O_2$. In agreement with the results of Yang _et al_. (9), those phenols which cochromatographed with the 3-hydroxybenzo(a)pyrene were 88 percent enriched in ^{18}O indicating the exclusive role of atmospheric oxygen in the cytochrome P-450-dependent monooxygenase reaction for the formation of this product. Mass spectral analysis (Figure 3) of the isolated 4,5-dihydrodiol and the 7,8-dihydrodiol provided no evidence for a molecule containing two atoms of ^{18}O (molecular weight of 290), but a 90% and 82% yield, respectively, of the dihydrodiols having a molecular weight of 288 - a result consistent with the introduction of one atom of atmospheric oxygen to form an epoxide followed by hydration with water from the reaction medium by epoxide hydrase.

Analysis of the 1,6-quinone and the 3,6-quinone (Figure 3) provided an unusual result; in both cases, approximately 40% of the molecules contained two atoms of ^{18}O while 40% of the molecules contained only one atom of ^{18}O. This result suggests that two pathways

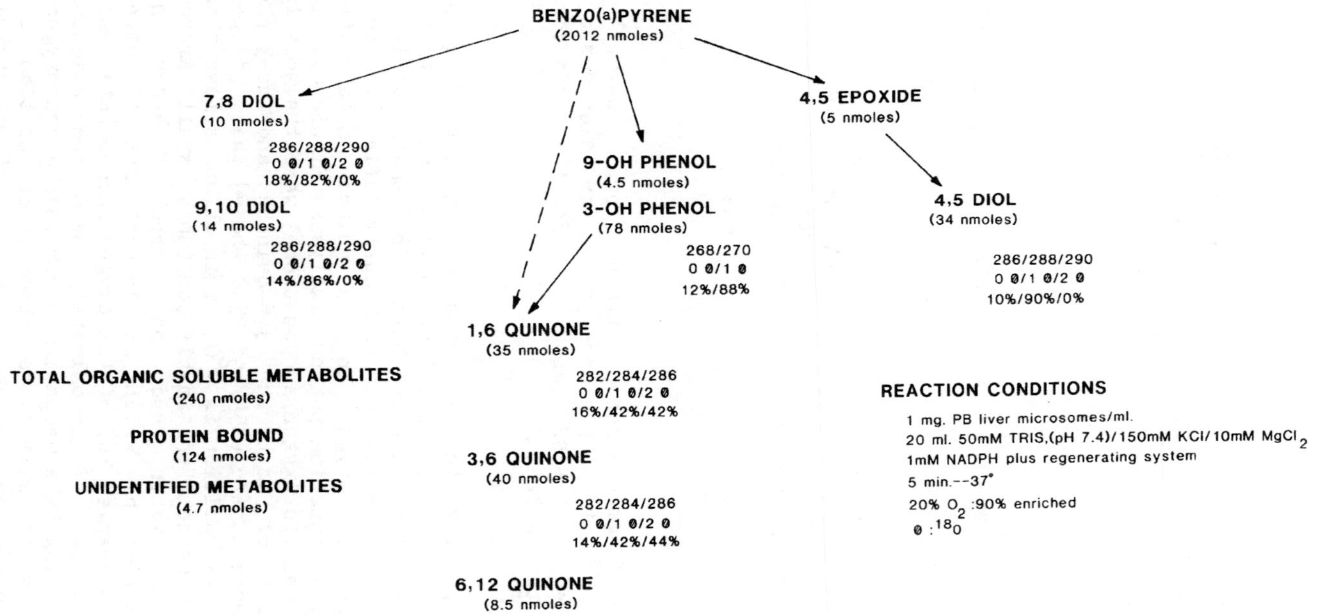

Figure 3. The metabolic balance for benzo(a)pyrene oxidation in the presence of liver microsomes from phenobarbital-pretreated rats with NADPH and $^{18}O_2$. The experimental conditions used are indicated on the figure. The reaction was terminated by the addition of ethyl acetate to the closed reaction vessel, the metabolites purified by hplc, and the properties of the isolated and purified compounds determined by mass spectrometry.

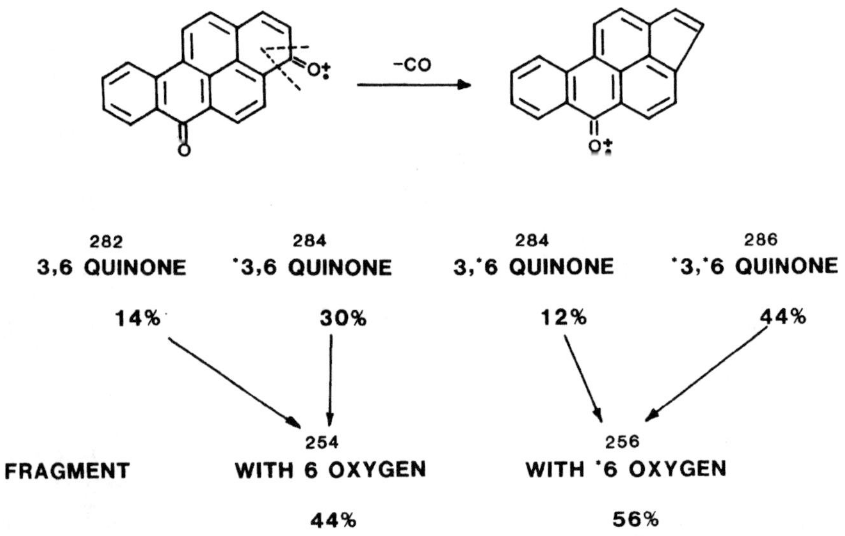

Figure 4. The mass fragmentation pattern for benzo(a)pyrene 3,6-quinone formed in the presence of liver microsomes from phenobarbital-pretreated rats, NADPH, and $^{18}O_2$. For the conditions of the experiment, refer to Figure 3.

of metabolism may be operative in the enzymatic formation of quinones. More than one-half of the quinone molecules formed appear to involve only monooxygenation reactions with the balance presumably derived from reactions involving free radicals (as previously defined). The exact location of the incorporated oxygen in the quinone metabolite was determined by mass fragmentation pattern analysis (Figure 4). It was noted that the carbon and its oxygen at position 3 were the most labile site for fragmentation of both the 3,6-quinone and the 3-phenol. Analysis of the fragments at m/e (254 and 256) showed that 78 percent of the oxygen at the C-3 position is ^{18}O. This value is less than the theoretical value of 90 percent presumably because a small amount of exchange occurs with water and the degree of error in the determination. At position 6, which is thought not to have any significant degree of exchange, 56% of the oxygen incorporated was ^{18}O. This provides a maximum limit of approximately 40% of the reaction at the 6 position occurring by a mechanism other than a monooxygenation reaction. These results establish that the major portion of the 3,6-quinone formed is not a consequence of an artifact resulting from the experimental procedure used, but instead represents a true enzymatic reaction catalyzed by the microsomal monooxygenase system. Thus, it appears that the formation of quinones and the redox couple between quinones and their hydroquinones (dihydroxy-compounds) may be more significant in the in vitro and in vivo metabolism of benzo(a)pyrene than previously considered.

D. Spatial Considerations of Tetrol Formation.

The oxidation of benzo(a)pyrene to reactive diol-epoxides is presumed to follow a sequence of reactions involving the initial oxidation of B(a)P to the 7,8-epoxide which is subsequently hydrated by the enzyme, epoxide hydrase, to form the 7,8-dihydrodiol. This dihydrodihydrodiol can be further oxidized to yield the 7,8-dihydrodiol-9,10-epoxide which can interact covalently with nucleic acids and proteins, thereby initiating a cascade of events resulting in the modification of a number of cellular characteristics. The question was posed whether there exists a structural interrelationship of the requisite enzymes (as bound to the microsomal membrane) in order to facilitate this metabolic transformation or whether the enzymes do not work in concert, but simply function as non-coordinated enzymatic reactions. Two possible alternatives for the spatial organization of the requisite enzymes, as they reside on the cytosol surface of the endoplasmic reticulum, are diagrammaticaly illustrated in Figure 5. In alternative A of Figure 5a, it is envisioned that a molecule of benzo(a)pyrene is bound to the active site of the hemoprotein, cytochrome P-450 (or P-448), where the initial reaction of oxygen interaction occurs. The resultant epoxide can then be visualized to migrate from this site to seek a molecule of the protein, epoxide hydrase, for its conversion to the 7,8-dihydrodiol. Presumably, the dihydrodiol molecule then must find an unoccupied molecule of cytochrome P-450 (P-448) for its subsequent oxidation to the 7,8-dihydrodiol-9,10-epoxide. In this case, a unique species of cytochrome P-450 (P-448) specifically dedicated to the oxidation of the dihydrodiol may exist (11) or the dihydrodiol may serve in competition with additional molecules of benzo(a)pyrene for a common species of hemoprotein. It should be realized that these reactions may occur on one location of the membrane or by intermembrane transfer. Another sequence of reactions which may occur is shown by alternative B (Figure 5b). In this case, the juxtoposition of cytochrome P-450 (P-448) and the epoxide hydrase facilitates an internal transfer of reactive molecules without the consequent release of intermediates and their dilution into the reaction medium. To test which alternative may be operative in the conversion of benzo(a)pyrene to the 7,8-dihydrodiol-9,10-epoxides (measured as tetrols), a series of experiments have been carried out to determine whether [^{3}H]-7,8-dihydrodiol, when added to a reaction mixture catalyzing the metabolism of [^{14}C]-benzo(a)pyrene, is in rapid equilibrium with a common pool of 7,8-dihydrodiol utilized for the formation of tetrols. This is diagrammatically shown in Figure 6. As summarized by the data

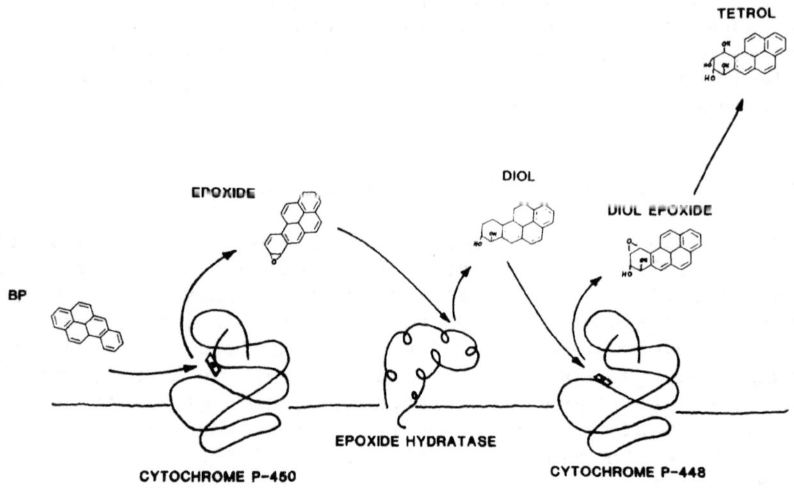

Figure 5a. The non-coordinated model for benzo(a)-pyrene-7,8-dihydrodiol-9,10-epoxide formation by rat liver microsomal cytochrome P-450.

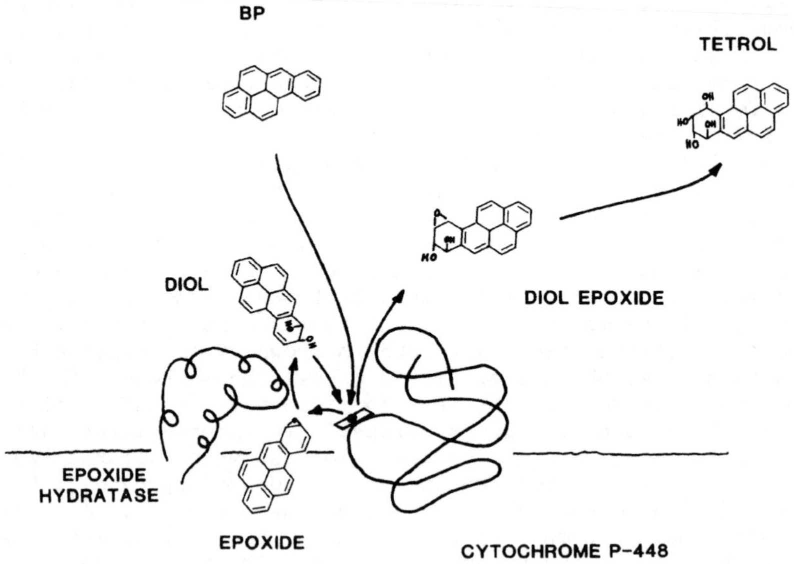

Figure 5b. The coordinated model for benzo(a)pyrene-7,8-dihydrodiol-9,10-epoxide formation by rat liver microsomal cytochrome P-450.

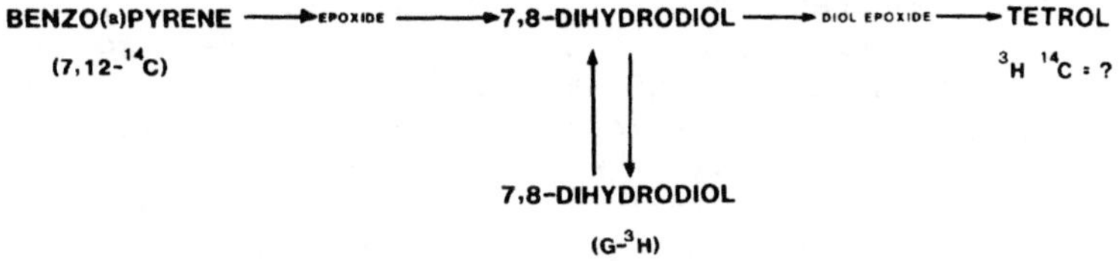

Figure 6. The reaction scheme of benzo(a)pyrene-7,8-dihydrodiol-9,10-epoxide formation and the possible free pool of 7,8-dihydrodiol.

Table I. The Metabolic Rates of Tetrol Formation

Substrate Used	Tetrol Formed[a]	
	^{14}C	^{3}H
	(pmoles/min/mg)	
^{14}C-B(a)P	56	-
^{3}H-Diol	-	0.1
^{14}C-B(a)P + ^{3}H-Diol	63	3.5
^{12}C-B(a)P + ^{3}H-Diol	-	3.1

a. The benzo(a)pyrene was incubated with 0.5 mg/ml liver microsomes from 5,6-benzoflavone-treated rats and 1.0 mM NADPH (with NADPH-regenerating system) in 0.15 M KCl, 10 mM $MgCl_2$, and 0.05 M Tris-HCl buffer, pH 7.4, for 10 min at 37°C. The ^{12}C- and ^{14}C-benzo(a)pyrene [B(a)P] concentration used was 7 μM and the ^{3}H-B(a)P-7,8-dihydrodiol [^{3}H-Diol] concentration was 37 nM. The samples were extracted with ethyl acetate and analyzed using a Toyosoda reverse phase column (octadecylsilane) with an isocratic solvent system (50% methanol in water at 1 ml/min flow rate). Authentic standards of the benzo(a)pyrene-7,8-dihydro-9,10-epoxide hydrolysis products (tetrols and triols) were added to the samples and the radioactivity which cochromatographed with the major tetrol [(7,10/8,9)-tetrol] was analyzed.

presented in Table I, it appears that only a small amount of the exogenously added [^{3}H]-7,8-dihydrodiol mixed with the [^{14}C]-7,8-dihydrodiol derived from the benzo(a)pyrene added to the reaction mixture. Under the experimental conditions employed, one can estimate that less than 10% of the potential pool of 7,8-dihydrodiol formed is in equilibrium with the added 7,8-dihydrodiol. This conclusion is based on the ratio of ^{3}H/^{14}C measured in the tetrols isolated by hplc. Measurements of the "carbon flux" through this reaction sequence clearly indicates that the conversion of benzo(a)pyrene to the tetrol via a 7,8-dihydrodiol intermediate is significantly faster than the rate of conversion of the exogenously added 7,8-dihydrodiol to tetrols. Based on these preliminary studies, one can postulate a sequence of events which favor a hypothesis of spatial organization of the enzymes involved in the metabolic conversion of benzo(a)pyrene to 7,8-dihydrodiol-9,10-epoxides. In order to eliminate a kinetic artifact due to the presence of sub-optimal concentrations of the 7,8-dihydrodiol used in Table I, a comparable experiment was carried out using a fifty-times greater chemical concentration of ^{3}H-7,8-dihydrodiol (2 μM). Similar results were obtained indicating a virtual independence of the chemical concentration of the exogenous dihydrodiol added on the kinetics of the reaction. Differences in the rates of metabolism for B(a)P and 7,8-dihydrodiol leads to a progressive accumulation of the 7,8-dihydrodiol during the reaction. These differences in the rates of metabolism may reflect the relative affinity of B(a)P and the 7,8-dihydrodiol for the cytochrome. From these studies, it is suggested that those molecules of benzo(a)pyrene that are transformed to diol-epoxides result from an organized relationship of the requisite enzymes in a manner that discourages the dissociation and migration of intermediates into a free pool of potentially reactive molecules.

E. Discussion

A study of benzo(a)pyrene metabolism offers many opportunities to gain a further insight into the mechanism of oxidative transformation of a variety of xenobiotics. The present report has focused on two aspects of benzo(a)pyrene metabolism which are currently under study in our laboratories. The enzymatic formation of quinone metabolites has not been fully explored and, indeed, the question of artifactual generation of quinones under _in vitro_ or _in vivo_ conditions remains unanswered. These studies, showing ^{18}O-incorporation into both oxygens of the 3,6-benzo(a)pyrene quinone, point to the enzymatic role of cytochrome P-450 (P-448) in this reaction. The presence of a significant amount of quinone product with oxygen incorporated from water suggests the existence of a second reaction mechanism operating, presumably, by free radical intermediates. Further study will be required to delineate the relative importance of these two mechanisms for quinone formation and the possible role of other additional pathways which remain to be elucidated. It is apparent, however, that quinones are real products resulting from an enzymatic process and that quinones may play a significant role in the regulation of the microsomal electron transport system and in the carcinogenic properties of benzo(a)pyrene metabolites.

A second question which is addressed relates to membrane structure and the organization of enzymes associated with the endoplasmic reticulum. The sequence of reactions which occur during the metabolic conversion of benzo(a)pyrene to the 7,8-dihydrodiol-9,10-epoxide of benzo(a)pyrene involves two steps of cytochrome P-450 (P-448)-catalyzed oxidations interposed by a hydration step catalyzed by epoxide hydrase. Our initial studies to evaluate the existence of an exchangable pool of the intermediate 7,8-dihydrodiol suggest that the requisite enzymes may be in a unique spatial relationship to facilitate the formation of the metabolite, benzo(a)pyrene 7,8-dihydrodiol-9,10-epoxide. The existence of an organized complex of membrane-bound proteins may influence the pattern of metabolites of benzo(a)pyrene formed *in vivo*. Future studies may reveal perturbations of this type of molecular organization and thereby explain, in part, the diversity of products of benzo(a)pyrene obtained using various experimental conditions and sources of the enzyme system.

Acknowledgements

The Authors are indebted to Dr. J. Napoli for his expertise in mass spectroscopy and to Mr. N. Chacos for his technical assistance. This research was supported by NIH Grant HL 19654 (RAP), NIH Grant GM 16488 (RWE), and Robert A. Welch Grant I-616 (RAP). RAP is a USPHS Research Career Development Awardee, HL 00255. Dr. Y. Saeki's permanent address is as follows: Laboratory of Toxicology, Department of Biochemistry, Shiga University of Medical Science, Seta, Ohtsu, Shiga 520-21, Japan.

References

1. Sims, P., and Grover, P.L.: 1975, Adv. Cancer Res. 20, pp. 165-274.
2. Remmer, H., and Merker, H.J.: 1965, Ann. N.Y. Acad. Sci. 123, pp. 79-86.
3. Conney, A.H.: 1967, Pharmacol. Rev. 19, pp.317-366.
4. Omura, T., Siekovitz, P., and Palade, G.E.: 1967, J. Biol. Chem. 242, pp. 2389-2396.
5. Gillette, J.R., Davies, D.C., and Sasame, H.A.: Ann. Rev. Pharmac. 12, pp. 57-84.
6. Capdevila, J., Estabrook, R.W., and Prough, R.A.: 1978, Biochem. Biophys. Res. Communs. 82, pp. 518-525.
7. Lesko, S., Caspary, W., Lorentzen, R., and Ts'o, P.O.P.: 1975, Biochem. 14, pp. 3978-3984.
8. Capdevila, J., Estabrook, R.W., and Prough, R.A.: 1978, Biochem. Biophys. Res. Communs. 83, pp. 1291-1298.

9. Yang, S.K., Roller, P.P., Fu, P.P., Harvey, R.G., and Gelboin, H.V.: 1977, Biochem. Biophys. Res. Communs. 77, pp. 1176-1182.
10. Capdevila, J., Estabrook, R.W., and Prough, R.A.: 1980, Arch. Biochem. Biophys. 200, pp. 186-195.
11. Wiebel, F.J., Selkirk, J.K., Gelboin, H.V., Haugen, D.A., Van Der Hoeven, T.A., and Coon, M.J.: 1975, Proc. Nat. Acad. Sci. USA 72, pp. 3917-3820.

STEREOCHEMICAL ASPECTS OF BENZO[A]PYRENE METABOLISM AND BIOCHEMICAL INDIVIDUALITY IN HUMAN CELLS

Joseph Deutsch*, Paul Okano and Harry V. Gelboin
*Department of Pharmaceutical Chemistry, The Hebrew University School of Pharmacy Jerusalem, Israel, and Laboratory of Molecular Carcinogenesis, NCI, National Institutes of Health, Bethesda, Maryland 20205 U.S.A.

ABSTRACT

Purified forms of rabbit liver microsomal cytochrome P-450, are active in the conversion of the (-) and (+) enantiomers of trans-7, 8-dihydroxy 7,8-dihydrobenzo[a]pyrene to stereoisomeric, highly reactive 7,8 dihydroxy-9,10-oxy-7,8,9,10-tetrahydrobenzo[a]pyrenes. Cytochrome P-450 LM4 has much higher catalytic activity than cytochrome P-450 LM2 with either enantiomer of the trans-7-8-diol. P-450 LM4 catalyzes the formation of diol epoxide I almost exclusively from the (-)trans-7,8-diol and forms more diol epoxide II than diol epoxide I from the (+)trans-7,8-diol. Thus P-450 LM4 is highly stereoselective for oxygenation at the 9,10 double bond, regardless of the absolute configuration of the hydroxyl groups at the 7 and 8 positions of the substrates.

BP and (-)trans-7,8-diol are metabolized by human monocytes and lymphocytes. The patterns of the metabolites formed by the two cells are different and some differences were observed in cells from different individuals. This suggests that these cells have a different complement of BP metabolizing enzymes which may relate to biochemical individuality in carcinogen susceptibility.

INTRODUCTION

Foreign compounds or xenobiotics to which we are all exposed include polycyclic hydrocarbons resulting from energy producing processes, chemicals from industry, pesticides, food additives, and a variety of drugs(1). The primary biological receptors which act as interfaces between this large variety of chemicals and man and other organisms are the microsomal mixed-function oxidases(2). These cytochrome P-450 containing enzyme systems, when coupled with other metabolically linked enzymes, provide the cells with an important pathway whereby the various xenobiotics can be converted to metabolites which are water soluble, allowing them to be safely excreted(3,4,5,22). These systems are thus highly beneficial and

B. Pullman, P. O. P. Ts'o and H. Gelboin (eds.), Carcinogenesis: Fundamental Mechanisms and Environmental Effects, 125–142.

necessary for organism survival. These apparently beneficial systems, however, have another face. They also catalyze the formation of metabolites which are toxic, mutagenic, teratogenic, and carcinogenic(6,7). Cytochrome P-450 exists in multiple forms and two of the cytochrome P-450's have been isolated in homogeneous form and three others have been partially purified(8,9). Previously we reported that the reconstituted systems containing the different cytochrome P-450's, NADPH cytochrome P-450 reductase and phospholipid are active in the metabolism of both benzo[a]pyrene and the conversion of (-)trans-7,8-diol metabolite to diol epoxides. These studies demonstrated that the different cytochrome P-450's possess regiosubstrate, and stereoselectivity in their catalytic activity(10).

With the introduction of the technique of HPLC, one can analyze and quantitate almost the entire BP metabolite pattern formed by cells under investigation(11). The various pathways of metabolism are summarized in Fig. 1. BP is converted into five different phenols, three quinones, and three dihydrodiols(23,24,29,12). Each of these compounds can be conjugated to water-soluble products and be excreted(13). The chart shows their conjugation with glutathione, sulfate and glucuronides. One of the primary metabolites, the 7,8-diol, can be further metabolized to 7,8 diol-9,10-epoxides of which there are two stereoisomeric forms (7). Both are highly reactive(14). They also are either hydrolyzed to tetrols and reduced to triols, or they react with macromolecules such as DNA(15,16). In fact, diol epoxide I has been found to be the most mutagenic metabolite of BP(7). It is also the major form found covalently bound to DNA which is isolated from cells treated in culture with BP(7). The activation pathway to diol epoxides involves several stereoselective enzymatic steps that yield primarily diol epoxide I, the anti stereoisomer, and lesser amounts of diol epoxide II, the syn stereoisomer (17,18). This brings us to the question of what regulates the amount of diol epoxides formed, relative to the other BP metabolites, which are primarily detoxification products. A primary factor is the activity of the cytochrome P-450 component of the mixed function oxidase system. The activity of other enzymes, such as epoxide hydratase and the conjugating enzymes, also affects the pathway of BP metabolism. Several laboratories have demonstrated that multiple forms of cytochrome P-450 are present in various animal tissues. We have examined the activity of several of the purified forms of cytochrome P-450 from rabbit liver and have found that they are regio- and stereoselective in terms of both substrate utilization and product formation (10). Six forms of cytochrome P-450 from rabbit liver microsomes have been resolved and defined by their electrophoretic mobilities by Dr. Minor Coon and his coworkers (8,9). They designate these different forms as LM1, LM2, LM3, LM4 and LM7. The two most highly purified forms are LM2 induced by phenobarbital and LM4 induced by either BP or β-naphthoflavone. The other forms are less pure. As could be seen in the Fig. 1, the activation pathways of benzo[a]pyrene to diol epoxides involves a first epoxidation at the 7,8-position followed by stereospecific hydration by epoxide hydratase to the (-)trans-7,8-diol. The diol is further epoxidized by the mixed

THE METABOLISM AND ACTIVATION OF BENZO[a]PYRENE

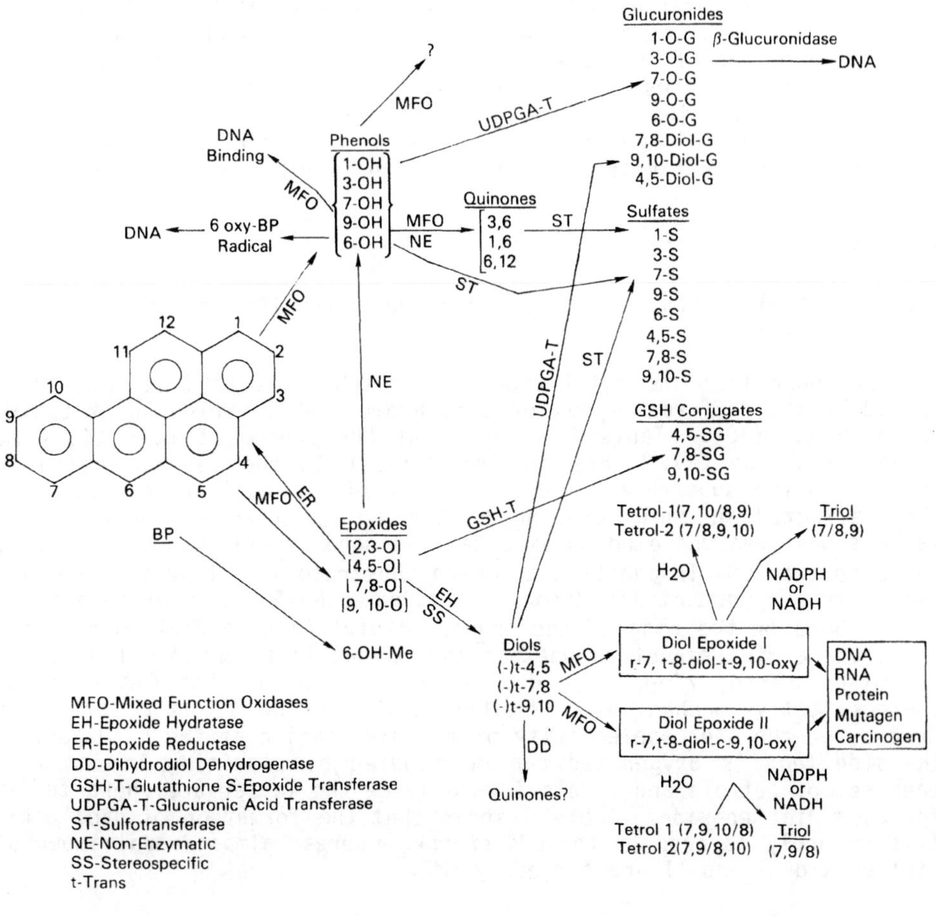

function oxidases to two stereoisomeric 7,8-diol-9,10 epoxides. Each isomer of the diol epoxides can be detected and identified by its tetrol hydrolysis and triol reduction products.

Studies from our laboratory and others have developed the conclusion that the 7,8-diol-9,10-epoxides are formed metabolically, are the most active metabolites in binding to DNA, and are super mutagens(7,14-17). Thus the diol epoxides may be the primary ultimate forms in BP carcinogenesis. An interesting aspect of this research has been the demonstration that the absolute stereochemistry of proximate and ultimate carcinogenic metabolites of BP plays a dominant role in the expression of biological activity by these compounds(19). Thus, in the metabolic sequence of BP to BP-7,8-oxide to 7,8-dihydrodiol and finally to 7,8-diol-9,10-epoxide, only one of the possible enantiomers of each metabolite displays high tumorigenic activity, and exhibit considerable stereoselectivity in their formation.

RESULTS AND DISCUSSION

Stereochemical aspects of benzy[a]pyrene metabolism by purified mixed function oxidase.

We found that the total amount of metabolites of (-)trans-7,8-diol formed by the different types of cytochrome P-450, varied quantitatively for each type(10). Table 1, shows that LM4 produce 5 to 8 times more metabolites than LM2. This is in contrast to the reverse being true when BP is the substrate. The metabolism of BP and (-)trans-7,8-diol is also reflected in their binding to DNA as is shown in Table 2. This Table shows that DNA binding is catalyzed more by LM4 than LM2. Further, we found that LM4 in addition to being substrate selective is also stereoselective in product formation. Cytochrome P-450 when oxygenating the double bond on the ring of the trans-7,8-diol to give diol epoxide shows a preference in respect to whether the oxygen is inserted below or above the plane of the ring. The diol epoxide I and II are formed by such stereoselective mechanisms. Whether diol epoxide I or II are formed depends on the stereoselectivity of the oxygenation at the 9,10 position. The side that is oxygenated can be studied by examining the amount of each set of tetrols and triols formed since the latter are characteristic for each diol epoxide. Table 3 shows that the formation of diol epoxide I is strongly favored by the LM4 enzyme, whereas almost equal amounts of diol epoxide I and II are formed by LM2.

For further study of the stereoselectivity of the cytochrome P-450 LM enzymes, the metabolism of both optical isomers of trans 7,8-diol were compared(20). The different forms of cytochrome P-450 might have a preference in respect to whether the oxygen is inserted below or above the plane of the ring when oxygenating a given position on the ring. Fig. 2 shows the two diol epoxides which could be obtained theoretically from the (-) and (+) trans-7,8-diol used as substrates. If we consider that a certain form of the cytochrome P-450, for instance, LM4, prefers

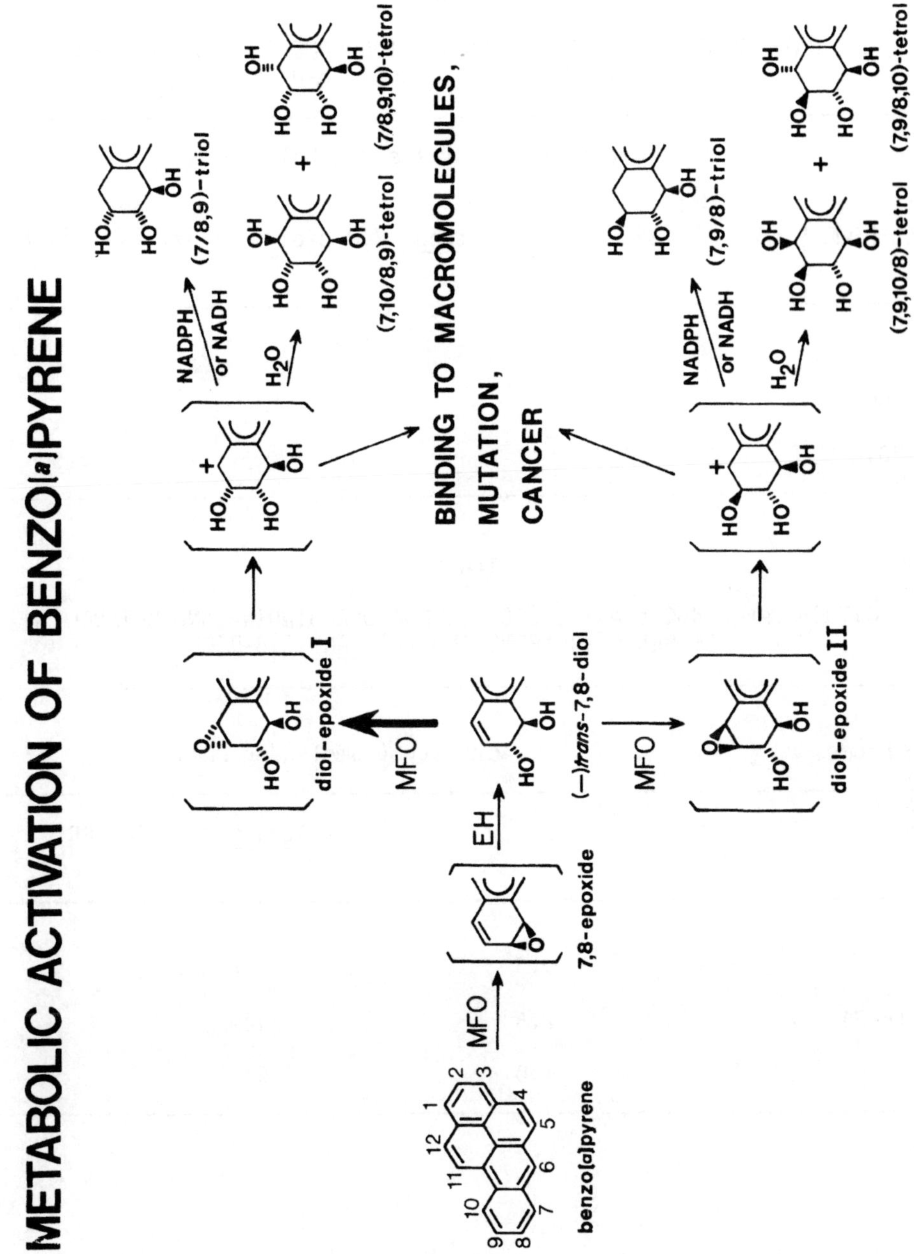
METABOLIC ACTIVATION OF BENZO[a]PYRENE
benzo[a]pyrene
MFO
7,8-epoxide
EH
(−)trans-7,8-diol
MFO
diol-epoxide I
MFO
diol-epoxide II
NADPH
or NADH
(7/8,9)-triol
H2O
(7,10/8,9)-tetrol
(7/8,9,10)-tetrol
NADPH
or NADH
(7,9/8)-triol
H2O
(7,9,10/8)-tetrol
(7,9/8,10)-tetrol
BINDING TO MACROMOLECULES,
MUTATION,
CANCER

TABLE 1

SUBSTRATE SELECTIVITY OF PURIFIED CYTOCHROME P-450's FOR BENZO[a] PYRENE AND (-)TRANS-7,8-DIOL

	Percent of substrate metabolized		
CYTOCHROME P-450	BP	(-)*trans*-7,8-diol	BP / (-)*trans*-7,8-diol
LM2	37.7	1.8	20.9
LM4(BNF)	2.7	8.6	0.3
LM4(PB)	4.8	14.5	0.3

TABLE 2

CYTOCHROME P-450's ARE SELECTIVE FOR DNA BINDING AND METABOLISM OF BENZO[a]PYRENE AND (-)TRANS-7,8-DIOL

Cytochrome	Activity (pmoles/mg DNA)	
	BP	(-)*trans*-7,8-diol BP
LM2	9.9	27
LM4 (BNF)	2.5	429
LM4	4.0	635

TABLE 3

STEREOSELECTIVITY OF DIOL EPOXIDE I AND II FORMATION FROM (-)TRANS-7,8-DIOL BY PURIFIED CYTOCHROME P-450's

Cytochrome P-450	Diol-epoxide I (pmol/nmol P-450/min)	Diol-epoxide II	Diol-epoxide I/ Diol-epoxide II
LM2	5	6	0
LM4(BNF)	117	10	11
LM4(PB)	178	15	11

TABLE 4

CATALYTIC ACTIVITY AND STEREOSELECTIVITY OF RABBIT P-450 LM2 AND P-450 LM4 IN OXYGENATION OF (-) AND (+) ENANTIOMERS OF BP-trans-7,8-DIOL

Metabolites of trans-7,8-diol	Activity (pmol products formed/min/nmol P-450 LM) LM4(PB)	LM4(BNF)	LM2
(-)trans-7,8-diol as substrate			
Diol-epoxide I	166	231	6
Diol-epoxide II	15	25	3
Ratio, I/II	11	9	2
(+)trans-7,8-diol as substrate			
Diol-epoxide I	7	9	3
Diol-epoxide II	12	23	1
Ratio, I/II	0.6	0.4	3

to insert the oxygen at the alpha position, (below the plane of the 9,10 double bond) independently of the stereochemistry of the hydroxyl groups, the ratio of diol epoxide I and II formed by such a stereoselective mechanism, should be reversed in the metabolism of the two trans-7,8-diol optical isomers. More diol epoxide I should be formed from the (-)enantiomer and more diol epoxide II from the (+) P-450 LM2, which yields the two diol epoxides almost to the same extents, we would anticipate no change in the ratio.

The catalytic activity of P-450 LM4 and LM2 toward the (-) and (+) enantiomers of trans-7,8-diol, as well as the degree of stereoselectivity of these catalysts in the oxygenation of the two optical isomers are compared in Table 4. The cytochrome LM preparations show large differences in their ability to metabolize the two enantiomeric forms of trans-7,8-diol, as judged by the total metabolites formed. It can be seen that the specificity for trans-7,8-diol in the reconstituted enzyme system resides almost exclusively in P-450 LM4, since this preparation from BP or BNF-induced microsomes is 20 to 30 times more active with the (-)enantiomer and 4 to 7 times more active with the (+) enantiomer than P-450 LM2. Both P-450 LM4 and LM2 exhibit higher activities with the (-) relative to the (+) optical isomer of the substrate, but this difference is more striking with P-450 LM4. Thus, P-450 LM4 from either source catalyzes the oxygenation of (-)trans 7,8-diol approximately ten times faster than that of (+)trans-7,8-diol, whereas the activity of P-450 LM2 is only twofold higher with the (-) than with the (+) enantiomer. Inspection of the metabolite ratios obtained with each enantiomer of the substrate reveals that the purified enzymes behave according to expectations; P-450 LM4 oxygenates the (-)trans-7,8-diol to give predominantly diol epoxide I. In contrast, about twice as much diol epoxide II as diol epoxide I is produced by P-450 LM4 when (+)trans7,8-diol serves as substrate. Thus, there is a marked reversal of the diol-epoxide I: diol-epoxide II ratio when P-450 LM4 is incubated with the (+) as compared to (-) diol. On the other hand, P-450 LM2 is devoid of stereoselectivity with respect to product formation, for it yields more diol epoxide I than diol epoxide II with either enantiomer of the trans-7,8-diol. The overall activity of P-450 LM2 with the enantiomeric diols is so small that the metabolite ratios shown at the right side of the slide, 1.8 and 3.0 with the (-) and (+) diol, respectively, are not believed to be significantly different.

The remarkable stereoselectivity of these cytochrome P-450 LM4 from rabbit liver microsomes is similar to that of 3-MC- inducible cytochrome P-448 purified from rat liver microsomes, though the latter preparation appears to be much more stereoselective in the conversion of the (+) trans-7,8-diol, giving diol epoxide II to diol epoxide I in a higher ratio. In contrast P-450 LM2 shows no stereoselectivity in oxygenation below the plane of the double bond of the trans-7,8-diol optical isomers. It should be noted that, unlike its poor overall activity toward BP-trans-7,8-diol, P-450 LM2 is the most active of the various forms of rabbit liver microsomal cytochrome P-450 in BP oxygenation.

AHH Activity and BP Metabolism in Man

In order to measure AHH activity(21) and BP metabolism in man, an easily obtainable source of human tissue is required. Tissues meeting this requirement, which we have used, include blood lymphocytes and monocytes. In our laboratory, interindividual and intraindividual variations in the basal and BA(benzo[a]anthracene) induced AHH activity were investigated in cultured monocytes from sets of monozygotic and dizygotic twins. The variation in BA-induced AHH activity among the various individuals show that the mean inducible values differ quite significantly, ranging from 4.26 to 17.69. Statistical analysis of this data suggests that the intraindividual variation in AHH activity is largely genetically determined(25). The data also suggested that the number of genes involved in AHH expression is relatively few in number. The relationship, however, of the total level of AHH activity to an individual's susceptibility to BP or other chemical carcinogen is still the subject of controversy today(12). Susceptibility to chemical carcinogens may be more related to the balance between different carcinogen metabolic pathways(26). In order to determine whether an individual's mixed-function oxidase activity could be characterized, we analyzed BP metabolite patterns in monocytes and lymphocytes, by using the HPLC technique. The amount of BP metabolism, however, was found to below in these cells. We modified existing procedures of adding an anti-oxidant to prevent oxidation of minute amounts of metabolites.

Comparison of BP Metabolism in BA-induced and Non-induced Human Cells (27-28)

Table 5 summarizes the pattern of BP metabolism in BA-induced and control monocytes from two different individuals. The formation of each BP metabolite is increased about 3-fold in BA-induced monocytes from Individual A as compared to control cells. The fold increase, however, of 7-OH and 9-OH formed in induced cells appears greater than the fold increase observed for other metabolites. Thus, the data suggest that induction preferentially induces enzyme systems forming 9-OH and 7-OH. The amount of 7,8-diol and 3-OH is increased to a lesser extent in the induced cells than are those of the other phenols. The 9,10-diol and 4, 5-diol are synthesized in low, barely detectable amounts in both BA-induced and control cells, making any comparison difficult. Formation of quinones varies from sample to sample, and differences in rate of formation in BA-induced and control cells are probably not significant. For Individual B, total BP metabolism is induced 3.6-fold by BA with a slight preferential synthesis of 7-OH and a slightly higher formation of 7-8-diol in control cells.

The summary of BP metabolism in BA-induced and control lymphocytes from the same two individuals is shown in Table 6. In Individual A, BA induction leads to a lower ratio of 7,8-diol and 3-OH and a higher ratio of 7-OH and 9-OH formation. In both monocytes and lymphocytes, the induction process appears to induce enzymes favoring 7-OH and 9-OH formation

TABLE 5

COMPARISON OF BP METABOLISM IN BA-INDUCED AND NON-INDUCED MONOCYTES

BP metabolite	Specific activity (f mol/min/10 :ells)					
	Individual A			Individual B		
	BA induced	Control	Ratio[a]	BA induced	Control	Ratio[a]
9,10-Diol	6(0.7)[b]	6(1.4)	1.0(0.5)	13(0.6)	7(1.2)	1.8(0.5)
4,5-Diol	4(0.5)	0(0.2)	(3.1)	65(3.2)	0(0)	
7,8-Diol	106(14.1)	47(15.8)	2.3(0.9)	205(10.1)	97(17.2)	2.1(0.6)
9-OH	265(35.2)	70(30.6)	3.8(1.1)	588(29.2)	181(32.1)	3.2(0.9)
7-OH	87(4)	20(10.1)	4.4(1.2)	256(12.6)	69(12.2)	3.7(1.0)
3-OH	212(28.6)	76(33.6)	2.8(0.8)	553(27.3)	197(34.8)	2.8(0.8)
1,6-Quinone	26(3.4)	6(1.5)	4.4(2.3)	108(5.3)	0(0)	
3,6-Quinone	44(5.8)	17(8.6)	2.6(0.7)	236(11.6)	14(2.5)	16.9(4.6)
6,12-Quinone						
Total	752(100)	242(100)	3.1	2,024(100)	565(100)	3.6
AHH	349	125	2.8	1,216	404	3.0

Duplicate samples of 5 to 6 x 10^6 frozen BA-induced or non-induced monocytes were incubated with [^{3}H]BP, and the BP metabolites were analyzed by HPLC. For the AHH assay, 4 x 10^6 monocytes were incubated, and the activity was determined. [a]BA-induced versus non-induced cells. [b]Numbers in parentheses, percentage of total metabolites.

TABLE 6

COMPARISON OF BP METABOLISM IN BA-INDUCED AND NON-INDUCED LYMPHOCYTES

BP metabolite	Specific activity (fmol/min/ 10^6 cells)					
	Individual A			Individual B		
	BA-induced	Control	Ratio[a]	BA-induced	Control	Ratio[a]
9,10-Diol	1(0.3)[b]	1(0.5)	1.0(0.5)	1(0.4)	0(0.2)	
4,5-Diol	3(1.2)	0(0.3)	(4.2)	23(6.2)	0(0)	
7,8-Diol	19(7.4)	13(12.3)	1.4(0.6)	20(5.2)	3(5.8)	6.4(0.9)
9-OH	61(23.3)	20(18.8)	3.0(1.2)	62(16.6)	9(16.6)	6.9(1.0)
7-OH	45(17.4)	11(10.4)	4.1(1.7)	36(9.7)	6(10.6)	6.4(0.9)
3-OH	95(36.4)	43(40.6)	2.2(0.9)	93(25.2)	24(44.1)	3.9(0.6)
1,6-Quinone	12(4.6)	6(5.5)	2.0(0.8)	43(11.6)	0(0)	
3,6-Quinone	25(9.4)	13(11.6)	2.0(0.8)	94(25.3)	12(22.7)	7.6(1.1)
6,12-Quinone						
Total	261(100)	107(100)	2.4	372(100)	54(100)	6.9
AHH	142	68	2.1	152	70	2.2

[a]BA-induced *versus* non-induced cell
[b]Numbers in parentheses, percentage of total metabolites

over those enzyme systems, favoring 3-OH and 7,8-diol formation.

Interindividual Differences in BP Metabolism

Table 7 shows a comparison of BP metabolism by BA-induced monocytes and lymphocytes which have been isolated from the same individual and a comparison of three different individuals. Total BP metabolism in BA induced monocytes is 2- to 5-fold higher than in BA-induced lymphocytes. The difference may be even greater since the BP metabolic activity in monocytes is more sensitive to freezing and thawing than it is in lymphocytes. One should note that, in this comparison, Individual A has the highest monocyte BP metabolic activity while Individual C has the highest lymphocyte activity. In the comparison of the rate of formation of individual BP metabolites, some common features emerge. The 9,10-diol and 7,8-diol are formed at about twice the rate as a percentage of total BP metabolites in monocytes compared to lymphocytes. The percentage of 9-OH formation is 30 to 80% higher in monocytes than in lymphocytes. Conversely, the rate of 3-OH formation is about 20% lower in monocytes than in lymphocytes. Quinone formation is also relatively higher in lymphocytes than in monocytes. In Individual A, the 7,8-diol: 4, 5 diol ratio is about 26:1 in monocytes vesus 6:1 in lymphocytes with the total diols comprising 15% of total BP metabolites in monocytes and about 9% in lymphocytes. Total phenol synthesis is about 76% in each case. For Individual A, the total phenol: diol ratios are 4.9 for monocytes and 8.7 for lymphocytes. The 7,8-diol: 4,5-diol ratios in Individual B are 3:1 in monocytes and about 1:1 in lymphocytes. The amount of diol formed is 14 and 12% of total BP metabolites, respectively. In this case, phenol formation is 69% of total BP metabolites in monocytes and 52% in lymphocytes with total phenol: diol ratios of 4.9 and 4.4 for monocytes and lymphocytes, respectively. This value for lymphocytes may be abnormally low since oxidation to quinones was higher than normal for these samples. In Individual C, the 7,8-diol: 4,5-diol ratios are about 5:1 and 3:1 for monocytes and lymphocytes, respectively. Total diols for Individual C represent 13 and 8%, respectively, of total BP metabolites for monocytes and lymphocytes. The phenols represent 83% of total BP metabolites in both these monocytes and lymphocytes. The total phenol: diol ratios are 6.5 and 10.7, respectively, for monocytes and lymphocytes of Individual C.

Quantitatively, the most striking interindividual difference seems to be in the relative amount of 4,5-diol formed by different individuals. This metabolite varied widely, and further work is needed to determine whether this parameter may indicate interindividual differences. In the individual with greater amounts of 4,5-diol formation, the amount of 9-OH was relatively reduced. In this individual, both the monocytes and lymphocytes formed relatively more 4,5-diol.

Both monocytes and lymphocytes can further metabolize the proximate carcinogen (-)trans-7,8-diol to 7,8-diol-9,10-epoxides. The data are summarized on Table 6. The results show that there is a preferential for-

TABLE 7

COMPARISON OF BP METABOLISM IN MONOCYTES AND LYMPHOCYTES FROM DIFFERENT INDIVIDUALS

BP metabolite	Individual A Monocyte	Individual A Lymphocyte	Ratio[a]
9,10-Diol	6 (0.7)[b]	1 (0.3)	6.0(2.7)
4,5-Diol	4 (0.5)	3 (1.2)	1.3(0.4)
7,8-Diol	106(14.1)	19 (7.4)	5.6(1.9)
9-OH	265(35.2)	61(23.3)	4.3(1.5)
7-OH	87(11.8)	45(17.4)	1.9(0.7)
3-OH	214(28.6)	95(36.4)	2.3(0.8)
1,6-Quinone	26 (3.4)	12 (4.6)	2.2(0.7)
3,6-Quinone	44 (5.8)	25 (9.4)	1.8(0.6)
6,12-Quinone			
Total	752 (100)	261 (100)	2.9

Specific activity (fmol/min/10^6cells)

Individual B Monocyte	Individual B Lymphocyte	Ratio[a]	Individual C Monocyte	Individual C Lymphocyte	Ratio[a]
13 (0.6)	1 (0.4)	13.0(1.8)	4 (0.4)	1 (0.2)	4.0 (1.8)
65 (3.2)	23 (6.2)	2.8(0.5)	22 (2.2)	9 (2.0)	2.4 (1.1)
205 (10.1)	20 (5.2)	10.2(1.9)	105 (10.3)	28 (5.6)	3.8 (1.8)
588 (29.2)	62 (16.6)	9.5(1.8)	353 (34.7)	128 (26.0)	2.8 (1.3)
256 (12.6)	36 (9.7)	7.1(1.3)	142 (13.9)	80 (16.2)	1.8 (0.9)
553 (27.3)	93 (25.2)	5.9(1.1)	350 (34.1)	200 (40.7)	1.7 (0.8)
108 (5.3)	43 (11.6)	2.5(0.5)	14 (1.4)	14 (2.8)	1.0 (0.5)
236 (11.6)	94 (25.3)	2.5(0.5)	31 (3.0)	33 (6.6)	0.9 (0.5)
2024 (100)	372 (100)	5.4	1021 (100)	493 (100)	2.1

The data for individuals A and B were taken from Tables 1 and 2 using the values obtained by the incubation of frozen BA-induced cells.

[a]BA-induced monocytes versus BA-induced lymphocytes.

[b]Numbers in parentheses, percentage of total metabolites.

TABLE 8

SUMMARY OF(-)trans-7,8-DIOL METABOLISM BY MONOCYTES AND LYMPHOCYTES: COMPARISON WITH BP METABOLISM AND AHH ACTIVITY

Individual A

	Monocytes			Lymphocytes		
(-)trans-7,8-Diol metabolites	BA induced	Control	Ratio[a]	BA induced	Control	Ratio
Tetrol 1-1	276	133	2.1	68	60	1.1
Tetrol 1-2	26	7	3.7	6	6	1.0
Triol 1	24	8	3.0	2	3	0.7
Tetrol 11-1	12	8	1.5	3	1	3.0
Tetrol 11-2	6	1	6.0	2	6	0.3
Total organic extractable	344	157	2.2	81	76	1.1
Aqueous	675	561	1.2	200	158	1.3
Total	1019	718	1.4	281	234	1.2
Total BP metabolites	752	242	3.1	261	107	2.4
AHH activity	349	125	2.8	142	68	2.1

Specific activity (fmol/min/10^6 cells)

Individual B

	Monocytes			Lymphocytes		
	BA induced	Control	Ratio	BA induced	Control	Ratio
	281	106	2.6	78	70	1.1
	16	4	4.0	10	1	10.0
	21	8	2.6	27	31	0.9
	8	7	1.1	3	5	0.6
	1	1	1.0	4	4	1.0
	327	126	2.6	122	111	1.1
	342	172	2.0	173	225	0.8
	669	298	2.2	295	336	0.9
	2024	565	3.6	372	54	6.9
	1216	404	3.0	152	70	2.2

Duplicate samples of 5 to 6 x 10^6 monocytes and 10 to 11 x 10^6 lymphocytes were incubated with (-)trans-7,8-[the 7,8-diol metabolites were analyzed by HPLC as described in "Materials and Methods". The nomenclature for the diol metabolites is as in the paper of Yang et al. (11). The data are from Table 1 for monocytes and Table 2 for lymphocytes. [a]BA-induced versus noninduced cells.

mation of diol epoxide I, the anti stereoisomer from (-)trans-7,8-diol, with about 20 times more diol epoxide I than diol epoxide II being formed in both cell types. There is, however, a large amount of material which remains in the aqueous phase that was not analyzed. This could represent material bound to cell DNA or protein or small molecules such as phosphate. One notable feature of 7,8-diol metabolism in monocytes and lymphocytes is that BA induction has relatively little effect on the level of 7,8-diol metabolism. The data in Table 7 shows that BP metabolism is increased to about 2-fold greater extent than total 7,8-diol metabolism. A similar result is seen when AHH activity is compared to the level of 7,8-diol binding to DNA. Table 8 shows that for monocytes, BA treatment induces AHH activity about 15-fold but only increases DNA binding of 7,8-diol about 2-fold. These results and other unpublished results with rat liver cells are consistent with our earlier studies with purified enzymes which showed that BP and 7,8-diol are predominantly metabolized by different forms of cytochrome P-450. The present studies suggest that the cytochrome P-450 forms metabolizing 7,8-diol is inducible by BA to a lesser extent than the forms metabolizing BP. There is, nonetheless, a sufficient quantity of enzyme in control cells to actively metabolize 7-8-diol.

Concluding Remarks

Figure 3 shows the alternative metabolic pathways to diol epoxide formation and its interaction with DNA. The figure shows that there are at least 14 pathways benzo[a]pyrene molecules can take which are alternate to diol epoxide-DNA adduct formation. These are controlled by the various forms of cytochrome P-450, epoxide hydratase and the three conjugating enzymes. An understanding of the regulation of these pathways may lead to understanding how individuals interact with polycyclic hydrocarbons and what factors govern their response in respect to activation and detoxification reactions. This may be a major factor of biochemical individuality in chemical carcinogenesis(26).

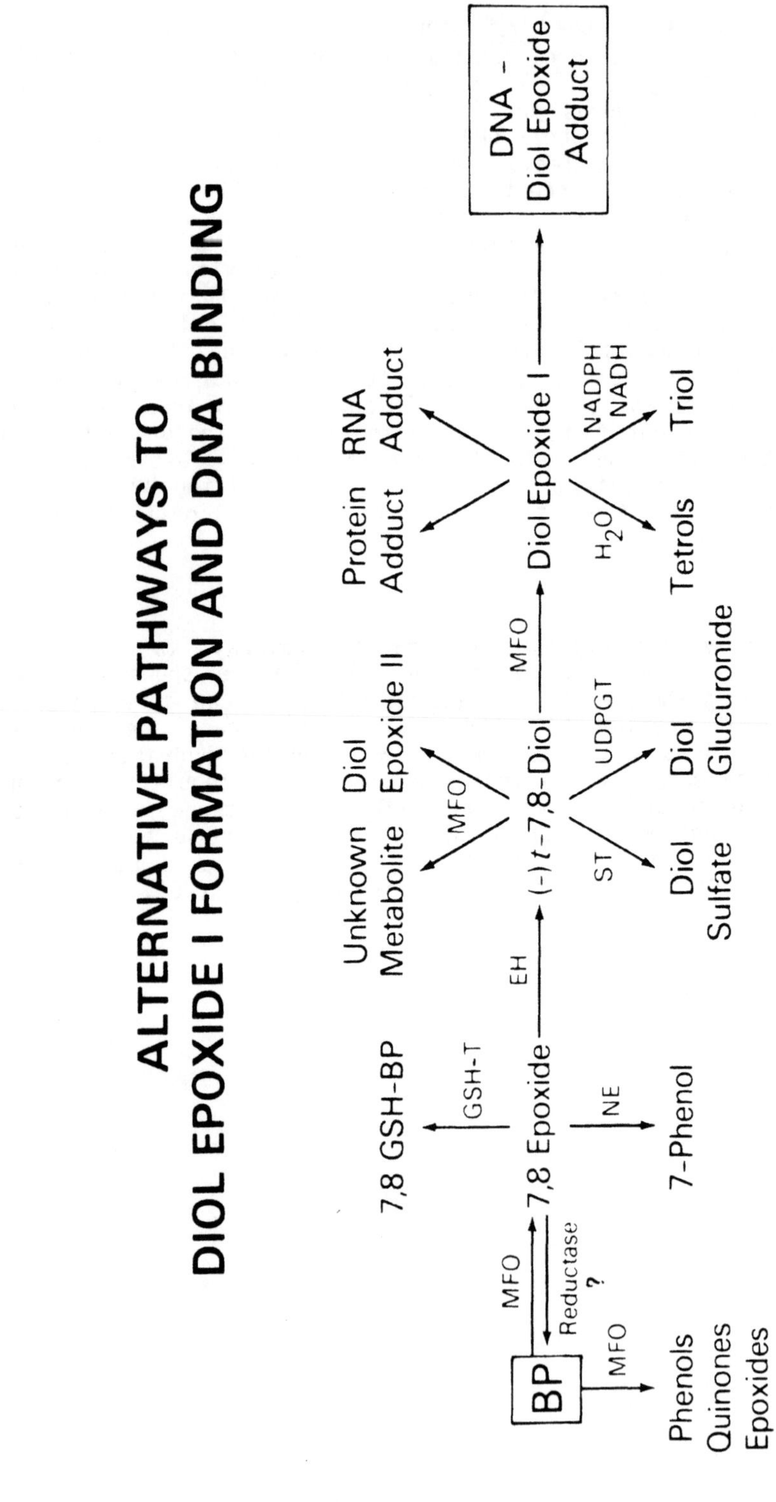
ALTERNATIVE PATHWAYS TO
DIOL EPOXIDE I FORMATION AND DNA BINDING
BP
MFO
Reductase ?
MFO
Phenols
Quinones
Epoxides
7,8 Epoxide
7,8 GSH-BP
GSH-T
NE
7-Phenol
EH
(-) t-7,8-Diol
Unknown Metabolite
Diol Epoxide II
MFO
ST
Diol Sulfate
UDPGT
Diol Glucuronide
MFO
Diol Epoxide I
Protein Adduct
RNA Adduct
H_2O
Tetrols
NADPH NADH
Triol
DNA - Diol Epoxide Adduct

References

(1) National Academy of Science, Washington, D.C.: 1972, Particulate Polycyclic Organic Matter. pp. 1-361.
(2) Gelboin, H.V.: 1967, Adv. Cancer Res. 10, pp. 1-81.
(3) Nemoto, N. and Gelboin, H.V.: 1976, Biochem. Pharmacol. 25, pp. 1221-1226.
(4) Nemoto, N., Gelboin, H.V., Habig, W.H., Ketley, J.N., and Jakoby, W.B.: 1975, Nature 25, p. 512.
(5) Cohen, G.M., Moore, B.P., and Bridges, J.W.: 1977, Biochem. Pharmacol. 26, pp. 551-553.
(6) Gelboin, H.V., Kinoshita, N., and Wiebel, F.J.: 1972, Fed. Proc. 31, pp. 1298-1309.
(7) Huberman, E., Sachs, L., Yang, S.K., and Gelboin, H.V.: 1976, Proc. Natl. Acad. Sci., USA 73, pp. 607-611.
(8) Haugen, D.A., van der Hoeven, T.A., and Coon, M.J.: 1975, J. Biol. Chem. 250, pp. 3567-3570.
(9) Haugen, D.A., and Coon, M.J.: 1976, J. Biol. Chem. 251, pp. 7929-7939.
(10) Deutsch, J., Leutz, J.C., Yang, S.K., Gelboin, H.V., Chiang, Y.L., Vatsis, K.P., and Coon, M.J.: 1978, Proc. Natl. Acad. Sci, USA 75, pp. 3123-3127.
(11) Yang, S.K., McCourt, D.W., Roller, P.P., and Gelboin, H.V.: 1976, Proc. Natl. Acad. Sci., USA 73, pp. 2594-2598.
(12) Sims, P.: 1970, Biochem. Pharmacol. 19, pp. 795-818.
(13) Conney, A.H., Miller, E.C., and Miller, J.A.: 1957, J. Biol, Chem. 228 pp. 753-766.
(14) Sims, P., Grover, P.L., Swaisland, A., Pal, K., and Hewar, A.: 1976, Nature, 252, pp. 326-328.
(15) Daudel, P., Duquesne, M., Vigny, P., Grover, P.L., and Sims, P.: 1975, FEBS Lett. 57, pp 250-253.
(16) Weinstein, I.B., Jeffery, A.M., Jennette, K.W., Blobstein, A.M., Harvey, R.G., Harris, C., Autrup, H., Kasai, H., and Nakanishi, K.: 1976, Science, 193, p. 592.
(17) Yang, S.K., and Gelboin, H.V.: 1976, Biochem. Pharmacol. 25, pp. 2221-2225.
(18) Thakker, D.R., Yagi, H., Akagi, H, Koreeda, M., Lu, A.Y.H., Levin, W., Wood, A.W., Conney, A.H., and Jerina, D.M.: 1977, Chem. Biol. Interaction. 16, pp. 281-300.
(19) Jerina, D.M., Yagi, H., Thakker, D.R., Karle, J.M., Mah, H.D., Boid, D.R., Gadaginamath, G., Wood, A.W., Buening, M., Chang, R.L., Levin, W., and Conney, A.H.: 1978, Advances in Pharmacology and Therapeutics. Vol. 9, p. 53. Ed. Cohen, Y. Pergamon Press Oxford.
(20) Deutsch, J., Vatsis, K.P., Coon, M.J., Leutz, J.C., and Gelboin, H.V.: 1979, Mol. Pharmacol. 16, pp. 1011-1018.
(21) Nebert, D.W. and Gelboin, H.V.: 1968, J. Biol. Chem., 243, pp. 6242-6249.
(22) Nebert, D.W. and Gelboin, H.V.: 1969, Arch. Biochem. Biophys. 134, pp. 76-89.
(23) Selkirk, J.K., Croy, R.G., and Gelboin, H.V.: 1976, Cancer Res., 36, pp.922-926.

(24) Yang, S.K., Gelboin, H.V., Weber, J.D., Sankaran, V., Fischer, D.L. and Engel, J.F.: 1977, Anal. Biochem., 78, pp. 520-526.
(25) Okuda, T., Vesell, E.S., Plotkin, E., Tarone, R., Bast, R.C., and Gelboin, H.V.: 1977, Cancer Res., 37, pp. 3904-3911.
(26) Gelboin, H.V.: 1977, N. Engl. J. Med., 297, pp. 384-387.
(27) Okano, P., Miller H.N., Robinson, R.C., and Gelboin, H.V.: 1979, Cancer Res., 39, pp. 3184-3193.
(28) Okano, P., Whitlock, J.P., and Gelboin, H.V.: In press, Ann. N.Y, Acad. Sci.
(29) Yang, S.K., Deutsch, J., and Gelboin, H.V.: 1978, Polycyclic Hydrocarbons and Cancer, Vol. 1, pp. 205-231. Eds. Gelboin, H.V. and Ts'o, P.O.P. Academic Press.

METABOLIC AND STRUCTURAL REQUIREMENTS FOR THE CARCINOGENIC POTENCIES OF UNSUBSTITUTED AND METHYL-SUBSTITUTED POLYCYCLIC AROMATIC HYDROCARBONS

Shen K. Yang, Ming W. Chou and Peter P. Fu[†]
Department of Pharmacology, School of Medicine, Uniformed Services University of the Health Sciences, Bethesda, Maryland 20014, and [†]National Center for Toxicological Research, Jefferson, Arkansas 72079, USA

ABSTRACT: Extensive metabolism studies of many non-K-region dihydrodiols of unsubstituted and methyl-substituted polycyclic aromatic hydrocarbons (PAHs) by mammalian drug-metabolizing enzymes indicate that the extent of formation of vicinal dihydrodiol-epoxides is determined by the conformations and geometric location of the adjacent double bond of the dihydrodiol precursors. Unsubstituted PAH dihydrodiols with one (in the case of cis-dihydrodiols) or both (in the case of trans-dihydrodiols) of the hydroxyl groups preferentially in quasi-equatorial conformations can be further metabolized to the vicinal dihydrodiol-epoxides. The trans-dihydrodiols with both of the hydroxyl groups preferentially or exclusively in the quasi-diaxial conformations can also be enzymatically epoxidized to form vicinal dihydrodiol-epoxides, but the extent of these formations is highly dependent on the geometric location of the adjacent double bond of the dihydrodiols. Quasi-diaxial trans-dihydrodiols with an adjacent double bond at the bay-regions are not or minimally metabolized to the vicinal dihydrodiol-epoxides. A methyl substituent at the hydroxylated carbons or at the adjacent double bonds of the quasi-diequatorial trans-dihydrodiols can substantially or completely inhibit the vicinal dihydrodiol-epoxide formations. These findings, along with the bay-region theory, provide the molecular basis for the observed carcinogenic potencies of many bay-region containing unsubstituted and methyl-substituted PAHs.

Polycyclic aromatic hydrocarbons (PAHs) are the most common particulate environmental pollutants and may be responsible for some cancer induction in man. The biological properties of PAHs, such as mutagenicity, carcinogenicity, and covalent binding to cellular macromolecules, require metabolic activation by the cytochrome P-450 containing drug-metabolizing enzyme systems. The metabolism of PAHs has been studied intensively in the past thirty years and the recent rapid progress in the understanding of their activation pathways is largely due to the recognition of benzo[a]pyrene (BaP) 7,8-diol-9,10-epoxides as the major carcinogenic and mutagenic metabolites (11,24,28,35,39,47,59, 60,75). These findings and the performance of perturbational molecular

B. Pullman, P. O. P. Ts'o and H. Gelboin (eds.), Carcinogenesis: Fundamental Mechanisms and Environmental Effects, 143–156.

orbital calculations by the method of Dewar (13) led to the development of the "bay-region" theory by Jerina and coworkers (27). Bay-region diol-epoxides of some unsubstituted PAHs, such as chrysene (34), benz[a]anthracene (BA) (45,58), dibenz[a,h]anthracene (DB[a,h]A) (5), and benzo[c]phenanthrene (9), have also been implicated as major carcinogenic and mutagenic metabolites. Methyl-substituted PAHs, such as 5-methylchrysene (21,22), 7-methylbenz[a]anthracene (7-MBA) (37), 7,12-dimethylbenz[a]anthracene (7,12-DMBA) (23,36,38,43), and 3-methylcholanthrene (3-MC) (46,62), are also believed to be metabolically activated at the bay-regions. However, bay-region containing PAHs are not all carcinogenic compounds (Fig. 1). It is possible that the bay-region vicinal diol-epoxides are not formed in vivo from these non-carcinogenic bay-region containing PAHs.

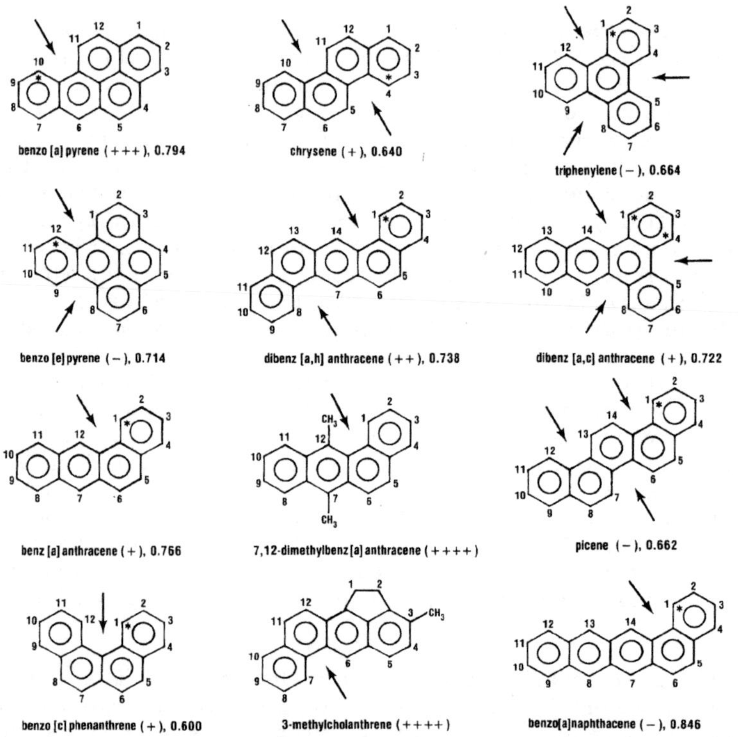

Fig. 1 Structure of some polycyclic aromatic hydrocarbons. Arrows indicate the location of bay region. The approximated relative carcinogenicity is indicated in the parenthesis ranging from noncarcinogenic (-) to highly carcinogenic (++++). The calculated $\Delta E_{deloc}/\beta$ values for the benzylic carbonium ions (at the positions indicated by "*") which are derived from the respective bay-region tetrahydro-epoxides are indicated for the unsubstituted PAHs.

It has been widely accepted that metabolic conversion of unsubstituted PAHs to highly mutagenic and carcinogenic bay-region vicinal diol-epoxides involves three enzymatic reactions: PAH → epoxide → dihydrodiol → bay-region diol-epoxide (18). Non-bay-region vicinal diol-epoxides are also enzymatically formed (6,55,67) but possess relatively weaker mutagenic and carcinogenic activities (33,45) although they may be the metabolite that binds to cellular DNA (55). For PAHs with a methylene bridge (such as 3-MC) or with methyl substituents (such as MBAs and 7,12-DMBA), initial hydroxylation reaction at the methylene carbons or methyl carbons may also play an important role in the activation pathways (7,46). PAH cis-diols, such as BA cis-1,2-diol and BaP cis-9,10-diol, have been found as bacterial oxidation products of the respective parent hydrocarbons (19). The bacterial PAH metabolites may be further transformed in the mammalian system to mutagenic and carcinogenic products. Thus, PAH cis-diols may also contribute to the incidence of cancer in man. Tentative identification of 3-MC cis-2a,3-diol and 3-MC cis-11,12-diol formed from the parent hydrocarbon by liver microsomal enzymes has been reported (51,53).

Examples are presented in this report to assess the structural basis for the metabolic conversion of non-K-region cis- and trans-diols to bay-region and non-bay-region vicinal diol-epoxides and their possible biological consequences. Since K-region diols do not possess a vicinal double bond, vicinal diol-epoxides cannot be formed. Only terminal benzo-rings (bay-region or non-bay-region) of PAHs are potentially capable of forming vicinal diol-epoxides.

DIHYDRODIOL CONFORMATIONS OF UNSUBSTITUTED AND METHYL-SUBSTITUTED POLYCYCLIC AROMATIC HYDROCARBONS

The conformation of trans- and cis-diols of unsubstituted and methyl-substituted PAHs can be defined and grouped into the following four classes (see refs. 9,17,20,30-32,50-53,76,77 and Table 1):

EE-DIOLS--Trans-diols with hydroxyl groups preferentially in quasi-equatorial conformations. The EE-diols include all unsubstituted PAH non-bay-region trans-diols, non-bay-region trans-diols with a methyl group at the adjacent double bond or other ring positions, and non-bay-region trans-diols formed at methyl-substituted double bond.

AA-DIOLS--Trans-diols with hydroxyl groups preferentially or exclusively in quasi-diaxial conformations. The AA-diols include all bay-region trans-diols and trans-diols that have a peri methyl group or a peri hydroxymethyl group.

AE_1-DIOLS--Cis-diols with one hydroxyl group in quasi-axial (or quasi-equatorial) and the other hydroxyl group in quasi-equatorial (or quasi-axial) conformations.

AE_2-DIOLS--cis-diols with the benzylic hydroxyl group preferentially or exclusively in quasi-axial conformation and the allylic hydroxyl group preferentially or exclusively in quasi-equatorial conformation (77).

EFFECT OF NON-K-REGION DIHYDRODIOL CONFORMATION ON THE METABOLIC FORMATION OF VICINAL DIHYDRODIOL-EPOXIDES

In considering the metabolic formation of vicinal diol-epoxides, each of the four classes of diols can be further subdivided depending on whether the vicinal double bond is located in the bay-region (BR) or in the non-bay-region (NBR). Methyl substituent peri to a diol creates a "bay-like" region, forcing the hydroxyl groups of both cis- and trans diols peri to the methyl group to adopt preferentially or exclusively quasi-axial conformations. The methyl group of the methyl-substituted PAHs thus creates a similar effect as the bay-region of the aromatic rings in determining the conformation of the diols that are peri to a methyl group. Examples of the structurally distinct types of non-K-region diols and the experimental evidence for their metabolism to vicinal diol-epoxides are shown in Table 2.

Table 1. Preferred Conformation of the trans-Dihydrodiols of Unsubstituted and Methyl- or Hydroxymethyl-substituted Benz[a]anthracene and Benzo[a]pyrene

Parent ** Hydrocarbon	Preferred Conformation of the t-diols* 1,2-	3,4-	5,6-	8,9-	10,11-	References
BA	AA	EE	EE	EE	EE	13,30,77
1-MBA	(AA)	EE	EE	EE	EE	6
2-MBA	(AA)	EE	EE	EE	(EE)	6
3-MBA	(AA)	(EE)	EE	EE	EE	6
4-MBA	(AA)	EE	AA	EE	EE	6
5-MBA	AA	AA	(EE)	EE	EE	6
6-MBA	(AA)	EE	(EE)	EE	EE	69
7-MBA	AA	EE	AA	AA	EE	50,68
8-MBA	AA	EE	EE	EE	EE	69,70
8-OHMBA	AA	EE	EE	EE	EE	6,68
9-MBA	(AA)	EE	EE	EE	EE	6
10-MBA	(AA)	EE	EE	EE	EE	6
11-MBA	(AA)	EE	EE	EE	EE	69
12-MBA	(AA)	EE	EE	EE	AA	69
7,12-DMBA	(AA)	EE	AA	AA	AA	8,51,52,69
7-OHM-12-MBA	(AA)	EE	AA	AA	AA	8,69
7-M-12-OHMBA	(AA)	EE	AA	AA	AA	8
7,12-diOHMBA	(AA)	EE	AA	AA	AA	8

	1,2-	2,3-	4,5-	7,8-	9,10-	11,12-	
BaP	(EE)	(EE)	EE	EE	AA	(AA)	48
6-MBaP	(EE)	(EE)	AA	AA	AA	(AA)	6

*Compounds in parenthesis have not been found as metabolites.
AA = quasi-diaxial, EE = quasi-diequatorial, and t = trans.
**In contrast to previous hypothesis (26) many metabolites are formed at the methyl-substituted aromatic double bonds.

Table 2. Different Types of Non-K-region Diols and Their Further Metabolism to Vicinal Diol-epoxides

Type of Diol	Vicinal Double Bond	Example (references in parenthesis)*
EE	NBR	BA t-8,9-diol (6,55,67); BA t-10,11-diol (6); 8-MBA t-8,9-diol (6A)**; 10-MBA t-8,9-diol (6B)**
EE	BR	Chrysene t-1,2-diol (34); BA t-3,4-diol (6,33,45); BaP t-7,8-diol (11,24,39,49,56,73); 7,12-DMBA t-3,4-diol (7,23,36,38,43,54); 5-methylchrysene t-1,2-diol and t-7,8-diol (22); 3-MC t-9,10-diol (54); 1-hydroxy-3-MC t-9,10-diol(46); DB[a,h]A t-3,4-diol (5)
AE_1	NBR	BA c-8,9-diol
AE_1	BR	BaP c-7,8-diol (16,63)
AE_2	NBR	BaP c-9,10-diol; 7-MBA c-8,9-diol
AE_2	BR	BeP c-9,10-diol; 5-MBA c-3,4-diol
AA	NBR	BaP t-9,10-diol (48); 7-MBA t-8,9-diol (6); 12-MBA t-10,11-diol (6); BA t-1,2-diol (6)
AA	BR	BeP t-9,10-diol (4,64)**; DB[a,c]A t-3,4-diol; 7,12-DMBA t-8,9-diol (6); 10-methylphenanthrene t-1,2-diol; 6-MBaP t-7,8-diol (6); 14-methyl-dibenzo[a,h]pyrene t-1,2-diol; 5-methyldibenzo-[a,i]pyrene t-3,4-diol; 3,11-DMC t-9,10-diol

*The metabolic formation of vicinal diol-epoxides is indicated by direct demonstration of their formations, by the high biological activities of the diols that could form bay-region diol-epoxides, or by the structural elucidation of the hydrocarbon-DNA adduct formation in vivo and/or in vitro. t = trans and c = cis.
**Non-detectable (6B,64) or trace (6A,48) amount of vicinal diol-epoxides was formed.

Detailed metabolism studies of BaP t-9,10-diol (48) and BeP t-9,10-diol (64) have led Wood, et al. (64) to suggest that all quasi-axial bay-region diol groups can shift metabolism away from vicinal diol-epoxide formation. However, our extensive metabolism studies of many non-K-region diols with rat liver microsomes indicated that the quasi-axial hydroxyl groups at the bay-region or peri to a methyl group do not necessarily shift the metabolism away from vicinal diol-epoxide formation. The extent of such diol-epoxide formation is highly dependent on the geometric location of the double bond adjacent to the diols and differs among different PAH diols. The evidence for the formation of vicinal diol-epoxides from the various types of diols indicated below was obtained by ultraviolet absorption spectral and mass spectral analyses of the HPLC isolated products formed from the in vitro rat liver microsomal incubations of the diols. Examples are summarized below.

Metabolism of EE-diols

The examples are: (a) Both syn- and anti-3,4-diol-1,2-epoxides were formed as metabolites of the racemic BA t-3,4-diol and were detected as 1,2,3,4-tetrols. (b) Both syn- and anti-8,9-diol-10,11-epoxides were formed as metabolites from the racemic BA t-8,9-diol. BA syn- and anti-8,9-diol-10,11-epoxides were stable and separable on the reversed-phase HPLC system. An anti-8,9-diol-10,11-epoxide was formed predominantly from BA (-)t-8,9-diol (8R,9R), whereas a syn-8,9-diol-10,11-epoxide was formed predominantly from BA (+)t-8,9-diol (8S,9S). (c) The 7-MBA t-10,11-diol, 12-MBA t-8,9-diol, and 4-MBA t-8,9-diol were each metabolized to the corresponding vicinal diol-epoxides. (d) A minor amount of 8,9-diol-10,11-epoxide was formed from the 8-MBA t-8,9-diol. The amount of formation is, however, considerably less than the 8,9-diol-10,11-epoxides formed from the racemic BA t-8,9-diol. (e) There was no detectable amount of 8,9-diol-10,11-epoxide formed from 10-MBA t-8,9-diol.

Metabolism of AA-diols at the Adjacent Non-bay-region Double Bond

The examples are: (a) The major metabolites that were formed from the racemic BA t-1,2-diol are syn- and anti-1,2-diol-3,4-epoxides. (b) Both syn- and anti-8,9-diol-10,11-epoxides were found as metabolites of 7-MBA t-8,9-diol. (c) Both syn- and anti-10,11-diol-8,9-epoxides were found as metabolites of 12-MBA t-10,11-diol.

Metabolism of AA-diols at the Adjacent Bay-region Double Bond

The examples are: (a) No detectable amount of 8,9-diol-10,11-epoxide was found from the HPLC analysis of in vitro rat liver microsomal incubation products of 7,12-DMBA t-8,9-diol. (b) A minor but detectable amount of metabolites with pyrene-like UV chromophores was formed from the in vitro rat liver microsomal metabolism of 6-MBaP t-7,8-diol.

Metabolism of AE_1-diols

BaP cis-7,8-diol is the only AE_1-diol whose metabolism has been studied. Unlike the racemic BaP t-7,8-diol whose metabolites are predominantly syn- and anti-7,8-diol-9,10-epoxides (49,73), the major metabolic products of BaP cis-7,8-diol are formed by metabolism at positions other than the 9,10 double bond, although a significant amount of the 7,8-diol-9,10-epoxides was formed and detected as tetrols (16). Metabolism studies of AE_2-diols have not yet been carried out.

The above evidence and other reports (48,49,64,73) thus indicate that the metabolic formation of vicinal diol-epoxides is primarily determined by the conformations of the hydroxyl groups and the geometric location of the double bond of the diol precursors. The most prominent structural features inhibitive to the enzymatic formation of vicinal diol-epoxides are: (i) quasi-diaxial conformations of the diols with the adjacent double bond located in the bay or "bay-like" region and (ii) quasi-diequatorial conformations of the diols with a methyl substituent at the hydroxylated carbons or at the adjacent double bond. In contrast to the previous suggestion (48,64), it is now clear that the diaxial hydroxyl groups alone do not determine the capability of the non-K-region diols to be further metabolized to vicinal diol-epoxides.

This discussion thus indicates the lack of a definitive rule for prediction of the capability and the extent of further metabolism of non-K-region diols to vicinal diol-epoxides. More detailed comparative and quantitative metabolism studies of the various types of diols (such as 1-5, 6-10, and 11-14) with different sources of carcinogen-metabolizing enzyme systems should provide a better understanding of the factors that determine the metabolic pathways of the diols.

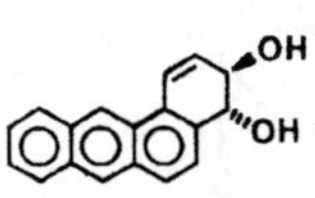
1
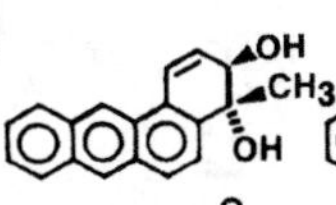
2
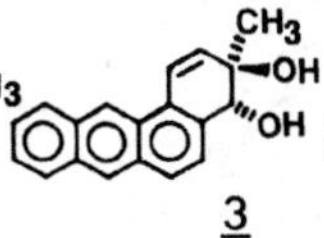
3
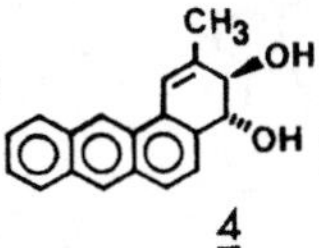
4
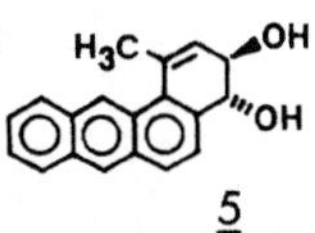
5
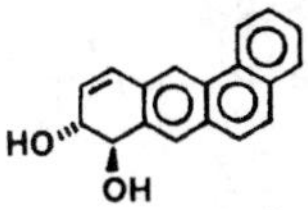
6
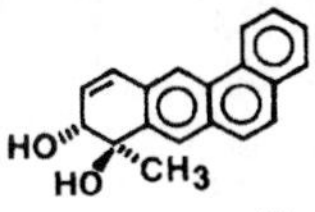
7
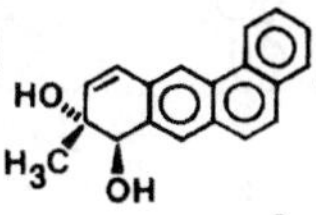
8
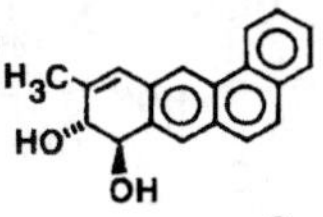
9
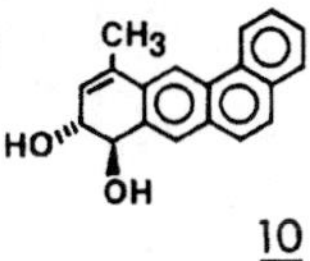
10
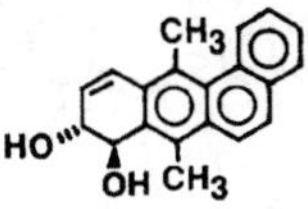
11
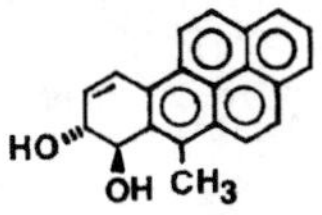
12
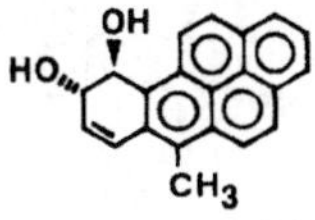
13
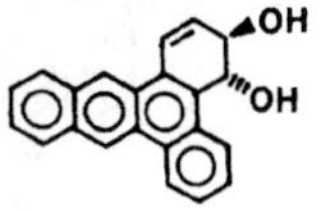
14

CHEMICAL REACTIVITY AND BIOLOGICAL ACTIVITY OF VICINAL DIHYDRODIOL EPOXIDES

Chemical Reactivity

Kinetic studies of the hydrolysis of BaP 7,8-diol-9,10-epoxides (23 and 24) (29,57,71) and the reaction of the BaP 7,8-diol-9,10-epoxides, BA 3,4-diol-1,2-epoxides, BA 8,9-diol-10,11-epoxides, and BA 10,11-diol-8,9-epoxides with either nitrothiophenolate or tert-butyl mercaptan (2,65) indicate that the chemical reactivity of non-K-region diol-epoxides for a given PAH increases with greater value of $\Delta E_{deloc}/\beta$ for the formation of benzylic carbonium ions. For a given pair of stereoisomeric diol-epoxides, the syn isomer is considerably more reactive (i.e., less stable) than the anti isomer (29,57,65,71). Because the hydroxyl and the epoxy groups in BeP 9,10-diol-11,12-epoxides (18 and 19) and in triphenylene 1,2-diol-3,4-epoxides are in the bay-regions, these "double bay-region" diol-epoxides are conformationally rigid molecules (66). The exact mechanism by which the axial hydroxyl group(s) affects the chemical reactivities of the conformationally rigid "double bay-region" diol-epoxides is not yet known. It is interesting to note that both BeP 9,10-diol-11,12-epoxides (18 and 19) are much less mutagenic than BeP 9,10,11,12-tetrahydro-9,10-epoxide (15) in the Ames direct mutagenicity test (66). The results thus suggest that the relative chemical reactivities are 15>18>19 (66) and these are consistent with the relative chemical reactivities of BaP tetrahydro-epoxides 20>23>24 (2,29,57,71). However, the axial hydroxyl groups are not necessarily detrimental to the chemical reactivity of a diol-epoxide. For example, the hydroxyl groups of BaP syn-7,8-diol-9,10-epoxide (23) are partially in diaxial conformations (65,66), yet it is an extremely reactive compound (2,29,57,71).

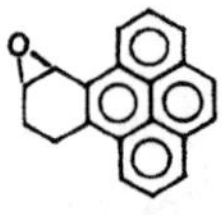
15
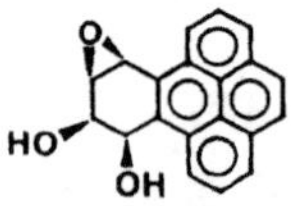
16
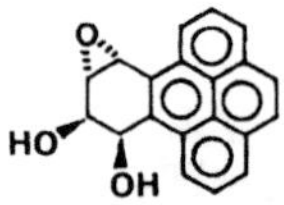
17
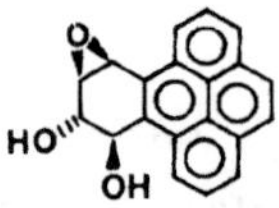
18
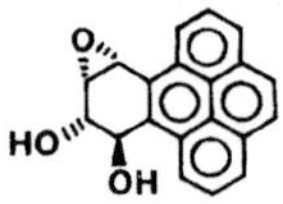
19
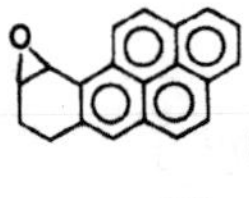
20
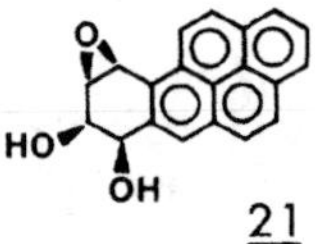
21
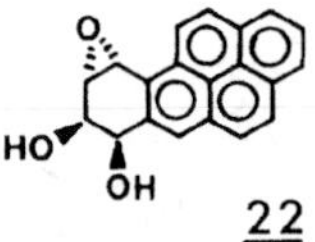
22
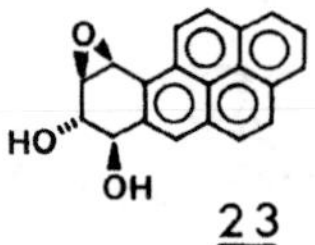
23
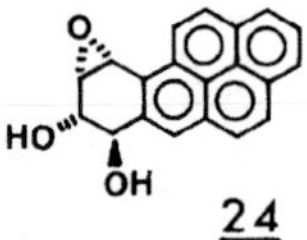
24
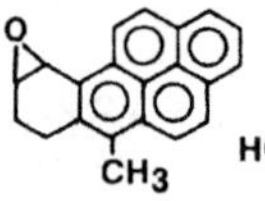
25
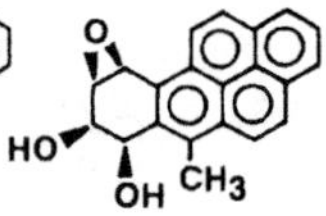
26
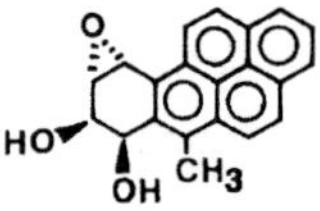
27
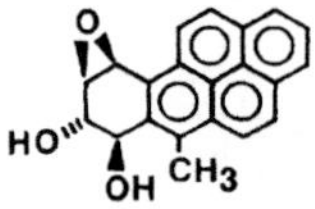
28
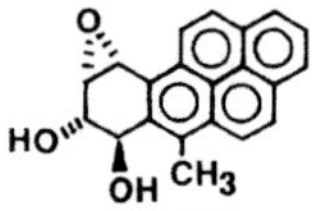
29

The diol-epoxides can react with nucelophiles by either the S_N1 or the S_N2 mechanism depending on the type of solvents and on the nucleophilicity of the nucleophiles. Knowledge of the relative chemical reactivities of the conformationally distinct molecules, such as compounds 15-29, and their reaction mechanisms are important toward a better understanding of the PAH-induced carcinogenesis.

Biological Activity

The relative mutagenic activity of bay-region PAH tetrahydro-epoxides in Salmonella typhimurium tester strains in the absence of rat liver S-9 fraction is generally parallel to their relative inherent chemical reactivities (e.g., 20>23>24 [59]). However, in many cases, the inherent chemical reactivities of PAH bay-region tetrahydro-epoxides and vicinal syn- and anti-diol-epoxides are not reflected by the related mutagenic activities in mammalian cells or by the tumorigenic activities on mouse skin (24,39,42,45,61). Thus the combination of the in vivo and in vitro testings of the diols (which are precursors of the vicinal diol-epoxides) and the knowledge of the diol metabolite formations and dispositions from the parent hydrocarbons in the target cells and tissues generally provide a more reliable assessment of the relative carcinogenic potencies of the PAH vicinal diol-epoxides.

METABOLIC AND STRUCTURAL REQUIREMENTS FOR THE CARCINOGENIC POTENCIES OF UNSUBSTITUTED AND METHYL-SUBSTITUTED POLYCYCLIC AROMATIC HYDROCARBONS

The first step in the metabolism of PAHs leading to the formation of vicinal diol-epoxides is the enzymatic epoxidation reaction catalyzed by the mammalian microsomal or nuclear aryl hydrocarbon epoxidase (AHE). The AHE activity from different tissues or cells toward different regions of a PAH molecule varies greatly. This phenomenon is termed "regioselectivity" and there is no rule to allow us to predict with any certainty the AHE's regioselectivity toward any particular PAH. The regioselectivity of AHE in the various tissues and cells is one of the major factors that determines the susceptibility of animals and humans to the carcinogenic effect(s) of a PAH. The epoxide intermediates formed may be reactive toward cellular macromolecules and may be responsible for the cytotoxic and carcinogenic effects of a PAH (1, 3,15,25,40). The enzymatically formed epoxide intermediates can be rearranged nonenzymatically to phenols and/or can be hydrated to trans-diols catalyzed by the mammalian microsomal or nuclear epoxide hydratase (EH). The extent of rearrangement/hydration is dependent on the stability of the epoxide intermediates and the endogenous EH activity. Some epoxide intermediates are very unstable (e.g., BaP 2,3-epoxide, [74]) and are rearranged nonenzymatically to phenolic products (10,25, 40). Some epoxide intermediates are more stable (e.g., some K-region epoxides) and only diols are formed from the metabolism of the parent hydrocarbons. The level of EH in the drug-metabolizing enzyme complex plays an important role in the relative amount of critical proximate carcinogenic diol metabolite(s). The epoxide intermediates may also form glutathione conjugates (25,40).

The stereospecificities of AHE and EH are important factors that ultimately determine the mutagenic and carcinogenic activities of a PAH. Thus the regioselectivity of AHE specifies the relative amounts of epoxide intermediates that are formed from the metabolism of a PAH. The stability, reactivity, and the extent of conjugation of the epoxide intermediates ultimately determine the amount of diols formed. The stereospecificity of the AHE and EH also determines the trans configuration and the ratio of the (+) and (-) enantiomers of a diol metabolite (12,72,75).

Non-K-region diols may be further metabolized by the AHE to form vicinal diol-epoxides. However, in addition to the regioselectivity of the AHE, the extent and the occurrence of this epoxidation reaction

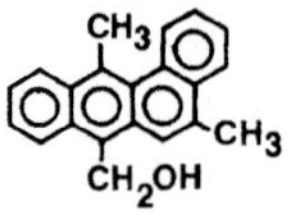
30

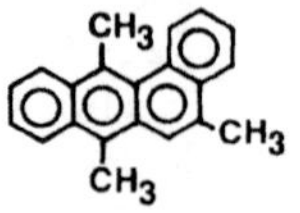
31

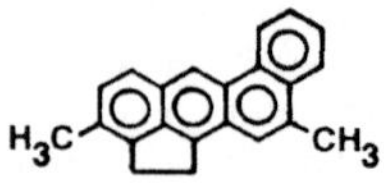
32

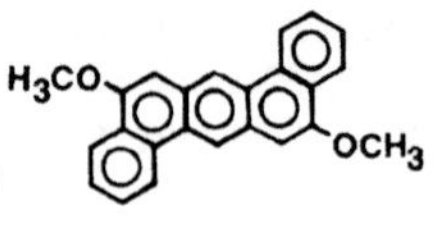
33

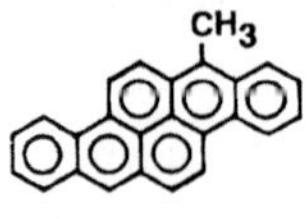
34

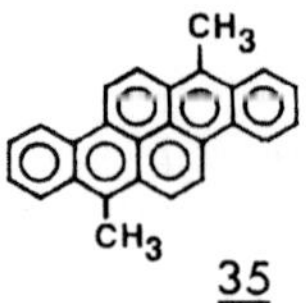
35

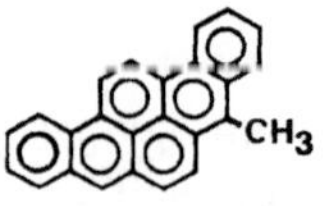
36

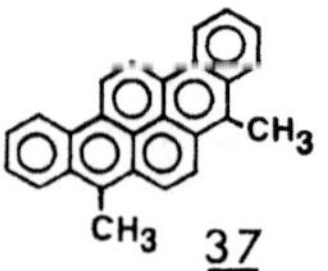
37

are also dependent on the conformation of the hydroxyl groups and the geometric location of the adjacent double bond of the diol. The relative amount of syn- and anti-diol-epoxides formed is dependent on the ratio of (+) and (-) enantiomeric diol precursors and the stereoselectivity of the AHE at the adjacent double bond. If the diol precursor adopts a preferentially or exclusively quasi-diaxial conformation and the adjacent olefinic double bond is at the bay or "bay-like" region, the extent of the epoxidation reaction by the AHE is substantially reduced or completely abolished. From the results of many PAHs studied, it is now established that all bay-region diols and diols with a peri methyl group or a hydroxymethyl group preferentially or exclusively adopt quasi-diaxial conformations, and the extent of further conversion of these diols to vicinal diol-epoxides is highly dependent on whether the adjacent double bond is located in the bay or "bay-like" region. Based on the assumption that bay-region vicinal diol-epoxides are the major ultimate carcinogenic metabolites, the weaker carcinogenicity of compounds 30-37 relative to the coresponding hydrocarbons without a peri methyl (or methoxy) group (14,44) can thus be attributed to the quasi-diaxial conformations of the critical proximate carcinogenic diol metabolites whose adjacent double bonds are all in the bay-regions. This concept is consistent with the finding that 7,12-DMBA t-8,9-diol is not metabolized to the 8,9-diol-10,11-epoxide (6) but it is not consistent with the higher observed carcinogenicity of 6-MBaP relative to that of BaP (14). HPLC analyses of the in vitro rat liver microsomal metabolites of 6-MBaP and BaP indicated that, under identical conditions, the amount of 6-MBaP t-7,8-diol formed from 6-MBaP is considerably lower than the amount of BaP t-7,8-diol formed from BaP. Furthermore, the extent of 7,8-diol-9,10-epoxide formation from BaP t-7,8-diol is much higher than that formed from 6-MBaP t-7,8-diol. Based on these observations, it is possible that 6-MBaP is actually less carcinogenic than BaP or that the higher carcinogenicity of 6-MBaP is primarily due to the formation of reactive metabolite(s) other than the bay-region 7,8-diol-9,10-epoxide(s).

Not all non-K-region quasi-diequatorial diols can be readily metabolized to vicinal diol-epoxides. For example, a small but detectable amount of 8,9-diol-10,11-epoxide was found as a rat liver microsomal metabolite of 8-MBA t-8,9-diol (7), but little if any 8,9-diol-10,11-epoxide was formed from the metabolism of 10-MBA t-8,9-diol (10). These results are in contrast to the abundance of 8,9-diol-10,11-epoxides formed from rat liver microsomal metabolism of 4-MBA t-8,9-diol or BA t-8,9-diol. Thus the methyl substituents at the hydroxylated carbons or at the adjacent olefinic double bond of a non-K-region diol do not alter the preferred quasi-diequatorial conformations of the hydroxyl groups but substantially reduce the ability of the diols to be further metabolized to vicinal diol-epoxides. These findings may be extended to interpret the extremely weak or noncarcinogenic activities of 1-,2-, 3-, 4-, and 5-MBAs and dibenz[a,c]anthracene and the weaker carcinogenicity of compounds 30-37 (14). But neither our findings nor the bay-region theory can, in the absence of other evidence, interpret the extremely weak or noncarcinogenic activities of 9-, 10-, and 11-MBAs although the quasi-diequatorial t-3,4-diols are each metabolites of the parent MBA (Table 1). Obviously more studies are required to fully understand the various carcinogenic potencies of many bay-region containing unsubstituted and methyl-substituted PAHs.

REFERENCES

1. Baird, W.M., Harvey, R.G. and Brookes, P.:1975, Cancer Res. 35: 54-57
2. Beland, F.A. and Harvey, R.G.:1976, J.C.S. Commun.: 84-85
3. Blobstein, S.H., Weinstein, I.B., Grunberger, D., Weisgras, J. and Harvey, R.G.: 1975, Biochemistry, 14: 3451-3458
4. Buening, M.K.,Levin,W., Wood, A.W., Chang, R.L., Lehr, R.E., Taylor, C.W., Yagi, H., Jerina, D.M. and Conney, A.H.: 1980, Cancer Res. 40: 203-206
5. Buening, M.K., Levin, W., Wood, A.W., Chang, R.L., Yagi, H., Karle, J.M., Jerina, D.M. and Conney, A.H.: 1979, Cancer Res. 39: 1310-1314
6. Chou, M.W., Fu, P.P. and Yang, S.K.: unpublished results
7. Chou, M.W. and Yang, S.K.: 1978, Proc. Natl. Acad. Sci., USA 75: 5466-5470
8. Chou, M.W. and Yang, S.K.: 1979, J. Chromatography 185: 635-654
9. Croisy-Delcey, M., Ittah, Y. and Jerina, D.M.: 1979, Tetrahedron Lett. 31:2849-2852
10. Daly, J.W., Jerina, D.M. and Witkop, B.: 1972, Experientia 28: 1129-1249
11. Daudel, P., Duquesne, M., Vigny, P., Grover, P.L. and Sims, P.: 1975, FEBS Lett. 57: 250-253
12. Deutsch, J., Leutz, J.C., Yang, S.K., Gelboin, H.V:, Chiang, Y.L. Vatsis, K.P. and Coon, M.J.: 1978, Proc. Natl. Acad. Sci., USA, 75:3123-3127
13. Dewar, M.J.S.: 1969, The Molecular Orbital Theory of Organic Chemistry, McGraw Hill, New York, pp. 214-217 and 304-306
14. Dipple, A.: 1976, in "Chemical Carcinogens", ACS Monograph Series

3, C.E. Searle (ed.), American Chemical Society, Washington, D.C., pp. 245-314
15. Dipple, A., Lawley, P.D. and Brookes, P.: 1968, Europ. J. Cancer 4:493-506
16. Fu, P.P., Beland, F.A. and Yang, S.K: unpublished results
17. Fu, P.P. and Harvey, R.G.: 1977, Tetrahedron Lett. 24:2059-2062
18. Gelboin, H.V. and Ts'o, P.O.P. (eds.): 1978, "Polycyclic Hydrocarbons and Cancer", Vols. 1 and 2. Academic Press, New York
19. Gibson, D.T., Mahadevan, V., Jerina, D.M., Yagi, H. and Yoh, H.J.C.:1975, Science 189:295-297
20. Harvey, R.G., Lee, H.M. and Shysmasundar, N.: 1979, J. Org. Chem. 44:78-83
21. Hecht, S.S., Loy, M. and Hoffmann, D.: 1976, in "Carcinogenesis", Vol. 1, R.L. Freudenthal and P.W. Jones (eds.), Raven Press, New York, pp. 325-340
22. Hecht, S.S., LaVoie, E., Mazzarese, R., Amin, S., Bodenko, V. and Hoffmann, D.: 1978. Cancer Res. 38:2191-2194
23. Huberman, E., Chou, M.W. and Yang, S.K.: 1979, Proc. Natl. Acad. Sci., USA 76:862-866
24. Huberman, E., Sachs, L., Yang, S.K. and Gelboin, H.V.: 1976, Proc. Natl Acad. Sci., USA 73:607-611
25. Jerina, D.M. and Daly, J.W.: 1974, Science 185: 573-582
26. Jerina, D.M. and Daly, J.W.: 1976, in "Drug Metabolism-From Microbe to Man", D.V. Parke and R.L. Smith (eds.), Taylor and Francis, Ltd., London, pp. 13-32
27. Jerina, D.M., Lehr, R.E., Yagi, H., Hernandez, O., Dansette, P.M. Wislocki, P.G., Wood, A.W., Chang, R.L., Levin, W. and Conney, A.H.: 1976, in: "In Vitro Metabolic Activation in Mutagenesis Testing". F.J. de Serres, J.R. Fouts, J.R. Bend and R.M. Philpot (eds.), Elsevier-North Holland Biomedical Press, Amsterdam, pp. 159-177
28. Kalpitulnik, J., Levin, W., Conney, A.H., Yagi, H. and Jerina, D.M.: 1977, Nature 266:378-380
29. Keller, J.W., Heidelberger, C., Beland, F.A. and Harvey, R.G.: 1976. J. Am. Chem. Soc. 98: 8276-8277
30. Lehr, R.E., Schaefer-Ridder, M. and Jerina, D.M.: 1977, Tetrahedron Lett: 539-542
31. Lehr, R.E., Schaefer-Ridder, M. and Jerina, D.M.: 1977, J. Org. Chem. 42:736-744
32. Lehr, R.E., Taylor, C.W., Kumar, S., Mah, H.D. and Jerina, D.M.: 1978, J. Org. Chem. 43:3462-3466
33. Levin, W., Thakker, D.R., Wood, A.W., Chang, R.L., Lehr, R.E., Jerina, D.M. and Conney, A.H.: 1978, Cancer Res. 38:1705-1710
34. Levin, W., Wood, A.W., Chang, R.L., Yagi, H., Mah, H.D., Jerina, D.M. and Conney, A.H.: 1978, Cancer Res. 38:1831-1834
35. Malaveille, C., Bartsch, H., Grover, P.L. and Sims, P.: 1975, Biochem. Biophys. Res. Commun. 66:693-700
36. Malaveille, C., Bartsch, H., Tierney, B., Grover, P.L. and Sims, P.: 1978, Biochem. Biophys. Res. Commun. 83:1468-1473
37. Malaveille, C., Tierney, B., Grover, P.L., Sims, P. and Bartsch, H.: Biochem. Biophys. Res. Commun. 75:427-433

38. Moschel, R.C., Baird, W.M. and Dipple, A.: 1977, Biochem. Biophys. Res. Commun. 76:1092-1098
39. Newbold, R.F. and Brookes, P.: 1976, Nature 261:52-54
40. Sims, P. and Grover, P.L.: 1974, Advan. Cancer Res. 20:165-274
41. Sims, P., Grover, P.L., Swaisland, A., Pal, K. and Hewer, A.: 1974, Nature 252:326-328
42. Slaga, T.J., Bracken, W.M., Viaje, A., Levin, W., Yagi, H., Jerina, D.M. and Conney, A.H.: 1977, Cancer Res. 37:4130-4133
43. Slaga, T.J., Gleason, G.L., DiGiovanni, J., Sukumaran, K.B. and Harvey, R.G.: 1979, Cancer Res. 39:1934-1936
44. Slaga, T.J., Gleason, G.L. and Hardin, L.: 1979, Cancer Lett. 7: 97-102
45. Slaga, T.J., Huberman, E., Selkirk, J.K., Harvey, R.G. and Bracken, W.M.: 1978. Cancer Res. 38:1699-1704
46. Thakker, D.R., Levin, W., Wood, A.W., Conney, A.H., Stoming, T.A. and Jerina, D.M.: 1978, J. Am. Chem. Soc. 100:645-647
47. Thakker, D.R., Yagi, H., Akagi, H., Koreeda, M., Lu, A.Y.H., Levin, W., Wood, A.W., Conney, A.H. and Jerina, D.M.: 1977, Chem.-Biol. Interac. 16:281-300
48. Thakker, D.R., Yagi, H., Lehr, R.E., Levin, W., Buening, M., Lu, A.Y.H., Chang, R.L., Wood, A.W., Conney, A.H. and Jerina, D.M.: 1978, Mol. Pharmacol. 14:502-513
49. Thakker, D.R., Yagi, H., Lu, A.Y.H., Levin, W., Conney, A.H. and Jerina, D.M.: 1976, Proc. Natl. Acad. Sci., USA 73:3381-3385
50. Tierney, B., Abercrombie, B., Walsh, C., Hewer, A., Grover, P.L. and Sims, P.: 1978, Chem.-Biol. Interac. 21:289-298
51. Tierney, B., Burden, P., Hewer, A., Ribeiro, O., Walsh, C., Rattle, H., Grover, P.L. and Sims, P.: 1979, J. Chromatography 176:329-335
52. Tierney, B., Hewer, A., MacNicoll, A.D., Gervasi, P.G., Rattle, H., Walsh, C., Grover, P.L. and Sims, P.: 1978, Chem.-Biol. Interac. 23:243-262
53. Tierney, B., Hewer, A., Rattle, H., Grover, P.L. and Sims, P.: 1978, Chem.-Biol. Interac. 23:121-135
54. Vigny, P., Duquesne, M., Coulomb, H., Tierney, B., Grover, P.L. and Sims, P.: 1977, FEBS Lett. 82:278-282
55. Vigny, P., Kindts, M., Duquesne, M., Cooper, C.S., Grover, P.L. and Sims, P.: 1980. Carcinogenesis 1:33-36
56. Weinstein, I.B., Jeffrey, A.M., Jennette, K.W., Blobstein, S.H., Harvey, R.G., Harris, C., Autrup, H., Kasai, H. and Nakanishi, K.: 1976, Science 193:592-595
57. Whalen, D.L., Ross, A.M., Yagi, H., Karle, J.M. and Jerina, D.M.: 1978, J. Am. Chem. Soc. 100:5218-5221
58. Wislocki, P.G., Kapitulnik, J., Levin, W., Lehr, R., Schaefer-Ridder, M., Karle, J.M., Jerina, D.M. and Conney, A.H.: 1978, Cancer Res. 38:693-696
59. Wislocki, P.G., Wood, A.W., Chang, R.L., Levin, W., Yagi, H., Hernandez, O., Dansette, P.M., Jerina, D.M. and Conney, A.H.: 1976, Cancer Res. 36:3350-3357
60. Wislocki, P.G., Wood, A.W., Chang, R.L., Levin, W., Yagi, H., Hernandez, O., Jerina, D.M. and Conney, A.H.: 1976, Biochem.

Biophys. Res. Commun. 68:1006-1012

61. Wood, A.W., Chang, R.L., Levin, W., Lehr, R.E., Schaefer-Ridder, M., Karle, J.M., Jerina, D.M. and Conney, A.H.: 1977, Proc. Natl. Acad. Sci., USA 74:2746-2750
62. Wood, A.W., Chang, R.L., Levin, W., Thomas, P.E., Ryan, D., Stoming, T.A., Thakker, D.R., Jerina, D.M. and Conney, A.H.: 1978, Cancer Res. 38:3398-3404
63. Wood, A.W., Levin, W., Lu, A.Y.H., Ryan, D., West, S.B., Yagi, H., Mah, H.D., Jerina, D.M. and Conney, A.H.: 1977, Mol. Pharmacol. 13:1116-1125
64. Wood, A.W., Levin, W., Thakker, D.R., Yagi, H., Chang, R.L., Ryan, D.E., Thomas, P.E., Dansette, P.M., Whittaker, N., Turujman, S., Lehr, R.E., Kumar, S., Jerina, D.M. and Conney, A.H.: 1979, J. Biol. Chem.. 254:4408-4415
65. Yagi, H., Hernandez, O. and Jerina, D.M.: 1975, J. Am. Chem. Soc. 97:6881-6883
66. Yagi, H., Thakker, D.R., Lehr, R.E. and Jerina, D.M.: 1979, J. Org. Chem. 44:3439-3442
67. Yang, S.K.: 1978, 7th Int. Pharmacology Congress, Paris. p. 652
68. Yang, S.K., Chou, M.W. and Fu, P.P.: 1980, in "Polynuclear Aromatic Hydrocarbons: Chemistry and Biological Effects", A. Bjorseth and P.W. Jones (eds.), Battelle Press, in press
69. Yang, S.K., Chou, M.W. and Roller, P.P.: 1979, J. Am. Chem. Soc. 101:237-239
70. Yang, S.K., Chou, M.W., Weems, H.B. and Fu, P.P.: 1979, Biochem. Biophys. Res. Commun. 90:1136-1141
71. Yang, S.K., McCourt, D.W. and Gelboin, H.V.: 1977, J. Am. Chem. Soc. 99:5130-5134
72. Yang, S.K., McCourt, D.W., Leutz, J.C. and Gelboin, H.V.: 1977, Science 196:1199-1201
73. Yang, S.K., McCourt, D.W., Roller, P.P. and Gelboin, H.V.: 1976, Proc. Natl. Acad. Sci. USA 73:2594-2598
74. Yang, S.K., Roller, P.P., Fu, P.P., Harvey, R.G. and Gelboin, H.V.: 1977, Biochem. Biophys. Res. Commun. 77:1176-1182
75. Yang, S.K., Roller, P.P. and Gelboin, H.V.: 1978, in "Carcinogenesis, Vol. 3: Polynuclear Aromatic Hydrocarbons", P.W. Jones and R.I. Freudenthal (eds.), Raven Press, New York, pp. 285-301
76. Zacharias, D.E., Glusker, J.P., Fu, P.P. and Harvey, R.G.: 1979, J. Am. Chem. Soc. 101:4043-4051
77. Zacharias, D.E., Glusker, J.P., Harvey, R.G. and Fu, P.P.: 1977, Cancer Res. 37:775-782

FORMATION AND INACTIVATION OF DNA-BINDING METABOLITES OF BENZO(a)PYRENE STUDIED WITH ISOLATED CELLS AND SUBCELLULAR FRACTIONS[1]

Bengt Jernström, Thomas M. Guenthner[2], Sigrun Hesse[3] and Sten Orrenius
Department of Forensic Medicine, Karolinska Institutet, Box 60 400, S-104 01 Stockholm, Sweden

The biotransformation of benzo(a)pyrene results in the formation of DNA-bound metabolites in various mammalian tissues. Products of 7,8-dihydro-7,8-dihydroxybenzo(a)pyrene and 9-hydroxybenzo(a)pyrene are the major DNA-binding species formed in different experimental systems. Both the extent and pattern of formation of DNA-bound benzo(a)pyrene metabolites varies however between different systems. The formation of DNA-bound metabolites of 9-hydroxybenzo(a)pyrene is favored by close contact between the site of generation of the binding species and target sites in DNA and counteracted by the presence of even weak nucleophiles in the incubation. In contrast, products of 7,8-dihydro-7,8-dihydroxybenzo(a)pyrene seem to be trapped primarily by enzyme-mediated glutathione conjugation. Our results underline the physiological importance of the nuclear envelope in the final activation of benzo(a)-pyrene to DNA-binding products.

It is now well established that the vast majority of chemical carcinogens require metabolic activation in order to express their carcinogenic property. As a result of biotransformation reactive electrophiles are formed which can act as ultimate carcinogens, presumably by covalent interaction with specific nucleophilic sites in DNA. Since the cell possesses protective systems, which can inactivate electrophilic metabolites and thereby prevent their accumulation and reaction with DNA, the formation of carcinogen - DNA adducts appears to be dependent on the relative rates at which these electrophiles are produced and inactivated.

During the last few years we have used freshly isolated mammalian cells, primarily hepatocytes, to study the metabolic activation of chemical carcinogens to DNA-binding products. Parallel studies have been performed with isolated subcellular fractions in order to compare formation of DNA-binding metabolites under different experimental conditions and elucidate the importance of various cellular protective mechanisms in preventing electrophilic carcinogen metabolites from reaching and interacting with nuclear DNA.

B. Pullman, P. O. P. Ts'o and H. Gelboin (eds.), Carcinogenesis: Fundamental Mechanisms and Environmental Effects, 157–165.

This communication summarizes some of our more recent studies on the formation and inactivation of DNA-binding products from benzo(a)-pyrene. The results indicate that cellular defence mechanisms influence the formation of DNA - benzo(a)pyrene adducts both quantitatively and qualitatively.

METABOLIC ACTIVATION OF BENZO(a)PYRENE

In mammalian tissues cytochrome P-450-linked monooxygenases metabolize benzo(a)pyrene (BP)[4] to various epoxides which are further converted to phenols, diones and dihydrodiols. Several of these metabolites may undergo a second cytochrome P-450-catalyzed monooxygenation to yield phenol and dihydrodiol epoxides (Sims et al, 1974; King et al, 1976a). In various biological systems metabolites of 9-hydroxybenzo(a)pyrene (9-OH-BP) and 7,8-dihydro-7,8-dihydroxybenzo(a)-pyrene (BP-7,8-diol) have been shown to be major DNA-binding products of BP (King et al, 1976a,b; Jernström et al, 1978a). The DNA-binding species formed from BP-7,8-diol have been identified as BP-7,8-diol-9,10-oxide isomers of which the (+)anti-BP-7,8-diol-9,10-oxide is considered the major ultimate carcinogenic form of BP (Sims et al, 1974; King et al, 1976b; Levin et al, 1976). DNA-binding products of the seemingly less carcinogenic 9-OH-BP have been tentatively identified as 9-OH-BP-4,5-oxide(s) (King et al, 1976a; Jernström et al, 1978b). Whereas susceptible sites in DNA for binding of reactive intermediates from BP-7,8-diol have been shown to involve the exocyclic and N-7 nitrogen atoms of guanine, the exocyclic nitrogen of adenine and phosphate, the sites involved in binding of 9-OH-BP intermediates are largely unknown.

The formation of these DNA-binding metabolites of BP requires a sequence of reactions which can occur in both the endoplasmic reticulum and the nuclear envelope. Precursors of the DNA-binding products may be formed in the endoplasmic reticulum and activated in the nuclear envelope. The close proximity of genetic material to the nuclear envelope may be particularly important in generating electrophilic metabolites which become bound to DNA.

Thus it appears that the formation of DNA-bound metabolites of BP in the cell depends on both the rate and site of generation of the binding species and their precursors, notably 9-OH-BP and BP-7,8-diol. However, it is also influenced by various inactivating pathways which may counteract the formation and accumulation of the reactive electrophiles. The precursors, 9-OH-BP and BP-7,8-diol, may form excretable conjugates with sulphate and glucuronic acid, and the ultimate DNA-binding metabolites may be trapped by interacting with cellular nucleophilic sites other than those in DNA. Thus, the generation of DNA-bound metabolites of BP in the cell is influenced by a number of factors including the rate and site of production of the reactive species and the activity of detoxifying metabolic pathways.

FORMATION OF DNA-BOUND METABOLITES OF BP IN VARIOUS EXPERIMENTAL SYSTEMS

The formation of DNA-bound metabolites of BP differs both quantitatively and qualitatively in different experimental model systems (Jernström et al, 1976 and 1978b and Guenthner et al, 1979). When liver microsomes from 3-methylcholanthrene(MC)-treated rat are incubated with BP and purified calf thymus DNA in the presence of NADPH and molecular oxygen, the generation of DNA-bound metabolites of BP is extensive, and products derived from 9-OH-BP predominate over those from BP-7,8-diol. On the other hand, incubation of hepatic nuclear fraction with BP and NADPH, in the absence or presence of microsomes, yields less DNA-bound products, and those derived from BP-7,8-diol dominate over those from 9-OH-BP. Finally, with hepatocytes from MC-treated rat only small amounts of DNA-bound BP metabolites are normally formed, and they are often almost exclusively derived from BP-7,8-diol (Guenthner et al, 1980). The latter observation may however vary with the experimental conditions, since the ratio between the two major DNA-bound BP products in hepatocytes has been found to be influenced by a number of factors including the composition of the medium, BP concentration and cell density in the incubation, the level of epoxide hydratase in the hepatocytes and the technique used for measuring DNA binding (Guenthner et al, 1980). The fluorescence technique employed in a previous study (Jernström et al, 1978b) measures total binding to intact DNA, including binding to phosphate, whereas the chromatographic technique used in a more recent investigation (Guenthner et al, 1980) measures specific binding to deoxyribonucleosides.

Table 1 compares the formation of metabolites of BP upon incubation with either hepatocytes, the microsomal fraction or the nuclear fraction isolated from MC-treated rats (Jernström et al, 1980). With hepatocytes non-polar metabolites increase during the early phase of incubation, then decrease as they are further metabolized, while polar metabolites increase continuously. Incubation of BP with microsomes or nuclei yield similar patterns of metabolites with relatively high proportions of BP-7,8-diol and 9-OH-BP; the latter is true also for hepatocytes at an early phase of incubation.

Incubation of hepatocytes isolated from MC-treated rat, but not from control rat, with BP results in the formation of DNA-bound BP metabolites (Table 2). The formation of bound products is inhibited by α-naphthoflavone but considerably increased in the presence of salicylamide or after pretreatment of the rat with diethylmaleate prior to cell isolation. The effect of α-naphthoflavone on DNA binding is associated with similar inhibition of production of organic-soluble BP metabolites, whereas salicylamide increases the concentration of such metabolites, presumably by interfering with their conjugation with sulphate and glucuronic acid. The effect of diethylmaleate treatment on formation of DNA-bound BP metabolites is presumably due to its effect on cellular glutathione (GSH) concentration, which is markedly lowered by administration of this agent prior to cell isolation. Thus,

Table 1. Formation of BP Metabolites by Hepatocytes, Microsomes and Nuclei from MC-Treated Rats

Metabolites	Hepatocytes 5 min	Hepatocytes 30 min	Microsomes 30 min	Nuclei 30 min
	Incubation time			
Total	100[a]	100[a]		
Organic-soluble	69	24		
Aqueous-soluble	31	76		
"Fraction I"	19[b]	63[b]	1[b]	1[b]
BP-9,10-diol	16	6	18	13
BP-4,5-diol	5	2	12	9
BP-7,8-diol	15	2	10	12
BP-1,6-dione	3	2	10	8
BP-3,6-dione	5	4	12	9
BP-6,12-dione	1	1	3	4
9-OH-BP	9	1	7	14
3-OH-BP	17	2	14	21

a/ Numbers represent % of total metabolites. b/ Numbers represent % of ethylacetate-extractable metabolites. Data from Jernström et al, 1980.

it appears from these results that the formation of DNA-bound metabolites of BP in hepatocytes is affected both by the removal of precursor, notably BP-7,8-diol, through conjugation with sulphate and glucuronic acid and by trapping of electrophilic BP metabolites through conjugation with glutathione.

Table 2. Effects of Inhibitors on Formation and Binding to DNA of BP Metabolites in Hepatocytes from MC-Treated Rat

Inhibitor	Metabolites: Total (nmol/1.5x10^6 cells)	Organic-soluble (nmol/1.5x10^6 cells)	Water-soluble (nmol/1.5x10^6 cells)	DNA-bound (pmol/mg DNA)
None	77	16	61	25
α-Naphthoflavone	16	8	8	10
Salicylamide	75	37	38	45
Diethylmaleate	73	24	49	57

Data from Jernström et al, 1980.

In order to further investigate factors influencing the formation of DNA-bound BP metabolites the series of experiments presented in

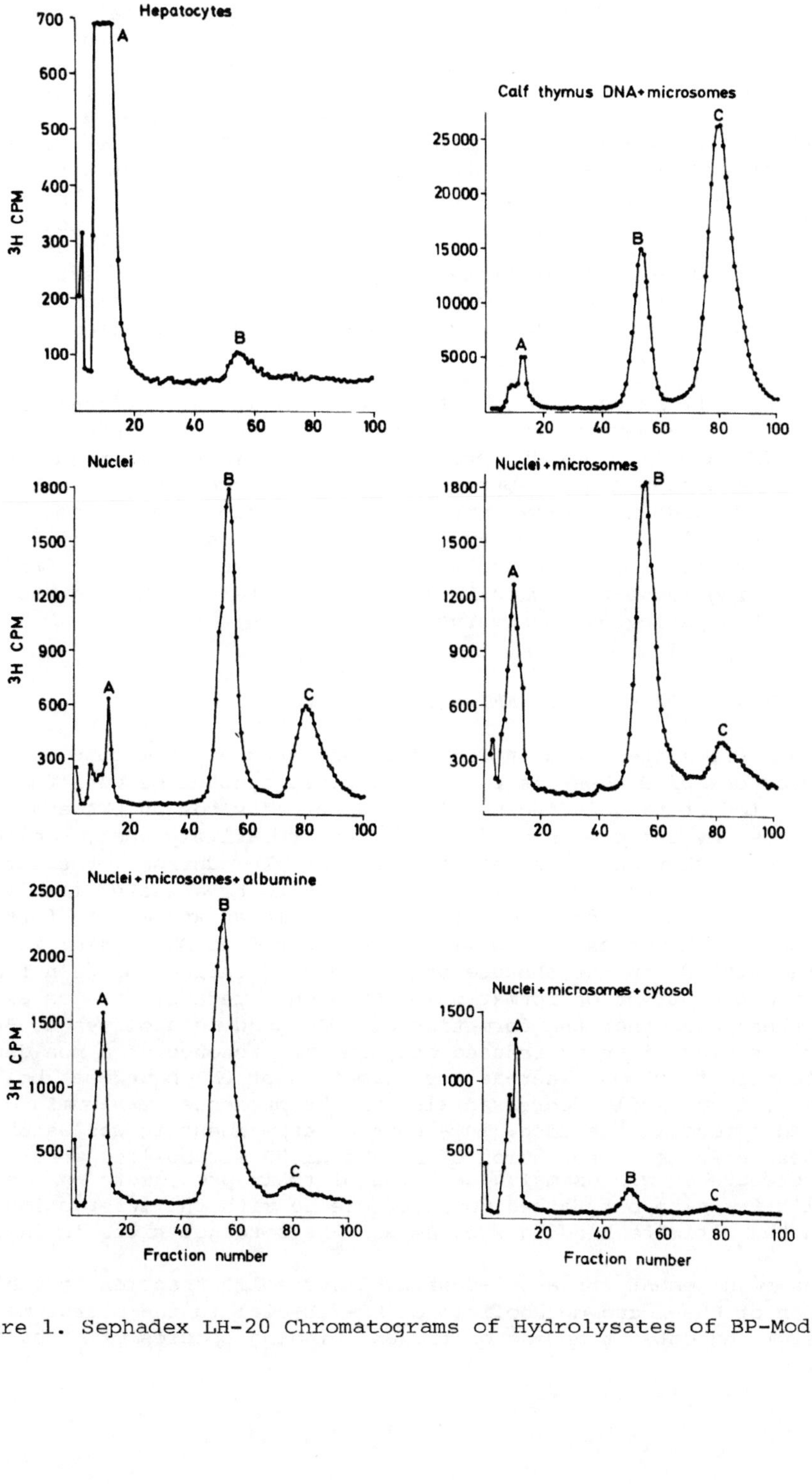

Figure 1. Sephadex LH-20 Chromatograms of Hydrolysates of BP-Modified DNA.

Figure 1 was performed. Both hepatocytes and hepatic subcellular fractions, and various combinations thereof, were incubated with BP, and modified DNA was analyzed by Sephadex LH-20 chromatography. The two peaks of interest, shown here as "B" and "C" reflect the binding of products of BP-7,8-diol and 9-OH-BP, respectively. In the case of peak "B", the binding species have been identified as BP-7,8-diol-9,10-oxides; the chromatographic system used in this study does not resolve the isomers of BP-7,8-diol-9,10-oxides (King et al, 1976b). In the case of peak "C", the binding species has not been as rigorously identified, but the available evidence has shown it to be 9-OH-BP-4,5-oxide(s) (King et al, 1976a). The identity of peak "A" has not been satisfactorily determined, but it may contain nuclease-resistent phosphotriesters, unhydrolyzed nucleotides and/or exchange-tritiated unmodified nucleosides.

Marked differences can be noted when the binding patterns obtained as a result of BP metabolism in the various incubation systems are compared (Figure 1). Peak "C" predominates only in the microsome - purified DNA system and is absent in the hepatocyte system. Peak "B", on the other hand, is prominent in all the incubation systems containing the nuclear fraction, except when the cytosolic fraction is also present. With both the nuclear, microsomal and cytosolic fractions present during incubation with BP a binding pattern much like that obtained with hepatocytes is observed.

INACTIVATION OF DNA-BINDING METABOLITES OF BP

The results presented in Figure 1 suggest that formation of DNA-bound products of 9-OH-BP is favored by close contact between the site of generation of the binding species and target sites in DNA and counteracted by the presence of other, even relatively weak, nucleophiles in the incubation (cf effect of albumin). In contrast, the formation of DNA-bound products of BP-7,8-diol appears to be affected mainly by factor(s) present in the cytosolic fraction. In an attempt to further elucidate this hypothesis, the microsome - purified DNA system was incubated with BP in the absence or presence of various protein fractions, and the effect on formation of DNA-bound metabolites was assayed. It was then found that the formation of DNA-bound metabolites of BP-7,8-diol is significantly reduced only in the presence of a non-denatured cytosolic fraction, whereas the formation of DNA-bound metabolites of 9-OH-BP is markedly decreased also in the presence of albumin or a denatured cytosolic fraction. In a control experiment it was ascertained that these effects on formation of DNA-bound BP metabolites were not due to effects on microsomal generation of their precursors by the cytosolic fraction but rather to interference with the interaction between the ultimate binding species and their target sites in DNA.

The requirement for a non-denatured cytosolic fraction to inhibit formation of DNA-bound metabolites of BP-7,8-diol in the microsome - DNA system suggested trapping of the DNA binding species of BP by conjugation with glutathione as a possible inhibitory mechanism. This hypothesis was further substantiated by comparing the effect of cytosolic fractions containing different concentrations of GSH on the binding of BP to DNA in the nuclear fraction. As shown in Table 3, lowering the level of hepatic GSH by pretreatment of the rat with diethylmaleate (DEM) prior to isolation of liver cytosol decreased the inhibitory effect of the cytosol on formation of DNA-bound metabolites of BP-7,8-diol; inhibition was restored upon addition of increasing concentrations of GSH to the incubate. Similar, although less marked, effects were observed in the case of formation of DNA-bound metabolites of 9-OH-BP in this system.

Table 3. Effect of Glutathione on DNA Binding of BP Metabolites Formed by Nuclei in the Presence of Cytosol

Additions	BP-7,8-diol metabolites	9-OH-BP metabolites
	pmol/mg DNA	
Nuclei + cytosol	0.4	1.4
Nuclei + DEM-cytosol[a/]	1.5	3.2
Nuclei + DEM-cytosol + 0.1 mM GSH	1.0	3.1
Nuclei + DEM-cytosol + 0.2 mM GSH	0.5	2.2

a/ DEM-cytosol represents cytosolic fraction with lowered GSH content due to pretreatment of the rat with diethylmaleate before isolation of liver cytosolic fraction. Data from Hesse et al, 1980.

The requirement for GSH and non-denatured cytosolic fraction to inhibit formation of DNA-bound metabolites of BP-7,8-diol by liver nuclei suggested that the binding species might be trapped by enzyme-mediated glutathione conjugation (Hesse et al, 1980). This was actually shown to be the case in the experiment depicted in Table 4, where the effect of purified glutathione S-transferase B on formation of DNA-bound metabolites of BP-7,8-diol in liver nuclei was investigated in the absence and presence of GSH. As shown in this Table, there was a GSH-dependent decrease in formation of DNA-bound BP metabolites in the presence of transferase. In the absence of transferase there was no apparent effect on DNA binding by GSH added in concentrations up to 1 mM. Further, in preliminary experiments the inhibitory effect of GSH and glutathione S-transferase B on the formation of DNA-bound metabolites of BP-7,8-diol has been shown to be associated with increased formation of glutathione conjugates of BP. From these results it appears that the interaction of binding species of BP with DNA is influenced by protective systems in the cell, and that enzyme-mediated glutathione conjugation is the main protective mechanism against forma-

Table 4. Effect of Glutathione and Glutathione S-Transferase B on the Binding of Products Formed During Nuclear Metabolism of BP-7,8-Diol

Additions	Metabolites bound (pmol/mg DNA) - Glutathione S-transferase B	+ Glutathione S-transferase B
Nuclei	51.8	50.7
Nuclei + 0.1 mM GSH	55.7	19.2
Nuclei + 0.2 mM GSH	55.7	17.8
Nuclei + 1.0 mM GSH	53.0	13.5

Data from Hesse et al, 1980.

tion of DNA-bound metabolites of BP-7,8-diol, the major ultimate carcinogenic form of BP.

CONCLUDING REMARKS

From these experiments it appears that the formation of DNA-bound metabolites of BP in the cell is markedly influenced both by the site of generation of the binding species and by protective systems present primarily in the cytosol. The extent and pattern of formation of DNA-bound BP metabolites in the microsome - purified DNA system suggest that binding - in particular binding of 9-OH-BP metabolites - is favored by close contact between the site of generation of the reactive species and target sites in DNA. Trapping of DNA-binding species of 9-OH-BP appears to occur by interaction with relatively non-specific nucleophiles, whereas the inactivation of DNA-binding products of BP-7,8-diol seems to require enzyme-mediated glutathione conjugation. These observations underline the physiological importance of final metabolic activation of BP in the nuclear envelope, since this mechanism would provide both close proximity between the site of generation of the binding species and target sites in DNA and minimal inactivation by glutathione conjugating systems present in the cytosol.

NOTES

[1]Supported by NIH (grant no 1RC1 CA26261-01) and the Swedish Cancer Society.

[2]Present address: Abt. Molekularpharmakologie, Pharmakologisches Institut der Universität Mainz, Obere Zahlbacherstrasse 67, D-6500 Mainz, Germany

[3]Present address: Department of Toxicology, Gesellschaft für Strahlen- und Umweltforsch., D-8042 Neuherberg/München, Germany

[4]Abbreviations used: BP, benzo(a)pyrene; BP-7,8-diol, 7,8-dihydro-7,8-dihydroxybenzo(a)pyrene; 9-OH-BP, 9-hydroxybenzo(a)pyrene; MC, 3-methylcholanthrene; GSH, glutathione, reduced form; DEM, diethylmaleate.

REFERENCES

Guenthner, T.M., Jernström, B., and Orrenius, S.: 1979, Biochem. Biophys. Res. Commun. 91, pp. 842-848.

Guenthner, T.M., Jernström, B., and Orrenius, S.: 1980, Carcinogenesis. Submitted for publication.

Hesse, H., Jernström, B., Martinez, M., Guenthner, T.M., Orrenius, S., Christodoulides, L., and Ketterer, B.: 1980, Biochem. Biophys. Res. Commun. In press.

Jernström, B., Vadi, H., and Orrenius, S.: 1976, Cancer Res. 36, pp. 4107-4113.

Jernström, B., Vadi, H., and Orrenius, S.: 1978a, Chem. Biol. Interactions, 20, pp. 311-321.

Jernström, B., Orrenius, S., Undeman, O., Gräslund. A., and Ehrenberg, A.: 1978, Cancer Res. 38, pp. 2600-2607.

Jernström, B., Guenthner, T.M., Dock, L., Svensson, S.-Å., and Orrenius, S.: 1980, In Coon, M.J., Estabrook, R.W., Gelboin, H.V., Gillette, J.R., and O'Brien, P.J. (Eds), Microsomes, Drug Oxidations and Chemical Carcinogenesis, Academic Press, New York. In press.

King, H.W.S., Thompson, M.H., and Brookes, P.: 1976a, Int. J. Cancer, 18, pp. 339-344.

King, H.W.S., Osborne, H.R., Beland, F.A., Harvey, R.G., and Brookes, P.: 1976b, Proc. Natl. Acad. Sci. USA. 73, pp. 2679-2681.

Levin, W., Wood, A.W., Yagi, H., Jerina, D.M., and Conney, A.H.: 1976, Proc. Natl. Acad. Sci. USA. 73, pp. 3867-3871.

Sims, P., Grover, P.L., Swaisland, A., Pal, K., and Hewer, A.: 1974, Nature, 252, pp. 226-228.

ENZYMIC CONTROL OF REACTIVE METABOLITES FROM AROMATIC CARCINOGENS

F. Oesch, J. C. S. Clegg, R. Billings, K. L. Platt, and H. R. Glatt
Institute of Pharmacology, University of Mainz,
D-6500 Mainz, FRG

ABSTRACT. Mutation and transformation in C3H 10T 1/2 mouse fibroblasts were coordinately induced by 4-nitroquinoline N-oxide and identically modulated by caffeine strongly suggesting mutation as one necessary step in the sequence of events ultimately leading to transformation. The enzymic control of reactive metabolites derived from aromatic carcinogens was then investigated using bacterial mutagenicity as an analytical tool. It was shown that the correlation of bacterial mutagenicity with carcinogenicity of BP and four major metabolites was substantially better when these compounds were activated by intact hepatocytes as compared to commonly used broken cell preparations which suggests that the relatively poor correlation between whole animal carcinogenicity and mutagenicity in the standard Ames assay of BP metabolites was due to differences in metabolism of the compounds in the two systems rather than to differences in the biological endpoints, *i.e.*, bacterial mutagenicity *versus* mammalian carcinogenicity. Addition and removal of cofactors or pure enzymes showed differential involvement in the control of the mutagenically reactive metabolites of the aromatic carcinogens investigated of monooxygenase, epoxide hydrolase, glutathione S-transferase, UDP-glucuronosyltransferase and sulfotransferase.

1. COORDINATE MUTATION AND TRANSFORMATION OF MOUSE FIBROBLASTS: INDUCTION BY NITROQUINOLINE OXIDE AND MODULATION BY CAFFEINE.

Comparisons of mutation and transformation by various agents in different systems, each suitable only for one of the two endpoints, suffer from the shortcoming that any non-correlation which may be observed cannot be attributed unambiguously to fundamental differences between the processes of mutation and transformation, since pharmacokinetic differences between the systems such as membrane permeability, activation and detoxification of test compound, or differences in repair capabilities, might also be responsible for such an effect. Analysis of correlations of active versus inactive compounds with respect to the two endpoints and comparisons of curves of the

B. Pullman, P. O. P. Ts'o and H. Gelboin (eds.), Carcinogenesis: Fundamental Mechanisms and Environmental Effects, 167–177.

dependence of the two responses on the dose in the same cell system provides stronger circumstantial evidence but still may not distinguish between the situation where mutation is a required step for transformation and the alternative, dependence of two separate events on the same or similar chemically reactive species. We therefore studied mutation and transformation in the same cell system and investigated whether caffeine, an agent known to inhibit post-replication repair of DNA, exerts a similar modulation on both endpoints. If transformation were not dependent on a mutation, modulation of the processing of DNA

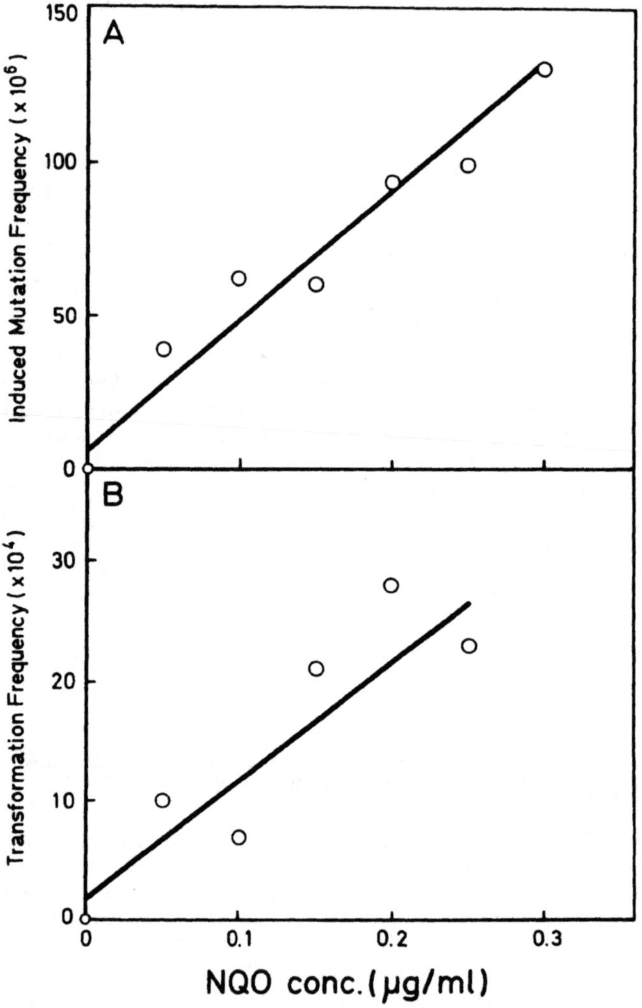

Fig. 1. Induction of (a) mutants resistant to ouabain and (b) transformants in C3H 10T 1/2 cells treated with NQO.

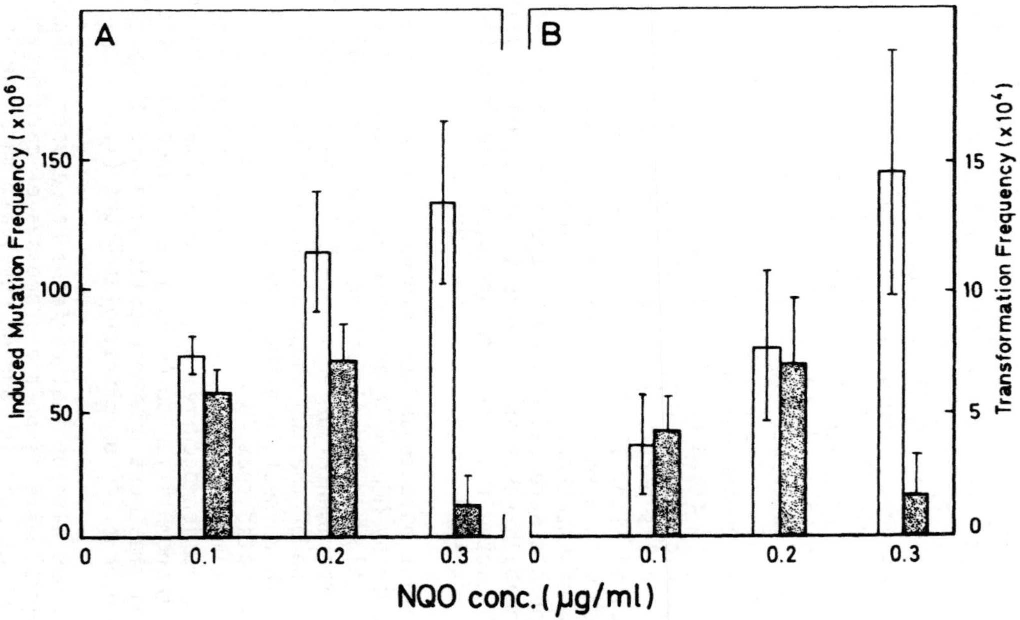

Fig. 2. Effect of caffeine on NQO-induced mutation and transformation. Panel A, mutation to ouabain resistance; panel B, transformation. Open bars, no added caffeine, solid bars, 1 mM caffeine 0-48 h after NQO treatment. The standard error between plates is indicated.

damage would not influence the response and it would be extremely unlikely that differential effects of caffeine would be the same for both endpoints.

Fig. 1 shows the characteristics of the response of C3H 10T 1/2 cells on exposure to 4-nitroquinoline-1-oxide (NQO). Cells in passages 9-13 were seeded in 175 cm^2 flasks (1.5 x 10^6 cells/flask) in 30 ml medium (Dulbecco modified MEM with 10% foetal calf serum, 100 U/ml penicillin, 100 μg/ml streptomycin) and allowed to attach and grow overnight. NQO (dissolved in DMSO) was added to give the desired concentration and incubation continued for 2 h at 37°C. The cell monolayer was then washed twice in medium, given fresh medium and incubated for 48 h to allow the expression of ouabain resistance. After this time the cells were trypsinised, counted and plated in three sets of dishes: 1) at 250,000 cells per 10 cm dish in medium containing 2 mM ouabain to detect ouabain-resistant mutants, stained after 14 days incubation (1); 2) at 300-400 viable cells per 10 cm dish in normal medium to detect foci of transformed cells (2), scored after 6 weeks with medium change twice a week until confluence and weekly thereafter; 3) at 70-100 viable cells per 10 cm plate in normal medium to determine survival, stained after 10-11 days. In the transformation assay plates were stained with Giemsa and types II and III foci were counted, using the criteria of Reznikoff *et al.* (3). The regression lines through the data points were calculated by the method of least squares. The frequency of ouabain-resistant mutants in untreated populations, which has been subtracted from the data shown, varied in different experiments between 2 x 10^{-6} and 8 x 10^{-6} per surviving cell.

Ouabain-resistant mutants and transformants were induced in a dose-dependent manner by NQO, with an apparently linear relationship between induced mutation or transformation frequency and NQO concentration. Within the experimental error inherent in the systems, the curves of the response of the two endpoints on the dose of NQO cannot be distinguished with respect to either shape or slope (Fig. 1). Pilot experiments had indicated that 48 h expression time for ouabain resistance gave a near maximum response over the range of NQO concentrations used here. Transformation occurred with an approximately 24-fold greater frequency than mutation to ouabain resistance, a ratio similar to that observed on treatment of these cells with N-acetoxyacetylaminofluorene, benzo(a)pyrene (4) or ultraviolet light (1).

It has been shown already that caffeine has diverse enhancing, inhibiting, or neutral effects on the processes of carcinogenesis and teratogenesis in intact animals, and on survival, mutagenesis and malignant transformation of cells in culture (5, 6, 7). These effects have frequently been attributed to its ability to inhibit certain pathways of DNA repair, possibly by interference with the supply of purine nucleotides (6). Fig. 2 shows the effects of addition of caffeine to the C3H 10T 1/2 mutation/transformation system. In order to obviate any influence of the compound on either the uptake or metabolism of NQO, or on the selection of ouabain-resistant mutants, caffeine was only present during the 48 h immediately following the end of NQO treatment, *i.e.*, during the expression time. This corresponds to the period in which active repair replication (as measured by unscheduled DNA synthesis) of NQO-induced damage takes place in Balb 3T3 (8) and C3H 10T 1/2 (this study, data not shown). Cells were prepared, treated with NQO, and washed as indicated above. During the subsequent 48 h expression time, fresh medium was given containing, where appropriate, 1 mM caffeine. Thereafter the cells were replated in caffeine-free medium to assay survival, mutation to ouabain resistance and transformation as indicated above. We initially established that exposure of NQO-treated cells to caffeine for 48 h, the procedure to be used in the transformation-mutation experiments, potentiated the toxicity of NQO without itself exerting significant toxicity, as has previously been shown in a number of rodent cell lines treated with a variety of agents (5, 6, 9). Although the reduction in plating efficiency mediated by caffeine was the same (about 60%) at each concentration of NQO tested, its effect on the induction of mutants and transformants was strongly dependent on the NQO concentration (Fig. 2). Caffeine had little or no effect on either mutation or transformation at an NQO concentration of 0.1 μg/ml. The inhibition of both responses was still not statistically significant ($P>0.05$, Student's t-test) at the intermediate concentration of 0.2 μg NQO/ml, whilst at a concentration of 0.3 μg NQO/ml a very strong inhibition was observed -- about 90% for both processes (Fig. 2). The possibility that the reduction in the NQO-induced mutation frequency is mediated through a caffeine-dependent increase in the time required for optimum expression of mutants has been ruled out by the observation that delaying replating in ouabain-containing medium up to 120 h after NQO exposure did not lead to any increase in the observed mutation frequency (data not shown).

The observations of Kakunaga (7), who demonstrated caffeine's ability to reduce NQO-induced transformation of another mouse fibroblast line, Balb 3T3, closely parallel the inhibitory effect of caffeine on the induction of transformation observed in these experiments. In his experiments the inhibition was also greater at higher NQO concentrations. He attributed the inhibitory effect to the caffeine-mediated suppression of post-replication repair, and the consequent repair of the lesions by an error-free excision pathway. On the other hand, the inhibition of mutation to ouabain resistance reported here is in contrast to recent results involving mutation at the *hgprt* locus (9), where it is clear that caffeine has no effect on mutation frequency in experiments where cells are respread to prevent density-dependent cross-feeding effects. Chang *et al.* have also reported (5) that addition of caffeine to UV-treated V79 chinese hamster cells during the period between irradiation and replating in selective medium causes an increase in the observed frequency of mutation to ouabain resistance. Satisfactory explanations of these differences, which may be related to the different mutagens, cell lines and genetic loci used, must await a more detailed understanding of the mechanisms of mutagenesis and the effect that caffeine has upon them. The important point to be made is that in the experiments described here, where mutation and transforma-

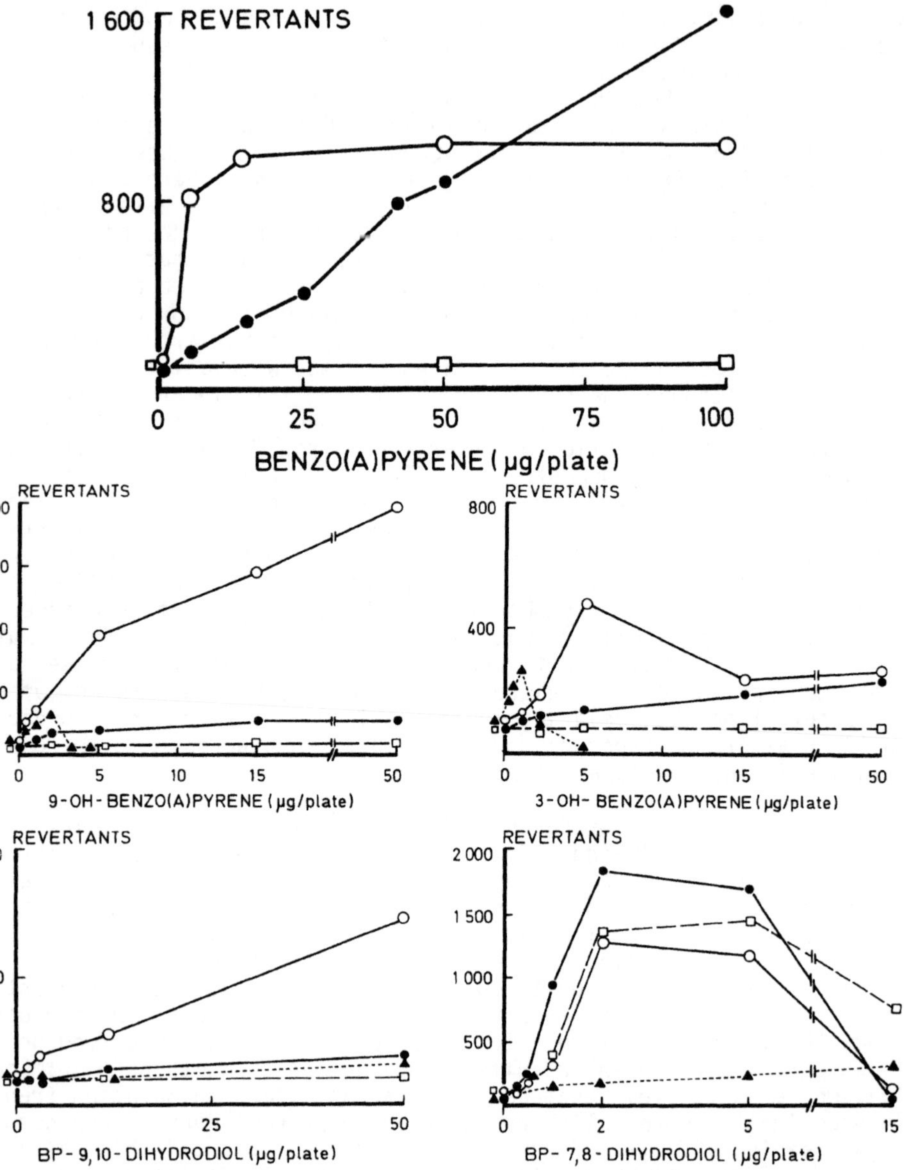

Fig. 3. Mutagenicity of benzo(a)pyrene and benzo(a)pyrene metabolites in *S. typhimurium* TA 100, directly (▲), in the presence of intact hepatocytes (●), homogenized hepatocytes (□) or homogenized hepatocytes and an NADPH-generating system (O). Values are means of three incubations. The variation coefficient was always smaller than 0.1, except for the highest two doses of benzo(a)pyrene with intact hepatocytes where it was about 0.4. Similar results were obtained in repeat experiments.

tion were measured concomitantly in the same cell system, there is a striking parallelism in the inhibitory effect of caffeine on both processes, and in its dependence on NQO dose. The parallel dose-response curves for the induction of mutation and transformation, albeit over a rather narrow range of concentrations, also argue against fundamentally different mechanisms for the two events. Somatic mutation theories of carcinogenesis have recently received strong support from the empirical correlation which is emerging between chemicals which are bacterial mutagens (in the presence of mammalian activating system) and those which are carcinogenic in animals (10, 11). The results reported here also suggest that mutation is a required step in the sequence of events ultimately leading to malignant transformation. The following studies on the control of reactive metabolites of aromatic carcinogens were performed using mutagenicity as an analytical tool.

2. ENZYMIC CONTROL OF REACTIVE METABOLITES FROM AROMATIC CARCINOGENS USING MUTAGENICITY AS AN ANALYTICAL TOOL.

2.1. Activation by whole cells versus cell homogenates.

The mutagenicity of benzo(a)pyrene (BP) and four major metabolites is shown in Fig. 3 as a function of the dose with four different metabolizing systems: (i) without any exogenous metabolizing system, (ii) with intact hepatocytes, (iii) with sonicated hepatocytes and (iv) with sonicated hepatocytes and an NADPH-generating system. BP (Fig. 3, upperst panel) was not mutagenic when tested directly or with homogenate without additional cofactors. It became strongly mutagenic when an NADPH-generating system was added to the homogenate or when whole hepatocytes were used. With homogenate, BP showed strong mutagenic effects at very low doses. At higher doses the mutagenic frequency was constant over a wide dose range, probably due to limited metabolic capacity, since no signs of toxicity were apparent. With whole hepatocytes, much higher doses of BP were required for induction of mutation than with homogenate and the mutagenic effect increased approximately linearly with the dose up to the highest concentration tested. At this high dose of BP, the mutagenic effect in the presence of hepatocytes was stronger than the maximum effect with NADPH-fortified homogenate. This result contrasts with that of Brouns *et al.* (12), who found mutagenicity of BP in the presence of hepatocytes only if glutathione was depleted. However, the highest BP dose used by these investigators was 25 μg. It should also be noted that at the highest two doses BP precipitated from the incubation mixture which may be the cause of the larger than normal variation of mutagenic effects at those doses.

3-OH-BP and 9-OH-BP (Fig. 3, middle panel) were mutagenic in the absence of a mammalian metabolizing system. The effects were not very strong, but they were found at low doses of test compound. At higher doses the two phenols were highly toxic, and all bacteria were killed at a dose of 5 μg. Direct mutagenic effects of hydroxybenzo(a)pyrenes have been previously described (10, 13, 14, 15). Interestingly, in the present study it was observed that addition of cell homogenate, without additional NADPH, completely prevented both mutagenic and visible toxic effects, even at doses of 50 μg. Since metabolism of the phenols is expected to be limited without additional NADPH, we suspect that derivatives formed spontaneously under the test conditions rather than the parent phenols themselves caused mutations and bacterial death in the absence of the homogenate. When an NADPH-generating system was added along with the cell homogenate, mutagenic effects were observed. At low doses, *e.g.*, at 5 μg per plate, the mutagenic effects were similar to those with BP. At higher doses, 9-OH-BP was more mutagenic than BP, whereas 3-OH-BP was somewhat toxic and the observable effect decreased with increasing dosage. The high mutagenicity of the two phenols, which is in agreement with other investigations which employed postmitochondrial supernatant from the liver of mice and rats (10, 14, 16), contrasts to tumor initiation and carcinogenicity studies. However, when tested for mutagenicity using unbroken mammalian cells for metabolism, the two phenols were very weak mutagens. However, the weak mutagenic effects were statistically significant and reproducible.

BP-9,10-dihydrodiol (Fig. 3, lower panel), at the highest dose tested, was very weakly mutagenic in the absence of a mammalian metabolizing system, but this mutagenicity was abolished by addition of liver cell homogenate without NADPH. This effect is similar to that observed with phenols of BP. Addition of an NADPH-generating system to the homogenate clearly resulted in activation of BP-9,10-dihydrodiol to a mutagen. Malaveille *et al.* (17), who used NADPH-fortified postmitochondrial fraction of rat liver, observed also an activation of BP-9,10-dihydrodiol. For equimutagenic effects, higher doses BP-9,10-dihydrodiol than of BP, 3-OH-BP or 9-OH-BP were required (Fig. 3). In the presence of intact hepatocytes, mutagenicity of BP-9,10-dihydrodiol was lower than in the presence of NADPH-fortified homogenate, but it was significantly mutagenic.

BP-7,8-dihydrodiol was weakly mutagenic in the absence of a mammalian metabolizing system. This mutagenicity is probably an artifact caused by the solvent dimethylsulfoxide. In assays conducted with freshly prepared solutions, the mutagenicity was substantially lower than when the compound was added from older solutions. In addition, when an acetone solution was used, no mutagenicity was observed in the absence of a mammalian metabolizing system (data not shown). However, acetone could not be used because BP added in acetone was not activated by isolated hepatocytes. Therefore, in subsequent experiments, BP-7,8-dihydrodiol and the other test compounds were dissolved in dimethylsulfoxide immediately, *i.e.*, a few minutes, before use. Under these conditions the direct mutagenicity was very low compared to that in the presence of a mammalian metabolizing system (Fig. 3, lower right panel). In the presence of homogenate, even without supplementation by an NADPH-generating system, BP-7,8-dihydrodiol was highly mutagenic. Addition of an NADPH-generating system had very little effect at low doses of BP-7,8-dihydrodiol. At higher doses, toxicity was potentiated. The apparent lack of an NADPH requirement for the activation of BP-7,8-dihydrodiol was somewhat surprising since it is assumed that this compound is activated by cytochrome P-450-dependent monooxygenases to bay region oxides (18, 19). Either the endogenous NADPH was sufficient for the activation of this compound, which is only used at very low doses in the mutagenicity test, or other mechanisms may also activate BP-7,8-dihydrodiol. Experiments with the monooxygenase inhibitor (20) α-naphthoflavone support the first hypothesis. α-Naphthoflavone reduced the mutagenicity of BP-7,8-dihydrodiol to the effects observed in the presence of boiled homogenate (data not shown).

2.2. Effect of cofactors for conjugation and of pure epoxide hydrolase on the homogenate-mediated mutagenicity.

Plausible reasons for the observed differences in mutagenicity of BP-derivatives with intact *versus* broken hepatocytes are differences in pharmacokinetics caused by permeability barriers and nucleophilic traps and differences in metabolism. We adopted the working hypothesis that many enzymes which operate in intact cells may not do so in the homogenate for reasons such as dilution of cofactors. Tests of this hypothesis by improvement of conditions in the homogenate for certain metabolic routes indeed markedly altered mutagenicity. The alterations were different for different test compounds and the results were then more similar to results found with intact cells. This was particularly true with 3-OH-BP, BP-9,10-dihydrodiol and with low doses of BP, which became very weak mutagens as is the case with whole hepatocytes. In contrast, BP-7,8-dihydrodiol, under the same conditions, remained a potent mutagen in agreement with its potency when activated by intact hepatocytes. 9-OH-BP, a potent mutagen with homogenate, but only a very weak mutagen with whole cells, was an exception, since its mutagenicity was reduced only slightly more than that of BP-7,8-dihydrodiol by the investigated mechanisms. It remains to be determined whether metabolic factors other than those modulated by the additions performed in this study are crucial for its inactivation in intact hepatocytes.

Differences in the activities of enzymes may not be the only reason for differences in mutagenicity between whole and broken cells. Pharmacokinetic aspects could also be important. Jones *et al.* (21) have shown that the distribution between hepatocytes and extracellular space is quite different for different BP metabolites. Other studies (22, 23) indicate that the localization of the target in relation to the site of generation of the reactive metabolite significantly effects mutagenic activities. It has also been shown (24) that metabolites of 9-OH-BP bind to hepatocyte DNA and may even be the major binding products when BP is used as precursor. This observation together with the high mutagenicity of 9-OH-BP in the presence of hepatocyte homogenate supplemented with an NADPH-generating system suggests that pharmacokinetic factors may be important for the lack of a high bacterial mutagenicity of 9-OH-BP activated by hepatocytes.

Only rarely are tumors in the liver caused by polycyclic aromatic hydrocarbons. Other organs are their main targets. Also, it is assumed that carcinogens usually are activated not in the liver, but in or close to the target cells. Therefore, the observed good correlation between hepatocyte-mediated mutagenicity and whole animal carcinogenicity is somewhat surprising. Above it has been discussed that metabolic inactivation may be the major (although probably not the only) reason for the differences between cell- or homogenate-mediated mutagenicity. Activities of enzymes such as epoxide hydrolase, with BP 4,5-oxide as substrate (25), glutathione S-transferase, also with BP 4,5-oxide (manuscript in preparation), and UDP-glucuronosyltransferase, with 3-OH-BP (26) appear to be present in all or nearly all organs and tissues of the rat, although with marked quantitative differences. The inactivation of the potent carcinogens and mutagens BP and BP-7,8-dihydrodiol was already less efficient in the intact hepatocyte and was further reduced by inhibitors of conjugation. Thus, an insufficient inactivation may easily occur in extrahepatic tissues, which usually show lower activities of the required enzymes (25, 26). Efficiency or failure of inactivation, depending upon the compound, and wide distribution of inactivating enzymes could be the reasons for the observed good correlation between hepatocyte-mediated mutagenicity and whole animal carcinogenicity even though the compounds mainly produce tumors in non-hepatic tissues.

ACKNOWLEDGEMENT. We thank Mrs. C. Künnecke from our institute for pure epoxide hydrolase, Dr. B. Ames, Berkeley for *S. typhimurium* TA strains, Dr. P. Sims, London for a generous gift of 9-OH-BP and the Bundesministerium für Forschung und Technologie for financial support.

REFERENCES.

1. Chan, G.L., and Little, J.B.: 1978, Proc.Natl.Acad.Sci.U.S.A.75, pp. 3363-3366.
2. Terzaghi, M., and Little, J.B.: 1976, Cancer Res.36, pp. 1367-1374.
3. Reznikoff, C.A., Bertram, J.S., Brankow, D.W., and Heidelberger, C.: 1973, Cancer Res.33, pp. 3239-3249.
4. Landolph, J.R., and Heidelberger, C.: 1979, Proc.Natl.Acad.Sci. U.S.A.76, pp. 930-934.
5. Chang, C.-C., Phillips, C., Trosko, J.E., and Hart, R.W.: 1977, Mutat.Res.45, pp. 125-136; Donovan, P.J., and DiPaolo, J.A.: 1974, Cancer Res.34, pp. 2720-2727.
6. Waldren, C.A., and Patterson, D.: 1979, Cancer Res.39, pp. 4975-4982.
7. Kakunaga, T.: 1975, Nature 258, pp. 248-250.
8. Ikenaga, M., and Kakunaga, T.: 1977, Cancer Res.37, pp. 3672-3678.
9. van Zeeland, A.A.: 1978, Mutat.Res.50, pp. 145-151; Myhr, B.C., and DiPaolo, J.A.: 1978, Chem.-Biol.Interactions 21, pp. 1-18;

McMillan, S., and Fox, M.: 1979, Mutat.Res.60, pp. 91-107.
10. McCann, J., Choi, E., Yamasaki, E., and Ames, B.N.: 1975, Proc. Natl.Acad.Sci.U.S.A.72, pp. 5135-5139.
11. McCann, J., and Ames, B.N.: 1976, Proc.Natl.Acad.Sci.U.S.A.73, pp. 950-954.
12. Brouns, R.-E., Bos, R.P., van Gemert, P.J.L., Yih-van de Hurk, E.W.M., and Henderson, P.T.: 1979, Mutat.Res.62, pp. 19-26.
13. Glatt, H.R., and Oesch, F.: 1976, Mutat.Res.36, pp. 379-384.
14. Owens, I.S., Koteen, G.M., and Legraverend, C.: 1979, Biochem. Pharmacol.28, pp. 1615-1622.
15. Wislocki, P.G., Wood, A.W., Chang, R.L., Levin, W., Yagi, H., Hernandez, O., Dansette, P.M., Jerina, D.M., and Conney, A.H.: 1976, Cancer Res.36, pp. 3350-3357.
16. Lubet, R.A., Capdevila, J., and Prough, R.A.: 1979, Int.J.Cancer 23, pp. 353-356.
17. Malaveille, C., Bartsch, H., Grover, P.L., and Sims, P.: 1975, Biochem.Biophys,Res.Commun.66, pp. 693-700.
18. Sims, P., Grover, P.L., Swaisland, A., Pal, K., and Hewer, A.: 1974, Nature 252, pp. 326-328.
19. Wood, A.W., Levin, W., Lu, A.Y.H., Yagi, H., Hernandez, O., Jerina, D.M., and Conney, A.H.: 1976, J.Biol.Chem.251, pp. 4882-4890.
20. Diamond, C. and Gelboin, H.V.: 1969, Science 166, pp. 1023-1025.
21. Jones, C.A., Moore, B.P., Cohen, G.M., Fry, J.R., and Bridges, J.W.: 1978, Biochem.Pharmacol.27, pp. 693-702.
22. Glatt, H.R. and Oesch, F.: 1977, Arch.Toxicol.39, pp. 87-96.
23. Kuroki, T. and Drevon, C.: 1978, Nature 271, pp. 368-370.
24. Jernström, B., Orrenius, S., Gräslund, A., and Ehrenberg, A.: 1978, Cancer Res.38, pp. 2600-2607.
25. Oesch, F., Glatt, H.R., and Schmassmann, H.U.: 1977, Biochem. Pharmacol.26, pp. 603-607.
26. Bock, K.W., v.Clausbruch, U.C., Kaufmann, R., Lilienblum, W., Oesch, F., Pfeil, H., and Platt, K.L.: 1980, Biochem.Pharmacol. 29, pp. 495-500.

AFLATOXIN-DNA INTERACTIONS: QUALITATIVE, QUANTITATIVE AND KINETIC FEATURES IN RELATION TO CARCINOGENESIS

Gerald N. Wogan, Robert G. Croy, John M. Essigmann and Richard A. Bennett
Department of Nutrition and Food Science, Massachusetts Institute of Technology, Cambridge, Massachusetts 02139

Aflatoxin B_1 (AFB_1) is covalently bound to calf thymus DNA by incubation in the presence of rat liver microsomes and appropriate cofactors. Under optimum conditions, a binding level of one AFB_1 residue per 18 DNA nucleotides can be attained. Acid hydrolysis of the modified DNA liberates a mixture of adducts, of which the major one, representing about 90% of bound AFB_1, is 2,3-dihydro-2-(N^7-guanyl) -3-hydroxyAFB_1 (I). A single i.p. injection of AFB_1 at a dose level of 1 mg/kg to rats produces 2 hours later maximum binding levels of 1 AFB_1 residue per 8.0×10^3 DNA nucleotides in liver and 1 per 8.4×10^4 in kidney. The N^7-guanyl adduct (I) represents the major DNA adduct in both organs (liver, 80%; kidney, 70%). Administration of an equitoxic dose (12 mg/kg) to mice produces 2 hour binding levels of 1 AFB_1 per 4.4×10^5 DNA nucleotides in liver and 1 per 1.4×10^5 in kidney. HPLC of DNA hydrolysates from mouse tissues shows qualitatively similar patterns of adducts as in rat tissues. Both contain at least 10 minor components which are as yet only partially identified. Two are probably derivatives of the major adduct, and two others are thought to be N^7-guanyl adducts of aflatoxin P_1 and M_1, respectively. The major adduct (I) is efficiently removed from rat liver DNA within 24 hours after dosing. However, several of the minor adducts are more stable and accumulate in DNA as a result of repeated dosing. Thus, the minor adducts could be functionally important in carcinogenesis. The major adduct (I) is excreted in urine of rats in dose-dependent manner, and measurement of urinary excretion could form the basis of a non-invasive method for monitoring aflatoxin exposure.

B. Pullman, P. O. P. Ts'o and H. Gelboin (eds.), Carcinogenesis: Fundamental Mechanisms and Environmental Effects, 179–191.

IDENTITY OF AFLATOXIN-DNA ADDUCTS

Aflatoxin B_1 (AFB_1) is converted metabolically to forms which bind covalently to DNA and other cellular macromolecules, with AFB_1-2,3 oxide representing a major reactive intermediate, possibly the ultimate carcinogenic form of AFB_1. When activation takes place *in vivo* in liver or other tissues of animals, or *in vitro* in the presence of DNA, the major DNA adduct formed is 2,3 dihydro-2-(N^7-guanyl)-3-hydroxy AFB_1 (AFB_1N^7-Gua) (Essigmann *et al* 1977; Lin *et al*, 1977; Croy *et al*, 1978). However, under both conditions, additional DNA adducts form in much smaller quantities. In order to evaluate fully the role of DNA modification in the initiation of carcinogenesis and mutagenesis, complete characterization of the remaining minor DNA adducts formed from AFB_1 will be required. We are attempting to carry out that identification to the limit of sensitivity imposed by technical capabilities, and the information currently available is presented in Figure 1.

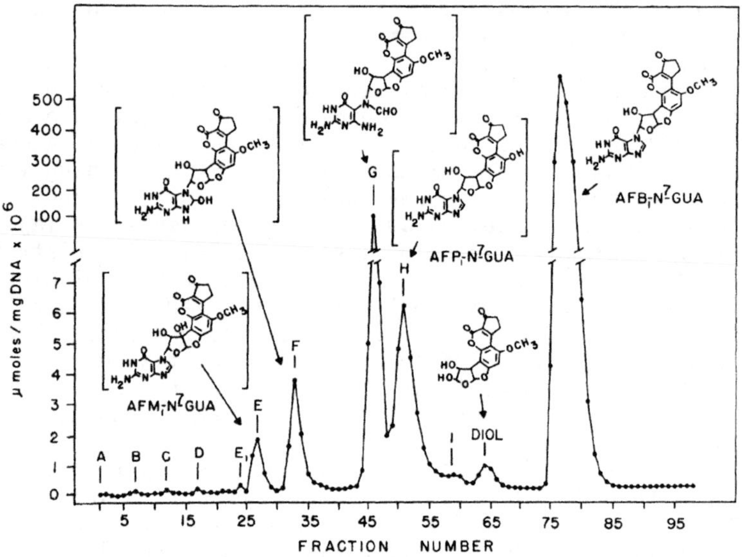

Figure 1. DNA adducts formed from AFB_1 in rat liver

This describes an HPLC chromatogram of hydrolysis products of DNA isolated from the liver of a rat injected with [^{3}H]-AFB_1 (0.6 mg/kg) and killed 2 hours later. A total of 12 peaks containing radioactivity can be identified. As noted above, the AFB_1-N^7-Gua adduct represents the major component; this is the only adduct which has been unambiguously identified. However, several additional structures can reasonably be proposed based on current evidence, which can be summarized as follows.

Two of the remaining adducts are postulated to be derivatives of AFB_1-N^7-Gua. Formation and structures of these compounds, designated peaks G and F were originally proposed by Lin et al (1977). They are thought to form as a consequence of the fact that substitution at the N^7-atom of guanine induces a strongly positive charge in the imidazole ring while the adduct is present in DNA. This in turn renders the bond linking the adduct to DNA labile to hydrolysis, especially under acid conditions, as is characteristic of the deoxyglycosidic bond of 7-substituted purines. However, in mildly alkaline conditions, OH^- attack occurs on C-8 of guanine, resulting in imidazole ring opening, producing the putative ring-opened derivative (Peak G). We have recovered a sufficient amount of this compound for extensive spectral studies, and all available evidence is consistent with the proposed structure. Peak F is consistently present concurrently with Peak G, and the possibility of its formation as an artifact of the acid hydrolysis of adducted DNA cannot be ruled out. Formation of these derivatives has important implications with respect to their possible functional significance, since it results in the stabilization of the deoxyglycosidic bond, reducing the facility with which depurination takes place, and stabilizing the adducts in DNA. Evidence discussed more fully below indicates that these products present relatively stable, persistent forms of the AFB_1-N^7-Gua adduct in rat liver.

Evidence from recent studies in our laboratories constitutes a basis for proposed structures of two additional adducts, both apparently derived from epoxidation of metabolites of AFB_1. The compound in Peak H of the chromatogram (Figure 1) has properties consistant with the structure of the N^7-guanyl derivative of aflatoxin P_1 (AFP_1), the O-demethylated derivative of AFB_1. Similarly, the adduct appearing in Peak E in Figure 1 is thought to be the N^7-guanyl adduct of aflatoxin M_1 (AFM_1).

Thus, structural evidence is available on 5 of the 11 adduct peaks consistantly observed in rat liver DNA after aflatoxin dosing. If, as seems reasonable, the AFP_1-N^7-Gua and AFM_1-N^7-Gua adducts are accompanied by derivatives resulting from N^7-Gua substitution (analagous to Peaks G and F formed from AFB_1-N^7-Gua), those derivatives could account for four of the remaining six unidentified minor adducts. We are currently engaged in further studies on the identities of these compounds.

DNA ADDUCTS AND ORGANOTROPISM IN THE RAT

Following identification of the major DNA adduct formed in the livers of rats injected with AFB_1 and with the availability of sensitive analytical methodology to detect and quantify minor DNA-bound AFB_1 derivatives, we have applied this experimental approach in evaluating possible relationships between qualitative and quantitative features of adduct formation and organotropic effects of AFB_1 in rats.

Among the noteworthy features of AFB_1 carcinogenesis in rats is the high degree of specificity with which it induces tumors and toxic lesions in the liver. Tissue specificity is not absolute, since the carcinogen induces kidney tumors in nearly the same incidence as liver tumors in one rat strain. Nonetheless, we have observed neither tumors nor other treatment-related histopathologic changes in kidneys of large numbers of Fischer rats treated with AFB_1 over a period of many years. Therefore, it seemed justifiable to compare liver and kidneys as representative of target and non-target tissues in this rat strain.

^3H-AFB_1 was injected into male Fischer rats at a dose of 1 mg/kg body weight, and the animals were killed two hours later. Nuclei were isolated from perfused livers and kidneys, and nucleic acids were in turn isolated from the nuclei and total DNA-bound radioactivity was determined. DNA was heat denatured and enzymatically hydrolyzed with nuclease P_1. The hydrolysate was analyzed by high pressure liquid chromatography (HPLC) and distribution of radioactivity in components of the chromatogram determined. The chromatogram for liver DNA is shown in Figure 1, and that for kidney was qualitatively very similar.

Levels of individual adducts present in the kidney DNA were only about 10% of their levels in liver, as indicated by the maximal concentration of AFB_1-N^7-Gua, i.e., 120 μmole/mg DNA x 10^{-7}. Six of the ten minor peaks identified in the liver (E,F,G,H,I, and diol) have been quantified in the kidney. These products represent a higher proportion of the adducts in the kidney (30%) than the liver (20%) with peak H increasing in the largest amount. Thus, it appears that virtually the same spectrum of DNA adducts is produced from AFB_1 in the non-target (kidney) as in the target tissue (liver). However, the total level of binding, and therefore of each of the adducts, was greatly reduced in the non-target tissue.

SPECIES SUSCEPTIBILITY AND AFB_1-DNA ADDUCT PATTERNS

A similar approach was used for investigating DNA adducts formed in tissues of the Swiss mouse, a species with a high level of resistance to the hepatocarcinogenic effects of AFB_1 (Wogan, 1973).

^3H-AFB_1 was injected into adult male Swiss mice at a dose of 12 mg/kg of body weight. As in the rat experiment described above, DNA was isolated from liver and kidneys, hydrolyzed and subjected to HPLC analysis.

The chromatogram of liver DNA in its main features was qualitatively similar to that of rat liver, in that the major components (AFB_1-N^7-Gua and peaks F, G, and H) were prominently represented. The most striking feature of the adduct pattern, however, was the very low level at which even the major adducts are present. The major component, AFB_1-N^7-Gua, has a concentration of only about 12 μmoles/mg DNA x 10^{-7} in the order of 0.1% of the level of the same adduct in rat liver DNA. The other lesser adducts were proportionately reduced. These results indicate that the mouse liver activates AFB_1 through the same pathways as rat liver and either has a very limited capacity for activation or possesses a very efficient pathway for removal of the activated derivative.

The adduct pattern seen in mouse kidney was also very similar in character to that of rat kidney, differing principally in the absence of several of the minor adduct peaks. However, the concentration of the major adducts was significantly higher in mouse kidney than in mouse liver. The major adduct attained a concentration approaching 80 μmoles/mg DNA x 10^{-7}, of comparable magnitude to the level of the same component in rat kidney (120 μmoles/mg DNA x 10^{-7}).

The main quantitative features of these experiments relating to tissue specificity and species responsiveness can be summarized as follows: in terms of total AFB_1 binding levels in DNA, the greatest difference occurred in adduct levels in rat and mouse liver. Rat liver DNA contained adducts at a total level of 1 AFB_1 residue per 8,000 nucleotides whereas the comparable value for mouse liver was 1 per 440,000. DNA from kidneys of the two species contained adducts at more similar concentrations, 1 AFB_1 per 80,000 nucleotides in the rat, and 1 per 140,000 in the mouse.

When these total binding levels are fractionated to take into account the levels of individual adducts in accordance with the HPLC data, the range of levels for individual components is very large, as illustrated in the data in Table 1.

TABLE 1. Quantitative Relationships Among Acid Hydrolysis Products of DNA Modified by AFB_1 *in vivo*

	Nucleotides/AFB_1 Residue x 10^3			
	Liver[a]		Kidney[b]	
HPLC Peak	Rat	Mouse	Rat	Mouse
A	4,900	--	--	--
B	5,600	--	--	--
C	4,400	17,000	--	--
D	2,400	38,000	--	--
E	430	35,000	5,300	
F	270	10,000	4,500	3,400
G	130	4,600	1,600	1,500
H,I	150	2,900	430	--
Diol	620	4,800	4,400	13,000
AFB_1-N^7-Gua	11	950	110	150

[a] Mean of 3 determinations. [b] Mean of 2 determinations.

The extremes of the range are indicated by the maximum concentration of AFB_1-N^7-Gua in the rat liver (1 adduct per 11,000 nucleotides) and the maximum level of peak D in mouse liver (1 AFB_1 residue per 38 x 10^6 nucleotides). At this time, evidence is insufficient to postulate whether any of these adducts have greater functional significance than others.

In summarizing these observations relating adduct formation to tissue specificity and species difference in response, differential ability to activate AFB_1 to its ultimate DNA binding form, the 2,3-epoxide, may play a role in determining the known tissue specificity for toxicity and carcinogenicity in the Fischer rat. Liver, the target tissue, and kidney, the non-target tissue, produce qualitatively similar patterns of DNA adducts, although there is a difference in total binding level in the two tissues of about 10-fold.

A comparable conclusion is supported by the experiments in mice. The great resistance of the liver to the toxic and carcinogenic effects of AFB_1 cannot be attributable to a total lack of metabolic competence for AFB_1 activation, since the spectrum of DNA adducts produced in mouse liver and kidney is generally similar to that present in rat tissues. However, in this case, the differential in total binding levels is very large (>100-fold) compared to the rat.

Thus, while the qualitative pattern of DNA adducts is not revealing with respect to mechanisms responsible for tissue susceptibility in either species, total binding level is strongly correlated with response suggesting that specific adducts levels may be determinants.

KINETICS OF DNA ADDUCT FORMATION IN RAT LIVER FOLLOWING A SINGLE AFB_1 DOSE

With the availability of sensitive analytical methodology to detect and quantify minor DNA-bound AFB_1 derivatives, it was of interest to apply this experimental approach in studying relationships between the kinetics of adduct formation and removal in rat liver under conditions of dosing which produce well characterized biological or biochemical effects in this tissue. Initial experiments involved a time-course study of the population of DNA adducts present 2 to 72 hours after a single sublethal dose of AFB_1 (0.6 mg/kg). The earliest sampling point (two hours after dosing) was selected on the basis of numerous previous investigations indicating that many biochemical changes induced by AFB_1 are maximal at that time.

^3H-AFB_1 was injected into male Fischer rats and the animals were killed 2 hours later. Nuclei were isolated from perfused livers, and nucleic acids were in turn isolated from the nuclei and total DNA-bound radioactivity was determined. DNA was heat denatured and enzymatically hydrolyzed with nuclease P_1. The hydrolysate was analyzed by high pressure liquid chromatography (HPLC) and distribution of radioactivity in components of the chromatogram determined.

In quantitative terms, the total level of binding was one AFB_1 residue per 11,000 nucleotides, and AFB_1-N^7-Gua, the major adduct, was present at a level of one adduct per 14,000 nucleotides. The next most numerous adducts were peaks G (1 per 160,000) and F (1 per 350,000). Peak B was the least abundant, with a frequency of one adduct per 1.3 x 10^8 nucleotides, a level which represents approximately the limit of sensitivity of the approach under conditions applied in these experiments.

Similar analysis of liver DNA of rats killed after 4, 12, 24, 48 and 72 hours after dosing revealed marked alterations in adduct levels, as shown in Figure 2.

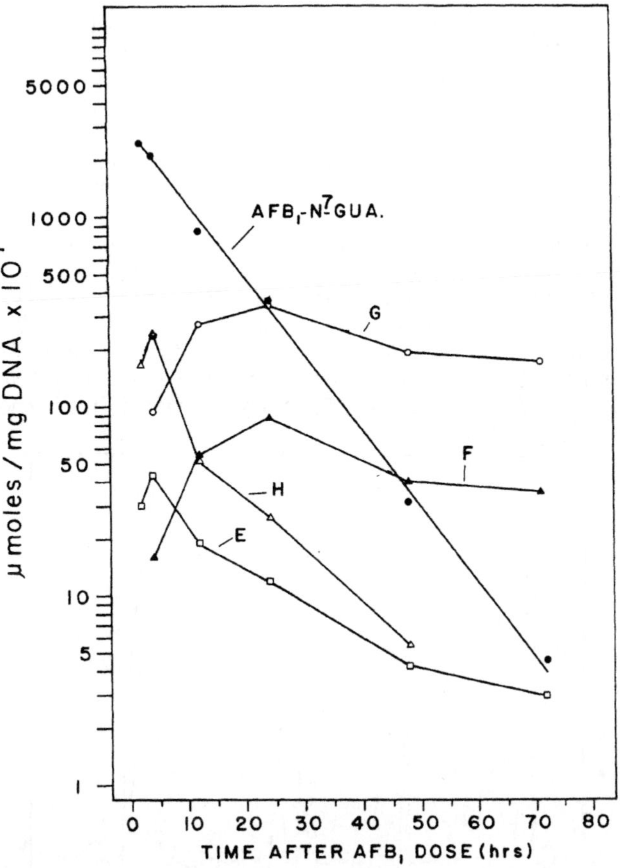

Figure 2. DNA adduct levels in rat liver following a single dose of AFB_1.

AFB_1-N^7-Gua levels declined linearly from an initial value of about 22,000 μmoles/mg DNA x 10^{-8} at 2 hours to 26 at 72 hours; these levels represent more than 90% and less than 1% of total bound radioactivity, respectively. Peaks H (the putative AFP_1-N^7-Gua) and E (AFM_1-N^7-Gua) declined with similar kinetics. In contrast, peaks G (the putative ring-opened derivative of AFB_1-N^7-Gua) and the related F remained at relatively constant levels after an initial increase during the 4 to 12 hour period. Thus, the greater stability predicted for the ring-opened form of the N^7-Gua adduct is substantiated by these observations.

RAT LIVER ADDUCT PATTERNS FOLLOWING REPEATED AFB_1 DOSING

A series of experiments was conducted in order to evaluate further the possible relationship of these observations to mechanisms of AFB_1 carcinogenesis in rat liver. It is well established that a single AFB_1 dose is not an efficient carcinogenic regimen so the relevance of the findings discussed above to initiating events in carcinogenesis is uncertain. However, a dosing regimen of small repeated doses can effectively induce hepatocellular carcinoma in every animal treated with appropriate total dose levels. We therefore conducted an investigation of DNA adducts in livers of rats dosed according to a schedule known to be an effective carcinogenic regimen.

During a 14 day period, 25 μg of ^3H-AFB_1 (0.125 mg/kg) was administered i.p. in 25 μl of DMSO to rats on days 0,1,2,3,4,7,8,9,10 and 11. Two rats were sacrificed each day and the DNA of their livers isolated, hydrolyzed, and analyzed chromatographically. With the exception of day 0, animals were not administered AFB_1 on the day of sacrifice.

Figure 3 shows the chromatographic profile of AFB_1-DNA hydrolysis products in rat liver, two hours after the injection of the initial 25 μg dose of AFB_1 (day 0).

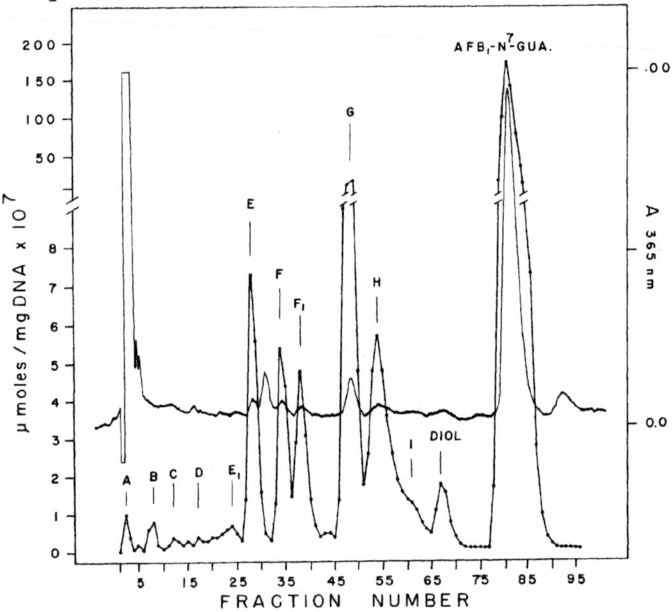

Figure 3. HPLC chromatogram of AFB_1-DNA adducts 24 hours after a single dose of 0.125 mg AFB_1/kg body weight. Upper curve A_{365}; lower [^3H].

The products identified in this figure are qualitatively identical to those isolated from the liver DNA of rats which were treated with 0.6 mg/kg (Figure 1), with the exception of peak F_1. The identity of this peak and reasons for its appearance in this experiment but not earlier ones are unknown. In quantitative terms, the levels of each peak range from 1.9×10^{-5} modifications/base (50,000 bases/modification) for AFB_1-N^7-Gua to 4.0×10^{-9} modifications/base (25×10^{-7} bases/modification), for peak D which represents less than 0.1% of the covalently bound material. After the two hour period, 7.3×10^{-5} moles of AFB_1 were covalently bound per milligram of DNA. Assuming 12 mg of DNA per liver, a total of 8.8 $\times10^{-4}$ moles of AFB_1 was covalently associated with liver DNA 2 hours after dosing. This is approximately 1% of the administered dose of 8.0 $\times10^{-2}$ moles of AFB_1.

The kinetics of removal of this material during the first 24 hour period were similar to those described previously for a 0.6 mg/kg body weight dose. After 24 hours, 88% of the covalently bound AFB_1 had been removed from DNA. The remaining 12% was found primarily in 3 peaks, F, G and AFB_1-N^7-Gua, containing 11, 51 and 34% respectively of this residual material. The remaining 4% was distributed among 7 other peaks representing 1% or less each. During the two week period the three major peaks account for greater than 90% of the AFB_1 derivatives hydrolyzed from DNA.

Figure 4 shows the level of G, F and AFB_1-N^7-Gua hydrolyzed from rat liver DNA during the two week period. Peak E is included in this figure to be representative of the relative levels of the minor AFB_1 hydrolysis products in relation to the three major adducts Peak G was present at the highest concentration in DNA and attained a relatively constant level of 1.8×10^{-5} μmoles/mg DNA at the end of the first 5 day dosing period. Its level did not significantly change with cessation of dosing on days 5 and 6, or during the second dosing period, days 7 through 11. A similar behavior was seen with peak F at a lower level of modification.

The amount of AFB_1-N^7-Gua remaining 24 hours after the AFB_1 administration showed a gradual decline during the first 5 day period. On day 6, 48 hours after cessation of dosing, a precipitous decline in the residual level was seen, which was further apparent on day 7, 72 hours after the last AFB_1 dose. Resumption of treatment on day 7, for a second 5 day period, resulted in an increase on day 8 to approximately 10 times the level seen on day 7. This level remained constant during this period until day 13, when dosing was again stopped and a second decline in the residual amount of AFB_1-N^7-Gua was apparent.

The level of peak E and other minor products, B, C, D, E, I, and diol, remained relatively constant throughout the entire period.

Figure 4 shows that most of the increase in the levels of the persistent derivatives, F and G, occurred within the first 5 day period. Relatively constant levels of these products were maintained over the remaining 9 days. A possible explanation for this pattern is derived from the examination of the kinetics of the precursor, AFB_1-N^7-Gua. During the first 5 day period a decrease in the amount of this product remaining after 24 hours is seen. In the subsequent 5 day period of AFB_1 administration, the residual level is constant, at a lower value than at any time during the first dosing period.

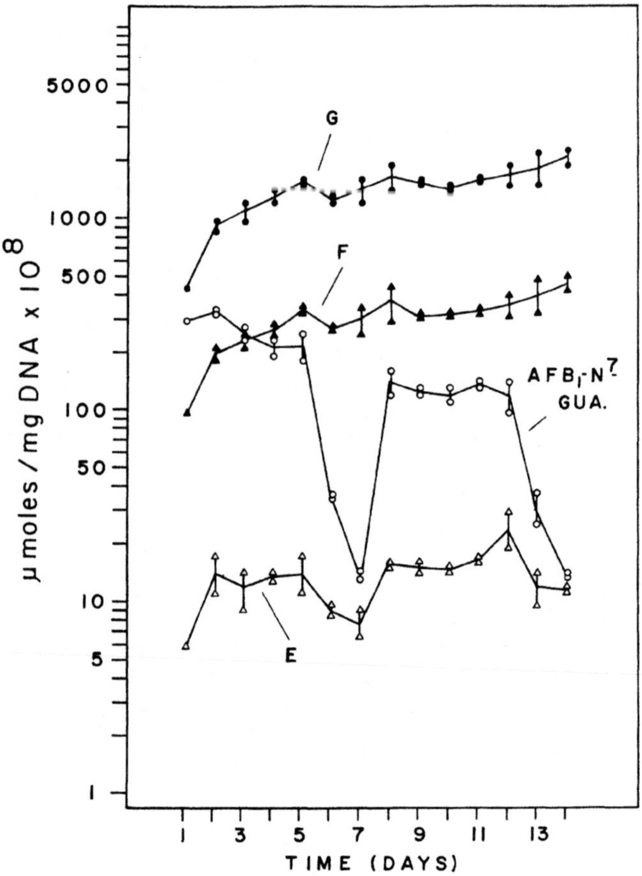

Figure 4. DNA adduct levels in liver of rats repeatedly dosed with AFB_1 on days 0-4 and 7-11. Animals were killed 24 hours after dosing.

Two processes could provide an explanation for these observations: an increased rate of removal of AFB_1-N^7-Gua from DNA; or a decreased level of modification of DNA by succeeding doses of AFB_1. Either of these mechanisms would reduce the residual level of AFB_1-N^7-Gua and the amounts of F and G produced in DNA by a given dose of AFB_1. Increased repair seems the less probable explanation since the half life of AFB_1-N^7-Gua calculated for the 24 hour period between days 5 and 6 is greater than its half-life during the first 24 hour period on day 0 (5.5 and 9.0 hours, respectively). The induction of other metabolic pathways which would limit AFB_1 activation or inactivate the 2,3-epoxide provides a more likely explanation, since these mechanisms would decrease the initial level of DNA modification. Induction of P_{450} mixed function oxidases by AFB_1 has been reported, and this phenomenon is well documented for other xenobiotics.

EXCRETION OF THE AFB_1-N^7-GUA ADDUCT IN URINE OF RATS

In view of the kinetics of removal of the major DNA adduct from rat liver DNA, we have explored the possibility that measurement of its excretion could provide the basis for a non-invasive method for quantitative measurement of AFB_1 body-burden as well as information on metabolic activation and repair capability. A suitably sensitive and quantitative method would have potential application for monitoring of human populations for aflatoxin exposure.

A method for isolation of AFB_1-N^7-Gua from urine was developed and subsequently applied to monitoring the excretion of this adduct from AFB_1-treated rats. Evaluation of the method for recovering AFB_1-N^7-Gua from spiked urine produced the data showing that recoveries ranged from a low of 69% at 25 ng adduct/sample to essentially quantitative at 200 ng adduct/sample. Analysis of variance of the data indicated that the recovery of adduct from these samples was linear for all but the smallest amount of adduct added (25 μg), where recovery deviated significantly from linearity. As a result, the method was shown to be applicable for accurately quantitating adduct in samples containing 50 ng adduct or more.

Preliminary investigations of the possible urinary excretion of adduct by rats injected with $[^{14}C]AFB_1$ indicated that a labeled compound chromatographically identical to authentic AFB_1-N^7-Gua was excreted by these animals (Figure 5). This compound was subsequently isolated from the urine of 15 rats injected i.p. with 1 mg/kg AFB_1 to give an estimated total of 3.6 μg of adduct. The UV spectra of the urinary compound in both acid and base are similar to those of the authentic compound for wavelengths greater than 300 nm, suggesting that the urinary compound contains aflatoxin. This observation is in accordance with the appearance of radiolabel in this compound following the injection of labeled AFB_1. The lack of bathochromic shift in base indicates that the putative urinary adduct is not a 2-hydroxy-aflatoxin derivative.

The presence of guanine in the putative urinary adduct was demonstrated by methylation of the adduct and subsequent perchloric acid hydrolysis of the methylated product. The methylation was performed under conditions which afford selective methylation of the imidazole nitrogen atoms of guanine, and the only detectable methylated guanine derivative formed from either authentic or putative urinary adduct was 9-methylguanine. This result is consistent with the assumption that the urinary compound contained guanine and that the quanine was substituted by the aflatoxin moiety at the 7-position at the time of methylation.

Taken together, the results of the methylation and the spectral analyses argue strongly that the urinary compound contains an AFB_1 derivative covalently bound to the N^7 atom of guanine. Thus, the urinary compound is identical to authentic AFB_1-N^7-Gua by these criteria.

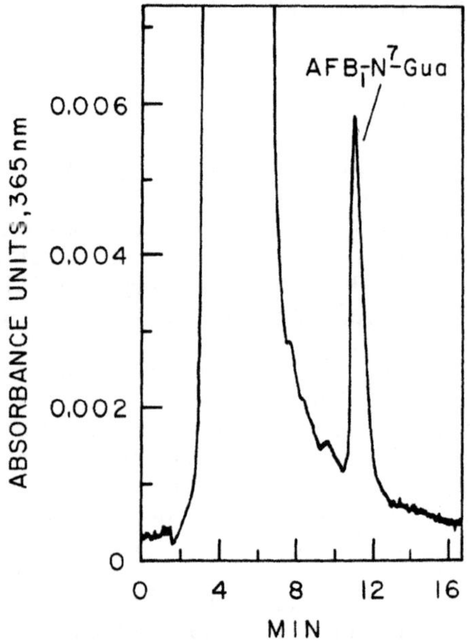

Figure 5. Chromatogram showing the final purification of AFB_1-N^7-Gua from the urine of a rat injected with AFB_1.

Determination of the amounts of adduct excreted by rats at various times following i.p. AFB_1 administration indicated that at the 1 mg/kg dose, 80% pf the total excretion of adduct occurred during a 58 hr period after dosing. The balance of adduct excretion occurred between 48 and 144 hr. In the case of the 0.25 mg/kg dose, no detectable levels (i.e., less than 5 ng) of adduct were observed at times later than 48 hr. Investigation of adduct excretion at shorter times indicated that the excretion of this compound occurs to an equal extent in the first and second days following AFB_1 administration.

Because the excretion of adduct was nearly complete in 48 hr, the amounts of adduct excreted by this time at various dose levels were used to generate the dose-response curve shown in Figure 6. These preliminary data indicate that an accurate dose-excretion plot could be constructed through which level of exposure could be estimated reliably.

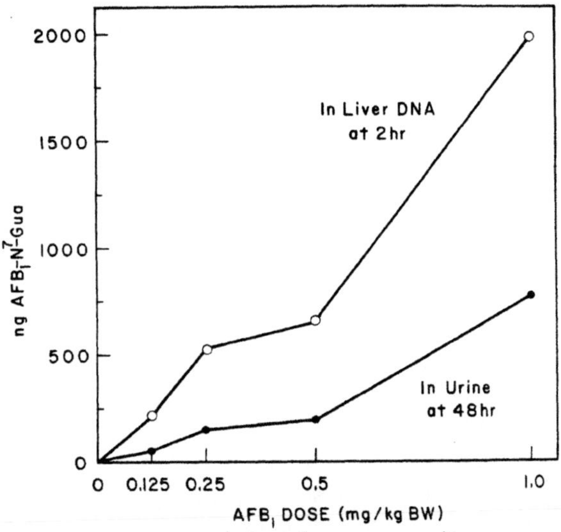

Figure 6. Dose-response relationships in 48-hour urinary excretion of AFB_1-N^7-Gua, and the corresponding levels of this adduct in liver DNA 2 hours after dosing of rats.

It is of interest that the amount of adduct excreted represented a relatively constant fraction (30-40%) of the maximum amount of adduct present in liver DNA. Thus, it is possible under the conditions used here to apply the method for estimation of AFB_1 body burden. It is doubtful, however, that the method would be applicable to human population studies without modification to significantly increase its sensitivity. We are currently exploring several alternative approaches to attain the necessary levels of sensitivity.

REFERENCES

Croy, R.G., Essigmann, J.M., Reinhold, V.N., and Wogan, G.N.: 1978, Proc. Natl. Acad. Sci. USA 75, pp. 1745-1749.

Essigmann, J.M., Croy, R.G., Nadzan, A.M., Busby, W.F., Jr., Reinhold, V.N., Buchi, G., and Wogan, G.N.: 1977, Proc. Natl. Acad. Sci. USA 74, pp. 1870-1874.

Lin, J.K., Miller, J.A., and Miller, E.C.: 1977, Cancer Res. 37, pp. 4430-4438.

Swenson, D.H., Miller, E.C., and Miller, J.A.: 1977, Cancer Res. 37, pp. 172-181.

Swenson, D.H., Miller, J.A., and Miller, E.C.: 1975, Cancer Res. 35, pp. 3811-3823.

Wogan, G.N.: 1973, IN: Methods in Cancer Research, Vol. VII, H. Busch, editor, Academic Press, New York, pp. 309-344.

STRUCTURAL MODIFICATIONS AND SPECIFIC RECOGNITION BY ANTIBODIES OF CARCINOGENIC AROMATIC AMINES BOUND TO DNA

Daune M.P., de Murcia G., Freund A.M., Fuchs R.P.P., Lang M.C., Leng M.* and Sage E.*
Department of Biophysics, Institut de Biologie Moléculaire et Cellulaire, 15 rue Descartes 67084 Strasbourg cédex, France
*Centre de Biophysique Moléculaire, 45045 Orléans cédex, France

ABSTRACT

Local structural modification of the native DNA molecule, induced by the covalent binding to the C-8 of guanine residue of acetylaminofluorene (nDNA-AAF), was studied previously by means of different physical and biochemical techniques. We present here the results obtained by using two other methods :
1) The local unwinding can be determined from the number of bound carcinogens necessary to transform the supercoiled circular form of DNA to the relaxed one. In the case of nDNA-AAF it amounts to about -20 degrees.
2) Detection and mapping of bound carcinogens can be realized by means of antibodies able to recognize specifically the structural features of the modified regions. With chromatin subunits (core particle and trinucleosomes) the mapping reveals the peculiar accessibility of the linker DNA.
This latter method appears to be applicable to other types of carcinogens and able, after some improvements, to solve several problems related to the initiation step of carcinogenesis.

INTRODUCTION

The "in vivo" modification of DNA by the hepatic carcinogen N-acetylaminofluorene (AAF) leads to the formation of a major amount of deacetylated adduct, the structure of which is still not completely elucidated (1) and of two acetylated adducts. These latter adducts are obtained when DNA is reacted "in vitro" with N-acetoxy-N-2-acetylaminofluorene (N-Aco-AAF). The first one which results from the covalent binding to the C-8 of guanine residue, is N-(2'-deoxyguanosin-8-yl)-2(acetylamino)fluorene (dGuo-C-8-AAF) and accounts for more than 80% of the acetylated adducts. The second one, which results from the binding to the amino group of G is 3-(2'-deoxyguanosin-N^2-yl)-2(acetylamino)fluorene (dGuo-N^2-AAF).

In order to determine the structural changes of "in vitro" modified DNA we have developped a series of experimental approaches. The situation

B. Pullman, P. O. P. Ts'o and H. Gelboin (eds.), Carcinogenesis: Fundamental Mechanisms and Environmental Effects, 193–205.

was rather favourable because of the presence of the large amount of C-8 adduct, which was practically the only one to induce all the observed changes. Evidences for a local desorganization of the double helix was given by melting profiles (2),kinetic studies of formaldehyde unwinding (3) and endonuclease S_1 digestion (4). The position of the fluorene ring relative to the helix axis as well as to the vicinal nucleotides was determined from linear dichroism measurements on electric field oriented DNA (5) and from circular dichroism spectra of model deoxytriribonucleotides (6). A model view of the locally modified DNA (as given in figure 1) illustrates the different features of the so-called "insertion denaturation" (2) or "base displacement" (7) model :

a) the partial insertion of the fluorene ring in place of guanine which is pushed outside,

b) the rupture of hydrogen bonds between G and C followed by a partial rotation of the cytosine toward the outside of the helix,

c) a local desorganization of the double-stranded helix which can extend on each side of the modified guanine to a number of base pairs depending strongly upon the temperature (3).

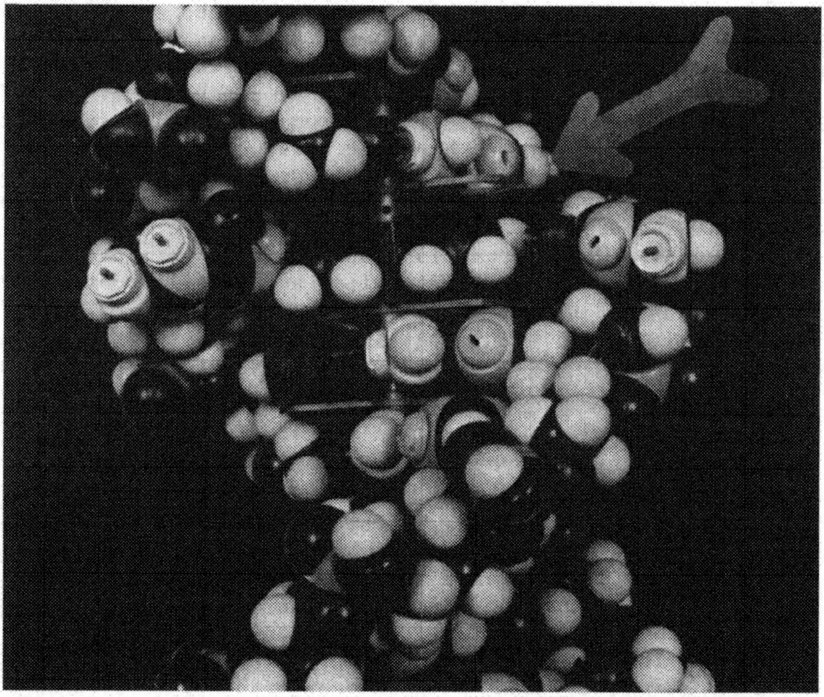

Fig. 1 - CPK building of the insertion-denaturation model. The arrow indicates the place of the fluorene ring bound to the C-8 of G. Guanine (left) and cytosine (right) have been rotated toward the outside.

On the opposite and because of unsuccessful attempts to obtain a major amount of N^2 adducts by modifying the conditions of the reaction, nothing is known about its local geometry, which is likely to be different from that of the C-8 adduct. The corresponding regions of modified DNA are indeed not recognized by Neurospora Crassa endonuclease, an enzyme which digests specifically single-stranded DNA (8). On the basis of molecular models (CPK models) one can only assume that the fluorene ring is lying in the small groove as illustrated in figure 2.

Fig.2 - CPK building of the assumed conformation of AAF bound to N^2 of G. As indicated by the arrow, the fluorene ring is lying in the small groove.

In an attempt to obtain more precise indication on the C-8 adduct we have developped recently two new different structural approaches : the determination of the tertiary structure of a covalently closed circular DNA, and the use of specific antibodies.

CIRCULAR DNA AND CONFORMATIONAL CHANGE

The supercoiled closed circular DNA has been used as a sensitive tool to detect and measure the binding of intercalative drugs and recently to detect the structural changes induced by the covalent binding of a carcinogen (9-10). According to Crick's notation (11), the linking number Lk, the writhing number Wr and the twisting number Tw are related by :

$$Wr = Lk - Tw$$

If Lk is assumed to remain constant for a given DNA molecule, then any change of the angle between two adjacent base-pairs will modify Tw and thus Wr. For example to a local unwinding ($\Delta Tw < 0$) will correspond an increase of Wr ($\Delta Wr > 0$). Since the initial value of Wr is negative, local unwinding will progressively lead to the relaxed circular form. As pointed out recently (12), the assumption of a constant value Lk is only true in the limit of an infinite shear modulus/Young modulus ratio. Practically this is never the case and the observed number c of constraint turns (12) is not equal to Wr.

With DNA extracted from plasmids like Col E_1 or pBr 322, the initial negative value of c is found to increase with the amount of bound carcinogen as followed by electron microscopy or electrophoresis in agarose gel (13). The process reflects a corresponding local unwinding of the DNA molecule and after a convenient scaling of gel electrophoresis, a linear relationship is obtained between c and the percentage of modified bases (figure 3), leading to an unwinding angle of about -20° per C-8 adduct.

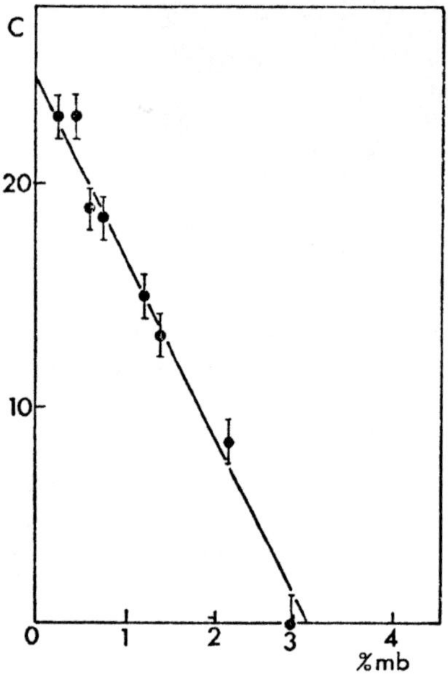

Fig. 3 - Linear variation of the number c of constraint turns of supercoiled Col E_1 DNA with the percentage of modified bases.

Such a structural change is likely to be correlated with trans-syn trans-conformation of guanine which was assumed in model ribodinucleotides (14). We have recently postulated that the conformational change of G is taking place during the process of nucleophilic substitution (15). In other words, the formation of a nitrenium ion during the approach of the electrophilic N-Aco-AAF and the rotation of the guanine moiety around the glycosydic bond must be considered as concerted events in the path of the nucleophilic substitution. A recent X-ray study of a minihelix of an alternating oligo d(G.C) (16) reveals the existence of a new double-stranded structure, called Z-form, in which guanine takes a syn conformation, leading to a better accessibility of C-8 than in the classical B-form. It can be shown indeed (17) that the AAF-modified poly d(GC). poly d(GC) is able to change its structure from a B to a Z form in presence of a lower amount of ethanol than the unmodified polynucleotide.

RECOGNITION BY SPECIFIC ANTIBODIES

Another manner to detect the local structural changes of modified DNA, was offered by using different classes of antibodies. In this paper we want to stress the importance of this new tool in the study of carcinogen-modified DNA. The following antibodies were successively used :

a) anti-C antibodies which are available for many years (18),

b) anti-Guo - AAF antibodies obtained from the bovine serum albumin-Guo - AAF conjugate as the antigen (19,20),

c) anti-nDNA - AAF antibodies elicited in rabbits immunized with native DNA substituted by AAF (21).

They provide a direct determination under the electron microscope of the number of modified regions in a given DNA. Moreover they can be used for the mapping of carcinogens bound to DNA.

A. Anticytosine (anti C) Antibodies

These antibodies were shown to react with denatured DNA but not with native DNA (18). The formation of a complex between antibodies and bases in a polynucleotide is possible only if the bases are not engaged in a base pair.

The reactivity of a modified DNA with antibodies directed against cytosine was determined by radioimmunoassay : one measures the inhibition of the precipitation of denatured $|^3H|$DNA by anti-C antibodies as a function of the amount of modified DNA which is added to the mixture as a competitor (22). The concentrations of inhibitor are expressed in moles of cytosine for denatured DNA and in moles of AAF-modified bases for nDNA-AAF. Since we expect to have one cytosine molecule accessible per each Guo-C-8-AAF adduct, this is the only manner to get a direct comparison of the number of accessible cytosine in both cases. In figure 4 the percentage of inhibition is plotted in function of the logarithm of inhibitor concentration. As shown in this figure, DNA-AAF is as good a competitor as denatured DNA itself ; we therefore conclude that a cytosine residue is accessible (i.e. non hydrogen bounded) for each AAF modified base.

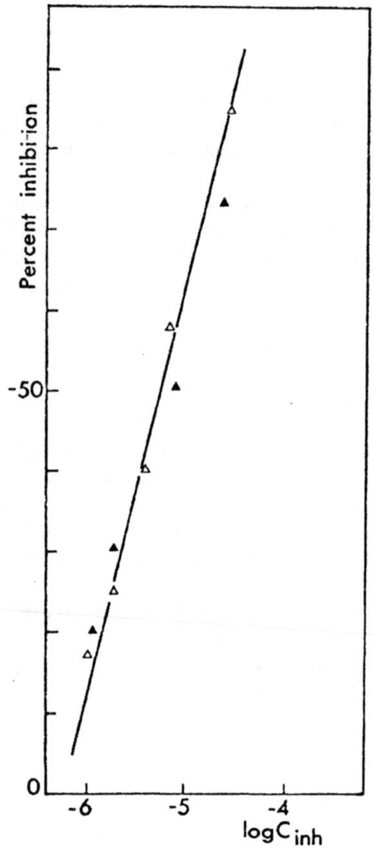

Fig. 4 - Inhibition of ^{3}H dDNA precipitation by anti-C antibodies.
Δ : competition with dDNA
▲ : competition with nDNA-AAF (5.5% modified bases)

B. Anti-modified-DNA Antibodies

1. Preparation - Native calf thymus DNA (nDNA) and corresponding denatured DNA (dDNA) were modified by the reaction with N-Aco-AAF according to a procedure previously described (5). After immunization of rabbits with nDNA-AAF, they were bled a week after the intravenous booster. The serum was applied on a Sepharose-dDNA-AAF column equilibrated with 0.15 M NaCl, 5 mM Tris-HCl pH 7.5, 0.1 mM EDTA (buffer I). The antibodies retained on the affinity column were eluted with 2 M acetic acid, neutralized and then dialysed against buffer I. They reacted with nDNA-AAF but also with dDNA. In order to get rid of antibodies directed against dDNA, a second purification step was made on a Sepharose-dDNA column on which about 15% of the proteins were retained. Unbound proteins were further purified on a Sephadex G-20 column and eluted in a

single peak. The interactions between these antibodies and several ligands were studied by radioimmunoassay. The affinity toward GMP-AAF, dDNA-AAF and nDNA-AAF was found to be identical. GMP, N-OH-AAF as well as nDNA and dDNA do not interact. The binding of antibodies to the modified monomer dGMP-AAF or to modified DNAs (dDNA-AAF and nDNA-AAF) induces the same perturbation of the optical properties of the fluorene ring, as revealed by U.V. spectroscopy and circular dichroism (22). Since the binding constant of the antibodies to the three antigens was the same, one can consider dGMP-AAF as the immuno-determinant group in agreement with the model of figure 1.

2. Visualization of modified regions of DNA - These antibodies as well as those obtained by using Guo-C-8-AAF coupled with bovine serum albumin (19,20) can be used to detect the modified regions of DNA under the electron microscope. In order to work with an homogenous population of DNA molecules, assays were made with Col E_1 DNA converted to linear molecules (form III) with the restriction enzyme Eco RI. This DNA was reacted with N-Aco-$|^{14}C|$AAF (50 mC/mM) to different extents ranging from 0 to 15 adducts per 10^4 nucleotides. It was then necessary, in a series of preliminary experiments to determine the optimal conditions for the isolation of the complex and a suitable method of spreading. The ionic strength was determined in such a way as to avoid any unspecific binding of antibodies to DNA (occurring in low ionic strength) and to keep a value as high as possible of the binding constant (which is a decreasing function of the ionic strength). The best value appear to be between 50 and 100 mM NaCl. For the antibody/antigen ratio, a similar balance has to be found. The ratio must be high enough to get a maximum number of antibodies bound to DNA, but low enough to avoid agregation phenomena due mainly to the bivalence of IgG and the related formation of intermolecular cross links. A ratio close to 10 was found to be the optimal value. After 30 min incubation at 20°C and pH 7.5, the excess antibodies were removed by gel filtration on a Sepharose 4B column. Then DNA molecules complexed with IgG were spread on positively charged carbon-coated grids (23). Pictures obtained by using the Siemens 101 Elmiskop are given in figure 5. The dark dots with an average diameter of 90 Å can be identified as bound IgG in a second labeling experiment in which ferritine labeled anti rabbit IgG are used (24). By using a Philips EM 301 electron microscope equipped with a video system[1], it was possible to count the number of bound IgG per DNA molecule and to determine an average value on a sample of about 120 molecules. When these average values are plotted against the number of modified guanine residues, a straight line was obtained the slope of which is equal to 0.4 (24). This low value is explained partly by the bivalence of the IgG which could recognize two antigens (and indeed several cross links are observed in the pictures), but also by the value of the binding constant leading to a partial dissociation of DNA-bound IgG during the chromatographic separation.

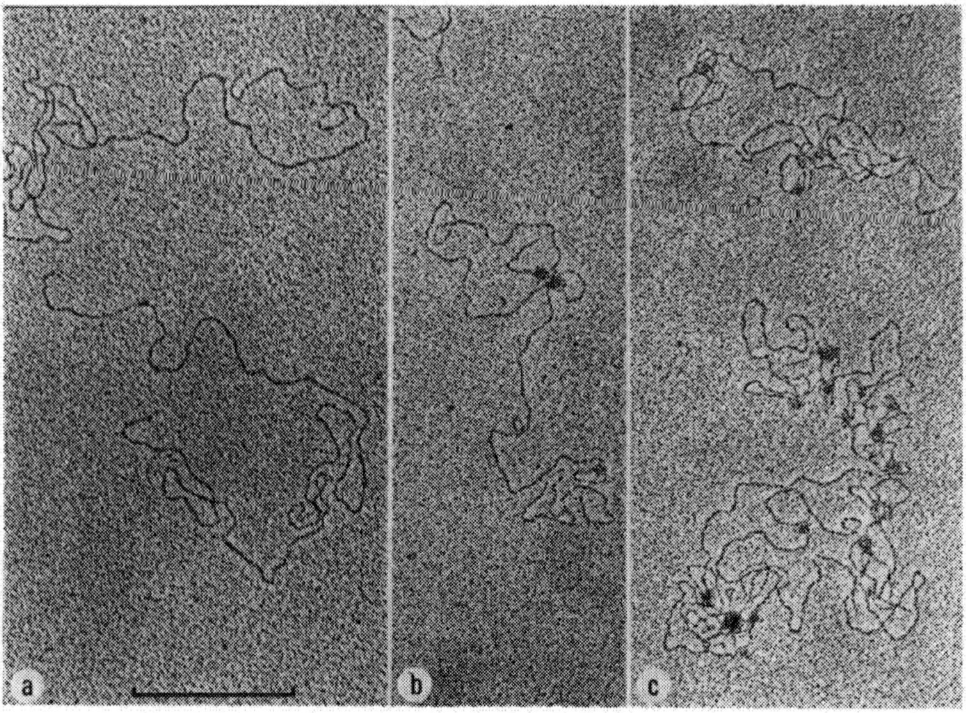

Fig. 5 - Electron micrographs of IgG binding to Col E_1 DNA form III. The bar represents 2000 Å. a) unmodified DNA b) 0.07% modified bases c) 0.15% modified bases (see text for details).

3. DNA mapping inside chromatin - In vivo binding of carcinogen has to take place on DNA inside chromatin and it was therefore interesting to use the antibodies as a probe of the accessibility of the DNA which depends on the structure of chromatin (25).

The first material to be used was the core particle. This subunit contains a stoechiometric amount of the four histones H2A, H2B, H3 and H4 assembled in an octamer and a DNA of 145 ± 3 bp. It was extracted from erythrocyte chromatin as previously described (26). Core particles were incubated with N-Aco-^{14}C-AAF at 37°C during 3 hours in a buffer containing 0.2 m EDTA, 4 mM sodium citrate, 10^{-4} M PMSF pH 7 and 10% ethanol. Unreacted carcinogen was removed by repeated extractions with ethylether. Following the incubation step, the structural integrity of the core particle was checked by circular dichroism and electron microscopy (figure 6a). The material was then deproteinized and DNA precipitated in ethanol. After redissolution, formation and isolation of the complex and spreading were made as indicated above. The percentage of modified bases (0.8%) was chosen in order to obtain an average of 2 sites per core particle. From inspection of fig. 6c and 6d, it is clear that only the extremities of the 145 bp DNA are labeled with IgG. Some cases of cross linking are visible. In the histogram (fig. 6e) is represented the distance between either a free end of DNA and the distal IgG, or two IgG.

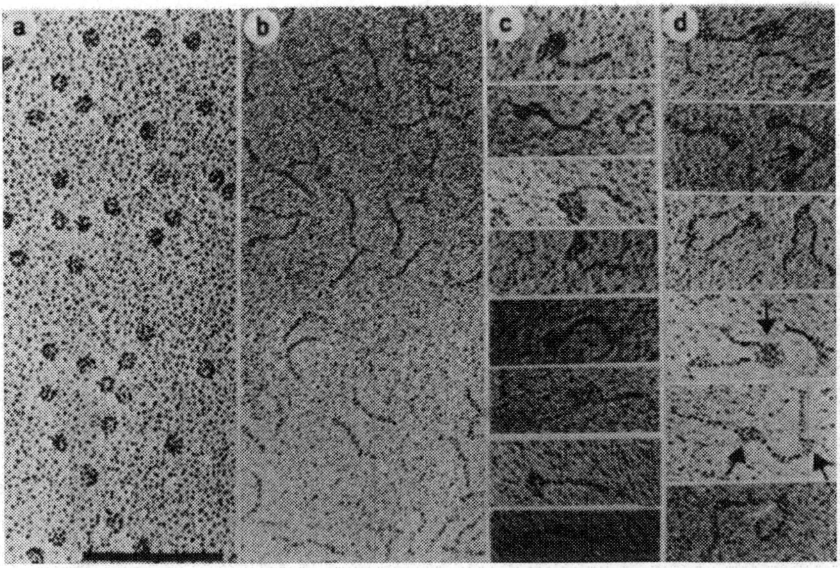

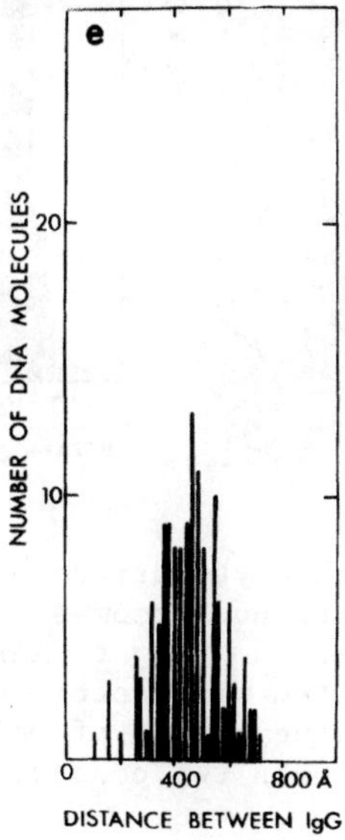

Fig. 6 - Electron micrographs of chicken erythrocyte core particles having 0.8% modified bases.
a) AAF-modified core particles
b) 145 bp AAF-DNA extracted from modified core particles
c) and d) 145 bp DNA extracted from modified core particles (see details in text) and reacted with specific IgG. Cross links between core particles are indicated by arrows in fig. 6d. The bar represents 1000 Å.
e) Histogram of the number of molecules in function of the distance between either a free end of DNA and the distal IgG or two IgG.

The distribution is centered on a length close to that corresponding to the DNA (493 Å). The width of the distribution reflects essentially the experimental error in the determination of such small lengths.

The same type of experiment was made with trinucleosomes which were isolated on a sucrose gradient after a partial digestion of erythrocyte chromatin. In these conditions, H1 and H5 histones are still present and the DNA has an average length of about 630 bp since in each nucleosome one finds a linker DNA of about 65 bp and the core DNA of 145 bp. The percentage of modified bases was equal to 0.36% which corresponds to an average number of 4 to 5 bound carcinogen molecules per trinucleosome. In figure 7 the distribution of IgG along the DNA extracted from trinucleosomes is apparently not random and indeed the histogram (figure 7d) indicates an average distance between two consecutive IgG of 740 ± 200 Å, compatible with the average repeat length of erythrocyte chromatin (27).

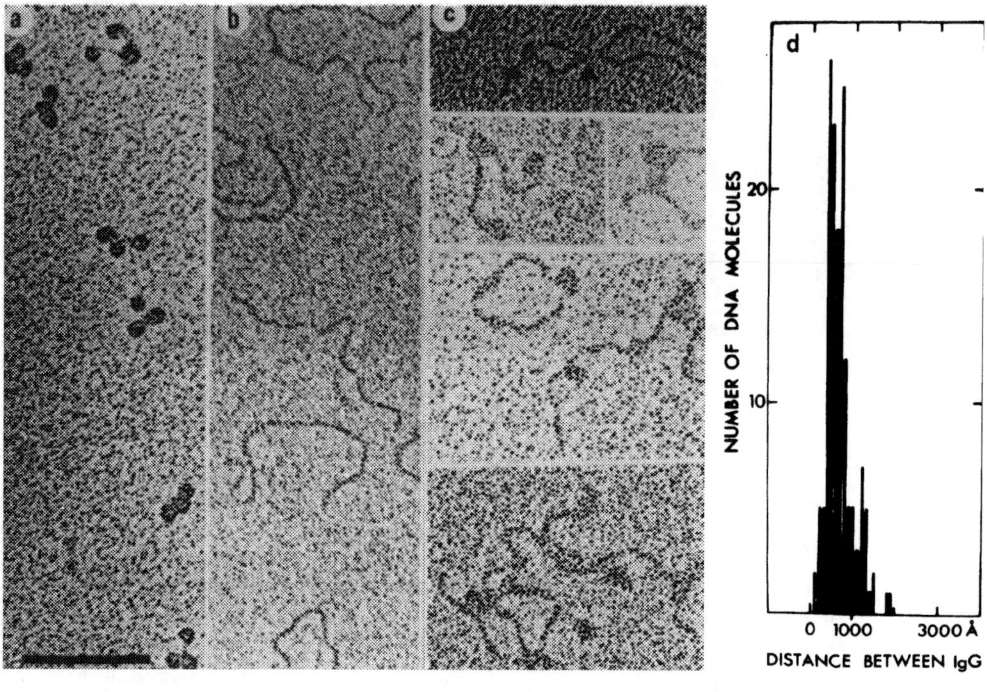

Fig. 7 – Electron micrographs of chicken erythrocyte trinucleosomes having 0.36% modified bases. a) AAF-modified trinucleosomes b) ∿ 630 bp AAF-DNA extracted from modified trinucleosomes. c) AAF-DNA extracted from modified trinucleosomes and reacted with specific IgG. The bar represents 1000 Å. d) Histogram of the number of molecules in function of the center-to-center distance between two consecutive IgG molecules.

As core particles are labeled only at their extremities, we conclude that in trinucleosomes, N-Aco-AAF is preferentially bound to the linker DNA.

CONCLUSION

Unwinding of supercoiled DNA as well as experiments with alternating poly d(G.C) point to the importance of the trans-syn transconformation of guanine to determine the local structure of the modified DNA. Unfortunately all the data give only indirect evidences, and structural studies on modified minihelices have to be made by using NMR or (and) X-ray crystallography.

The use of antibodies directed specifically against DNA modified by a covalently bound carcinogen has allowed us to present some new interesting experiments which can be considered as preliminary results. In fact, a large development of this promising technique could be predicted in the future and we want to conclude in terms of prospective.
1) At first, it is necessary to obtain a better stoichiometry that is a better correspondance between the number of sites and the number of bound IgG. Two possible ways to improve the technique are :

a) Monovalent Fab must be used in place of bivalent IgG, in order to avoid the formation of intramolecular cross links. Their presence not only modifies the apparent stoichiometry but also makes more difficult any mapping or distance measurement. Since Fab are too small to be easily visible, coupling with anti-rabbit goat Fab or with an avidine-biotin complex are necessary.

b) A higher association constant between the antibody and the antigenic site would also improve the stoichiometry in view of the range of concentration in which are made the experiments. This is possible by increasing the antigenic power of the carcinogen by means of some chemical modifications, or by developing monoclonal antibodies.

2) Results obtained with core particles and trinucleosomes have of course to be extended to chromatin, or at least to long fragments of 50 to 100 nucleosomes. In many respects, the hexanucleosome appears as an intermediary step because its structure can be considered as reflecting already the superhelical structure of chromatin fiber. The role of histone H1 or of HMG non histone chromosomal proteins on the accessibility of DNA could therefore be checked at the level of this subunit. DNAse I which was reported to digest selectively the active genes of chromatin (29,30) was also shown to release initially a fraction of DNA enriched in AAF modified nucleotides, when liver chromatin is extracted at early time after injection of the carcinogen to rats (31). Antibodies could therefore be used to characterize the repartition of modified guanines in active and inactive chromatin respectively.

3) The existence of mutagenic hot-spots in bacteria is well documented and in view of the close relationship between the initiation step of carcinogenesis and mutagenesis, the question of sequence specificity to a given carcinogen can be raised. A careful study of the mapping by antibodies of a carcinogen-modified DNA could in principle solve the problem.

If specific regions are detected their sequence analysis would offer new insight in the mechanism of carcinogenesis.
4) Present antibodies are directed against Guo-C-8-AAF. Beside this major adduct and the N^2 derivative, other minor adducts have been found but their structure is not elucidated. The present state of knowledge of repair mechanisms seems to point out the important role of these minor adducts which could escape to any enzymatic system of repair. Specific antibodies against these minor adducts would be of the utmost importance to detect their presence in vivo.

REFERENCES

1. Westra, J.G., and Visser, A. : 1979, Cancer Letters 8, pp. 155-162
2. Fuchs, R., and Daune, M. : 1972, Biochemistry 11, pp. 2659-2666
3. Fuchs, R.P.P., and Daune, M.P. : 1974, Biochemistry 13, pp. 4435-4440
4. Fuchs, R.P.P. : 1975, Nature 257, pp. 151-152
5. Fuchs, R.P.P., Lefèvre, J.F., Pouyet, J., and Daune, M.P. : 1976, Biochemistry 15, pp. 3347-3351
6. Lefèvre, J.F., Fuchs, R.P.P., and Daune, M.P. : 1978, Biochemistry 17, pp. 2561-2567
7. Weinstein, I.B., and Grunberger, D. : 1974, in Model Studies in Chemical Carcinogenesis, Ts'o and Di Paolo ed., N.Y., Marcel Dekker, pp. 217-235
8. Yamasaki, H., Pulkrabek, P., Grunberger, D., and Weinstein, I.B.: 1977, Cancer Res. 37, pp. 3756-3760
9. Drinkwater, N.R., Miller, J.A., Miller, E.C., and Yang, N.C. : 1978, Cancer Res. 38, pp. 3247-3255
10. Lang, M.C.E., Freund, A.M., de Murcia, G., Fuchs, R.P.P., and Daune, M.P. : 1979, Chem.-Biol. Interactions 28, pp. 171-180
11. Crick, F.H.C. : 1976, Proc. Natl. Acad. Sci. USA 73, pp. 2639-2643
12. Le Bret, 1979 : Biopolymers 18, pp. 1709-1725
13. Deleys, R.J., and Jackson, D.A. : 1975, Biochem. Biophys. Res. Comm. 69, pp. 446-453
14. Nelson, J.H., Grunberger, D., Cantor, C.R. and Weinstein, I.B. : 1971, J. Mol. Biol. 62, pp. 331-346
15. Daune, M.P., Fuchs, R.P.P., and Leng, M. : 1980, in press
16. Wang, A.H.J., Quigley, G.J., Kolpak, F.J., Crawford, J.L., Von Boom, J.H., Van der Marel, G., and Rich, A. : 1979, Nature 282, pp. 680-686
17. Sage, E., and Leng, M. : 1980, in press
18. Erlanger, B.F., and Beiser, S.M. : 1964, Proc. Natl. Acad. Sci. USA 52, pp. 68-74
19. Poirier, M.C., Yuspa, S.H., Weinstein, I.B., and Blobstein, S. : 1977, Nature 270, pp. 186-188
20. Guigues, M., and Leng, M. : 1979, Nucl. Acid. Res. 6, pp. 733-744
21. Leng, M., Sage, E., Fuchs, R.P.P., and Daune, M.P. : 1978, FEBS Letters 92, pp. 207-210
22. Sage, E., Fuchs, R.P.P., and Leng, M. : 1979, Biochemistry 18, pp. 1328-1332
23. Dubochet, J., Ducommun, M., Zollinger, M. and Kellenberger, E. : 1971, J. Ultrastruct. Res. 35, pp. 147-167

24. de Murcia, G., Lang, M.C.E., Freund, A.M., Fuchs, R.P.P., Daune, M.P., Sage, E., and Leng, M. : 1979, Proc. Natl. Acad. Sci. USA 76, pp. 6076-6080
25. Wilhelm, F.X., Wilhelm, M.L., and Metzger, G. : 1978, in "Chromatin Structure and Function from Nucleosome to Nuclei", NATO Advanced Study Institute Series, Nicolini, C., ed., Plenum Press, N.Y., vol. 2, pp. 781-801
26. Erard, M., Das, G.C., de Murcia, G., Mazen, A., Pouyet, J., Champagne M., and Daune, M. : 1979, Nucl. Acid. Res. 6, pp. 3231-3253
27. Wilhelm, M.L., Mazen, A., and Wilhelm, F.X.: 1977, FEBS Letters 79, pp. 404-408
28. Paoletti, J., Magee, B.B., and Magee, P.T. : 1977, Biochemistry 16, pp. 351-357
29. Garel, A., and Axel, R. : 1976, Proc. Natl. Acad. Sci. USA 73, pp. 3966-3970
30. Weintraub, H., and Groudine, M. : 1976, Science 193, pp. 848-856
31. Metzger, G., Wilhelm, F.X., and Wilhelm, M.L. : 1977, Biochem. Biophys. Res. Comm. 75, 703-710

[1] We are deeply grateful to P. Oudet and P. Chambon for using their E.M. equipment.

IMMUNOLOGICAL DETECTION AND QUANTIFICATION OF DNA COMPONENTS STRUCTURALLY MODIFIED BY ALKYLATING CARCINOGENS (ETHYLNITROSOUREA)

M.F. Rajewsky, R. Müller, J. Adamkiewicz, and W. Drosdziok
Institut für Zellbiologie (Tumorforschung),
Universität Essen (GH), Hufelandstrasse 55,
D-4300 Essen 1, Federal Republic of Germany

ABSTRACT

The reaction products of alkylating N-nitroso carcinogens in DNA have to a large extent been identified. Thus in the case of N-ethyl-N-nitrosourea (EtNU), a highly potent mutagen and pulse-carcinogen (with a developmental stage-dependent, neural tissue-specific tumorigenic effect in rats), about a dozen different DNA ethylation products are known. Of the various sites of ethylation in DNA, the O^6-position of deoxyguanosine (dGuo) is of particular interest, since the carcinogenic potency of different types of alkylating agents increase with the relative amount of O^6-alkyl-dGuo they produce in the DNA of a given target tissue, and since the rate at which cells of different tissues are able to enzymatically remove this product from their DNA appears to be inversely correlated with their risk of malignant conversion. Thus rat brain cells, contrary to the cells of other rat tissues, are unable to eliminate from their DNA the O^6-ethyl-dGuo produced by EtNU. To facilitate quantification of alkylation products at low concentrations in the DNA of small amounts of tissues and cells, we have, as a first step, developed high-affinity "conventional" and monoclonal antibodies specifically directed against O^6-ethyl-dGuo, O^6-butyl-dGuo, and O^4-ethyl-2'-deoxythymidine. When used in a radioimmunoassay, the antibodies for O^6-ethyl-dGuo (affinity constants, $\sim 10^{10}$ l/mol), for example, will detect this product at an O^6-ethyl-dGuo/dGuo molar ratio of $\sim 3 \times 10^{-7}$ in a sample of 100 μg of DNA. The data underline the potential of immunological methods for the sensitive quantification of DNA components structurally modified by non-radioactive (e.g., environmental) carcinogens.

INTRODUCTION

Structural modifications of target cell DNA are primary events in the multistage process of malignant transformation and tumorigenesis by chemical carcinogens (1). Such alterations of DNA structure result from molecular interactions of DNA with the reactive, generally

B. Pullman, P. O. P. Ts'o and H. Gelboin (eds.), Carcinogenesis: Fundamental Mechanisms and Environmental Effects, 207–218.

electrophilic derivatives ("ultimate carcinogens"; 2) of the respective parent compounds, generated in vivo either by enzyme-catalyzed metabolic "activation" or via non-enzymatic decomposition. Of the various classes of carcinogenic chemicals, the alkylating N-nitroso compounds have been particularly well characterized with respect to their reaction products in DNA (1,2,3).

A representative of the latter group is the highly mutagenic (4, 5) ethylating agent N-ethyl-N-nitrosourea (EtNU; 6) which, due to its selective and developmental stage-dependent neuro-oncogenic effect in rats (6,7), has become one of the favorite model carcinogens for the study of the process of chemical oncogenesis (7,8,9,10). The reactive derivative of EtNU, an ethyldiazonium ion, is produced in vivo via rapid, heterolytic decomposition of the parent compound (half-life, < 8 min; 11). Ethylation of DNA by EtNU occurs at about a dozen nucleophilic sites (3,8,11,12,13). Both in the DNA of intact cells and in purified DNA in vitro, ~80 % of all reaction products are formed by ethylation on oxygen atoms (O^6-ethylguanine, ~7 %; O^2-ethylthymine; O^4-ethylthymine; O^2-ethylcytosine; and ethylphosphotriesters, ~60 %) (8,12,13). The remaining products result from ethylation on nitrogens (7-ethylguanine, ~11 %; 3-ethylguanine; 1-ethyladenine; 3-ethyladenine; 7-ethyladenine; 3-ethylcytosine; and 3-ethylthymine) (8,12,13). Most of these products have now also been synthesized in the laboratory (1,13,14,15, 16). The presence in cellular DNA of O^6-alkylguanine may be of particular relevance for carcinogenesis, since on the one hand it is a potentially mutagenic DNA lesion (17), and on the other the capacity of different rat cells and tissues to enzymatically remove this product from DNA appears to be inversely correlated with their risk of malignant conversion (8,10,12). A selective persistence of O^6-alkylguanine has also been found in Xeroderma pigmentosum cells of complementation groups A and C (18,19,20). These cells are repair-defective for DNA damage induced by ultraviolet light, and are characterized by a high risk of malignant transformation (21).

The detection of carcinogen-modified DNA constituents, particularly at low levels of DNA modification, and the kinetic analysis of their enzymatic elimination from DNA require highly sensitive methods. Thus far, radiochromatographic techniques have been used to analyze DNA for the presence of alkylation products (1, 8,12,13). The sensitivity of these methods is however, limited by the specific radioactivity of the radiolabeled carcinogens applied. Furthermore, the relatively large amounts of DNA required for analysis preclude measurements in small numbers of cells (e.g., embryonic and fetal tissues of laboratory animals, bioptic tissue samples, etc.). Lastly, since radiolabeled carcinogens are required for these techniques, their use is restricted to experiments on laboratory animals and cell cultures, and does not permit the analysis of DNA from (e.g., human) tissues exposed to non-radioactive carcinogens (e.g., from the environment).

In view of the high specificity of antibodies in recognizing minor alterations of molecular structure, we have begun to investigate systematically the potential of immunological methods for the quantification of alkylation products in the DNA of cells exposed to non-radioactive N-nitroso carcinogens. In principle, the development of antibodies for this purpose can be approached in two ways. If the product in question is to be measured in DNA hydrolysates (nucleosides), then the respective (synthesized) alkyl-nucleoside would be expected to be the preferred immunogen (i.e., the hapten in the conjugate with a carrier protein). If, on the other hand, the alkylation product should be recognizable as a constituent of intact DNA (e.g., in the nuclei of individual cells), then carcinogen-treated DNA (i.e., DNA containing all the different alkylated structures produced by the respective carcinogen) might be better suited for immunization. In the former case, the choice between "conventional" antisera (especially if affinity-purified antibodies are obtained from the antiserum with the use of specific immunosorbents) and monoclonal, hybridoma-produced antibodies (22,23) may be somewhat less important (although antibody-secreting hybridomas permit large-scale production of pure antibody). However, the latter case (i.e., "shotgun" immunization with a multi-componential immunogen) calls for the production of monoclonal antibodies which would then each be specific for a distinct type of structural DNA modification. As a first step, we have used as haptens several alkylated nucleosides. We describe here the properties of some high-affinity "conventional" antisera and monoclonal antibodies used for the quantification by radioimmunoassay (RIA) or enzyme-linked immunosorbent assay (ELISA) of O^6-ethyl-2'-deoxyguanosine (O^6-EtdGuo), O^6-n-butyl-2'-deoxyguanosine (O^6-BudGuo), and O^4-ethyl-2'-deoxythymidine (O^4-EtdThd).

MATERIALS AND METHODS

Animals

Rats of the inbred BDIX strain (24) were used for <u>in vivo</u> application of EtNU. Rabbits (Graue Riesen), BDIX-rats, and Balb/c mice were used for immunization.

Ethylnitrosourea

EtNU (Roth) twice recrystallized from methanol was used.

Synthesis of Alkylated Nucleic Acid Constituents

O^6-EtGua, O^6-ethylguanosine (O^6-EtGuo), O^6-EtdGuo, O^6-ethyl-2'-dGMP, 7-ethyl-2'-deoxyguanosine, 7-ethylguanine, and O^6-methylguanosine (O^6-MeGuo), were synthesized as described (14,16). O^6-methyl-2'-deoxyguanosine (O^6-MedGuo) and O^6-BudGuo were prepared by alkylation of 2'-deoxyguanosine (dGuo) with diazomethane and diazobutane, respectively, as described for O^6-EtdGuo in Ref. 16. The tracers for the RIA, O^6-methyl- [8,5'-^{3}H]-2'-dGuo, O^6-ethyl-[8,5'-^{3}H]-2'-dGuo, and O^6-n-butyl- [8,5'-^{3}H]-2'-dGuo, were synthesized from [8,5'-^{3}H]-dGuo by the same procedure. O^4-EtdThd and O^4-ethyl-thymine riboside (O^4-EtrThd) were synthesized by ethylation of 2'-deoxythymidine and thymine riboside, respectively, with diazoethane. The products were purified by thin-layer chromatography on silica gel (16) and Sephadex G10 chromatography. The tracer (RIA) O^4-ethyl-[6-^{3}H]-2'-deoxythymidine was synthesized from [6-3H]-2'-deoxythymidine in the same way. O^6-n-butylguanosine (O^6-BuGuo) was synthesized by Dr. R. Saffhill (Paterson Laboratories, Manchester, England; collaborative program of the European Medical Research Council on "Early Changes in Chemical Carcinogenesis").

Nucleoside-Protein Conjugates

O^6-EtGuo, O^6-BuGuo, and O^4-EtrThd were coupled to the carrier proteins keyhole limpet hemocyanin (KLH; Calbiochem) or horseshoe crab hemocyanin (HCH; Worthington), respectively (14,16,25).

Enzymatic Hydrolysis of DNA

DNA isolated from rat tissues (either treated with EtNU in vivo or untreated) by a modified Kirby method (12) or by a hydroxylapatite adsorption technique (16), or DNA exposed to EtNU in vitro (14), was enzymatically hydrolyzed to nucleosides with DNase I (EC 3.1.4.5; Boehringer), snake venom phosphodiesterase (EC 3.1.4.5.; Boehringer) and grade I alkaline phosphatase (AP; EC 3.1.3.1; Boehringer), as described in Ref. 16. Adenosine deaminase (EC 3.5.4.4; Boehringer; 0.3 units/ml) was used to convert deoxyadenosine (dAdo) to deoxyinosine (dIno) in the DNA hydrolysates prior to analysis. This reaction is complete after 5 min (20°C) and does not lead to measurable dealkylation of O^6-EtdGuo and O^6-BudGuo (in contrast to O^6-MedGuo). dGuo concentrations in DNA hydrolysates were determined as described (16) or by peak integration after separation of DNA hydrolysates by high pressure liquid chromatography (HPLC).

Immunization Procedures

A mixture of 500 µg of a nucleoside-protein conjugate in 0.5 ml of phosphate buffered saline (PBS) and 0.5 ml of aluminium hydroxide (Alugel S, Serva) was stirred for 1 hr at 4°C and emulsified in 1 ml of complete Freund's adjuvant. Rabbits were immunized by injection into the hind foot pads and about 50 intracutaneous (i.c.) sites. Eight weeks later, the rabbits were boosted by the same procedure. After another 8-week interval, the animals received a second booster by intramuscular injection of 500 µg of conjugate in 1 ml of PBS emulsified in 1 ml of complete Freund's adjuvant. Sera were collected 2 weeks later. Immunization of 6-8 weeks old female Balb/c mice was carried out by i.c. injection (20 µg of conjugate in 12.5 µl PBS and 12.5 µl aluminium hydroxide, emulsified in 25 µl complete Freund's adjuvant) into about 10 sites. Twelve and 24 weeks later, the mice were boosted by the same procedure. After another 2 week-interval, the animals received a third intraperitoneal (i.p.) booster injection, and spleen cells for hybridisation were collected 3 days thereafter. Female rats (12-15 weeks old) were immunized by i.c. injection into about 20 sites with an emulsion of 50 µg of conjugate in 50 µl of PBS, 50 µl of aluminium hydroxide, and 100 µl of complete Freund's adjuvant. Five weeks later, the animals were boosted by the same procedure. After another 8 weeks, a second i.p. booster injection was given, and spleen cells were collected 3 days thereafter.

Myeloma Cell Lines

The HAT (hypoxanthine/aminopterine/thymidine)-sensitive mouse myeloma cell line P3-X63-Ag 8.653 (26), a clonal subline of P3-X63-Ag 8 (22), was used for hybridisation with mouse spleen cells. The P3-X63-Ag 8.653 clone, which does not express immunoglobulin heavy or light chains, was obtained from K. Rajewsky (Institut für Genetik, Universität Köln, Germany). For hybridisation with BDIX-rat spleen cells, we used the HAT-sensitive rat myeloma cell line Y3-Ag.1.2.3 (27) which secretes κ-chains. This rat myeloma, obtained from Dr. C. Milstein (MRC Laboratory of Molecular Biology, Cambridge, England), is an interesting alternative to mouse myelomas due to the possibility of producting about 10 times more antibody in a hybridoma-bearing rat as compared to a mouse.

Cell Fusion, Culture, and Cloning of Monoclonal Antibody-Secreting Hybridoma Cells

Spleen cells from immunized Balb/c mice or BDIX-rats were fused with P3-X63-Ag 8.653 cells or Y3-Ag.1.2.3 cells, respectively, using polyethylene glycol (PEG 4000; Roth) as a fusion reagent, and cultured in multi-well tissue culture plates (Costar No. 3524; 1-2 x 10^6 cells/well; RPMI 1640-HAT medium for mouse cells, and Dulbecco modified Eagle's-HAT medium with 1 mM sodium pyruvate for rat cells, both supplemented with 5 x 10^{-5} M mercaptoethanol and 20 % fetal bovine serum) (28). After about 2 weeks, culture supernatants were tested for the presence of specific antibodies by ELISA. Cells from positive cultures were cloned and recloned without aminopterin in the presence of Balb/c mouse or BDIX-rat "feeder" spleen cells, respectively. Positive clones were maintained in cell culture, and aliquots injected i.p. into Pristan-pretreated Balb/c mice or BDIX-rats, respectively, for growth and antibody production in the ascites. Isotope analysis of mouse x mouse hybridoma-produced monoclonal antibodies was carried out by ELISA, using alkaline phosphatase-labeled goat anti-mouse isotope antibodies kindly donated to us by A. Radbruch (26).

Antibody Concentrations and Affinity Constants

Antibody concentrations in sera (rabbits) and cell culture or ascites fluid (mice, rats), respectively, and antibody affinity constants for the respective alkylated deoxyribonucleoside, were calculated from the data obtained by competitive RIA (29).

Isolation of Antibodies by Specific Hapten-Immunosorbents

Hapten was coupled to epoxy-activated Sepharose 6B (Pharmacia), the immunosorbent then incubated with rabbit antiserum, mouse or rat ascitic fluid, or culture supernatants, respectively, and the antibodies eluted at acid or alkaline pH (16).

Enzyme-Linked Immunosorbent Assay (ELISA)

The competitive ELISA was performed in 96-well microtiter plates (type M 129 A; Greiner) coated with bovine serum albumin conjugates of the respective haptens, as described (16). The wells were incubated for 1 hr at 37°C with a conjugate of an alkyl-deoxyribonucleoside-specific antibody with AP (AbAP-conjugate), premixed either with Tris-buffered saline (calibration standard) or with the inhibitor (test sample) (16). In the case of cell cultures to be assayed for monoclonal antibody activity, culture supernatant was added to duplicate wells (with and without 0.15 mM nucleoside) together with AP-labeled rabbit-anti-mouse IgG or goat-anti-rat IgG, respectively, and incubated at 37°C for 3 hr. After removing the incubation mixture, the wells were extensively washed and a 10 mM solution of p-nitrophenyl phosphate (phosphatase substrate; Sigma) was added. After incubation for 1 hr at 37°C, the plates were screened for specific wells (i.e., wells without added nucleoside: yellow color; wells containing nucleoside: colorless), or absorption at 405 nm was determined spectrophotometrically and the degree of inhibition of AbAP binding to the solid phase calculated according to Ref. 16.

Radioimmunoassay (RIA)

The competitive RIA, a modified Farr assay (30), was carried out as described (16). Each sample contained in a total volume of 100 µl of Tris-buffered saline (with 1 % bovine serum albumin [w/v] and 0.1 % bovine IgG [w/v]), ~10^3 dpm of tracer, antiserum at a dilution giving 50 % binding of tracer in the absence of inhibitor, plus varying amounts of inhibitor. After incubation for 2 hr at room temperature (equilibrium), 100 µl of a saturated ammonium sulphate solution (pH 7.0) were added, and after 10 min the samples were centrifuged for 3 min at 10.000 x g. Radioactivity in 150 µl supernatant was measured by liquid scintillation spectrometry. The degree of inhibition of tracer-antibody binding was calculated as described (16).

RESULTS AND DISCUSSION

The characteristics of 3 antisera (specific for O^6-EtdGuo, O^6-BudGuo, and O^4-EtdThd) and 2 monoclonal antibodies (specific for O^6-EtdGuo) are presented in Figure 1 and Tables 1-4. As indicated by

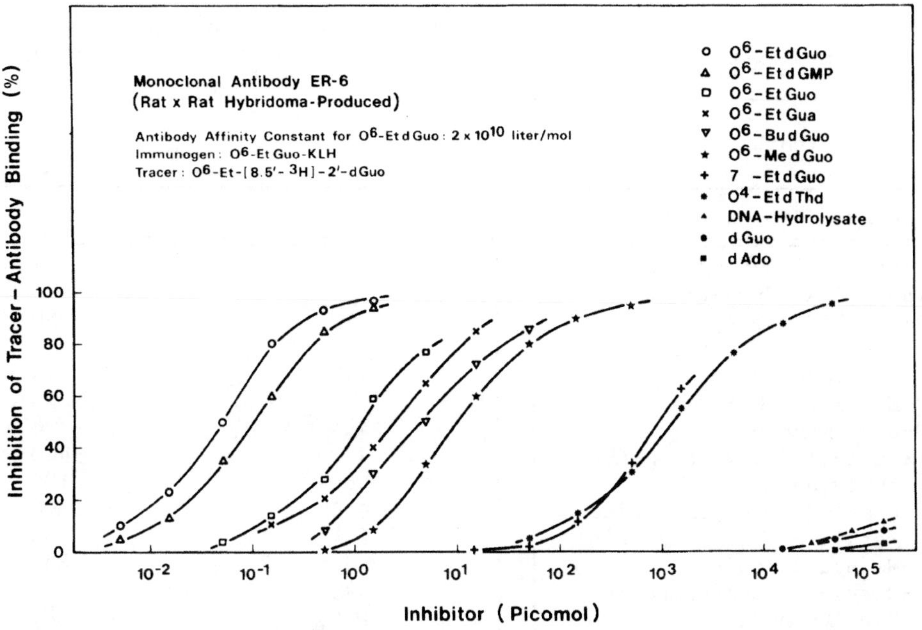

Figure 1. Inhibition of tracer-antibody binding by various natural and alkylated nucleic acid components in the competitive RIA. ER-3 (see Table 4). - DNA hydrolysate: Mixture of dGuo, dIno, dCyd, and dThd, at the molar ratios for rat DNA.

their high antibody affinity constant (up to 2×10^{10}/l mol) and specificity, a low-dose, slow immunogen-release immunization scheme, and the use of a carrier protein phylogenetically unrelated to the proteins of the immunized species, are suitable conditions for the production of such antibodies. As can be seen from the data for the inhibition of tracer-antibody binding by alkylated and normal nucleic acid components in the RIA, the antibodies show very low degrees of cross-reactivity with normal 2'-deoxyribonucleosides. The antibodies recognize to varying relative extents both the alkylated base and the sugar moiety. When used in the RIA, the antibodies of antiserum E-3, and the monoclonal antibodies EM-9 and ER-6, will detect O^6-EtdGuo at an O^6-EtdGuo/dGuo molar ratio of $\sim 3 \times 10^{-7}$ in a sample of 100 µg of hydrolyzed DNA. The antibodies of antiserum E-3 (Table 1) recognize O^6-EtdGuo about one order of magnitude less well in single-stranded

Table 1. Radioimmunoassay for O^6-ethyl-2'-deoxyguanosine: Inhibition of tracer-antibody binding by natural and alkylated nucleic acid components. Rabbit-produced, affinity-purified antiserum E-3. Antibody affinity constant for O^6-ethyl-2'-deoxyguanosine: 1.8×10^{10} liter/mol. Immunogen: O^6-ethylguanosine-KLH. Tracer: O^6-ethyl-(8,5'-^{3}H)-2'-deoxyguanosine (spec. activity: 27 Ci/mmol). SS-DNA: Single-stranded DNA. DS-DNA: Double-stranded DNA.

ANTISERUM E-3	Amount of Inhibitor Required for 50 % Inhibition of Tracer-Antibody Binding	
Inhibitor	pMol	Multiple of O^6-EtdGuo
O^6-Ethyl-2'-Deoxyguanosine	.05	1
O^6-Ethyl-2'-Deoxyguanosine in SS-DNA (EtNU-treated)	.65	13
O^6-Ethyl-2'-Deoxyguanosine in DS-DNA (EtNU-treated)	13	250
O^6-Ethylguanosine	.15	3
O^6-Ethyl-2'-dGMP	.70	14
O^6-Ethylguanine	1.4	28
7-Ethyl-2'-Deoxyguanosine	170	3400
7-Ethylguanine	4000	8×10^4
O^4-Ethyl-2'-Deoxythymidine	140	2800
O^6-Methyl-2'-Deoxyguanosine	2.4	48
O^6-Methylguanosine	12	240
O^6-Butyl-2'-Deoxyguanosine	.45	9
O^6-Butylguanosine	2.0	40
2'-Deoxyguanosine	3×10^4*	~ 5×10^6
2'-Deoxyadenosine	6×10^4*	~ 1×10^7
2'-Deoxyinosine	6×10^4*	~ 1×10^7
2'-Deoxypyrimidines	1×10^5**	> 3×10^7
DNA-Hydrolysate***	5×10^4*	~ 1×10^7

* : < 10 percent inhibition at this concentration. ** : 10 percent inhibition at this concentration. *** : Mixture of dGuo, dIno, dCyd, and dThd, at the molar ratios for rat DNA.

DNA (comparable to O^6-ethyl-2'-dGMP), and about 2 orders of magnitude less well in double stranded DNA. However, the ability to recognize a given alkylation product in DNA may vary for different antibodies, and could be greater if O^6-EtdGuo-containing DNA is used as a hapten (see Introduction). In comparison with the RIA, the ELISA was about 5 times less sensitive (16). It has, however, the advantage of being carried out more readily in large-scale experiments.

Table 2. Radioimmunoassay for O^6-n-butyl-2'-deoxyguanosine: Inhibition of tracer-antibody binding by natural and alkylated nucleic acid components. Rabbit-produced antiserum E-20. Antibody affinity constant for O^6-butyl-2'-deoxyguanosine: 1 x 10^{10} liter/mol. Immunogen: O^6-butylguanosine-HCH. Tracer: O^6-butyl-(8,5'-^3H)-2'-deoxyguanosine (spec. activity: 22 Ci/mmol).

ANTISERUM E-20	Amount of Inhibitor Required for 50 % Inhibition of Tracer-Antibody Binding	
Inhibitor	pMol	Multiples of O^6-BudGuo
O^6-Butyl-2'-Deoxyguanosine	.04	1
O^6-Butylguanosine	.04	1
O^6-Methyl-2'-Deoxyguanosine	500	1.4 x 10^4
O^6-Ethyl-2'-Deoxyguanosine	7	200
2'-Deoxyguanosine	5 x 10^4*	> 1 x 10^8
2'-Deoxyinosine	5 x 10^4*	> 1 x 10^8
DNA-Hydrolysate**	3 x 10^5*	> 1 x 10^8

* : < 10 percent inhibition at this concentration. ** : Mixture of dGuo, dIno, dCyd, and dThd, at the molar ratios for rat DNA.

Table 3. Radioimmunoassay for O^4-ethyl-2'-deoxythymidine: Inhibition of tracer-antibody binding by natural and ethylated nucleic acid components. Rabbit-produced antiserum E-30. Antibody affinity constant for O^4-ethyl-2'-deoxythymidine: 4 x 10^8 liter/mol. Immunogen: Mixture of O^4-ethylthymidine-HCH and O^4-ethylthymidine-KLH. Tracer: O^4-ethyl-(6-^3H)-2'-deoxythymidine (spec. activity: 17 Ci/mmol).

ANTISERUM E-30	Amount of Inhibitor Required for 50 % Inhibition of Tracer-Antibody Binding	
Inhibitor	pMol	Multiples of O^4-EtdThd
O^4-Ethyl-2'-Deoxythymidine	.6	1
O^4-Ethylthymine Riboside	.6	1
2'-Deoxythymidine	5 x 10^4*	> 1 x 10^7
2'-Deoxycytidine	5 x 10^4*	> 1 x 10^7
DNA-Hydrolysate**	3 x 10^5*	> 5 x 10^6

* : < 10 percent inhibition at this concentration. ** : Mixture of dGuo, dIno, dCyd, and dThd, at the molar ratios for rat DNA.

Table 4. Radioimmunoassay for O^6-ethyl-2'-deoxyguanosine: Inhibition of tracer-antibody binding by natural and alkylated nucleic acid components. Mouse x mouse hybridoma-produced, monoclonal antibody EM-9 (left columns); rat x rat hybridoma-produced monoclonal antibody ER-6 (right columns). Antibody affinity constants of EM-9 and ER-6: 8,7 x 10^9 and 2,0 x 10^{10} liter/mol, respectively. EM-9 antibody isotype: γ_1; κ. Immunogen: O^6-ethylguanosine-KLH. Tracer: O^6-ethyl-(8,5'-^{3}H)-2'-deoxyguanosine (spec. activity: 27 Ci/mmol).

MONOCLONAL ANTIBODIES (Mouse x Mouse Hybridoma-Produced; EM-9, Rat x Rat Hybridoma-Produced; ER-6)	Antibody EM-9		Antibody ER-6	
	Amount of Inhibitor Required for 50 % Inhibition of Tracer - Antibody Binding			
Inhibitor	pMol	Multiples of O^6-EtdGuo	pMol	Multiples of O^6-EtdGuo
O^6-Ethyl-2'-Deoxyguanosine	.06	1	.04	1
O^6-Ethylguanosine	1.3	22	.88	22
O^6-Ethyl-2'-dGMP	.34	5.7	.09	2.2
O^6-Ethylguanine	23	383	2	50
7-Ethyl-2'-Deoxyguanosine	600	1.0×10^4	840	2.1×10^4
O^4-Ethyl-2'-Deoxythymidine	3420	5.7×10^4	1080	2.7×10^4
O^6-Methyl-2'-Deoxyguanosine	3.8	63	6.9	173
O^6-Methylguanosine	80	1333	260	6500
O^6-Butyl-2'-Deoxyguanosine	5.2	87	3.3	82
O^6-Butylguanosine	125	2083	56	1400
2'-Deoxyguanosine	4×10^4*	3×10^6	1×10^5**	$\sim 2 \times 10^7$
2'-Deoxyadenosine	1×10^5**	$> 3 \times 10^7$	1×10^5**	$\sim 3 \times 10^7$
2'-Deoxyinosine	1×10^5**	$\sim 2 \times 10^7$	1×10^5**	$> 3 \times 10^7$
2'-Deoxypyrimidines	1×10^5**	$> 3 \times 10^7$	1×10^5**	$> 3 \times 10^7$
DNA-Hydrolysate***	1×10^5*	1×10^7	1×10^5*	$\sim 1 \times 10^7$

* : 10 percent inhibition at this concentration.
** : $<$ 10 percent inhibition at this concentration.
*** : Mixture of dGuo, dIno, dCyd, and dThd, at the molar ratios for rat DNA.

We are presently trying to expand the panel of high-affinity antibodies against various DNA alkylation products including O^6-MedGuo. For the latter product only less affine antibodies have been reported to date (31,32; C. Bordet and R. Montesano, personal comm.). Generally, the data thus far obtained underline the potential of immunological methods for the sensitive quantification of DNA components structurally modified by non-radioactive (e.g., environmental) carcinogens, and for kinetic studies on the persistence (or rate of elimination) of such lesions from cellular DNA.

ACKNOWLEDGEMENTS

The authors express their appreciation to Ch. Knorr, I. Spratte, and B. Wirth for expert technical assistance. This work was supported by Grant 1 R01 CA 20017 awarded by the National Cancer Institute, DHEW, and by the Deutsche Forschungsgemeinschaft (Schwerpunkt "Biochemie des Nervensystems"; Ra 119/11).

REFERENCES

1. Grover, P.L. (ed.): 1979, "Chemical Carcinogens and DNA". Boca Raton, Florida: CRC Press.
2. Miller, E.C. and Miller, J.A.: 1976, in "Chemical Carcinogens" (C.E. Searle, ed.), ACS Monograph No. 173, pp. 737-762. Washington, D.C.: American Chemical Society.
3. Rajewsky, M.F.: 1980, in "Molecular and Cellular Aspects of Carcinogen Screening Tests" (R. Montesano, H. Bartsch, and L. Tomatis, eds.), IARC Scientific Publication No. 27, pp. 41-54. Lyon: International Agency for Research on Cancer.
4. Russel, W.L., Kelly, E.M., Hunsicker, P.R., Bangham, J.W., Maddux, S.C., and Phipps, E.L.: 1979, Proc. Natl. Acad. Sci. USA 76, pp. 5818-5819.
5. Vogel, E. and Natarajan, A.T.: 1979, Mutation Res. 62, pp. 51-100.
6. Ivankovic, S. and Druckrey, H.: 1968, Z. Krebsforsch. 71, pp. 320-360.
7. Rajewsky, M.F.: 1980, in "Advances in Neuroblastoma Research" (A.E. Evans, ed.), pp. 127-134. New York, N.Y.: Raven Press.
8. Rajewsky, M.F., Augenlicht, L.H., Biessmann, H., Goth, R., Hülser, D.F., Laerum, O.D., and Lomakina, L.Ya.: 1977, in "Origins of Human Cancer" (H.H. Hiatt, J.D. Watson and J.A. Winsten, eds.), Cold Spring Harbor Conf. on Cell Prolif., Vol. 4, pp. 709-726. Cold Spring Harbor, N.Y.: Cold Spring Harbor Laboratory.
9. Laerum, O.D., Haugen, Å., and Rajewsky, M.F.: 1979, in "Neoplastic Transformation in Differentiated Epithelial Cell Systems <u>in vitro</u>" (L.M. Franks and C.B. Wigley, eds.), pp. 190-201. London-New York: Academic Press.
10. Rajewsky, M.F.: 1980, Arch. Toxicol., Suppl. 3, pp. 229-236.
11. Goth, R. and Rajewsky, M.F.: 1972, Cancer Res. 32, pp. 1501-1505.
12. Goth, R. and Rajewsky, M.F.: 1974, Proc. Natl. Acad. Sci. USA 71, pp. 639-643.
13. Singer, B., Bodell, W.J., Cleaver, J.E., Thomas, G.H., Rajewsky, M.F., and Thon, W.: 1978, Nature (Lond.) 276, pp. 85-88.

14. Müller, R. and Rajewsky, M.F.: 1978, Z. Naturforsch. 33c, pp. 897-901.

15. Müller, R. and Rajewsky, M.F.: 1979, in "Short Term Tests for Prescreening of Potential Carcinogens" (L. Santi and S. Parodi, eds.), pp. 70-75. Genova: Istituto Scientifico per lo Studio e la Cura dei Tumori.

16. Müller, R. and Rajewsky, M.F.: 1980, Cancer Res. 40, pp. 887-896.

17. Loveless, A.: 1979, Nature (Lond.) 223, pp. 206-207.

18. Goth-Goldstein, R.: 1977, Nature (Lond.) 267, pp. 81-82.

19. Altamirano-Dimas, M., Sklar, R., and Strauss, B.: 1979, Mutation Res. 60, pp. 197-206.

20. Bodell, W.J., Singer, B., Thomas, G.H., and Cleaver, J.E.: 1979, Nucleic Acids Res. 6, pp. 2819-2829.

21. Cleaver, J.E. and Bootsma, D.: 1975, Ann. Rev. Genet. 9, pp. 19-38.

22. Köhler, G. and Milstein, C.: 1975, Nature (Lond.) 256, pp. 495-497.

23. Melchers, F., Potter, M., and Warner, N.L. (eds.): 1978, "Lymphocyte Hybridomas". Curr. Top. in Microbiol. and Immunol., Vol. 81, Berlin-Heidelberg-New York: Springer.

24. Druckrey, H.: 1971, Arzneim.-Forsch. 21, pp. 1274-1278.

25. Erlanger, B.F. and Beiser, S.M.: 1964, Proc. Natl. Acad. Sci. USA 52, pp. 68-74.

26. Kearny, J.F., Radbruch, A., Liesegang, B., and Rajewsky, K.: 1979, J. Immunol. 123, pp. 1548-1550.

27. Galfré, G., Milstein, C., and Wright, B.: 1979, Nature (Lond.), 277, pp. 131-133.

28. Lemke, H., Hämmerling, G.J., Höhmann, C., and Rajewsky, K.: 1978, Nature (Lond.) 271, pp. 249-251.

29. Müller, R.: 1980, J. Immunol. Meth., in the press.

30. Farr, R.S.: 1978, J. infect. Dis. 103, pp. 239-262.

31. Briscoe, W.T., Spizizen, J., and Tan, E.M.: 1978, Biochemistry 17, pp. 1896-1901.

32. Kyrtopoulos, S.A. and Swann, P.F.: 1978, Abstr. 19th Ann. Gen. Meeting Brit. Assoc. Cancer Res., Brit. J. Cancer 38, p. 170.

CHARACTERISTICS OF STAGES OF HEPATOCARCINOGENESIS

Henry C. Pitot
McArdle Laboratory for Cancer Research
The Medical School, University of Wisconsin
Madison, Wisconsin 53706, U.S.A.

Abstract

Hepatocarcinogenesis induced by chemicals may be delineated into at least two distinct stages which are analogous to those of initiation and promotion as previously defined in epidermal carcinogenesis in the mouse. Evidence is presented that the immediate clonal progeny of initiated hepatocytes can be distinguished as foci by one of four histochemical techniques. 85% of the total identifiable foci could be scored as γ-glutamyl peptidase positive together with one other histochemical parameter. With phenobarbital used as a promoting agent, incomplete carcinogens or "pure" initiators of hepatocytes have been identified. After initiation by a subcarcinogenic dose of diethylnitrosamine 24 hours following 70% partial hepatectomy, 2,3,7,8-tetrachlorodibenzo-p-dioxin (TCDD) and butylated hydroxy anisole (BHA) were found to promote the formation of both foci and hepatomas. In particular, TCDD is a millionfold more efficient, whereas BHA is significantly less efficient than phenobarbital as promoting agents.

That the natural history of the carcinogenic process in skin can be divided into several distinct stages has been known for almost 40 years (1). That such separable steps exist during the pathogenesis of other histogenetic types of neoplasms was not readily apparent until this last decade (2,3). While many experimental oncologists have accepted the concept that the natural developmental histories of neoplasms arising from different tissues are probably basically similar, a number still do not equate the natural history of neoplastic development in skin with that in other cell types. The purpose of this presentation is to demonstrate experimental evidence that the process of carcinogenesis in two diverse tissues, epidermis and liver, is biologically almost completely analogous insofar as the experiments have progressed to date.

B. Pullman, P. O. P. Ts'o and H. Gelboin (eds.), Carcinogenesis: Fundamental Mechanisms and Environmental Effects, 219–233.

STAGES IN EPIDERMAL CARCINOGENESIS

Among the earliest investigations which provided experimental evidence demonstrating a two-stage mechanism for carcinogenesis in skin was that of Rous and his associates (4). In their studies the application of coal tar to the ears of rabbits followed by wounding in the affected area resulted in the appearance of neoplasms along the edge of the wound. Rous coined the terms "initiate" to designate the process resulting from the application of coal tar and "promote" for that resulting from the wounding. Subsequent studies by many investigators (5-8) confirmed and refined these studies replacing the technique of wounding with that of chemical treatment during the process of promotion. The chemical most commonly employed in earlier studies was croton oil, a local irritant. Subsequent investigations demonstrated that neither irritation itself nor cell proliferation was absolutely necessary for the second stage to occur (7). A number of the distinctive characteristics of each of the two processes, initiation and promotion, as reviewed by Boutwell (7), may be seen in Table 1 (1).

TABLE 1

CHARACTERISTICS OF STAGES OF INITIATION AND PROMOTION IN SKIN CARCINOGENESIS

Initiation	Promotion
Irreversible, with 'memory'	Reversible, at least in early stages
Initiated cells and immediate progeny not usually identifiable	Promoted neoplasm seen grossly
'Pure' initiation (incomplete carcinogen) causes irreversible change but not neoplasm unless promoter applied	Promoting agents not carcinogenic but may promote fortuitously initiated cells (i.e., background)
Dependent on cell cycle and, for many chemicals, on the metabolism of the cell	Modulated by diet, hormonal, environmental, and related factors

One of the most important characteristics of the initiated cell is that its induced properties are not lost with time. The action of the initiating agent is to evoke an irreversible change in a number of cells of the target tissue. On the other hand, tumor promotion is not unalterable, but rather may actually be reversed under appropriate circumstances (9). Most known carcinogenic agents are capable of inducing both the stage of initiation and that of promotion. These agents are termed "complete carcinogens". Some agents, however, are capable only of initiating the process of carcinogenesis in specific tissues. One of the earliest examples of such agents, termed "incomplete carcinogens" or "pure" initiating agents, was urethane, which was capable of inducing neoplasms in the liver and lung but not in the skin unless the tissue was subsequently treated with multiple applications of croton oil (10).

An important experimental drawback of the study of the two-stage process of carcinogenesis in mouse epidermis is that the initiation of cells may only be confirmed by the ultimate production of neoplasms. In fact, if the tumors produced are clonal in origin, as some investigations suggest (11), then the process of initiation in the skin would seem to be extremely inefficient since only a few neoplasms develop relative to the total cell population of the epidermal area treated by the initiating agent.

Although initiation results in an irreversible cellular change, the efficiency of the process is dependent on the cell cycle (12) as well as the capability of the target tissue to metabolize the chemical to its ultimate carcinogenic form if necessary (13). Agents that enhance the process of initiation have been termed co-carcinogens, whereas those that inhibit the process of carcinogenesis are anti-carcinogens (14). On the other hand, promotion may be markedly affected by environmental factors, as shown by many investigators (9,15-17) who have demonstrated effects of nutrition, hormonal status, and age on the efficiency of tumor promotion.

Since tumor promotion depends on the repeated application of the promoting agent for an extended period, it is perhaps not surprising to find that the repeated application of promoting agents may give rise to a very low incidence of neoplasms, usually only after a greatly extended time period of treatment. Although some investigators (18) have suggested that tumor promoters are thus weak complete carcinogens, one would certainly expect that efficient promoting agents would cause the expression of initiated cells resulting from the action of ambient environmental conditions such as dietary contaminants (19), background radiation, and minor contaminants of the promoting agent itself. At the present time, however, it is not possible to distinguish absolutely between an efficient promoting agent and a weak complete carcinogen by any demonstrated experimental methodology.

STAGES IN HEPATOCARCINOGENESIS

That the process of hepatocarcinogenesis occurred in specific, separable, and characteristic stages was first implied by the early studies of several investigators (20,21) who demonstrated that one of the earliest morphologically demonstrable lesions occurring in the process of hepatocarcinogenesis was the hyperplastic nodule. Histologic evidence was presented suggesting a transition from the hyperplastic nodule into hepatocellular carcinoma (20). However, later investigations indicated that many hyperplastic nodules regressed or disappeared after the animal was removed from the carcinogenic stimulus (22). This latter finding is analogous to the papillomas of the skin produced following treatment with initiating and promoting agents or complete carcinogens (9).

Later studies by Friedrich-Freksa and his associates (23) demonstrated the occurrence of another lesion, the enzyme-deficient or altered island or focus. Such foci have now been demonstrated to occur following the administration of a number of carcinogens. Furthermore, other studies showed that these foci did not disappear following removal of the carcinogenic stimulus (24) and that they were probably clonal in origin (25).

A number of studies by many authors (26-28) including the Millers, who were among the pioneers in the study of chemically induced hepatic carcinomas, demonstrated that hepatocarcinogenesis could be modulated by a variety of means. These included dietary changes as well as hormonal changes within the organism. However, no attempt was made to determine systematically whether such effects were co-carcinogenic, anti-carcinogenic, or the action of promoting agents. The first demonstration that hepatocarcinogenesis may occur in stages was reported by Peraino and his associates in 1973 (29). Following a short feeding of 2-acetylaminofluorene at a level of 0.02% in the diet to weanling rats, the animals were placed on diets containing 0.05% phenobarbital for 5 months or more. Although a low level of hepatic carcinomas occurred in those animals not continuously receiving the drug, 100% of animals fed phenobarbital following initiation by 2-acetylamino-fluorene developed hepatomas. Since these pioneering studies of Peraino, a number of other agents have been demonstrated to be effective in the promotion of hepatocarcinogenesis. A list of these is seen in Table 2 (1).

It is of interest that at least four of the agents in the table have also been shown to induce hepatomas after quite prolonged feeding, usually for periods of 18 to 14 months. On the other hand, if these agents are administered simultaneously with or previous to the initiating agent (2-acetylaminofluorene, diethyl-nitrosamine, etc.), then carcinogenesis is inhibited.

Table 2

PROMOTING AGENTS IN HEPATOCARCINOGENESIS (1)

Promoter	Dosage regimen	Initiating agents effective with:	Species
Phenobarbital	0.05% in diet continuous for 4-5 months	2-Acetylaminofluorene	Rat
	Same 9-11 months	-	Mouse
	Same 12-24 weeks	Diethylnitrosamine	Rat
	500 ppm in diet for 19 weeks	Diethylnitrosamine	Rat
	0.05% in diet for 5-7 months	Diethylnitrosamine	Rat
	Same 12-14 weeks	3'-Methyl-4-(dimethyl-amino)azobenzene	Rat
Dichlorodiphenyltrichloroethane (DDT)	112 mg/kg daily intraperintoneal injection	2-Acetylaminofluorene	Rat
Butylated hydroxytoluene	0.5% in diet	2-Acetylaminofluorene	Rat
Polychlorinated biphenyl(s)	400 ppm in diet 6 months	3'-Methyl-4-(dimethyl-amino)azobenzene	Rat
	1000 ppm in diet 8 weeks	2-Acetylaminofluorene	Rat
3-(3,5-dichlorophenyl)-5,5'-dimethyloxazoline-2,4-dione	2500 ppm in diet 8 weeks	2-Acetylaminofluorene	Rat
Estradiol-17-phenyl propionate and estradiol benzoate	645 μg/100 g body wt.	Nitrosomorpholine	Rat

Following these studies of Peraino and his associates (29) as well as those of Friedrich-Freksa and of Scherer and Emmelot (23,24), we have developed a system involving the initiation of hepatocytes by a single dose of diethylnitrosamine (10 mg/kg) following partial hepatectomy with promotion by phenobarbital after the Peraino protocol (30). However, the promoting agent was not administered until 2 months following initiation. Utilizing this technique, we were able to delineate clearly two distinct stages in the process of hepatocarcinogenesis (Table 3).

Table 3

PROMOTION OF DEN-INDUCED ENZYME-ALTERED FOCI (EAF) BY PHENOBARBITAL FOLLOWING PARTIAL HEPATECTOMY (P.H.)

Treatment	EAF/cm^3 liver	% Liver volume occupied by EAF	No. animals with carcinomas
None	0	0	0/5
P.H.	11 ± 6	0.004	0/5
0.05% Phenobarbital	54 ± 17	0.02	0/5
0.05% Phenobarbital + P.H.	109 ± 22	0.1	0/5
DEN + P.H.	254 ± 65	0.4	0/6
DEN + P.H. + Phenobarbital	874 ± 137	11.3	10/12

As can be seen from the table, the administration of phenobarbital for 6 months results in no enzyme-altered foci or carcinomas. Phenobarbital administration following partial hepatectomy is associated with a relatively small number of enzyme-altered foci but no carcinomas. On the other hand, the administration of a single dose of diethylnitrosamine following partial hepatectomy gives rise to a significant number of enzyme-altered foci but no carcinomas during the 8-month period of the experiment, in confirmation of earlier studies of Scherer and Emmelot (24), who demonstrated that this regimen resulted in no hepatomas even after 24 months. Diethylnitrosamine/partial hepatectomy followed by the feeding of phenobarbital results in a 3.5-fold increase in enzyme-altered foci as well as hepatocellular carcinomas in most of the animals. Furthermore, it is only when the initiating agent is followed by administration of the promoter, phenobarbital, that a substantial volume of the liver is replaced by the altered cellular populations.

CHARACTERISTICS OF ENZYME-ALTERED FOCI IN RELATION TO THE STAGES OF HEPATOCARCINOGENESIS

As indicated in Table 1, the demonstration of initiated cells during the process of epidermal carcinogenesis occurs only "after the fact" by the appearance of neoplasms. That the enzyme-altered foci in our experiments arise from initiated cells has substantial experimental support.

As seen from the studies of Scherer and Emmelot (24), enzyme-altered foci induced by diethylnitrosamine are stable for periods up to 24 months in the liver of the treated animal. In our experiments (31), enzyme-altered foci resulting from a combination of diethylnitrosamine and phenobarbital treatment also show no disappearance after at least 4 months following the cessation of phenobarbital treatment subsequent to that of diethylnitrosamine/partial hepatectomy. That such foci are probably clonal in origin has been reported by Scherer and Hoffmann (25).

Although earlier studies on enzyme-altered foci were largely limited to determination of single enzyme alterations, our investigations (30) utilized serial sections with the monitoring of three enzymes, glucose-6-phosphatase (G6P), canalicular ATPase (ATPase) and gamma-glutamyl transpeptidase (GGT). In this way, by utilization of three biochemical variables it was possible to determine whether all enzyme-altered foci were identical with respect to these parameters. Our studies demonstrated that such was not the case but rather that all possible combinations of biochemical alterations existed in the phenotypes, although those abnormally expressing gamma-glutamyl transpeptidase were clearly in the majority in the initiated and promoted animals. As we have pointed out, such phenotypic heterogeneity of enzyme-altered foci is quite analogous to the phenotypic heterogeneity characteristic of fully developed hepatocellular carcinomas (30). Other markers for putative "preneoplastic" cell populations in liver have also been employed. Watanabe and Williams demonstrated foci in the liver of animals given 2-acetylaminofluorene which were resistant to the accumulation of iron pigment following iron-loading (32). However, a number of these foci disappeared on removal of the carcinogenic stimulus from the diet. Farber and his associates (33) have demonstrated the presence of the "PN" antigen in hyperplastic nodules, and studies in our laboratory (unpublished observations) have shown that enzyme-altered foci also contain this antigen, which has now been shown to cross-react completely with the microsomal enzyme, epoxide hydratase (34). Since no single biochemical marker of enzyme-altered foci produced a maximum of the total number of foci scored, we undertook to determine the effect of iron-loading and the lack of iron pigment formation in enzyme-altered foci as an additional variable in order to determine if such a procedure would increase the total number of foci scored in concert with the three enzymatic markers already used by us. These data are seen in Table 4.

Table 4

PERCENT DISTRIBUTION OF PHENOTYPES OF FOCI BY USE OF FOUR MARKERS

PHENOTYPE	PERCENT DISTRIBUTION OF				
	All Markers	All Markers with GT	All Markers with AP	All Markers with GP	All Markers with Fe
AP*	10.1		28.7		
GT	46.5	62.0			
GP	2.8			17.3	
FE	6.7				23.8
AP,GT	6.3	8.4	17.9		
AP,GP	1.7		4.8	10.5	
AP,Fe	2.5		7.1		8.9
GT,GP	2.0	2.7		12.3	
GT,Fe	5.2	7.0			18.4
GP,Fe	0.8			4.9	2.8
AP,GT,GP	2.3	3.1	6.5	14.2	
GT,GP,Fe	0.7	0.9		4.3	2.5
AP,GT,Fe	6.4	8.5	18.2		22.7
AP,GP,Fe	0.5		1.4	3.1	1.8
AP,GT,GP,Fe	5.4	7.2	15.3	33.3	19.1
% of total of all markers:	100.0	74.8	35.2	16.2	28.2
Absolute number of foci:	1597 ± 490	1197	564	260	453
		GT only	GT + AP	GT + AP + GP	All markers
Cumulative total %:		74.8	89.8	93.4	100.0

*AP, ATPase-deficient; GT, GGT-positive, GP; G6P-deficient; FE, iron pigment-deficient.

As can be seen from the table, the addition of a fourth marker, that of the lack of accumulation of iron pigment, does not increase the maximal number of foci scored by more than 8%. The phenotypic heterogeneity of foci is further emphasized in that all possible phenotypes are demonstrable in the liver of an animal given diethylnitrosamine after partial hepatectomy, followed by phenobarbital, although approximately three-fourths of the foci are positive (scorable) by gamma-glutamyl transpeptidase. The data in this table thus demonstrate that more than 80% of the foci demonstrable by a combination of these four markers can be scored by a combination of gamma-glutamyl transpeptidase positivity in combination with one of the other markers employed.

Substantial evidence for a sequential relationship of enzyme-altered foci to hepatocellular carcinomas has been demonstrated. Laishes and Farber (35) have demonstrated the transplantability of cells capable of forming enzyme-altered foci in the liver of a suitably prepared, partially hepatectomized recipient, beginning with cells of an animal given a relatively large dose of diethylnitrosamine. Scherer and Emmelot presented evidence that enzyme-altered foci were directly related to hepatocellular carcinomas, although not every focus developed into a frank neoplasm during the life span of the animal (36). Such would not be expected since Pugh and Goldfarb (37) demonstrated that the thymidine-labeling of cells of enzyme-altered foci was different for the various phenotypes when the three enzyme markers employed by Pitot were used (30). In addition, Rabes and his associates have demonstrated that foci of cells histologically compatible with early carcinomas and exhibiting a relatively high rate of thymidine labeling arise within some enzyme-altered foci (38). Finally preliminary evidence from our laboratory (A.E. Sirica and H.C. Pitot, unpublished data) has shown that cells cultured from livers of animals containing numerous enzyme-altered foci may be transplanted into the livers of partially hepatectomized rats prepared by the method of Laishes and Farber (35) and give rise to foci exhibiting many of the histologic and histochemical characteristics of enzyme-altered foci.

IDENTIFICATION OF INCOMPLETE CARCINOGENS AND PROMOTING AGENTS IN THE TWO-STAGE PROCESS OF HEPATOCARCINOGENESIS

Published studies from this laboratory have shown that at least two agents not considered to be hepatocarcinogenic except under special circumstances can be shown to be capable of initiating both hepatocellular carcinomas and enzyme-altered foci following treatment with the promoting agent, phenobarbital. 2-Methyldimethylaminoazobenzene has been shown to be weakly carcinogenic only under the circumstance of a partial hepatectomy performed during the feeding of the agent (39). In collaboration with Dr. Kitagawa and the Drs. Miller, we demonstrated that a relatively short-term feeding of this chemical in the absence of a partial hepatectomy resulted in almost no hepatic lesions, while the short-term feeding followed by the feeding of 0.05% phenobarbital for 12 months resulted in hepatocellular carcinomas in all animals so treated (40). In addition, the single administration of dimethylbenzanthracene (10 mg/kg) to rats within 24 hours following a partial hepatectomy and subsequent dietary administration of phenobarbital for 8-12 months resulted in a very high level of enzyme-altered foci, comparable to those seen after the diethylnitrosamine/partial hepatectomy-phenobarbital feeding regimen, and, in most animals, hepatocellular carcinomas (31). Under this regimen animals not receiving the promoting agent showed relatively few enzyme-altered foci (100 EAF/gm liver) and no carcinomas. In addition, in unpublished experiments, the administration of the mutagenic agent, proflavin, at a level of 0.06% for 6 weeks together with a partial hepatectomy and followed by phenobarbital feeding for 6 months resulted in a slight but significant increase in the number of enzyme-altered foci over control levels. The administration of the mutagen in the absence of the promoting agent showed few if any enzyme-altered foci in this relatively short-term experiment, although feeding of the compound for 2 years did cause the appearance of some relatively large enzyme-altered foci, but no significant number of carcinomas (41). Thus it would appear that this latter agent may be a weak incomplete carcinogen, demonstrable only in the presence of an efficient promoting agent such as phenobarbital.

Table 2 lists a number of agents found to have promoting action on hepatocarcinogenesis by a protocol similar to that described by Peraino. We have now tested two compounds for their promoting action in the diethylnitrosamine/partial hepatectomy system (30). The first of these is the environmental toxin, 2,3,7,8-tetrachlorodibenzo-p-dioxin, a contaminant of a number of organic industrial synthetic processes. This agent is quite toxic in a number of animal species (42). Its action appears to be mediated by a receptor protein by a mechanism quite analogous to steroid hormones (43). Its toxicity is not related to mutagenicity or a covalent interaction with DNA, since none of these mechanisms appears to be active in the action of this compound (44). Yet at least one extensive investigation (45) demonstrated that low doses of the dioxin administered to rats for 2 years or more did result in hepatic and tracheal carcinomas in some animals. On these bases we undertook to test the promoting action of the compound in our system.

The data of Table 5 clearly demonstrate the promoting action of TCDD following initiation by diethylnitrosamine/partial hepatectomy. In this table two different doses of TCDD were employed, and it is interesting to note, although no difference in the number of enzyme-altered foci occurred when TCDD was used to promote diethylnitrosamine initiation, it is only at the higher dose that carcinomas were produced and that a very high percentage of the liver volume

Table 5

PROMOTING EFFECT OF 2,3,7,8-TETRACHLORODIBENZO-P-DIOXIN (TCDD) ON HEPATOCARCINOGENESIS BY A SINGLE DOSE OF DIETHYLNITROSAMINE (DEN) AFTER PARTIAL HEPATECTOMY (P.H.)[+]

Group Number	Treatment after P.H.	No. rats	EAF/cm^3 liver	EAF as % liver volume	Rats with carcinoma
1	DEN	4	346 ± 65	5.0	0
2	TCDD (low dose)	5	46 ± 15	0.1	0
3	TCDD (high dose)	5	76 ± 20	0.1	0
4	Phenobarbital	6	138 ± 40	0.1	0
5	DEN + TCDD (low dose)	5	1582 ± 300	7.8	0*
6	DEN + TCDD (high dose)	7	1280 ± 40	35.0	5**
7	DEN + Phenobarbital	4	1510 ± 185	5.0	2

[+]Female rats (200 g) were intubated where shown with DEN. 7 days later administration of TCDD (injected subcutaneously) or phenobarbital (0.05% in the diet) was begun and continued for 28 weeks, at which time the animals were sacrificed and the livers examined. The low and high doses of TCDD were 0.14 and 1.4 µg/kg/2 weeks respectively. DEN was given at a dose of 10 mg/kg. See text for further details.

[++]EAF = enzyme-altered foci.

*Three rats showed "neoplastic nodules".

**One other rat showed "neoplastic nodules".

was occupied by the combination of carcinomas and enzyme-altered foci. The other noteworthy point is that, at the doses employed, TCDD appeared to be one million times more effective than phenobarbital on a molar basis as a promoting agent in this sytem.

The data of this table clearly demonstrate that the administration of TCDD enhances the growth but not the number of enzyme-altered foci when administered as a promoting agent. In further support of the promoting action of TCDD is the fact that on a proportional basis almost three times as many foci at the higher promoting dose of TCDD exhibited all three enzymatic abnormalities compared with those at the lower dose. It is this phenotype that Pugh and Goldfarb (37) demonstrated as exhibiting the highest level of thymidine labeling.

In preliminary experiments we have also demonstrated that the widely used preservative, butylated hydroxyanisole, acts as a promoting agent when it is used in the place of TCDD or phenobarbital in the protocol of Table 5. However, this substance is significantly less efficient as a promoting agent than is phenobarbital.

COMPARISON OF THE TWO-STAGE PROCESS OF CARCINOGENESIS IN LIVER AND SKIN

Pitot and Sirica have reviewed the analogies between the stages of carcinogenesis in liver and those during epidermal carcinogenesis (1). Table 6 shows a comparison of the characterisitics of initiation and promotion during these two processes. While the reader is referred to the original paper for a detailed discussion of this comparison, it is important to note that most of the characteristics of these processes seen in epidermal carcinogenesis also apply to liver. An interesting difference is the fact that initiated cell populations can be identified early during hepatocarcinogenesis but not in skin. There are, however, two critical experiments yet to be done during hepatocarcinogenesis to complete the analogy between the two processes. These are a demonstration of the reversibility of promotion during hepatocarcinogenesis as it has been shown in the skin (9) and a clear demonstration of the modulation of the process of promotion during hepatocarcinogenesis, since earlier studies on modulation of the entire process of tumor formation in the liver did not take into account the two stages involved.

Nevertheless, at our present state of knowledge it is quite apparent that hepatocarcinogenesis, like epidermal carcinogenesis is a multi-stage phenomenon. Furthermore, as we have pointed out earlier (1), it becomes of considerable importance to take into account the two-stage phenomenon and the action of initiators, promoters, and complete carcinogens in any determination of the relative human risk of such compounds in our environment.

TABLE 6

COMPARISON OF TWO-STAGE PROCESS OF EPIDERMAL AND HEPATOCELLULAR CARCINOGENESIS IN THE RAT

Epidermis	Hepatocyte
Initiation:	
Irreversible	Irreversible
Initiated cells and their progeny not usually identifiable	Enzyme-altered foci probably represent progeny of initiated cells
'Pure' initiator (incomplete carcinogen) causes no discernible change unless promoter applied	Incomplete carcinogens induce enzyme-altered foci
Dependent on cell cycle and, for many chemicals, on the metabolism of the cell	Markedly enhanced by cell replication. Promoting agents administered with or before initiator inhibit initiation
Promotion:	
Reversible, at least in early stages	Reversibility not tested, but does require certain time of promotion for full effect
Papilloma resulting from promotion may regress	Many hyperplastic nodules resulting from complete carcinogen administration regress on removal of carcinogenic stimulus
Promoting agents may promote cells initiated by ambient environmental factors	Promoting agents do promote cells initiated by ambient environmental factors
Promotion modulated by diet, hormonal and other environmental factors	Modulation of promotion not yet studied directly, but hepatocarcinogenesis is modulated by hormonal and dietary factors

Bibliography

1. Pitot, H.C., and Sirica, A.E.: 1980, Biochim. Biophys. Acta 605, pp. 191-215.
2. Hicks, R.M., Wakefield, J.St.J., and Chowaniec, J.: 1973, Nature, 243, pp. 347-349.
3. Peraino, C., Fry, R.J.M., Staffeldt, E., and Kisieleski, W.E.: 1973, Cancer Res. 33, pp. 2701-2705.
4. Rous, P., and Kidd, J.G.: 1941, J. Exp. Med. 73, pp. 365-390.
5. Mottram, J.C.: 1944, J. Pathol. Bacteriol. 56, pp. 181-187.
6. Berenblum, I., and Shubik, P.: 1947, Br. J. Cancer 1, pp. 379-391.
7. Boutwell, R.K.: 1974, Crit. Rev. Toxicol. 2, pp. 419-443.
8. Hecker, E.: 1971, Methods Cancer Res. 6, pp. 438-484.
9. Boutwell, R.K.: 1964, Prog. Exp. Tumor Res. 4, pp. 207-250.
10. Berenblum, I., and Haran-Ghera, N.: 1957, Br. J. Cancer 11, pp. 77-87.
11. Iannaccone, P.M., Gardner, R.L., and Harris, H.: 1978, J. Cell Sci. 29, pp. 249-269.
12. Iversen, O.H.: 1974, Excerpta Med. Int. Congress Series 321, Characterization of Human Tumours, Vol. 1 (W. Davis and C. Maltone (eds.), American Elsevier Publ. Co., Inc., N.Y., pp. 21-29.
13. Miller, J.A., and Miller, E.C.: 1971, J. Natl. Cancer Inst. 47, pp. v-xiv.
14. Hecker, E.: 1976, Z. Krebsforsch. 86, pp. 219-230.
15. Boutwell, R.K., Brush, M.K., and Rusch, H.P.: 1949, Cancer Res. 9, pp. 747-752.
16. Sporn, M.B., Dunlop, N.M., Newton, D.L., and Smith, J.M.: 1976, Fed. Proc. 35, pp. 1332-1338.
17. Van Duuren, B.L., Smith, A.C., and Melchionne, S.M.: 1978, Cancer Res. 38, pp. 865-866.
18. Roe, F.J.C., and Clack, J.: 1964, Br. J. Cancer 17, pp. 596-604.
19. Edwards, G.S., Fox, J.G., Policastro, P., Goff, U., Wolf, M.H., and Fine, D.H.: 1979, Cancer Res. 39, pp. 1857-1858.
20. Farber, E., and Ichinose, H.: 1959, Acta Unio Intern. Contra Cancrum 15, pp. 152-153.
21. Goldfarb, S., and Zak, F.G.: 1961, J. Am. Med. Assoc. 178, pp. 729-731.
22. Teebor, G.W., and Becker, F.F.: 1971, Cancer Res. 31, pp. 1-3.
23. Friedrich-Freksa, S.H., Gössner, W., and Börner, P.: 1969, Z. Krebsforsch. 72, pp. 226-239.
24. Scherer, E., and Emmelot, P.: 1975, Eur. J. Cancer 11, pp. 689-696.
25. Scherer, E., and Hoffmann, M.: 1971, Eur. J. Cancer 7, pp. 369-371.
26. Bielschowsky, F.: 1944, Br. J. Exp. Pathol. 25, pp. 90-95.

27. Miller, E.C., Miller, J.A., Kline, B.E., and Rusch, H.P.: 1948, J. Exp. Med. 88, pp. 89-98.
28. Weisburger, J.H., and Weisburger, E.K.: 1963, Clin. Pharmacol. Ther. 4, pp. 110-129.
29. Peraino, C., Fry. R.J.M., Staffeldt, E., and Kisieleski, W.E.: 1973, Cancer Res. 33, pp. 2701-2705.
30. Pitot, H.C.: 1979, In: The Induction of Drug Metabolism, R.W. Estabrook and E. Lindenlaub, eds. F. K. Schattauer Verlag, New York, pp. 471-483.
31. Pitot, H.C., Barsness, L., Goldsworthy, T., and Kitagawa, T.: 1978, Nature 271, pp 456-458.
32. Williams, G.M., and Watanabe, K.: 1978, J. Natl. Cancer Inst. 61, pp. 113-121.
33. Lin, J.-C., Hiasa, Y., and Farber, E.: 1977, Cancer Res. 37, pp. 1972-1981.
34. Levin, W., Lu, A.Y.H., Thomas, P.E., Ryan, D., Kizer, D.E., and Griffin, M.J.: 1978, Proc. Natl. Acad. Sci. USA 75, pp. 3240-3243.
35. Laishes, B.A., and Farber, E.: 1978, J. Natl. Cancer Inst. 61, pp. 507-512.
36. Scherer, E., and Emmelot, P.: 1975, Eur. J. Cancer 11, pp. 145-154.
37. Pugh, T.D., and Goldfarb, S.: 1978, Cancer Res. 38, pp. 4450-4457.
38. Rabes, H.M., Scholze, P., and Jantsch, B.: 1972, Cancer Res. 32, pp. 2577-2586.
39. Warwick, G.P.: 1967, Eur. J. Cancer 3, pp. 227-233.
40. Kitagawa, T., Pitot, H.C., Miller, E.C., and Miller, J.A.: 1979, Cancer Res. 39, pp. 112-115.
41. Bioassay of Proflavine for Possible Carcinogenicity. N.C.I. Carcinogenesis Technical Report Series No. 5, N.I.H., Bethesda, 1977.
42. Poland, A., Greenlee, W.F., and Kende, A.S.: 1979, Ann. N.Y. Acad. Sci. 320, pp. 214-230.
43. Poland, A., Glover, E., and Kende, A.S.: 1976, J. Biol. Chem. 251, pp. 4936-4946.
44. Poland, A., and Glover, E.: 1979, Cancer Res. 39, pp. 3341-3344.
45. Kociba, R.J., Keyes, D.G., Beyer, J.E. Carreon, R.M., Wade, C.E., Dittenber, D.A., Kalnins, R.P., Frauson, L.E., Park, C.N., Barnard, S.D., Hummel, R.A., and Humiston, C.G.: 1978, Toxicol. Appl. Pharmacol. 46, pp. 279-303.

MECHANISMS OF GENETIC SUSCEPTIBILITY TO CANCER

D.G. Harnden
Department of Cancer Studies, University of Birmingham, U.K.

A consideration of clinical features as well as cellular, cytogenetic and biochemical studies on patients susceptible to cancer suggests that a simple correlation between exposure to an environmental agent, DNA damage, mutation and cancer may not be an adequate explanation for the observed cancer incidence patterns.

Many inherited conditions where there is a susceptibility to develop cancer are described in man and these have been widely reviewed (for example, Mulvihill, 1975; Harnden, 1976). In addition, there are many families where there is an elevated familial incidence of cancer, even though there is no clear pattern of inheritance. Remarkably little is known of the mechanisms underlying these genetically determined predispositions to cancer, even though over the past few years many groups of workers have begun to study these families. A number of groups of conditions have been recognised which share common features which could suggest the underlying mechanism. For example, many of the inherited immunological deficiencies have an increased incidence of cancer (Spector et al, 1978); certain patients with hormone imbalance have a high risk of specific cancers (Klinefelter syndrome, XY females); patients with a susceptibility to benign hyperplasias such as neurofibromatosis, polyposis coli and naevoid basal cell carcinoma syndrome, have a specific risk of cancers of these particular tissues. There is also a group of diseases in which there is either an abnormality of DNA repair or replication, or a demonstrable sensitivity, in vivo or in vitro, to the effects of an environmental agent.

It is possible to suggest model schemes which would accommodate all the different mechanisms that these groups of diseases suggest, but for the topic which is the subject of this symposium the most relevant is the latter group of diseases which appear to involve an interaction of an environmental carcinogen with DNA as a critical step in carcinogenesis. These diseases are listed in Table 1.

B. Pullman, P. O. P. Ts'o and H. Gelboin (eds.), Carcinogenesis: Fundamental Mechanisms and Environmental Effects, 235–244.

Table 1. Diseases with either Spontaneous Chromosome Breakage or Susceptibility to an environmental agent.

	Inheritance	Cancer Susceptibility	Spontaneous Chromosome Damage	Immune Defect	Clinical Sensitivity	In Vitro Sensitivity[2]
Ataxia telangiectasia	AR	Lymphoid and others	+(clones)	+	X-rays	X-rays, bleomycin
Xeroderma pigmentosum	AR	Skin	-	-	UV	UV + long patch repair stimulators
Fanconi anaemia	AR	Leukaemia	Breaks exchanges	±	-	Cross linking agents
Bloom syndrome	AR	Lymphoid and others	Homologous exchanges; SCEs	+	UV	UV(±)
Cockayne syndrome	AR	-	-	-	UV	UV
Progeria	AR	Sarcomas meningiomas	-	-	-	-
Retinoblastoma	AD[1]	Retinoblastoma and osteosarcoma	-	-	X-ray sarcoma induction	?X-rays
Basal cell naevus syndrome	AD	Skin and medulloblastoma	±	-	X-ray carcinoma induction	-
Scleroderma	?Multi-factorial	Lung	+	-	?	?Transmissible agent
Incontinentia pigmenti	?AD	?Leukaemia	+	-	?	?

1. A small number have a constitutional deletion of chromosome 13.
2. Either by cell survival or chromosome damage.

For the present purpose I want to concentrate on two of these in particular.

1. ATAXIA TELANGIECTASIA (AT)

In the recessively inherited disease ataxia telangiectasia there is clear evidence of susceptibility to a variety of cancers (Harnden and Taylor, 1978; Spector et al, 1978). These are mainly lympho-reticular neoplasms, but there are also reports of cancers of the stomach and of other organs (Table 2). Amongst the complex of clinical abnormalities of this disease, there are two features in particular which may be associated with the susceptibility to cancer, namely immunodeficiency and chromosome breakage. There is a defect of cell mediated immunity in all patients and, in some cases, there is also a low level of IgA. Progression towards neoplasia may be more probable because of reduced surveillance. However, reservations have recently been expressed about the whole concept of immune surveillance and the increased incidence of lymphoid tumours may be a reflection of a special sensitivity of the lymphoid cells of these patients rather than of a defective recognition of neoplastic foci. Furthermore, AT is unlike the other immune deficiency diseases in that all the others show a susceptibility only to tumours of the reticulo-endothelial system. There are also many features of AT which are not shared with the other immunodeficiency syndromes and which seem likely to be associated with the cancer susceptibility in these patients. There is a spontaneous occurrence of chromosome aberrations in peripheral lymphocytes and in other cells (Harnden, 1974). More-over, clones of cytogenetically marked cells occur in the peripheral blood lymphocyte population (McCaw et al, 1975). In the vast majority of cases the chromosomal rearrangements in these clones involve an exchange point on chromosome 14 at a specific point on the long arm (band 14q12). Breakage and exchange may occur specifically at this point, but it seems more likely that the population of cells with this particular type of exchange is selected out because it has a proliferative advantage. We do not know that this population reflects a potentially malignant line of cells, but in a very small number of cases it has been shown that a neoplasm has arisen from within the clone of cytogenetically marked cells. The suggestion that a specific clone of abnormal cells is selected out has important implications for the consideration of DNA carcinogen interactions. It would suggest that of the many reactions that occur only a very tiny minority will be of any significance in the initiation of neoplasia.

Another line of enquiry leads to the same conclusion. Patients with AT are unusually sensitive to the damaging effects of ionising radiation at the clinical level. Several patients being treated for neoplasms have died as a result of their exposure to X-rays (Cunliffe et al, 1974). This sensitivity to ionising radiation can also be demonstrated in vitro where the cells can be shown to be sensitive to the lethal effects of radiation (Taylor et al, 1975). Obviously events that are lethal are not the events which lead to the initiation of neoplasia. However, radiosensitivity can also be demonstrated at the chromosomal level (Taylor et al, 1976; Cox et al, 1978). AT cells exposed to ionising radiation show an increase in aberrations of specific kinds, in particular of fragments, chromatid breaks and chromatid rearrangements. Heddle et al (1978) has also demonstrated a radiosensitivity of the same magnitude using the micronucleus test. Taylor (1978) has argued that this pattern of damage suggests that there is a defect in repair of specific DNA lesions. Since we know that rejoining of double and single strand breaks following radiation treatment is normal in these patients using classical gradient centrifugation techniques, it seems therefore that any defect in repair must relate to a specific class of lesions that represents only a minority of the breaks that are formed. This conclusion is consistent with the observations on spontaneous aberrations.

There are now several observations on DNA repair in these patients. Paterson et al (1976) has suggested that there is a defect in the ability of cells from some AT patients to recognise and remove lesions in DNA which specifically detected using a Micrococcus luteus γ endonuclease under anoxic conditions. He found, however, that other AT patients were proficient in this function. On the other hand Inoue et al (1977 and personal communication) has shown that AT cells from all the patients that he has studied lack a function which enhances incorporation of labelled precursors into irradiated DNA in an in vitro synthetic system. He suggests that this could represent an exonuclease activity which primes breakage sites for the attachment of DNA polymerase. Recently, Edwards et al (1980) have confirmed this basic observation using both fibroblasts and lymphoid cell lines from AT patients.

These two results appear, at first, to be in conflict, but since we know little of the enzyme pathways that are involved in the repair of damage induced by ionising radiation it is possible that these two suggested functions represent two related activities both of which are important in repairing radiation induced DNA lesions.

It has also been shown that AT cells are sensitive to the cell killing and chromosome damaging effects of bleomycin (Taylor et al, 1979; Lehman, 1979) a radiomimetic drug which is believed to break internucleosomal DNA selectively.

Studies with other DNA damaging agents have given conflicting results (Hoar and Sargent, 1976; Lehman and Arlett, 1977). It is clear that further studies on the precise nature of the repair defect in AT patients are required, but results achieved so far on this disease could suggest that more than one repair pathway may be involved.

Table 2.

Comparison of Cancer Incidence. Number of cases of a specific cancer expressed as a percentage of all cases.

	Normal children <15 years (814 cases) [1]	AT cases (76 cases) [2]	Children irradiated at <10 years with >100 rads (33 cases) [3]
Leukaemia	33	17	45
Brain Tumours	22	4	3
Neuroblastoma	6	0	0
Wilm's Tumour	6	0	0
Bone Tumours	6	0	0
Non-Hodgkin's Lymphoma	6	49	3
Hodgkin's Disease	4	9	0
Retinoblastoma	5	0	0
Rhabdomyosarcoma	4	0	9
Histiocytosis X	2	0	0
Yolk Sac Tumours	2	0	0
Liver Tumours	1	1	0
Thyroid Tumours	1	0	15
Stomach Tumours	0	7	9
Skin Tumours	0	4	6
Ovarian Tumours	0	4	0
Pancreatic Tumours	0	0	3
Miscellaneous or Unspecified	5	5	6

1. Data from Report of U.K. Northern Region Co-ordinating Committee for Malignant Disease in Childhood (1968-1977).

2. Data from Harnden and Taylor (1978)

3. Data from Jablon et al (1971)

Paterson and his colleagues (1979) have suggested that a proportion of heterozygotes are intermediate between AT homozygotes and controls in both cell survival and repair replication following γ irradiation under anoxic conditions. Under oxic conditions, however, he does not find any difference between AT heterozygote cell lines and controls. This latter result has also been obtained in our laboratory (Taylor - personal communication), but studies under anoxic conditions have not been carried out.

In an epidemiological study Swift et al (1976) have found more cases of cancer amongst relatives of AT patients than would have been expected by chance and Paterson et al (1979) has suggested that this finding of increased sensitivity in heterozygotes might be linked to this observation of increased cancer incidence in relatives. However, since tissues are normally well oxygenated and since no difference is found under oxic conditions this hypothesis may not be well founded. The whole idea that AT heterozygotes represent a very substantial number of cancer susceptible people in the normal population is at present speculative.

If we assume that the increased radiosensitivity in AT patients is the cause of their cancer susceptibility we might expect to find in AT patients a range of cancers similar to those that are found in excess in patients known to have been exposed to high doses of ionising radiation such as the atomic bomb survivors in Hiroshima or Nagasaki and the population of radiation exposed ankylosing spondylitis patients.

Table 2 shows a comparison of the relative cancer incidence in a population of normal children (under 15 years of age) in the U.K., all the reported cancers in AT patients and the cases of cancer reported to have occurred in a fifteen year period in persons who were exposed as children to ionising radiation following the dropping of the atomic bombs in Japan in 1945. Although the figures are only approximations and based on very different numbers of cases, a number of striking differences are immediately apparent. As compared with the normal children, the AT patients show a striking excess of non-Hodgkin's lymphoma. There is also a relative deficiency of both leukaemias and brain tumours while the cancers of stomach, skin and ovary must constitute an excess even though the numbers are small. If the AT cancers are due to a generalised susceptibility to radiation one might expect the distribution of cancers to be roughly comparable to that in subjects who have been irradiated as children. This is clearly not so. The irradiated children show a relative excess of leukaemias and thyroid cancers and possibly also of soft tissue sarcomas, stomach cancers and skin cancers. (The apparent increase in stomach cancers in the irradiated children may be simply a reflection of the high incidence of stomach cancer in Japan since there were just as many cases found in the local control population as in the irradiated population). It can be concluded from these figures that the increased incidence of cancer is not simply due to their increased sensitivity to natural radiations. The interpretation of the increased incidence appears to be more complex than it appears to be at first sight.

2. RETINOBLASTOMA

Retinoblastoma is a cancer of the neural retina which occurs in children under 10 years of age. It may be sporadic, in which case it is most often unilateral or it may be familial in which it is frequently bilateral. In some pedigrees it is clearly inherited as an autosomal dominant trait, although it is not uncommon for the penetrance to be incomplete. A small group of retinoblastoma cases have a deletion of the short arms of chromosome 13; the band 13q14 is always deleted (Lele et al, 1963; Harnden and Taylor, 1979). These children often have other malformations and it is probably better to consider these cases as a malformation syndrome of which retinoblastoma is one feature. The majority of retinoblastoma patients have a normal karyotype using present techniques. Patients with retinoblastoma also tend to develop osteosarcomas, particularly of the long bones (Kitchin and Ellsworth, 1974), but they also develop osteosarcomas of the orbit following radiotherapy at a frequency greater than would have been expected following comparable doses of radiation to a normal individual (Sagerman et al, 1969).

Weichselbaum and his colleagues have found cell lines derived from some cases of retinoblastoma to be radiosensitive as measured by cell survival techniques. They suggest on the basis of a series of deletions (Nove et al, 1979) that there are two genes close together on the long arms of chromosome 13, one involved in the production of retinoblastoma, the other concerned with radiosensitivity. In our laboratory we have failed to find any radiosensitivity in cells cultured from either sporadic or familial cases of retinoblastoma (Taylor and Morten - personal communication) using cell survival techniques. We find that cell lines derived from normal individuals show variation in sensitivity and cells from retinoblastoma patients fall within that range of normal variability.

Knudson (1971) has suggested that retinoblastoma may be due to two sequential mutations and, on the basis of observations of age of onset, that in sporadic cases both mutational events occur spontaneously while in familial cases the retinoblastoma mutation is itself the first of these two events. While this idea is attractive in many ways and does fit with current ideas on the role of somatic mutation in the initiation of cancer, there are a number of difficulties. For example, as Matsunaga (1979) points out, it is hard to reconcile the idea that the second mutation in familial cases is random with the observation that tumours tend to occur simultaneously in both eyes. This is much more consistent with the idea that there is some critical point in differentiation at which some event occurs for which the correct functioning of the retinoblastoma gene is necessary. Failure of the gene function could lead to the establishment of populations of potentially malignant cells. It is still hard to see, however, how only a few cells come to be affected out of all the many cells at risk in the retina and it may be that the concept of a second mutation would have to be combined with the idea of a critical developmental stage to give a model which could explain the curious incidence pattern of this disease.

In some families the gene remains silent for several generations only to reappear more or less simultaneously in several different lineages. Such pedigrees are consistent with the idea that the mutation when it occurs may be subject to the effect of control genes and only when these are deleted out can expression occur. Such a concept could also explain the poor penetrance in some pedigrees. Genetic phenomena of this type have been reported in other instances notably the expression of melanisation in Xiphophorus species. Such pedigrees are hard to reconcile with the two mutation hypothesis, but suggest a much more complex mechanism.

CONCLUSION

In both these rare diseases there are a number of curious features which suggest that it might be premature to make a simple link between clinical sensitivity to a carcinogenic agent, the induction of DNA damage, mutation and the emergence of a clinical cancer.

REFERENCES

Cox, R., Hosking, G.P. and Wilson, J. 1978, Arch. Dis. Child, 53, pp.386-390.

Cunliffe, P.N., Mann, J.R., Cameron, A.H., Roberts, K.D. and Ward, H.W.C. 1974, Brit. J. Radiology, 48, pp.374-376.

Edwards, M.J., Taylor, A.M.R. and Duckworth, G. 1980, J. Biochem, (In press).

Harnden, D.G. 1974, In Chromosomes and Cancer, ed. J. German. John Wiley & Sons Inc, New York.

Harnden, D.G. 1976, In Scientific Foundations of Oncology, eds. T. Symington and R.L. Carter. Heinemann, London.

Harnden, D.G. and Taylor, A.M.R. 1978, In Mutagen Induced Chromosome Damage in Man, eds. H.J. Evans and D.C. Lloyd. Edinburgh University Press, pp.52-61.

Harnden, D.G. and Taylor, A.M.R. 1979, In Advances in Human Genetics, eds. H. Harris and K. Hirschhorn. Plenum Press, New York, pp.1-70.

Heddle, J.A., Benz, R.D. and Countryman, P.I. 1978, In Mutagen Induced Chromosome Damage in Man, eds. H.J. Evans and D.C. Lloyd. Edinburgh University Press, pp.191.

Hoar, D.I. and Sargent, P. 1976, Nature, 261, pp.590.

Inoue, T., Hirano, K., Yokoiyama, A., Kada, T. and Kato, H. 1977, Biochem. Biophys. Acta, 479, pp.497.

Jablon, S., Tachikawa, K. and Belsky, J.L. 1971, Lancet, i, pp. 927-932.

Kitchin, F.D. and Ellsworth, R.M. 1974, J. Med. Genet., 11, pp.244-246.

Knudson, A.G. 1971, Proc. Natl. Acad. Sci (Wash), 68, pp.820.

Lehman, A.R. and Arlett, C.F. 1977, Biochem. Soc. Trans., 5, pp.1199.

Lele, K.P., Penrose, L.S. and Stallard, H.B. 1963, Ann. Hum. Genet., 27, 171.

McCaw, B.K., Hecht, F., Harnden, D.G. and Teplitz, R.L. 1975, Proc. Nat. Acad. Sci. (USA), 72, pp.2071-2075.

Matsunaga, E. 1979, J. Nat. Cancer Inst., 63, pp.933-939.

Mulvihill, J.J. 1975, In Persons at High Risk of Cancer, ed. J.F. Fraumeni. Academic Press, New York.

Nove, J., Little, J.B., Weichselbaum, R.R., Nichols, W.W. and Hoffman, E. 1979, Cytogenetics & Cell Genetics, 24, pp.176-184.

Paterson, M.C., Anderson, A.K., Smith, B.P. and Smith, P.J. 1979, Cancer Research, 39, pp.3725-3734.

Paterson, M.C., Smith, B.P., Lohman, P.H.M., Anderson, A.K. and Fishman, L. 1976, Nature, 260, pp.444-447.

Sagerman, R.H., Cassady, J.R., Tretter, P. and Ellsworth, R.M. 1969, Am. J. Roentgenol, 105, pp.529-535.

Spector, B.D., Perry, G.S. and Kersey, J.H. 1978, Clin. Immunol. Immunopathol, 11, pp.12-29.

Swift, M., Stolman, L., Perry, M. and Chase, M. 1976, Cancer Research, 36, pp.209-215.

Taylor, A.M.R. 1978, Mutat. Res., 50, pp.407-418.

Taylor, A.M.R., Harnden, D.G., Arlett, C.F., Harcourt, S.A., Lehman, A.R., Stevens, S. and Bridges, B.A. 1975, Nature, 258, pp.427.

Taylor, A.M.R., Metcalfe, J.A., Oxford, J.M. and Harnden, D.G. 1976, Nature, 260, pp.441.

Taylor, A.M.R., Rosney, C.M. and Campbell, J.B. 1979, Cancer Res., 39, pp.1046.

Weichselbaum, R.R. 1978, Proc. Natl. Acad. Sci. (USA), 75, pp.3962-3964.

CONSTITUTIVE UNCOUPLING OF PATHWAYS OF GENE EXPRESSION THAT CONTROL GROWTH AND DIFFERENTIATION AND THE MOLECULAR MECHANISM OF CARCINOGENESIS

Leo Sachs
Department of Genetics, Weizmann Institute of Science
Rehovot, Israel

ABSTRACT. Chemical carcinogens and tumor promoters have pleiotropic effects. Tumor initiators can produce a variety of mutations and tumor promoters can regulate a variety of physiological molecules that control growth and differentiation. Some of these mutations and regulation of the appropriate molecules can initiate and promote the sequence of changes required for the transformation of normal into malignant cells. After this sequence of changes, some tumors can still be induced to revert with a high frequency from a malignant to a non-malignant phenotype. Results obtained from the analysis of regulation of growth and differentiation in normal and leukemic myeloid cells, the phenotypic reversion of malignancy by induction of normal differentiation in myeloid leukemia, and the blocks in differentiation defective leukemic cell mutants, have been used to propose a general molecular model for the origin and development of malignancy. This model states that malignancy originates by changing specific pathways of gene expression required for growth from inducible to constitutive. This changes the requirements for growth and uncouples growth from differentiation. Constitutive expression of other specific pathways then causes blocks in differentiation and the further development of malignancy.

PLEIOTROPIC EFFECTS OF TUMOR INITIATORS AND PROMOTERS AND THE PHENOTYPIC REVERSION OF MALIGNANCY

Results obtained on the effects of chemical carcinogens and tumor promoters on cultured cells have shown that tumor initiators and tumor promoters have pleiotropic effects. The pleiotropic effects of carcinogens, which interact directly with DNA, include the ability to induce cell transformation (Berwald and Sachs, 1963; Chen and Heidelberger, 1969; DiPaolo, Donovan and Nelson, 1969) and mutations of various genes in mammalian cells (Huberman and Sachs, 1976; Huberman, Sachs, Yang and Gelboin, 1976; Barrett and Ts'o, 1978a; Landolph and Heidelberger, 1979). This includes mutations of genes which code for specific functions that are presumably not essential for this initiation of tumor formation. The pleiotropic effects of

B. Pullman, P. O. P. Ts'o and H. Gelboin (eds.), Carcinogenesis: Fundamental Mechanisms and Environmental Effects, 245–256.

tumor promoters, which do not seem to interact directly with DNA, include induction of plasminogen activator (Wigler and Weinstein, 1976), prostaglandin (Levine and Hassid, 1977), Epstein-Barr virus (Zur Hausen, Bornkmann, Schmidt and Hecker, 1979), type C virus (Lotem and Sachs, 1979), inhibition of cell differentiation (Yamasaki, Fibach, Nudel, Weinstein, Rifkind and Marks, 1977; Rovera, O'Brien and Diamond, 1977), induction of cell differentiation (Sachs, 1978; Miao, Fieldsteel and Fodge, 1978), and promotion of cell multiplication (see Lotem and Sachs, 1979). The pleiotropic effects of tumor promoters can be explained by the ability of these compounds to regulate a variety of molecules that can influence growth and differentiation. The final results obtained then depend on which molecules are being regulated in the treated cells (Lotem and Sachs, 1979).

The change of normal into malignant cells produced by initiation and subsequent promotion (Berenblum, 1975) involves a sequence of genetic changes (Huberman, Salzberg and Sachs, 1968; Barrett and T'so, 1978b), including specific chromosome changes (Sachs, 1974; Azumi and Sachs, 1977; Rowley, 1977; Klein, 1979). Evidence has, however, been obtained with various types of tumors including sarcomas (Sachs, 1974), myeloid leukemias (Sachs, 1974; 1978), and teratocarcinomas (Mintz and Illmensee, 1975; Dewey, Martin, Martin and Mintz, 1977), that malignant cells have not lost the genes that control normal growth and differentiation. This was first shown in sarcomas by the finding that it was possible to reverse the malignant to a non-malignant phenotype with a high frequency in cloned sarcomas cells (Rabinowitz and Sachs, 1968; 1970a; 1970b; Sachs, 1974). The comparison of sarcomas with myeloid leukemias then showed that reversion of the malignant phenotype can be achieved by 2 mechanisms. In one mechanism, found with sarcomas (Rabinowitz and Sachs, 1968; 1970a; Yamamoto, Rabinowitz and Sachs, 1973; Sachs, 1974), this reversion was obtained by chromosome segregation resulting in a change in the balance of specific chromosomes. This reversion of malignancy by chromosome segregation, with a return to the normal gene balance required for expression of the non-malignant phenotype, occured without hybridization between different types of cells. Reversion of malignancy associated with chromosome changes has also been found after hybridization between different types of cells (Klein, Frieberg, Wiener and Harris, 1973; Wiener, Klein and Harris, 1974; Ringertz and Savage, 1976).

PHENOTYPIC REVERSION OF MALIGNANCY BY INDUCTION OF NORMAL CELL DIFFERENTIATION

In addition to this reversion of malignancy by chromosome segregation, another mechanism of reversion has been found in myeloid leukemia (Sachs, 1974; 1978). In this second mechanism, reversion to a non-malignant phenotype was also obtained in certain clones with a high frequency, but in contrast to sarcomas this reversion was not associated with chromosome segregation. Phenotypic reversion of malignancy in these leukemic cells was obtained by induction of the normal sequence of cell differentiation by the physiological inducer of differentiation

(Sachs, 1974; 1978; Paran, Sachs, Barak and Resnitzky, 1970; Fibach, Landau and Sachs, 1972; Fibach and Sachs, 1974; 1975; Lotem and Sachs, 1974; 1977a; 1977b; 1978; Liebermann, Hoffman-Liebermann and Sachs, 1980). Myeloid leukemic cells also have specific chromosome changes compared to normal cells (Hayashi, Fibach and Sachs, 1974; Azumi and Sachs, 1977; Rowley, 1977; Lotem and Sachs, 1977a). The chromosome changes in these leukemic cells thus seem to involve changes in genes other than those involved in the induction of normal differentiation. Induction of normal differentiation in teratocarcinoma cells (Mintz and Illmensee, 1975; Dewey, Martin, Martin and Mintz, 1977) may also be due to this mechanism. Studies on the phenotypic reversion of malignancy by induction of normal cell differentiation, can provide an insight into the mechanism of differentiation and the origin and development of tumors. I shall summarize below our results on the controls that regulate growth (proliferation) and differentiation in myeloid leukemic cells and indicate how these may serve as a general model to explain the origin and development of malignancy.

MGI AND THE CONTROL OF GROWTH AND DIFFERENTIATION IN NORMAL AND LEUKEMIC MYELOID CELLS

The culture of normal hematopoietic cells *in vitro*, has made it possible to study the controls that regulate the proliferation and differentiation of normal and leukemic myeloid cells (Sachs, 1974; 1978). We first showed (Ginsburg and Sachs, 1963; Pluznik and Sachs, 1965), as was then confirmed by others (Bradley and Metcalf, 1966), that normal myeloid precursor cells cultured with a feeder layer of other cell types can form colonies of granulocytes and macrophages *in vitro*. We also found that the formation of these colonies is due to secretion, by cells of the feeder layer, of a specific inducer that induces both the formation of colonies and the differentiation of cells in these colonies to macrophages or granulocytes (Pluznik and Sachs, 1965; 1966; Ichikawa, Pluznik and Sachs, 1966). After we first detected its presence in culture supernatants (Pluznik and Sachs, 1966; Ichikawa, Pluznik and Sachs, 1966), this protein inducer has been referred to by a number of names including mashran gm (Ichikawa, Pluznik and Sachs, 1967), macrophage and granulocyte inducer (MGI) (Landau and Sachs, 1971), colony-stimulating factor (CSF) (Metcalf, 1969) and colony-stimulating activity (CSA) (Austin, McCulloch and Till, 1971). This protein inducer can be produced and secreted by various normal and malignant cells in culture and *in vivo* (Sachs, 1974; 1978; Pluznik and Sachs, 1965; 1966; Ichikawa, Pluznik and Sachs, 1966; 1967; Paran, Ichikawa and Sachs, 1968; Landau and Sachs, 1971; Mintz and Sachs, 1973; Weiss and Sachs, 1978; Cline and Golde, 1979; Douer and Sachs, 1979).

Experiments with normal myeloblasts, the normal counterpart of myeloid leukemic cells, have shown that these normal cells require the normal macrophage and granulocyte inducing protein MGI for cell viability, proliferation and differentiation to mature macrophages or granulocytes (Fibach and Sachs, 1976; Sachs, 1978). There are, however, clones of myeloid leukemic cells that can be induced to differentiate normally to mature macrophages or granulocytes by the physiological inducer MGI, but that no longer require MGI for cell viability and proliferation. These clones, which in mice have been found in 3 different inbred strains (Sachs, 1974; 1978; Ichikawa, 1969; Fibach, Landau and Sachs, 1972; Fibach, Hayashi and Sachs, 1973; Ichikawa, Maeda and Horiuchi, 1976; Lotem and Sachs, 1977a), are referred to as MGI^+D^+ (D^+ for differentiation to mature cells). This loss in the requirement of MGI for cell viability and proliferation then allows these leukemic cells, unlike the normal cells, to multiply in the absence of MGI. It has been suggested (Sachs, 1978) that this loss, either partial or complete, is the origin of myeloid leukemia. Other clones of myeloid leukemic cells that can also proliferate without MGI, are either partially (MGI^+D^-) or almost completely (MGI^-D^-) blocked in their ability to be induced to differentiate by MGI (Sachs, 1974; 1978; Fibach, Hayashi and Sachs, 1973; Lotem and Sachs, 1974; 1977a; Liebermann, Hoffman-Liebermann and Sachs, 1980; Hoffman-Liebermann and Sachs, 1978; Symonds and Sachs, 1979). These clones, which have specific chromosome changes (Hayashi, Fibach and Sachs, 1974; Azumi and Sachs, 1977), have been considered as further stages in the development of myeloid leukemia (Sachs, 1978).

UNCOUPLING OF THE CONTROLS FOR GROWTH AND DIFFERENTIATION IN MYELOID LEUKEMIA

Proliferation in the normal cells is regulated at 2 control points. The first control is that which requires MGI to produce more cells that can then differentiate, and the second control is the stopping of cell multiplication that occurs in the programme of terminal differentiation in the mature cells. In the presence of MGI, proliferation of the normal cells is followed by initiation of differentiation and termination of the differentiation programme, and it is assumed that these 3 processes are coupled in the normal cells. In the MGI^+D^+ leukemic cells, the first proliferation control is uncoupled from the process that initiates differentiation, but the coupling between the initiation of differentiation and the stopping of proliferation in the mature cells is still maintained. In the MGI^+D^- leukemic cells, both the first proliferation control and the terminal proliferation control have been uncoupled from the initiation of differentiation. In the final stage, MGI^-D^-, uncoupling of the first proliferation control is associated with almost no inducibility for differentiation by MGI (Table 1). The origin and development of myeloid leukemia is thus associated with this uncoupling of controls, presumably due to the chromosome changes found in the leukemic cells that can produce changes in gene balance (Sachs, 1974; 1978; Azumi and Sachs, 1977), or the integration of appropriate viruses at specific sites in the cellular genome (Liebermann and Sachs, 1979). The results presented below can explain the mechanism of this uncoupling.

TABLE 1. Uncoupling of controls for proliferation and differentiation in myeloid leukemia

Cell type	MGI induction of: Initial proliferation[a] (A)	Partial differentiation (B)	Terminal differentiation (C)	Coupling or uncoupling of different controls
Normal myeloblasts	+	+	+	Coupling of A, B and C
Leukemic $MGI^{+}D^{+}$	—	+	+	Uncoupling of A from B and C
Leukemic $MGI^{+}D^{-}$	—	+	—	Uncoupling of B from A and C
Leukemic $MGI^{-}D^{-}$	—	—	—	Uncoupling of A and blocking of B and C

[a] Initial proliferation before differentiation requires addition of MGI to normal myeloblasts, whereas the 3 types of leukemic cells can proliferate without adding MGI.

CONSTITUTIVE GENE EXPRESSION AND THE ORIGIN AND DEVELOPMENT OF MYELOID LEUKEMIA

Since the normal but not the leukemic myeloblasts require MGI for cell viability and proliferation (Sachs, 1978), the molecular changes required for cell viability and proliferation that have to be induced in the normal cells are constitutive in the leukemic cells. This suggests that the origin of myeloid leukemia is due to a change from an induced to a constitutive expression of genes that control cell viability and proliferation. Studies on changes in the synthesis of specific proteins in normal myeloblasts, $MGI^{+}D^{+}$, $MGI^{+}D^{-}$ and $MGI^{-}D^{-}$ leukemic clones at different times after treatment with MGI, using two dimensional gel electrophoresis (Liebermann, Hoffman-Liebermann and Sachs, 1980), have provided evidence at the level of protein synthesis that supports this explanation and also indicate the relationship between constitutive gene expression and uncoupling of other controls. The leukemic cells were found to be constitutive for changes in the synthesis of specific proteins that were only induced in the normal cells after treatment with MGI. These constitutive changes, which included the constitutive presence of some proteins and the constitutive absence of other proteins, can be divided into 2 classes. One class consisted of changes in 14 specific proteins that were induced by MGI in the normal cells and were constitutive in all the 8 leukemic clones studied derived from 4 different tumors. The other class consisted of changes in other specific proteins that were induced by MGI in normal and $MGI^{+}D^{+}$ leukemic cells, but were constitutive in $MGI^{+}D^{-}$ and $MGI^{-}D^{-}$ leukemic cells. In this second class there were more constitutive changes in MGI-D- than in MGI+D- cells (Liebermann, Hoffman-Liebermann and Sachs, 1980).

The protein changes during the normal proliferation and differentiation of myeloblasts seem to be induced by MGI as a series of parallel multiple pathways of gene expression (Liebermann, Hoffman-Liebermann and Sachs, 1980). It can be assumed, that the normal developmental programme that appears to couple proliferation and differentiation in normal cells requires a synchronous initiation and progression of these multiple parallel pathways. The presence of constitutive gene expression for some pathways can be expected to produce asynchrony in the developmental programme. Depending on the pathways involved, this asynchrony could then result in an uncoupling of the controls for proliferation and differentiation and produce different blocks in the ability to be induced for and to terminate the differentiation process. The suggestion derived from these results (Sachs, 1978; Liebermann, Hoffman-Liebermann and Sachs, 1980) is, therefore, that myeloid leukemia originates by a change that produces certain constitutive pathways of gene expression and that this uncouples the normal requirement for MGI for proliferation and differentiation. This can be followed by constitutive expression of other pathways resulting in the uncoupling of other controls and thus interfering with the normal programme of terminal differentiation (Fig. 1).

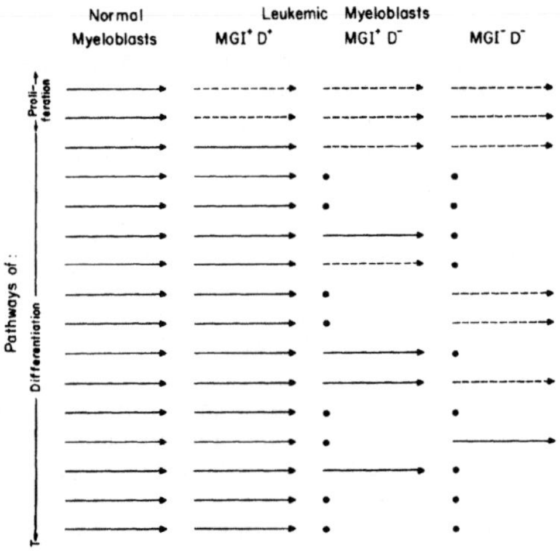

Fig. 1. Schematic relationship between MGI induced and constitutive changes in gene expression and the control of cell proliferation and differentiation in normal and different types of leukemic myeloblasts. ——→, MGI induced changes; ---→, constitutive changes; •, blocked for induction of changes by MGI; T, terminal stage of differentiation.

The MGI independent proliferation of myeloid leukemic cells appears to proceed through stages (Collins, Gallo and Gallagher, 1977), probably starting from a small decrease to finally a complete loss of the requirement for MGI for cell viability and proliferation. The study of these stages, which may be due to stages in cell susceptibility to different molecular forms of MGI (Lotem, Lipton and Sachs, 1980), would allow a further dissection of the constitutive pathways associated with the initial uncoupling of the requirement for MGI for proliferation. Myeloid leukemic cells show both a clonal difference in inducibility by the physiological inducer MGI, and a clonal difference in inducibility for differentiation by steroid hormones and by non-physiological inducers such as actinomycin D, dimethylsulfoxide, or tumor promoting phorbol esters (Sachs, 1974; 1977; 1978; Fibach, Hayashi and Sachs, 1973; Lotem and Sachs, 1974; 1975; 1977a; 1978; 1979; 1980; Ichikawa, Maeda and Horiuchi, 1976; Krystosek and Sachs, 1976; 1977; Honma, Kasukabe, Okabe and Hozumi, 1977; Maeda and Sachs, 1978; Weiss and Sachs, 1978; Hoffman-Liebermann and Sachs, 1978; Symonds and Sachs, 1979; Oshima, Yamada and Sugimura, 1979). Erythroleukemic cells that do not respond to the physiological inducer of erythroid differentiation, also show clonal differences in their response to non-physiological inducers (Friend, 1978; Marks and Rifkind, 1978). The difference in response to steroids and non-physiological inducers, like the difference in response to the physiological inducer MGI, may also be due to constitutive expression of specific pathways in the non-responding cells.

ORIGIN AND DEVELOPMENT OF MALIGNANCY IN OTHER TYPES OF TUMORS AND EXPRESSION OF SPECIALIZED FUNCTIONS IN MALIGNANT CELLS

This suggestion for the origin and development of myeloid leukemia may be applicable to malignant tumors derived from other types of normal cells, whose viability, proliferation and differentiation is induced by other physiological inducers. Different tumor cells can also express certain functions that are induced in normal development and yet the cells are malignant. These functions include the production of myeloma proteins (Potter, 1972; Natvig and Kunkel, 1973), foetal proteins, hormones and other specialized products (Weinhouse, 1972; Norgaard-Pederson and Axelsen, 1978; Potter, 1978; Lehmann, 1979; Lotem and Sachs, 1979; Siden, Baltimore and Clark, 1979; Unkeless, Kaplan and Plutner, 1979). This could include cases where some of the tumor cells behave like $MGI^{+}D^{+}$ cells and are induced for expression of these products by the appropriate physiological inducer. It can, however, also be suggested that these products, including myeloma and foetal proteins, which are induced by the appropriate physiological inducer during normal development, are constitutive in the malignant cells and this prevents the cells proceeding to terminal differentiation. The synthesis of these products may indeed be an indication of constitutive changes in these malignant cells and this can be experimentally tested. The model of specific constitutive pathways of gene expression that uncouple the controls for proliferation and differentiation in myeloid leukemia, may thus be a general model to explain the origin and development of malignancy and the expression of specialized functions of normal development in different types of tumors.

REFERENCES

Austin, P.E., McCulloch, E.A., and Till, J.E.: 1971, J. Cell. Physiol. 77, pp. 121-134.

Azumi, J., and Sachs, L.: 1977, Proc. Natl. Acad. Sci. USA 74, pp. 253-257.

Barrett, J.C., and Ts'o, P.O.P.: 1978a, Proc. Natl. Acad. Sci. USA 75, pp. 3297-3301.

Barrett, J.C., and Ts'o, P.O.P.: 1978b, Proc. Natl. Acad. Sci. USA 75, pp. 3761-3765.

Berenblum, I.: 1975, in Cancer, ed. Becker, F.F. (Plenum, New York),Vol. 1, pp. 323-344.

Berwald, Y., and Sachs, L.: 1963, Nature 200, pp. 1182-1184.

Bradley, T.R., and Metcalf, D.: 1966, Aust. J. Exp. Biol. Med. Sci. 44, pp. 287-300.

Chen, T.T., and Heidelberger, C.: 1969, Int. J. Cancer 4, pp. 166-178.

Cline, M.J., and Golde, D.W.: 1979, Nature 277, pp. 177-181.

Collins, J.S., Gallo, R.C., and Gallagher, R.E.: 1977, Nature 270, pp. 347-349.

Dewey, M.J., Martin, D.W.Jr., Martin, G.R., and Mintz, B.: 1977, Proc. Natl. Acad. Sci. USA 74, pp. 5564-5568.

DiPaolo, J.A., Donovan, P., and Nelson, B.: 1969, J. Natl. Cancer Inst. 42, pp. 867-876.

Douer, D., and Sachs, L.: 1979, J. Immunol. 122, pp. 2473-2477.

Fibach, E., Hayashi, M., and Sachs, L.: 1973, Proc. Natl. Acad. Sci. USA 70, pp. 343-346.

Fibach, E., Landau, T., and Sachs, L.: 1972, Nature New Biology 237, pp. 276-278.

Fibach, E., and Sachs, L.: 1974, J. Cell. Physiol. 83, pp. 177-185.

Fibach, E., and Sachs, L.: 1975, J. Cell. Physiol. 86, pp. 221-230.

Fibach, E., and Sachs, L.: 1976, J. Cell. Physiol. 89, pp. 259-266.

Friend, C.: 1978, Harvey Lectures 72, New York, Academic Press, pp. 253-281.

Ginsburg, H., and Sachs, L.: 1963, J. Natl. Cancer Inst. 31, pp. 1-40.

Hayashi, M., Fibach, E., and Sachs, L.: 1974, Int. J. Cancer 14, pp. 40-48.

Hoffman-Liebermann, B., and Sachs, L.: 1978, Cell 14, pp. 825-834.

Honma, Y., Kasukabe, T., Okabe, J., and Hozumi, M.: 1977, J. Cell. Physiol. 93, pp. 227-235.

Huberman, E., and Sachs, L.: 1976, Proc. Natl. Acad. Sci. USA 73, pp. 188-192.

Huberman, E., Sachs, L., Yang, S.K., and Gelboin, H.V.: 1976, Proc. Natl. Acad. Sci. USA 73, pp. 607-611.

Huberman, E., Salzberg, S., and Sachs, L.: 1968, Proc. Natl. Acad. Sci. USA 59, pp. 77-82.

Ichikawa, Y.: 1969, J. Cell. Physiol. 74, pp. 223-234.

Ichikawa, Y., Maeda, M., and Horiuchi, M.: 1976, Int. J. Cancer 17, pp. 789-797.

Ichikawa, Y., Pluznik, D.H., and Sachs, L.: 1966, Proc. Natl. Acad. Sci. USA 56, pp. 488-495.

Ichikawa, Y., Pluznik, D.H., and Sachs, L.: 1967, Proc. Natl. Acad. Sci. USA 58, pp. 1480-1486.

Klein, G.: 1979, Proc. Natl. Acad. Sci. USA 76, pp. 2442-2446.

Klein, G., Frieberg, S., Wiener, F., and Harris, H.: 1973, J. Natl. Cancer Inst. 50, pp. 1259-1268.

Krystosek, A., and Sachs, L.: 1976, Cell 9, pp. 675-684.

Krystosek, A., and Sachs, L.: 1977, J. Cell. Physiol. 92, pp. 345-352.

Landau, T., and Sachs, L.: 1971, Proc. Natl. Acad. Sci. USA 68, pp. 2540-2544.

Landolph, J.R., and Heidelberger, C.: 1979, Proc. Natl. Acad. Sci. USA 76, pp. 930-934.

Lehmann, F.G.: Carcino-embryonic proteins. Elsevier, Amsterdam, 1979.

Levine, L., and Hassid, A.: 1977, Biochem. Biophys. Res. Commun. 79, pp. 477-484.

Liebermann, D., Hoffman-Liebermann, B., and Sachs, L.: 1980, Develop. Biol. in press.

Liebermann, D., and Sachs, L.: 1979, Proc. Natl. Acad. Sci. USA 76, pp. 3353-3357.

Lotem, J., Lipton, J., and Sachs, L.: 1980, Int. J. Cancer, in press.

Lotem, J., and Sachs, L.: 1974, Proc. Natl. Acad. Sci. USA 71, pp. 3507-3511.

Lotem, J., and Sachs, L.: 1975, Int. J. Cancer 15, pp. 731-740.

Lotem, J., and Sachs, L.: 1977a, Proc. Natl. Acad. Sci. USA 74, pp. 5554-5558.

Lotem, J., and Sachs, L.: 1977b, J. Cell. Physiol. 92, pp. 97-108.

Lotem, J., and Sachs, L.: 1978, Proc. Natl. Acad. Sci. USA 75, pp. 3781-3785.

Lotem, J., and Sachs, L.: 1978, Int. J. Cancer 22, pp. 214-220.

Lotem, J., and Sachs, L.: 1979, Proc. Natl. Acad. Sci. USA 76, pp. 5158-5162.

Lotem, J., and Sachs, L.: 1980, Int. J. Cancer, in press.

Maeda, S., and Sachs, L.: 1978, J. Cell. Physiol. 94, pp. 181-186.

Marks, P., Rifkind, R.A.: 1978, Ann. Rev. Biochem. 47, pp. 419-448.

Metcalf, D.: 1969, J. Cell. Physiol. 74, pp. 323-332.

Miao, R.N., Fieldsteel, A.H., and Fodge, D.W.: 1978, Nature 274, pp. 271-272.

Mintz, B., and Illmensee, K.: 1975, Proc. Natl. Acad. Sci. USA 72, pp. 3585-3589.

Mintz, U., and Sachs, L.: 1973, Blood 42, pp. 331-340.

Natvig, J.B., and Kunkel, H.G.: 1973, Adv. Immunol. 16, pp. 1-54.

Norgaard-Pederson, B., and Axelsen, N.H.: 1978, Scand. J. Immunol. Suppl. No. 8 Vol. 8, pp. 1-683.

Oshima, G., Yamada, M., and Sugimura, T.: 1979, Biochem. Biophys. Res. Commun. 90, pp. 158-163.

Paran, M., Ichikawa, Y., and Sachs, L.: 1968, J. Cell. Physiol. 72, pp. 251-254.

Paran, M., Sachs, L., Barak, Y., and Resnitzky, P.: 1970, Proc. Natl. Acad. Sci. USA 67, pp. 1542-1549.

Pluznik, D.H., and Sachs, L.: 1965, J. Cell. Comp. Physiol. 66, pp. 319-324.

Pluznik, D.H., and Sachs, L.: 1966, Exp. Cell. Res. 43, pp. 553-563.

Potter, M.: 1972, Physiol. Rev. 52, pp. 632-719.

Potter, V.R.: 1978, Brit. J. Cancer 38, pp. 1-23.

Rabinowitz, Z., and Sachs, L.: 1968, Nature 220, pp. 1203-1206.

Rabinowitz, Z., and Sachs, L.: 1970a, Nature 225, pp. 136-139.

Rabinowitz, Z., and Sachs, L.: 1970b, Int. J. Cancer 6, pp. 388-398.

Ringertz, N.R., and Savage, R.E.: 1976, Cell Hybrids, New York, Academic Press.

Rovera, G., O'Brien, T.G., and Diamond, L.: 1977, Proc. Natl. Acad. Sci. USA 74, pp. 2894-2898.

Rowley, J.D.: 1977, Proc. Natl. Acad. Sci. USA 74, pp. 5729-5733.

Sachs, L.: 1974, Harvey Lectures 68, New York, Academic Press, pp. 1-35.

Sachs, L.: 1977, Isr. J. Med. Sci. 13, pp. 654-665.

Sachs, L.: 1978, Nature 274, pp. 535-539.

Siden, E.J., Baltimore, D., and Clark, D.: 1979, Cell 16, pp. 389-396.

Symonds, G., and Sachs, L.: 1979, Somat. Cell Genet. 5, pp. 931-944.

Unkeless, J.C., Kaplan, G., Plutner, H. et al.: 1979, Proc. Natl. Acad. Sci. USA 76, pp. 1400-1404.

Weinhouse, S.: 1972, Cancer Res. 32, pp. 2007-2016.

Weiss, B., and Sachs, L.: 1978, Proc. Natl. Acad. Sci. USA 75, pp. 1374-1378.

Wiener, F., Klein, G., and Harris, H.: 1974, J. Cell. Sci. 15, pp. 177-183.

Wigler, M., and Weinstein, I.B.: 1976, Nature 259, pp. 232-233.

Yamamoto, T., Rabinowitz, Z., and Sachs, L.: 1973, Nature New Biology 243, pp. 247-250.

Yamasaki, H., Fibach, E., Nudel, V., Weinstein, E.B., Rifkind, R.A., and Marks, P.A.: 1977, Proc. Natl. Acad. Sci. USA 74, pp. 3451-3455.

Zur Hausen, H., Bornkmann, G.W., Schmidt, R., and Hecker, E.: 1979, Proc. Natl. Acad. Sci. USA 76, pp. 782-785.

AMPLIFICATION OF CARCINOGENESIS BY NON-CARCINOGENS - DITERPENE TYPE PROMOTERS AND MODELS OF ENVIRONMENTAL EXPOSURE

E. Hecker
Institut für Biochemie
Deutsches Krebsforschungszentrum
Im Neuenheimer Feld 280, D-6900 Heidelberg 1
Federal Republic of Germany

SUMMARY

In the last 15 years the cocarcinogens of the polyfunctional diterpene ester type originating from plants were established molecularly and biologically as non-carcinogenic amplifiers of carcinogenesis. Thus the long suspected concept of amplification of carcinogenesis by non-carcinogens was finally verified. In addition the new principle of "cryptic" cocarcinogens was detected in the form of higher esters of diterpenes. Besides for investigations in molecular cytology and molecular mechanisms of carcinogenesis, cocarcinogens of this type may be used in models of environmental exposure to investigate the role of cocarcinogens in general as possible carcinogenic risk factors of the human environment. Three model situations are devised and investigated: (i) Several polyfunctional diterpenes were detected to be present in roots of Croton flavens L. and identified as highly active irritants and cocarcinogens of the promoter type. The utilization of parts of this plant according to local habits may be suspected therefore to be responsible for the unusually high incidence of esophageal cancer on Curacao. (ii) Many but not all species of the Euphorbiaceae and Thymelaeaceae families utilized as ornamentals were shown to contain irritant and cocarcinogenic diterpene esters. Provided intensive contacts are avoided with skin and/or by ingestion of plant parts, it is concluded that most likely an actual carcinogenic risk does not exist by maintenance of single species. However, it is called for epidemiological investigations of a possible risk in persons involved professionally in raising and handling of mass cultures of species of Euphorbiaceae and Thymelaeaceae containing irritants. (iii) It was shown that nectar collected by honey bees from certain species of Euphorbiaceae may contain irritant polyfunctional diterpenes of the promoter type. - The actual new findings i - iii suggest that in the etiology of human tumors besides solitary carcinogens, the classical first order carcinogenic risk factors, diterpene ester type co-carcinogens of the promoter type - non-carcinogenic amplifiers of carcinogenesis - may contribute. As compared to solitary carcinogens cocarcinogens of whatever type may be classified as second order carcinogenic risk factors.

B. Pullman, P. O. P. Ts'o and H. Gelboin (eds.), Carcinogenesis: Fundamental Mechanisms and Environmental Effects, 257–272.

INTRODUCTION

Stimulated by previous clinical observations the existence of solitary carcinogens (for proposals of a standardized terminology see Hecker 1976a-d) was verified experimentally for the first time by Yamagiwa and Ichikawa in 1914. Meanwhile in animal experiments about 700 solitary carcinogens were detected, most of them of exogenous origin. Since about 1950 it was learned additionally that solitary carcinogens present in the human environment may not solely be of industrial origin and therefore man-made, but occur also naturally as products of the plant kingdom. This important aspect of environmental carcinogenesis was covered by International Symposia e.g. by the UICC in 1968 and by the Princess Takamatsu Cancer Research Fund in 1979 (see references). Of all experimentally detected solitary carcinogens about 30 are considered solitary carcinogenic for human beings (Tomatis et al. 1978). In terms of the carcinogenic risk imposed on the human environment solitary carcinogens of whatever origin may be classified as first order carcinogenic risk factors (Hecker 1978b, 1979).

In most countries all over the world every 5th death is caused by cancer. The incidence of cancer is even higher: throughout his life about every 3rd to every 4th human being is inflicted. On this background it appears a disenchanting fact that up to now despite the large number of solitary carcinogens known experimentally the etiology of only a few human cancers is known beyond doubt: e.g. for the so-called "occupational cancers". Yet, according to data from WHO (1977) occupational cancers account for only 1 - 5 % of all cancers. Therefore, indeed, the carcinogenic risk factors involved in the generation of the majority and even of the most frequently occuring human cancers, such as for example mammary carcinoma, carcinoma of the stomach, carcinoma of the colon etc. are unknown. As a consequence, in the last about ten years the idea began to develop that, for many human cancers, the simple "unifactorial" relation between cause and consequence called solitary carcinogenesis may be the exception rather than the rule. The idea of "multifactorial" etiologies of human cancers was particularly stimulated when the existence of a new class of carcinogenic factors was established experimentally: they are non-carcinogenic by themselves yet capable to strongly amplify carcinogenesis "initiated" by subcarcinogenic exposure to solitary carcinogen(s). Therefore these amplifiers of carcinogenesis were collectively called "cocarcinogens" (Shear 1938). In the course of life of the individual human being it is only too likely that, depending on its individual "life style" (Higginson 1978), exposures to various well defined carcinogenic factors - solitary and co-carcinogens - may occur simultaneously or consecutively. Therefore, indeed, in the etiology of human cancer combination effects of more than one carcinogenic risk factor(s) collectively called syncarcinogenesis, may prevail and thus come more close to every day reality than solitary carcinogenesis. Since per se multifactorial effects are of a considerable degree of complexity, syncarcinogenesis calls for experimental investigation in form of thoughtfully selected "prototype processes" as model systems (see footnotes a,b at the end of this article).

POLYFUNCTIONAL DITERPENES: MOST ACTIVE, YET PER SE NON-CARCINOGENIC AMPLIFIERS OF CARCINOGENESIS

Amplification of carcinogenesis as the result of combination effects of more than one carcinogenic risk factor (i.e. syncarcinogenesis) attains increasing attention in experimental oncology in recent years (see footnote a.). Those prototype processes of syncarcinogenesis representing amplification of carcinogenesis by non-carcinogens deserve particular interest since they define a new class of carcinogenic risk factors: the cocarcinogens (see footnote b,c).

Fig. 1

UNI - AND MULTIFACTORIAL PROCESSES OF CARCINOGENESIS

For Cocarcinogenesis the Prototype Initiationpromotion is described

Exposure of Organism	Result (Tumors)	Processes of Carcinogenesis
■ ■ ■ ■ ■	+	Solitarycarcinogenesis
■	0	–
□ □ □ □	0	–
■ □ □ □ □	+	Cocarcinogenesis

■ = Solitarycarcinogen (Initiator) □ = Cocarcinogen (Promoter)

Already from the early results of experimental oncology it was postulated that exposure of the host to even a single small dose of a solitary carcinogen may produce in the cells of any target tissue one or more permanent (irreversible) biochemical lesion(s). During chronic exposure such lesions may sum up (Fig.1, solitary carcinogenesis) yielding, within the lifespan of the organism exposed, the phenotypical manifestation of cancer (summation theory, Druckrey and Küpfmüller, 1948, Druckrey, 1967). This theory implies that the biochemical lesion(s) imposed by exposure to small dose(s) of (or for a short time interval to) a solitary carcinogen must not necessarily yield cancer within the lifespan of an organism (Fig.1, line 2). Rather, it was postulated that subcarcinogenic exposure of this kind may generate an irreversible latent disposition for cancer at the genetic or molecular level (somatic mutation hypothesis, Bauer 1928, 1963). That this truly may be the case was shown experimentally using a basic fraction of creosote oil as a "co-carcinogen" (Shear, 1938). Thus an up to then unknown toxicologic quality of action was detected called cocarcinogenic activity, i.e. amplification of carcinogenesis by a per se non-carcinogenic factor. In fact, in the ideal experiment, cocarcinogens do not produce tumors - not even by chronic exposure of the target tissue (Fig.1, third line). In the fourties this concept was worked out (Berenblum 1941 a,b, Mottram 1944, Berenblum and Shubik 1947) using mouse skin as the target tissue and croton oil as cocarcinogen: after "initiation" with a "subcarcinogenic dose" of a solitary carcinogenic polycyclic aromatic hydrocarbon (PAH, Fig.1, line 2) tumors were generated experimentally if exposure to croton oil follows to "promote" initiation (Fig.1, line 4). This prototype process of cocarcinogenesis was called initiation (or tumor) promotion (see footnote b.). While Berenblum, Mottram and Shubik have used croton oil as an initiation promoter, it was shown by Pound and Bell (1962) and by Shinozuka and Ritchie (1967) that croton oil exhibits also an initiation modifying activity (see footnote c.). In contrast to the "initiating" effects of initiators, those of promoters were postulated to be reversible (e.g. Boutwell 1964).

Although studied intensively for about twenty years the cocarcinogenic quality of action (i.e. amplification of carcinogenesis by non-carcinogens) was not fully accepted in experimental oncology. Some investigators even denied entirely the existence of this non-classic toxicologic quality (for a review of arguments see Hecker 1972, 1976a-d). It became fully established only in the late sixties, when the active principles of croton oil were isolated as molecularly uniform factors (Fig.2) and identified chemically as phorbol-12,13-diesters (Hecker, van Duuren). They then were characterized biologically and beyond doubt as per se non-carcinogenic amplifiers of initiation in mouse skin i.e. as true initiation promoters (for reviews see Hecker 1971, Hecker and Schmidt 1974). Subsequently also the initiation modifying property of croton oil was assigned to its phorbol-12,13-diesters (van Duuren et al. 1971, Goerttler and Löhrke 1976b). In more recent experiments initiation promotion by phorbol-12,13-diesters was demonstrated to take place in organs of mice other than skin (Goerttler and Löhrke 1976a, Goerttler et al. 1979) and in organs of species other than mice if initiation is done properly (Goerttler et al. 1980a,b).

Fig.2

COCARCINOGEN (PROMOTER) CROTON OIL FACTOR A_1 (TPA), ITS DITERPENOID PARENT ALCOHOL PHORBOL AND CORRESPONDING CRYPTIC COCARCINOGEN (CRYPTIC PROMOTER)

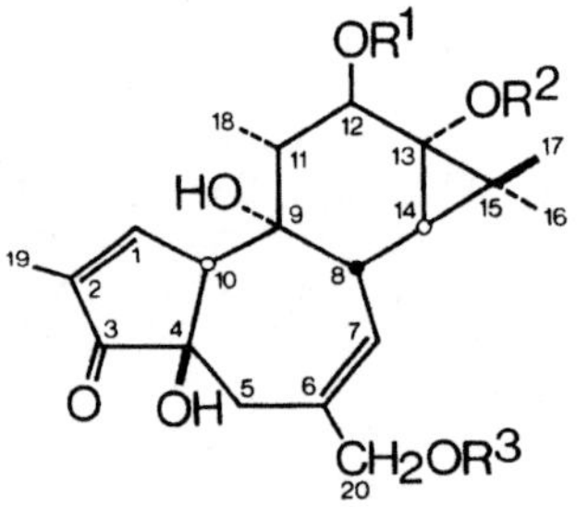

Phorbol:	$R^1 = R^2 = R^3 = H$	inactive
Croton Oil Factor A_1 (TPA):	$R^1 = CO(CH_2)_{12}CH_3$; $R^2 = COCH_3$; $R^3 = H$	promoter
TPA-20-ester:	R^1 = and R^2 as in TPA; $R^3 = COR'$	cryptic promoter

The typical structure of the prototype of the new class of promoters, croton oil factor A1,12-O-tetradecanoylphorbol-13-acetate (TPA), is shown in Fig.2. Some related structures of this prototype will be introduced below as possible risk factors in human cancer. In fact and quite in contrast to solitary carcinogens such as initiating PAH, the cocarcinogenic phorbol-12,13-diesters do not require metabolic activation (Hecker 1978b). Also, altogether "ultimate promoters", they do not exhibit mutagenic activities in bacteria (Soper and Evans 1977) and yeast (Hecker et al. 1978, Hecker 1979) and do not show initiating activities in mouse skin (Hecker et al. 1978). Further, again in contrast to "ultimate initiators" (activated metabolites of PAH), they do not exhibit alkylating activities (Hecker et al. 1978) and do not bind covalently to the fraction of soluble proteins of mouse skin (Traut et al. 1971). Thus, as far as the parameters investigated are concerned the mechanism of action of solitary carcinogens and of cocarcinogens of the phorbol ester type is entirely different. Such differences have been found with respect to many more of their biological and biochemical effects (e.g. Hecker et al. 1979b) and establish the existence of the new quality of action as per se reversible entity.

In addition to the naturally occuring phorbolester type cocarcinogens a variety of structurally different promoters of mouse skin of industrial origin is known yet, of much lower biological activity (e.g. Hecker 1972). Furthermore, also many pharmaceutical drugs of diverse structures have been detected to exhibit cocarcinogenic activity primarily in liver (e.g. Kunz et al. 1978, Kunz et al. in press).

Fig.3

THE DITERPENE PARENTS OF THE BIOLOGICALLY ACTIVE PRINCIPLES OF EUPHORBIACEAE AND THYMELAEACEAE, trivial names, structures and stereochemistry of parent hydrocarbons and prototype polyfunctional parent alcohols

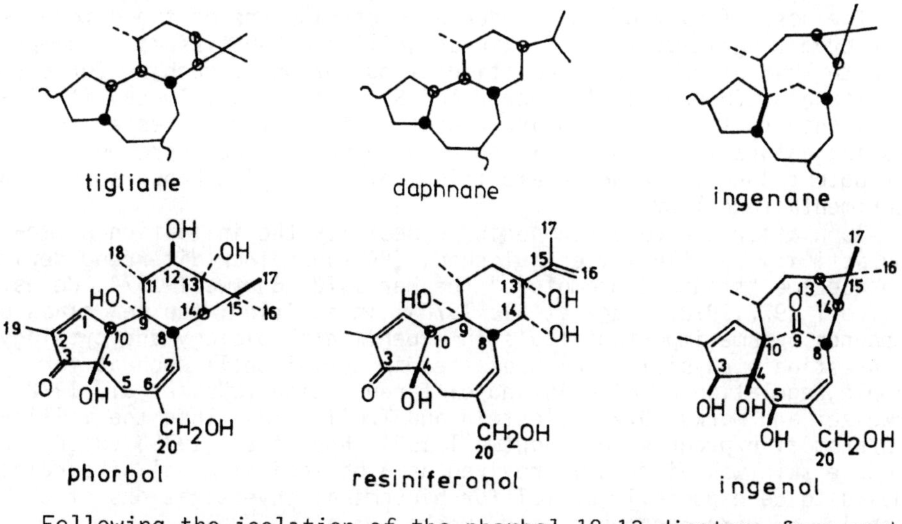

Following the isolation of the phorbol-12,13-diesters from croton oil it was shown that cocarcinogens of the promoter type with similar

structures are widely distributed in plant species which belong to the botanical families of the Euphorbiaceae and of the Thymelaeaceae (e.g. Hecker 1976e). Chemically they were identified as polyfunctional diterpenes with either one of three typical carbon skeletons: the tetracyclics tigliane and ingenane and the tricyclic daphnane (Fig.3). Up to now, including the prototype parent alcohols phorbol, resiniferonol and ingenol (Fig.3), more than 18 tri- and tetracyclic polyfunctional diterpene parent alcohols and a multiplicity of molecularly uniform partial esters thereof are known. Most of them have been characterized biologically as skin irritants and many, but not all of them also show initiation promoting activities. As a rule, their parent alcohols are neither irritant and practically non-promoting in mouse skin (Hecker 1976e, 1978b). Under certain circumstances, however, they may be active as weak promoters of internal organs (e.g. Armuth et al. 1979).

Besides the highly active polyfunctional diterpene esters derived from parent alcohols such as phorbol, resiniferonol ingenol (Fig.3) and others, many of the plant materials investigated, including croton oil, often but not always contain further diterpene ester entities called "cryptic" irritants (and promoters). They were identified as 20-esters of irritants and promoters of the tigliane and ingenane type. For example, of the entire phorbol content of croton oil about 50 % revealed to be phorbol-12,13,20-triesters (Fig.2). Upon testing cryptic irritants and promoters show neither impressive irritant nor promoting activities. Yet their estergroup in position 20 may be split selectively, either by chemical or encymatic means i.e. "activated" (Hecker 1976e, 1978b) to yield the corresponding "ultimate" promoters. Thus the 20-esters of irritants and promoters represent naturally occuring "cryptic" amplifiers of carcinogenesis i.e. "cryptic co-carcinogens". They deserve particular attention as concerns environmental cocarcinogenesis (see below).

The dose of many of the exogenous cocarcinogens of the diterpene ester promoting initiation of skin as well as other tissues is comparable to that of proliferation stimulating hormones, such as for example estrogens (Hecker 1971, Hecker and Schmidt, 1974). To-day it is an established fact that the cocarcinogens of the diterpene ester type occuring naturally in Euphorbiaceae and Thymelaeaceae represent the most potent non-carcinogenic amplifiers of initiation known so far in experimental oncology.

Soon after the cocarcinogenic, especially the initiation promoting activity was finally established, TPA, and other diterpene derivatives became "standard promoters" (Hecker 1972, Boutwell 1974, Weinstein and Troll 1977,1978, Slaga et al. 1978). In addition, the new class of compounds became important tools in experimental biology and cytology to investigate physiological processes in normal cells such as for example regulation of mitosis and differentiation (Süss et al. 1972, Schweizer and Marks 1977, Weinstein and Troll 1978). Thus the initiation/promotion/progression (Boutwell 1964, Roe et al. 1972) experiment in mouse skin was vigorously revived as a biological model of carcinogenesis to be especially useful for biochmical investigations of the mecahnism of carcinogenesis (e.g. Hecker et al. 1979b).

MODELS OF ENVIRONMENTAL EXPOSURE TO TUMOR PROMOTERS OF THE DITERPENE ESTER TYPE

Based upon clinical and experimental data the existence of cocarcinogens - non-carcinogenic amplifiers of carcinogenesis - was suspected for a long time. Already Virchow was considering in depth the relationship between chronic irritation and cancer (see Bauer 1963). Also, soon after the estrogenic hormones were isolated, their possible role as endogenous cocarcinogens was suspected and investigated (Butenandt 1949, 1950, Kaufmann et al. 1949, Dontenwill 1966). However, in the absence of conclusive evidence for corresponding initiators, other researchers considered these hormones just solitary carcinogens. An unequivocal final clarification of this problem, which is of well known actuality in connection with the pill (Upton 1978, WHO 1978), is still open. Yet, recent findings on the activation of latent Epstein-Barr-Virus by the cocarcinogens of the diterpene ester type in cell culture (zur Hausen et al. 1978, 1979) throw an interesting new light on this problem.

Altogether, as compared to solitary carcinogenesis, it is much more difficult to provide conclusive evidence for multifactorial etiologies of cancer and in particular for cocarcinogenesis as exemplified by the long period of time which was required to finally establish this nonclassical toxicologic quality. As a consequence of the rather complex situation regarding the possible contribution of cocarcinogens to the overall carcinogenic load of the human environmental, models of environmental exposure are highly desireable for more detailed evaluation. Here again the investigation of promoters of the diterpene ester type may pave the way towards a better understanding also of the environmental aspects of cocarcinogenesis. As a lead to identify possible environmental models populations at risk of direct or indirect exposure to cocarcinogens of the diterpene ester type may be defined, based upon the well known utilization of plants or plant parts of species of the Euphorbiaceae and Thymelaeaceae families.

Thus, for example, direct exposures may be suspected by use of plants or plant parts as drugs in officinal and in folk medicine - such as for example croton oil and many others. In fact quite a few drugs of this origin are recorded as officinal drugs in the pharmacopeias of many countries up until today. They may be a source of an iatrogenic risk. Also, many species of these botanical families are grown as ornamental plants in family homes, gardens and parks allover the world. Others are being used directly or as raw materials to produce bulk food (e.g. oils and fats, starch) or fine food (e.g. certain beverages, honey, chewing gum). Further, diverse Euphorbiaceae are of technological importance to produce, for example, timber and rubber or according to most recent concepts, gasoline. Indirect exposure of human beings may result from use of certain sources of protein: in some countries species of Euphorbiaceae and Thymelaeaceae are being used in veterinary medicine and residual amounts may be retained in milk and in meat. Further they are used to bulk feed cattle or may occur casually as contamination in food of cattle, especially during dry seasons. In primitive fishing many Euphorbiaceae and also several species of Thymelaeaceae are used as highly active fish poisons.

These and other utilizations of Euphorbiaceae and Thymelaeceae may provide model situations to investigate the carcinogenic risk possibly involved. To do this it is necessary to know in the first place the polyfunctional diterpenes contained in the plant(s) or plant part(s) and to follow-up their appearance in the (processed) product(s) finally consumed. If the outcome of such investigations is positive it would be desireable to demonstrate the definite existence of the risk indicated by epidemiological investigations. Even legislative action might be desirable before final results of long term epidemiological investigations are available. Thus, already some time ago, based upon our previous results (Hecker 1971, Hecker and Schmidt 1974) croton oil was eliminated from the German Pharmacopoeia as an officinal drug. Further, since 1978, by legislative act of the health authorities of the Federal Republic of Germany, croton oil is no longer available without written prescription by an approved medical doctor.

Concerning the contents of polyfunctional diterpenes of Euphorbiaceae and Thymelaeaceae the results of our ongoing screening program (Hecker 1976e, 1978b) allow the following generalizations to be made:

1. Species of the plant families of the Euphorbiaceae and the Thymelaeaceae do not necessarily contain skin irritant activity and/or polyfunctional diterpenes.
2. In case any species contains irritant activity it is most likely represented by polyfunctional diterpene esters.
3. Presence of skin irritant polyfunctional diterpene esters does not necessarily mean initiation (or tumor) promoting activity.

Therefore, for risk/benefit evaluations every particular case of utilization of Euphorbiaceae or Thymelaeaceae by human beings has to be investigated separately as to the presence and chemical nature of irritant and promoting polyfunctional diterpenes.

At present three possible model situations are being investigated: (i) a certain plant species utilized in every day life on Curacao - that Caribbian island with particularly high rate of esophageal cancer, (ii) ornamental plants of Euphorbiaceae and Thymelaeaceae and (iii) the possibility of polyfunctional diterpenes to occur in nectar collected by bees from Euphorbiaceae species to produce honey.

(i). In the native population of the island of Curacao carcinoma of the esophagus ranks high above average as established by Dutch and American pathologists since about 1920. By interview of the local population an American economic botanist, Dr. Julia F. Morton, detected that plant parts from the attractively smelling bush Croton flavens L., growing abundantly on the island, are being used in every day life for a variety of purposes (Morton 1971). Thus the roots of this bush are used as a substitute of chewing gum. Its leaves are used as an insecticide or as an insect repellant and also as a cheap detergent for dish washing. Also they are being kept in the mouth to cope painful irritations. Further, the young leaves and the tips of twigs are used to prepare a "bush tea" as an every day beverage. Based upon the local habits unraveled by inquires a causal relation between them and the high rate of carcinoma of the esophagus on the island was considered possible (Morton 1971). Consequently extracts of Croton flavens were investigated for solitary carcinogenic activity in mice, rats and hamsters. However, negative results were obtained (O'Gara et al., 1971).

Fig.4

TYPICAL CROTON FACTORS FROM ROOTS OF CROTON FLAVENS L. AND THEIR PARENT ALCOHOLS

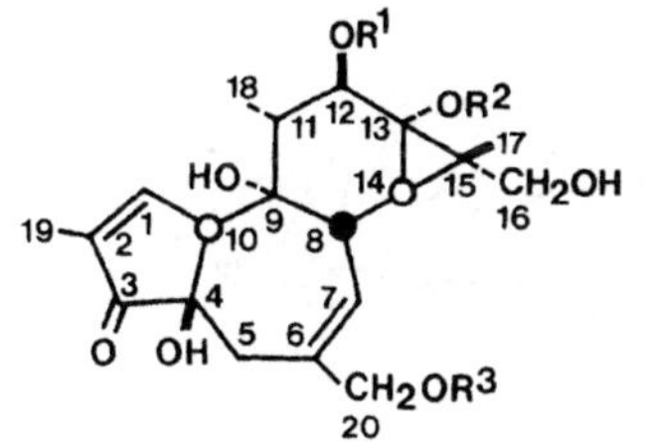

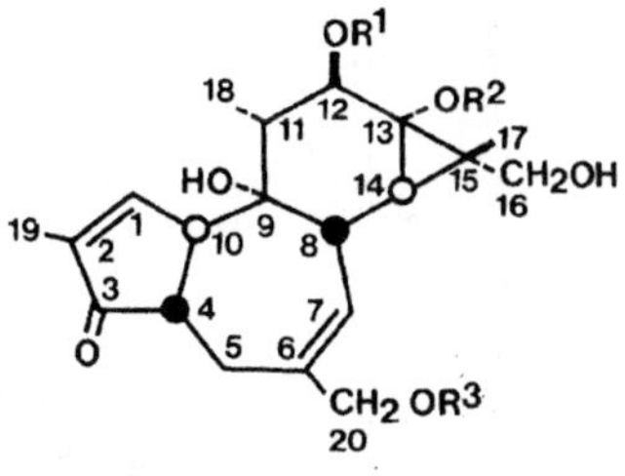

Croton Factors F_1 and F_1'

16-hydroxyphorbol:	$R^1 = R^2 = R^3 = H$	inactive
Factor F_1:	R^1 = hexadecanoyl R^2 = acetyl $R^3 = H$	promoter
Factor F_1':	$R^1 - R^2$ as in F_1 R^3 = decanoyl	cryptic promoter

Croton Factors F_2 and F_2'

4-deoxy-16-hydroxyphorbol:	$R^1 = R^2 = R^3 = H$	inactive
Factor F_2:	R^1 = hexadecanoyl R^2 = acetyl $R^3 = H$	promoter
Factor F_2':	$R^1 - R^2$ as in F_2 R^3 = decanoyl	cryptic promoter

In 1973 we began to investigate the plant parts of Croton flavens utilized on Curacao with respect to irritants and cocarcinogens of the initiation promoter type. Extracts of both the roots as well as the young leaves and tips of twigs were found to exhibit considerable irritant activity in our standard assay on the mouse ear (Hecker and Weber, 1977).

From the extract of the roots six molecularly uniform new croton factors F1-F3 and F'1-F'3 were isolated (Fig.4). They were identified as polyfunctional diterpene esters of the tigliane type: the croton factors F1-F3 representing the structural type of "free" irritants and the factors F'1-F'3 representing the structural type of "cryptic" irritants. As polyfunctional parent alcohol, the factor F1 and the corresponding cryptic factor F'1 contain 16-hydroxyphorbol (Fig.4). The croton factors F3 and F'3 differ from F1 and F'1 only in the ester function R1 which is tetradecanoyl in the former instead of hexadecanoyl. The croton factors F2 and F'2 are corresponding esters of 4-deoxy-16-hydroxyphorbol (Fig.4). Of the croton factors F tested on the back skin of mice for promoting activity in our standard assay F1 and F2 revealed strong promoters. Their activity is qualitatively and quantitatively comparable with that of croton oil factor A1 or TPA (Fig.5). Hence they may be considered as "ultimate" promoters. The "cryptic" types F', more difficult to detect as factors of potential carcinogenic risk in actual tests, may be a source of "ultimate" promoter(s) to be released by esterases in target tissues (Hecker and Weber, 1977). - In ongoing investigations we are now concerned with the contents of the constituents of the extract from young leaves and tips of twigs used for the "bush tea" (Lutz and Hecker, 1979) and with the tea itself as prepared according to the orginal recipe.

Fig.5

COCARCINOGENIC (PROMOTING) ACTIVITY OF CROTON FACTORS F_1 AND F_2 IN MOUSE SKIN COMPARED WITH THAT OF CROTON OIL FACTOR A_1 (TPA)

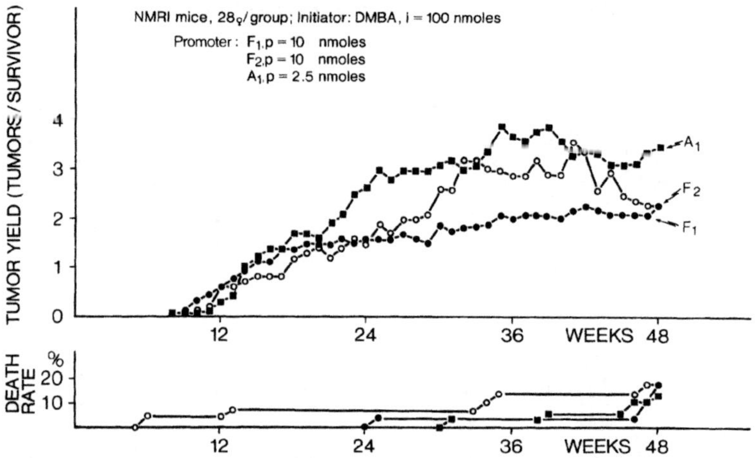

Since per se cocarcinogens of the promoter type do not produce cancer, for environmental models of this kind hypotheses as to the possible nature of the initiator(s) involved have to be provided. With respect to the local conditions of Curacao such hypothesis may postulate exposure to PAH by drinking water contamined with oil (see Weber 1976). It seemingly is supported by recent experimental evidence: by intragastric administrations to mice tumors may be produced in their forestomach in initiation/promotion experiments using DMBA/TPA (Goerttler et al. 1979) and in the stomach of rats using N-methyl-N-nitro-N-nitrosoguanidine (MNNG)/croton oil (Sugimura et al. 1978). Thus the utilization of croton flavens on Curacao with its high incidence of esophageal cancer and the diterpene ester type promoters already demonstrated to be present in roots of this plant indicates that cocarcinogens of the promotertype may not be neglegible as carcinogenic risk factors. However, as compared to solitary carcinogens, the classical first order carcinogenic risk factors, for obvious reasons the degree of carcinogenic risk involved is different: cocarcinogens are considered as second order carcinogenic risk factors (e.g. Hecker 1978b, 1979).

(ii). How about the possible carcinogenic risk involved in maintenance of ornamental plants of Euphorbiaceae and Thymelaeaceae such as, for example, the most popular Poinsettia (Euphorbia pulcherrima)? In view of the generalizations 1-3 derived out of our screening program (see above) and with respect to so many ornamental plants which we already investigated in detail, it was pointed out (Hecker, 1978a) that, according to present knowledge, maintaining and handling of single species of such plants most likely does not mean an actual risk of cancer if only intensive contacts with the skin or ingestion is avoided. Prevention measures of this kind are generally valid and have been exercised at all times in connection with poisonous ornamentals, the beauty of which we enjoy although we know that they are toxic. However, we call for epidemiological investigations concerning a possible occupational risk of professional gardeners and other persons who are maintaining and handling mass cultures of species from the plant families under consideration (Hecker, 1978a).

(iii). Certain species of Euphorbiaceae and Thymelaeaceae are known to be attractive sources of nectar for honey bees. Thus, according to the literature, Euphorbiaceae species occuring in Africa and in North America yield respectable crops of unpalatable honey. Indeed, it was possible to obtain from Africa honey harvested almost exclusively from one of the species under consideration, Euphorbia coerulescens Haw.. Its latex was shown to contain irritant polyfunctional diterpenes. Thus, the honey was also investigated to see, if and to what extent polyfunctional diterpenes contained in the latex may show up in the nectar secreted by the plants in nectar glands and collected by honey bees. This research program is undertaken as a joint venture of our laboratory with the Institut für Bienenforschung in Celle, Germany (Director: Dr. H.J. Dustmann).

Fig.6

CHEMICAL STRUCTURES, YIELDS AND IRRITANT DOSE 50 OF THE POLYFUNCTIONAL DITERPENES ISOLATED FROM HONEY OF EUPHORBIA COERULESCENS, HARVEST: 1977/78

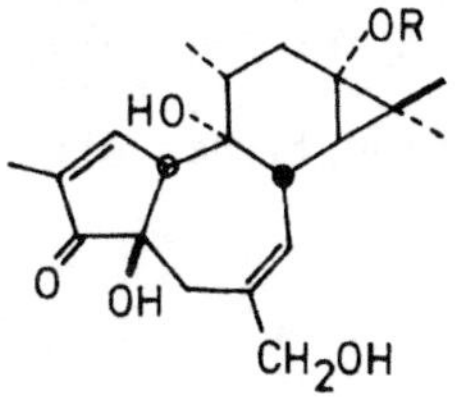

R = H: 12-deoxyphorbol (8,2 ppm), ID_{50} >> 500 µg/ear

R = CO CH=CH $(CH_2)_6$ CH_3: 13-O-decenoyl-12-deoxyphorbol (0,6 ppm), ID_{50} 0,1µg/ear.

R = CO $(CH_2)_{10}$ H_3: 13-O-dodecanoyl-12-deoxyphorbol (0,1 ppm), ID_{50} 0,14µg/ear.

Our present results are summarized in Figs. 6 and 7. From Fig. 6 it may be seen that at least three different polyfunctional diterpenes are present in the honey. The main constituent (8.2 ppm) was identified as the polyfunctional diterpene parent alcohol 12-deoxyphorbol, which is practically inactive as an irritant (and as a promoter). In addition, yet in smaller amounts, the irritant esters of this parent alcohol, 12-deoxyphorbol-13-decenoate (o,6 ppm) and -13-dodecanoate (0,1 ppm) were isolated (Fig.6). The promoting activities of 13-esters of 12-deoxyphorbol, prepared independently by partial synthesis, are presented in Fig.7 in comparison to croton oil factor A1 (TPA). For example, with 4 times the single dose p of TPA, the 12-desoxyphorbol-13-decanoate exhibits definite promoting activity as do all the other esters shown. The former ester is almost identical with the 12-deoxyphorbol-13-decenoate isolated

Fig.7

INITIATION PROMOTING ACTIVITY OF SOME PARTIAL SYNTHETIC ESTERS OF 12-DEOXYPHORBOL AS COMPARED TO CROTON OIL FACTOR A_1 (TPA)

NMRI mice, 28 ♀/group, initiator: DMBA, i = 100 nM.

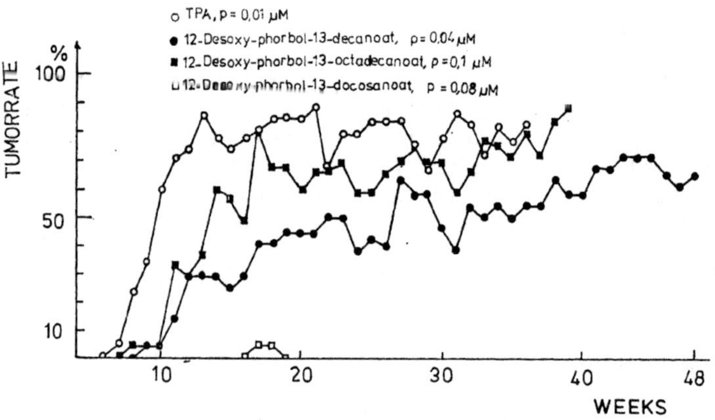

from the honey (Hecker et al. 1979). Most recently the non-irritant polyfunctional diterpene parent alcohol, ingenol, was isolated from honey of Euphorbia seguieriana Neck, but promoting esters and quantitative data are lacking so far (Uphadhyay et al. 1979). Altogether, these results call for more detailed investigations of the possibility of honey to contain promoters of the polyfunctional diterpene ester type (see Ott and Hecker, 1980).

CONCLUSION

As a consequence of the ever increasing scientific knowledge concerning the occurence of solitary carcinogens in the human environment private and governmental organizations, including WHO, have developed means to detect and even more difficult, measures to exclude solitary carcinogens from the environment of human beings. For this purpose basic data on carcinogenic activities of the classical first order carcinogenic risk factors - the solitary carcinogens - are being collected since many years (e.g. Tomatis et al. 1978). They are used as scientific background for risk/benefit evaluations on which measures of prevention can be based. According to experience, however, rather long ranges of time are required to implement in human societies scientific knowledge in general and that on carcinogenic risk factors in particular. Now, that the long suspected role in the human environment of co-carcinogens as second order carcinogenic risk factors has been verified environmental models using promoters of the diterpene ester type. Hence, it would be necessary to start systematic recording and evaluation of environmental cocarcinogens of what ever origin and nature in a similar way as in case of solitary carcinogens. A small but first step in this direction is the "File on Drug Interactions" which has been developed by the Schweizer Apothekerorganisation. It is available also for the German and Austrian organisations of pharmaceutists.

Footnotes:

a. The latent disposition for cancer imposed on a target tissue by subcarcinogenic exposure to solitary carcinogen(s) (e.g. Fig. 1, line 2) may become manifested within life exspectancy of the organism also by subsequent exposure of the target tissue to another solitary carcinogen of identical organotropy and in doses which per se may be also "subcarcinogenic". To differentiate this important prototype process of amplification of carcinogenesis as an alternative to cocarcinogenesis it was proposed to call it "pluricarcinogenesis" (see Hecker 1976a-d). In recent years pluricarcinogenesis has been investigated experimentally to considerable extent (Likachev, 1968, Schmähl 1976). - Some researchers call processes of pluricarcinogenesis just "syncarcinogenesis" (e.g. Schmähl, 1976) neglecting cocarcinogenesis as an alternative prototype process with a different degree of carcinogenic risk.

b. It was proposed that in accordance with the classical terminology used by Shear 1938 amplification of carcinogenic processes involving non-carcinogenic amplifiers be called collectively "co-carcinogenesis" (Hecker 1976a-d, 1978b). More specifically that particular prototype process of cocarcinogenesis in which exposure to the subcarcinogenic dose of the solitary carcinogen is followed by the co-carcinogen (as represented by Fig. 1, line 4) may be called initiation (or tumor) promotion. (Hecker 1978b) and contrasted to initiation modification (see footnote c.).

c. Initiation modification may be considered an alternative prototype process of cocarcinogenesis (Hecker 1976, 1978b) in which alteration of initiation is achieved by exposure of the target tissue to a subcarcinogenic dose of a solitary carcinogen (initiator) preceded by or accompanied with exposure to a co-carcinogen which may or may not be a promoter. If the tissue is promoted subsequently, an amplification of the carcinogenic response in terms of an increase of tumor rate or tumor yield may be achieved as compared to initiation alone followed by promotion. Similarly, in alternative processes of initiation modification inhibition of initiation may be accomplished (Hecker 1976a-d, 1978b). - Some researchers prefer to restrict the term "cocarcinogenesis" as introduced by Shear 1938 to the process of initiation stimulation (e.g. van Duuren et al. 1971, 1973).

REFERENCES

Armuth V, Berenblum I, Adolf W, Opferkuch HJ, Schmidt R, Hecker E (1979) J Cancer Res Clin Oncol 95; 19-28

Bauer KH (1928) Mutationstheorie der Geschwulstentstehung, Springer Verlag, Berlin

Bauer KH (1963) Das Krebsproblem, 2. Aufl., Springer, Berlin Göttingen Heidelberg

Berenblum I (1941a) Cancer Res. 1; 44-48
Berenblum I (1941b) Cancer Res. 1; 807-814
Berenblum I, Shubik P (1947) Brit J Cancer, 1; 383-391
Boutwell RK (1964) Progr. exp. Tumor Res. 4; 207-250
Boutwell RK (1974) CRC Critical Reviews in Toxicology 2; 419-443
Butenandt A (1949) Verhandlungen der Dt.Ges.f.innere Med., 55.Kongreß, Wiesbaden, 342-364
Butenandt A (1950) Dt. Med. Wochenschrift 75; 5-7
Dontenwill W (1966) Hdb. der experimentellen Pharmakologie, Bd. XVI/13; Springer, Berlin Heidelberg New York, 74-204
Druckrey H, Küpfmüller K (1948) Dosis und Wirkung, Editio Cantor KG, Aulendorf/Württ.
Druckrey H (1967) In: Truhaut R (ed.) Potential Carcinogenic Hazards from Drugs, UICC Monographs Series Vol 7; Springer, Berlin Heidelberg New York, 60-77
van Duuren BL, Blazej T, Goldschmidt BM, Katz C, Melchione S, Sivak A (1971) J. Nat. Cancer Inst. 46; 1039-1044
van Duuren BL, Katz C, Goldschmidt BM (1973) J. Nat. Cancer Inst. 51; 703-705
O'Gara RW, Lee C, Morton JF (1971) J. Nat. Cancer Inst. 46; 1131-1137
Goertller K, Löhrke H (1976a) Virchows Arch. A Path. Anat and Histol. 376; 117-132
Goerttler K and Löhrke H (1976b) Exp.Path.,12; 336-341
Goerttler K, Löhrke H, Schweizer J, Hesse B (1979) Cancer Res. 39; 1293-1297
Goerttler K, Löhrke H, Schweizer J, Hesse B (1980a) Cancer Res. 40; 155-161
Goerttler K, Löhrke H, Schweizer J, Hesse B (1980b) Virchows Arch. A Path. Anat. and Histol. 385; 181-186
Hecker E (1971) In: Busch H (ed.) Methods in Cancer Research, Vol. 6; 439-484, Academic Press, New York London
Hecker E (1972) Z. Krebsforsch. 78; 99-122
Hecker E (1976a) Internat. J. Cancer 18; 122-129 (b) Z. Krebsforsch. 86; 219-230. (c) GANN 67; 471-481. (d) Bull. World Health Organ. 54; 1-10
Hecker E (1976e) Pure and Applied Chem. 49; 1423
Hecker E (1978a) Cocarcinogenesis in occupational cancer. Symp. No. 26, XII. Internat. Cancer Congress, Buenos Aires, Oct. 5-11, In: Adv. in Med. Oncology, Research and Education, Vol. 3. Epidemiology
Hecker E (1979) In: Miller EC, Miller JA, Hirono T, Sugimura T, Takayama S (eds.) Naturally occuring carcinogens - mutagens and modulators of carcinogenesis. Japan Scientific Societies Press, Tokyo, 263-286
Hecker E, Schmidt R (1974) Chem.Org.Nat.Prod. 31; 377-467
Hecker E, Weber J (1977) 7th Internat. Symposium on the Biological Characterization of Human Tumors, Budapest, 13-15 April Abstracts p. 52; see also Experientia (Basel) 34; 679-682 (1978)
Hecker E, Friesel H, Marquardt H, Schmidt R, Siebert D. (1978) Symposium No. 2, XII. Internat. Cancer Congress, Buenos Aires, Oct. 5-11; Adv. in Med. Oncology, Research and Education, Vol.1, Carcinogenesis, Margison GP (ed.) 1979, Pergamon Press Oxford and New

Hecker E (1979) In: Miller EC, Miller JA, Hirono T, Sugimura T, Takayama S (eds.) Naturally occuring carcinogens - mutagens and modulators of carcinogenesis. Japan Scientific Societies Press, Tokyo, 263-286
Hecker E, Schmidt R (1974) Chem.Org.Nat.Prod. 31; 377-467
Hecker E, Weber J (1977) 7th Internat. Symposium on the Biological Characterization of Human Tumors, Budapest, 13-15 April Abstracts p. 52; see also Experientia (Basel) 34; 679-682 (1978)
Hecker E, Friesel H, Marquardt H, Schmidt R, Siebert D. (1978) Symposium No. 2, XII. Internat. Cancer Congress, Buenos Aires, Oct. 5-11; Adv. in Med. Oncology, Research and Education, Vol.1, Carcinogenesis, Margison GP (ed.) 1979, Pergamon Press Oxford and New York, 207-212; J.Cancer Res. Clin. Oncology, in press
Hecker E, Adolf W, Opferkuch HJ, Ott HH, Weber G (1979a) Internat. Symposium on Environmental Carcinogenesis, Bombay, Dec. 9-11, Abstracts p 29. Proceedings J. Cancer Res. Clin. Oncology, in press
Hecker E, Pyerin W, Friesel H, Oberender H. Schmidt R (1979b) Internat. Symposium on Environmental Carcinogenesis, Bombay, Dec. 9-11. Abstracts 2728; Proceedings J. Cancer Res. Clin. Oncol., in press
Higginson J (1978), UICC Bulletin Cancer 16; 67
Kaufmann C, Müller HA, Butenandt A und Friedrich-Freska H (1949) Z. Krebsforsch. 56; 482-542
Kunz W, Appel KE, Rickart R, Schwarz M, Stöckle G (1978) In: Remmer H, Bolt HM, Bannasch P, Popper H (eds.) Primary Liver Tumors, Press Ltd, Lancaster UK, 261-283
Kunz W, Appel KE, Jones GRN, Rickart R. Schwarz M, Stöckle G (in press) Arch. Toxicol. supplement, Proceedings of the European Society of Toxicol.
Likhachev AY (1968) Vopr. Onkol. 14; 114-124
Lutz D, Hecker E (1979) VII. Internat. Symposium on the Biological Characterization of Human Tumors, May 8-11, Book of Abstracts
Morton JF (1971) Econ. Bot. 25; 457-463
Mottram JC (1944) J. Path. Bacteriol. 56; 181-187
Ott HH, Hecker E (1980) Japanese-German Workshop on Chemical, Viral and Environmental Carcinogenesis, Deutsches Krebsforschungszentrum Heidelberg May 20-23
Pound AW, Bell Jr (1962) Brit. J. Cancer 16; 690-695
Princess Takamatsu Cancer Research Fund (1979) 9th Internat. Symposium: Naturally occuring carcinogens - mutagens and modulators of carcinogenesis, Jan. 23-25, Tokyo, Japan; Proceedings see loc. cit. Hecker 1979
Roe FJC, Carter RJ, Mitschley BCV, Peto R, Hecker E (1972) Int. J. Cancer 9; 264-273
Schmähl D (1976) (Experimental Results) Oncology 33:73
Schweizer J, Marks F (1977) Cancer Res. 37; 4195-4201
Shear MJ (1938) Amer. J. Cancer 33; 499-537, dort S. 532
Shinozuka H, Ritchie AC (1967) Int. J. Cancer 2; 77-84
Sivak A (1979) Biochem. Biophys. Acta 560; 67-89
Slaga TJ, Sivak A, Boutwell RK (eds.) (1978) Carcinogenesis Vol. 2, Mechanism of Tumor Promotion and Cocarcinogenesis, Raven Press, New York

Soper CB, Evans FB (1977) Cancer Res. 37; 2487-2491
Sugimura T, Kawachi T, Nakayasu N, Matsukura N, Takayama S (1978) Symposium Biological Effects of Phorbol Esters in Cell Culture Systems, Cold Spring Harbour Laboratory, May 11-14, Abstracts of Papers 30
Süss R, Kreibich G, Kinzel V (1972) Europ. J. Cancer 8; 299-304
Tomatis L, Agathe C, Bartsch H, Huff I, Montesano T, Saracci R, Walker E, Wilbourn I (1978) Cancer Res. 38; 877-885
Traut M, Kreibich G und Hecker E (1971) In: Lettré H and Wagner G (eds.) Aktuelle Probleme der Cancerologie III, Springer Berlin Heidelberg New York, 91-96
UICC, International Symposium "Carcinogens of PLant Origin" (1968), National Institutes of Health, Bethesda, Maryland, USA, see Cancer Res. 28; 2233-2396; see also IARC Monograph series "Evaluation of Carcinogenic Risk of Chemicals to Man, Vol. 10, Internat. Agency for Research on Cancer, Lyon, 1976
Upadhyay RR, Eslampanah S, Daudi A (1979) IV. Asian Cancer Conference, Dec. 4-8, Bombay, Abstracts 15
Upton AC (1978) National Cancer Program (USA), Special Communication: Cancer and estrogen use, May 22
Weber J (1976) Dissertation, Univ. Heidelberg
Weinstein B, Troll W (1977) Cancer Res. 37; 3461-3463
Weinstein B, Troll W (1978) (eds.) Symposium Biological Effects of Phorbol Esters in Cell Culture Systems, Cold Spring Harbour Laboratory, May 11-14, Abstracts of Papers
WHO Chronicle (1977) 31; IARC: past and present developments, 239-245, see p 243
WHO Scientific Group (1978) Steroid contraception and the risk of neoplasia. WHO Technical Report Series No. 619
Yamagiwa K, Ichikawa K (1914) GANN, Tokyo, 8; 11-15; see: Collected Papers on Artificial Production of Cancer by Prof. K. Yamagiwa (1965) 1-9. Maruzen Comp. Ltd. Tokyo, Japan
Zur Hausen H, O'Neill FB, Freese UK and Hecker E (1978) Nature 272; 373-374
Zur Hausen H, Bornkamm GW, Schmidt R und Hecker E (1979) Proc. Ntl. Acad. Sci. (USA) 76; 782-785

Reduction of Human Exposure to Environmental N-Nitroso-Carcinogens. Examples of Possibilities for Cancer Prevention.

R. Preussmann, B.Spiegelhalder, G.Eisenbrand
Institute of Toxicology and Chemotherapy,
German Cancer Research Center, Heidelberg,
Federal Republic of Germany

Summary

Characteristic examples of recent achievements in reduction of exposure to environmental N-nitroso compounds are given. Elucidation of the causes of such exposure has led to suggestions for preventive measures, as is shown for nitrosamines in beer, for occupational exposure in the rubber industry and some other areas.

Investigations of the environmental occurrence of chemical carcinogens and of corresponding human exposure are often misunderstood or misinterpreted. Usually such research is undertaken neither to cause trouble to regulatory agencies or industry, nor are they to cause fear of cancer in the general population. The aim of such work is to collect data on potential health risks for the human population, to pinpoint emission sources and determine exposure levels and the causes thereof. The ultimate aim is always elimination or reduction of human exposure to such carcinogens, or primary prevention

B. Pullman, P. O. P. Ts'o and H. Gelboin (eds.), Carcinogenesis: Fundamental Mechanisms and Environmental Effects, 273–285.

of human cancer. In view of the slow progress in the treatment of manifest human cancer this approach seems more important than ever.

Nevertheless preventing exposure as the theoretically best way to reduce the incidence of cancer is often thought to be "obviously not practical" (e.g. Slaga, 1980). Although it is certainly true that intervention may also be effective in later stages of the multistage process of carcinogenesis, such as in the promotion phase (Weinstein et al., 1979) we maintain that the most important step in cancer prevention is intervention at the initial stage, the transformation of normal cells by DNA damaging agents.

We will present some recent data from research in environmental N-nitroso carcinogens showing that the identification of exposure sources leads to investigations into the causes of such exposure and finally to suggestions for preventive measures. In the examples given the suggested technological changes seem to be neither complicated nor expensive and can be effected without great technical problems or great costs. In such cases it is not even necessary to have difficult risk-benefit-evaluations before effecting proposed solutions of the problem in question.

Environmental N-nitroso compounds

Human exposure to carcinogenic N-nitroso compounds is twofold: by exposure to preformed nitrosamines from the environment as well as by endogenous formation of such compounds from precursors in the human body. The latter problem, unique for N-nitroso compounds, is dealt with during the meeting by Dr. Tannenbaum. We will concentrate therefore on exogenous exposure. Table 1 gives the most important sources of preformed environmental nitrosamines known to date.

Table 1: Human exposure by preformed N-nitroso compounds and potential prevention measures

"LIFE-STYLE"	OCCUPATIONAL
FOOD	RUBBER INDUSTRY
TOBACCO	LEATHER TANNERIES
DRUGS	CUTTING FLUIDS
COSMETICS	HYDRAULIC FLUIDS
(RUBBER)	PESTICIDE PRODUCTION AND APPLICATION

→ TOTAL EXPOSURE ←

↑

→ REDUCTION ←

TECHNOLOGICAL IMPROVEMENT (MALT DRYING)	EXCHANGE OF NITROSATING CHEMICAL (NITROSODIPHENYLAMINE)
REDUCTION OF NITRITE IN CURING SALT	REDUCTION OF NO_X POLLUTION
LEGAL REGULATIONS	USE OF "SAFE" AMINES
	VENTILATION AT WORKING PLACES

"Life-style" exposure

In the "lifestyle" exposure food and tobacco seem to be the most important sources of human contact. In our own survey on the occurrence of volatile nitrosamines in food on the German market, in which more than 3,000 commercially available samples were analyzed, it was shown that there was an average daily intake of 1.1 µg N-nitrosodimethylamine (NDMA) and of 0.1–0.15 µg N-nitrosopyrrolidine (NPYR) per capita in 1978. For NDMA 64% of the total daily intake resulted from the consumption of beer and 10% from meat products (Preussmann et al., 1979; Spiegelhalder et al., 1980).

The investigation of the origin of NDMA formation in beer as the most important source of exposure was done by analyzing every stage of the entire brewing process together with all the different ingredients. Invariably we found that malt was practically the only source of NDMA contamination. Therefore a wide variety of different malt samples was analyzed thereafter and it was found that NDMA is formed during the kilning (drying) of the green malt. The extent of nitrosamine formation clearly depended on the type of heating technique used: Indirect heating systems led to NDMA concentration in malt of <1 µg/kg. Much higher values were obtained with direct heating systems, which are predominant in many countries: The use of oil burners resulted in concentrations of 0.5-10 µg/kg NDMA and malt dried directly with gas burners had 15-80 µg/kg NDMA. For dark malts, which are dried at higher temperatures (up to 105 °C vs. 80 °C for pale malts), the corresponding concentrations were 2-8 µg/kg with indirect heating, 6-10 µg/kg with oil burners and 80-320 µg/kg with gas burners.

In cooperation with experts it was then found out that sulfur burning ("sulfuring") i.e. SO_2-treatment during drying significantly reduced NDMA formation.

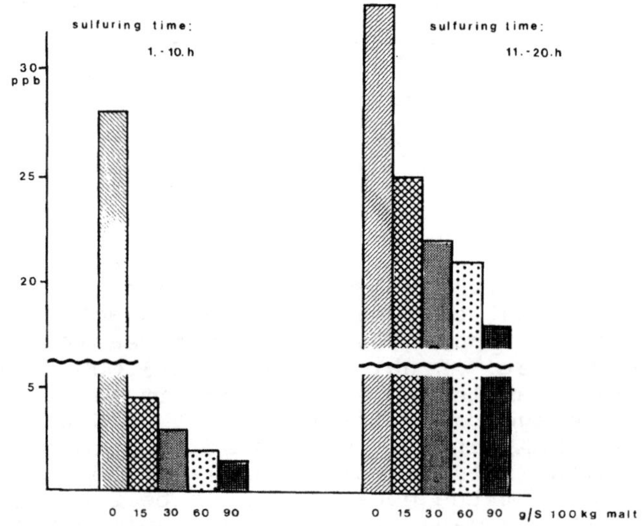

Fig. 1: Reduction of NDMA in malts during sulfuring

As shown in Fig. 1, this method is very effective, but only when applied during the first ten hours, when the malt is still wet. In the later stage sulfuring is much less effective.

Since NOx is a precursor in NDMA formation and high combustion temperatures (usually from 1500-1800 ^{o}C) yield high reaction rates between oxygen and nitrogen, a decrease in NDMA formation can also be achieved by lowering the flame temperature. Excess air for combustion seems to be the most economic way to reduce flame temperature and NO_x synthesis. In a new type of burner developed on this principle the resulting air had only 0,05-0,1 mg NO_x/m^3 as compared with 1-4 mg/m^3 in conventional burners. Accordingly malt dried with these burners contains only 1-3 ug NDMA/kg , a 15-30 fold reduction of NDMA concentration.

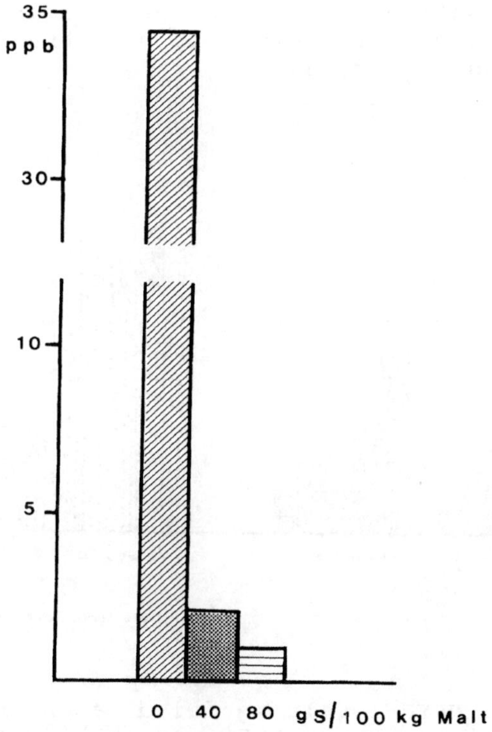

Fig. 2: Reduction of NDMA in "Rauchbier"

Using these methods of reduction of NDMA formation during kilning, it is now possible to produce malt with low NDMA content even with direct heating techniques. Beer produced with this kind of malt shows significantly lower NDMA contamination. An example is given in Fig. 2: "Rauchbier", a speciality in some areas of Germany, which had the highest contamination in our investigations(Spiegelhalder et al., 1979), was reduced from a 40-70 µg/kg level to only 1-2 µg/kg. Similar effects were seen in other types of beer.

The contamination of nitrite treated meat and meat products, mainly with NDMA and NPYR, is more difficult to deal with, because nitrite as the main source of nitrosating agents in such products cannot be eliminated totally at present, because of the botulism problem. However, Havery et al. (1978) have shown for bacon that a reduction of nitrite levels (down to 125 mg/kg nitrite initially added) and increased levels of ascorbate (550 mg/kg) as nitrosation inhibitors resulted in a substantial reduction of NPYR-levels in some brands of fried bacon, as shown in Fig. 3.

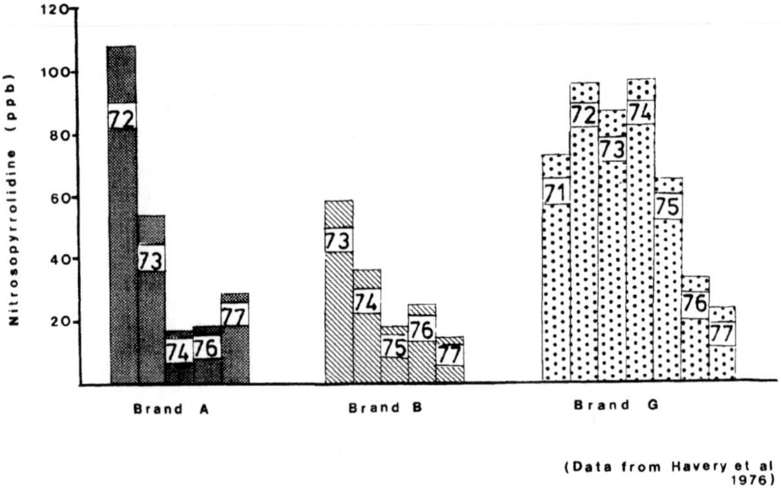

Fig. 3: Reduction of NPYR-levels in fried bacon in a six-year period between 1972 and 1977 (data from Havery et al., 1978).

Further refinement in the use of nitrite in meat curing might bring further progress, but a definite solution of the problem probably will only be possible if other preservatives become available as complete or at least partial substitute for nitrite.

A thorough investigation of the occurrence of volatile nitrosamines in tobacco smoke has been presented by Brunnemann et al. (1977). Concentrations found in mainstream-smoke of commercial cigarettes ranged between 6-65 ng/cigarette for NDMA, with corresponding values for N-nitrosomethylethylamine of 0.4-8, for N-nitroso-diethylamine 1-4, for NPYR 5-34 and for N-nitroso-nornicotine, a tobacco-specific nitrosamine, 40-250 ng/cig.; Side-stream smoke concentrations were 10-40fold higher. A smoker of 20 cigarettes per day is exposed by inhalation to considerably higher quantities of several nitrosamines as compared to exposure from food. The authors have also shown that the use of cellulose acetate filters in cigarettes reduces effectively and selectively nitrosamine concentration in mainstream smoke. Some relevant data are summarized in Fig. 4.

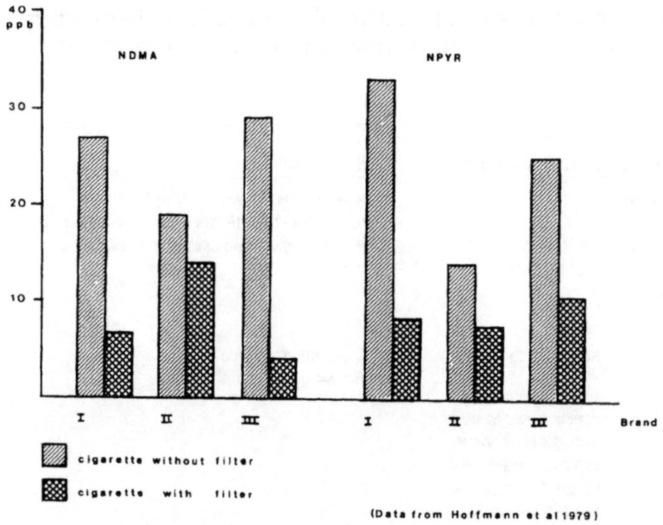

Fig. 4: Reduction of nitrosamine concentration in cigarette smoke by cellulose acetate filters in different brands (data adapted from Brunnemann et al., 1977 and Hoffmann et al, 1979)

No systematic investigations are available on the effects of other adsorbents in filters; we are certain that further improvements can be obtained by such studies.

Occupational exposure

Data on occupational exposure to different nitrosamines are of very recent origin and as yet by no means representative. They clearly demonstrate, however, that certain industries have a serious nitrosamine problem. Exposure has been shown to vary considerably in regard to the amount of nitrosamines found in the working places in different industries or occupations as well as to the nitrosamine in question. In the latter regard, NDMA, N-nitrosomorpholine, N-nitrosodiethanolamine and several other dialkyl- and cyclic nitrosamines are of relevance. Several problem areas already identified are listed in Table 1 (Spiegelhalder and Preussmann, 1980).

We have data from all areas indicated, but our analytical program has been concentrated recently on nitrosamines in the rubber and tire industry. Knowledge and data available up to date are summarized in Table 2.

Table 2: Nitrosamines in rubber and tire industry. Precursors, maximum concentrations and exposures

AMINE CHEMICALS (ACCELERATORS) USED	+	NITROSATING AGENTS
ALKYL CARBAMATES		NITROSO DIPHENYLAMINE (RETARDER)
ALKYL THIURAMES		NITROGEN OXIDES IN AIR FROM FORK-LIFT TRUCKS
SECONDARY AMINE DERIVATIVES (MORPHOLINE, PIPERIDINE, DICYCLOHEXYLAMINE)		NO_x FROM OTHER COMBUSTION PROCESSES

↓

NITROSAMINES FORMED	MAX. CONCENTRATION IN WORKING AREA ($\mu g/m^3$)
NITROSODIMETHYLAMINE	140
NITROSODIETHYLAMINE	1
NITROSODIPROPYLAMINE	2
NITROSODIBUTYLAMINE	1
NITROSOPIPERIDINE	2
NITROSOPYRROLIDINE	1
NITROSOMORPHOLINE	20

MAX. CALCULATED EXPOSURE	1700 µG NDMA/8 HRS
AVERAGE CALCULATED EXPOSURE	50 µG NDMA + NMOR/8 HRS

Amine precursors are mainly polymerisation accelerators. Some typical representatives are shown in Fig. 5

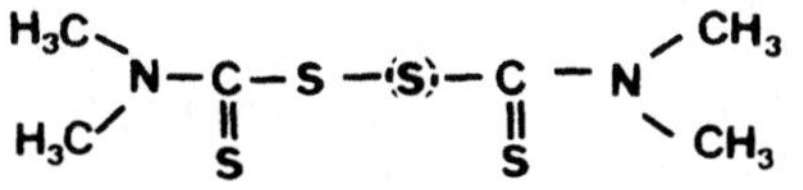

TETRAMETHYL-THIURAM-DISULFIDE
-MONOSULFIDE

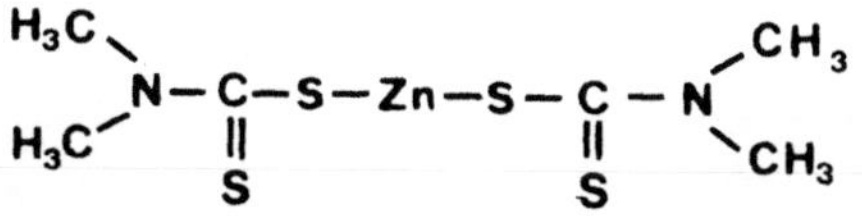

ZINC-DIMETHYL-DITHIOCARBAMATE

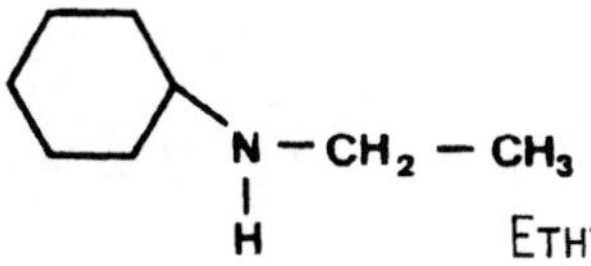

ETHYLCYCLOHEXYLAMINE

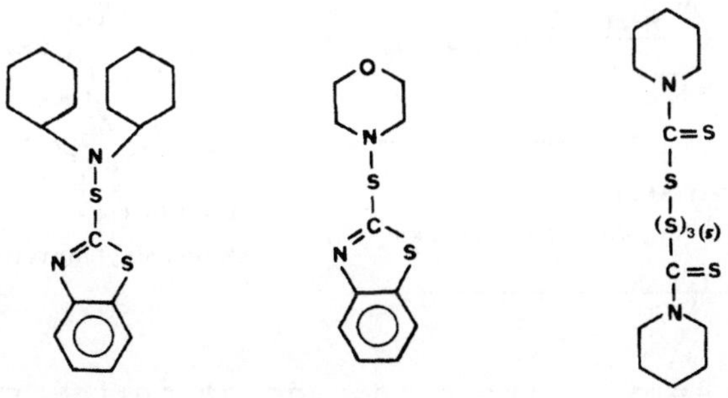

DICYCLOHEXYL-2-BENZOTHIAZOL-SULFENAMIDE

MORPHOLINYL-2-BENZOTHIAZOL-SULFENAMIDE

DIPENTAMETHYLENE-THIURAM-TETRA(HEXA)-SULFIDE

Fig. 5: Polymerisation accelerators used in the rubber industry

Most of the compounds are labile and release the corresponding secondary amine, especially during heating in the process. It should be mentioned that according to our investigations most of these accelerators as technical products are already contaminated with the corresponding nitrosamine in the range between 10-5,000 µg/kg. Such concentrations, however, can by no means explain the air contamination measured.

Nitrosation of amines very likely is mainly mediated either by the decomposition of N-nitrosodiphenylamine, used as retarder, or by N,N'-dinitrosopentamethylenetetramine, used as a foaming agent. Very important also are nitrogen oxides in the indoor air, and here one major source are fork-lift trucks, which are powered by propane or butane gas and have a high NO_x emission (Fig. 6).

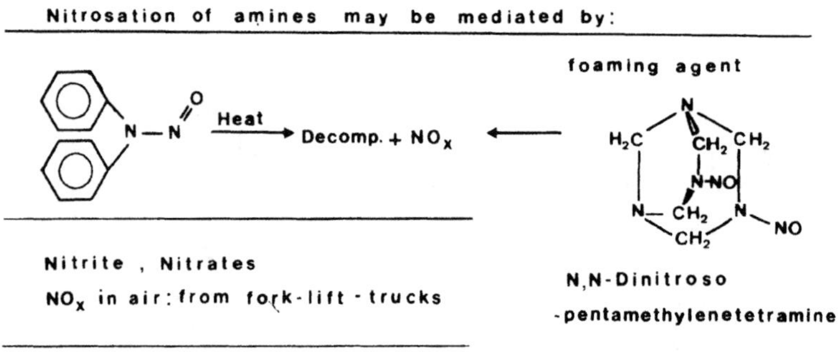

Fig. 6: Sources of nitrosating agents in the rubber industry

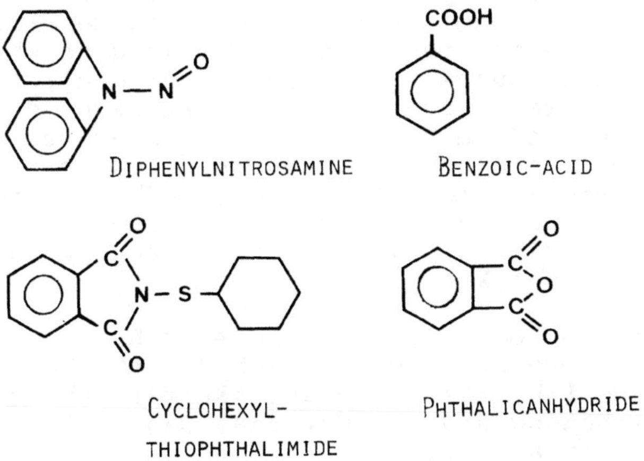

Fig. 7: Polymerisation retarders used in the rubber industry

Fig. 7 shows that there are also other retarders used in this industry, which in contrast to N-nitrosodiphenylamine cannot serve as nitrosation sources.

The maximum air concentrations that we found in several rubber factories (Tab. 2) compare well with the data obtained by Fajen et al. (1979) for N-nitrosomorpholine (NMOR). These authors, however, only found 0-0.5 $\mu g/m^3$ NDMA, while our data go up to 140 $\mu g/m^3$ for this component. From our data an average daily exposure of about 50 µg NDMA + NMOR can be calculated.

The likelihood of the proposed scheme of nitrosamine formation in the rubber industry is confirmed by our results from measurements in one factory, where neither diphenylnitrosamine nor fork-lift trucks powered by gas are used: Indoor air in this factory has almost no detectable nitrosamines.

This clearly shows the way for future preventive measures: Elimination of N-nitrosodiphenylamine, exchange of gas-powered fork-lift trucks by electrically-driven ones and further ways of reducing NO_x in the air.

Accelerators used should be nitrosamine-free and should not liberate amines in the process, which yield potentially carcinogenic nitrosamines.

In summary, we hope to have shown in the case of environmental nitrosamines that there are several areas of existing exposure to these potent carcinogens where knowledge of the causes of exposure can result in preventive measures, leading to greatly reduced exposure. Both direct and indirect costs of such measures would be relatively small and no insurmountable technological difficulties are apparent to prevent such measures. Therefore, even without a precise estimate of the risk for man, such measures for primary cancer prevention have been or should be undertaken.

Acknowledgements: Our experimental work has been supported by the Deutsche Forschungsgemeinschaft, the Federal Ministry of Health, the Bundesanstalt für Arbeitsschutz und Unfallverhütung and the National Cancer Institute, USA (Contract No. 1CP65796).

References

1. Brunnemann, K.D., Yu, L. and Hoffmann, D., 1977. Cancer Res. 37, pp. 3218-3223.

2. Fajen, M.J., Carson, G.A., Rounbehler, D.P., Fan, T.Y., Vita, R., Goff, U.E., Wolf, M.H., Edwards, G.S., Fine, D.H., Reinhold, V., Biemann, K., 1979. Science 205, pp. 1262-1264.

3. Havery, D.C., Fazio, T. and Howard, J.W., 1978. In? Environmental Aspects of N-Nitroso Compounds. IARC Sci. Publ. No. 19, IARC Lyon, pp. 305-310.

4. Hoffmann, D., Adams, J.D., Piade, J.J. and Hecht, S.S., 1979. Paper presented at the 6th Meeting on Environmental N-Nitroso Compounds, Budapest, Oct. 1979. IARC Sci. Publ., in press.

5. Preussmann, R., Eisenbrand, G. and Spiegelhalder, B., 1979. In: Environmental Carcinogenesis. Elsevier/ North Holland Biomed. Press, Amsterdam, pp. 51-71.

6. Slaga, T.J., 1980. In: Modifiers of Chemical Carcinogenesis. Raven Press, New York, p. V.

7. Spiegelhalder, B., Eisenbrand, G. and Preussmann, R., 1979. Fd. Cosmet. Toxicol. 17, pp. 29-31.

8. Spiegelhalder, B., Eisenbrand, G. and Preussmann, F., 1980. Oncology, in press.

9. Spiegelhalder, B. and Preussmann, R., 1980. Paper given at 15. German Cancer Congress, March 1980, Munich, Germany.

10. Weinstein, I.B., Lee, L.S., Fisher, P.B., Muffson, A. and Yamasaki, H., 1979. In: Naturally Occurring Carcinogens, Mutagens and Modulators of Carcinogenesis. Univ. Park Press, Baltimore, pp. 301-313.

ENDOGENOUS CARCINOGENESIS: NITRATE, NITRITE AND N-NITROSO COMPOUNDS

Steven R. Tannenbaum, Laura C. Green, Katherine R. de Luzuriaga, Guido Gordillo, Lori Ullman and Vernon R. Young
Department of Nutrition and Food Science, Massachusetts Institute of Technology, Cambridge, Massachusetts 02139

Abstract: Man is exposed to carcinogens in both his internal and external environment. In the case of N-nitroso compounds the endogenous burden may greatly exceed environmental exposure. Endogenous synthesis may occur from nitrite preformed or added to foods, or from nitrite formed endogenously from nitrate. Nitrate may originate from food or water, but there is also an important component introduced via endogenous synthesis not of microbial origin. The distribution and metabolism of nitrate is discussed in relation to formation of nitrite, synthesis of N-Nitroso compounds, and the etiology of gastric cancer.

Nitrate has been identified in several locales as a component of the environment possibly associated with greater than ordinary risk of certain cancers (Zaldivar and Robinson, 1973; Hill et al., 1973; Cuello et al., 1976). Nitrate is a stable chemical species, readily converted to nitrites.

A carcinogenic agent may result from the reaction of nitrite with a nitrogen-containing compound to form an N-nitroso compound. Exposure to these N-nitroso compounds might occur in two different ways. The first possibility is through preformed compounds in the environment, particularly in food and water (Preussmann, this volume). The second possibility is through formation of these compounds in the body by the reaction of ingested nitrite with an appropriate nitrogen compound.

A summary of human exposure to one important N-nitrosamine, dimethylnitrosamine (NDMA), is shown in Table 1. The endogenous exposure is an estimate based upon measurements of NDMA in human physiological fluids and a mathematical model for metabolism (Tannenbaum, 1980). This is a theoretical estimate, and requires substantial further investigation.

B. Pullman, P. O. P. Ts'o and H. Gelboin (eds.), Carcinogenesis: Fundamental Mechanisms and Environmental Effects, 287–296.

Table 1. Human Dimethylnitrosamine Exposure

Source	Amount	Route
cigarettes	5 ng/cigarette	air
auto interior	0.3 μg/M^3	air
workplace	up to ppm	air
beer	5 μg/L	oral
bacon	1 μg/kg	oral
Scotch whiskey	1 μg/L	oral
endogenous	0 to 700 μg/day	unknown

Evidence that N-nitroso compounds are formed in the body from amine precursors is based on: (a) their chemical detection in gastric juices _in vitro_ and in the mammalian or human stomach _in vivo_; and (b) the observation of acute toxic and carcinogenic effects as well as damage to cellular macromolecules after simultaneous administration of nitrite and various amines and amides (Magee _et al_., 1976). Likely reaction sites in the body are the stomach because of the reaction-promoting acidic conditions, but consideration must also be given to the mouth, esophagus, intestinal tract, bladder, and urinogenital tract because under normal or pathological conditions, they also contain bacteria which can reduce nitrate to nitrite, and which catalyze nitrosation.

The relevance of these traces of N-nitroso compounds to a possible risk to man has yet to be established. However, a large variety of animal species and organs are susceptible to their carcinogenic action and human liver samples are capable of forming alkylating and mutagenic metabolites. The exceedingly difficult question of whether some human cancer or transmissable genetic damage could be attributed to these groups of carcinogens has not so far been answered, although some studies have examined the possible role of N-nitrosamine intake in the etiology of a number of human cancers. With this in mind we can now consider factors which influence these endogenous processes through the study of nitrate metabolism and its consequences for nitrosamine formation.

METABOLISM OF NITRATE

Little is known about the metabolism of nitrate in man. Nitrate is formed in the oral cavity by bacterial reduction of nitrate, and the normal concentration range in indiviuals who do not consume large amounts of nitrate is 6-10 mg/l (Tannenbaum _et al_., 1974). When foods that contain nitrate, such as certain roots and leafy vegetables, are consumed, salivary nitrate and nitrite levels are elevated to hundreds of mg/l (Speigelhalder et al., 1976; Tannenbaum et al., 1974). The nitrate and nitrite contents of saliva in Japan is about 3-fold higher than in the U.S. as a result of higher dietary nitrate intake (Kawabata et al., 1979). This is of particular interest when viewed in the light of the approximately 10-fold higher incidence of gastric cancer in Japan compared to the U.S.

In our laboratory experiments have been carried out with single and multiple doses of nitrate given in vegetable juice or in water. We have carried out experiments with $^{15}NO_3$ to measure pharmacokinetic parameters, and also conducted long-term experiments on low-nitrate diets to determine patterns of apparent endogenous nitrate formation. All of these studies have been carried out in man and in the rat. It is worth noting that certain important physiological differences exist between man and the rat, particularly in the ability to concentrate nitrate into saliva, which is lacking in the rat (Cohen and Myant, 1959). On the other hand, the rat probably concentrates nitrate into the secretory flow of the small bowel, which may be lacking in man (Pastan, 1957). The need to develop a model for prediction of nitrate concentration in various body compartments as a function of time following dose is acute, because it has been shown that salivary nitrite is proportional to salivary nitrate (Spiegelhalder et al., 1976) as a result of microbial action. More recently, we have also shown that when gastric pH is above 5, gastric nitrite is proportional to gastric nitrate concentration, also as a result of microbial action (Tannenbaum et al., 1979). Under normal conditions, nitrite in the stomach would probably arise only from the swallowing of saliva (Klein et al., 1978).

The quantitative relationship between salivary nitrate and nitrite has been discussed in some detail by Spiegelhalder et al. (1976). Their work suggests that a minimum oral intake of approximately 50 mg $NaNO_3$ is required before excess nitrate appears in saliva, but that above that level of intake a linear proportionality exists between nitrate and nitrite. However, since salivary flows are not constant with time the total amount of nitrite formed from a given dose of nitrate is not presently known.

An overall model for metabolism of nitrate and nitrite is shown below:

Model for Metabolism of Nitrate and Nitrite

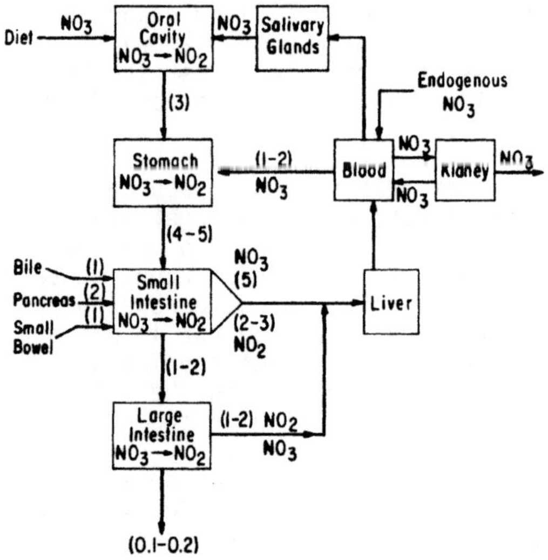

Numbers in parentheses represent flow (liters/day).

The various processes of conversion are shown for each compartment, and the relative flow across the compartment is also indicated. In this model, nitrate is ingested via the gastrointestinal tract, which serves as a "depot". It is absorbed into the blood and distributed throughout the body into extracellular water. It is cleared from the bloodstream by the kidneys into urine and by the salivary glands and the parietal glands, which returns nitrate to the gastrointestinal tract. Nitrate also flow into the small bowel via the bile, pancreatic juice, and intestinal secretion. Additional nitrate is apparently supplied to the body via de novo endogenous synthesis. Also, nitrite is destroyed via chemical reactions. In this model nitrate flow to and from each compartment is governed either by passive diffusion, or by active transport in the case of the salivary glands. In the kidney nitrate is filtered and resorbed. We have little information on the characteristics of systems such as the parietal glands, bile, etc.

As a result of these processes, nitrate is distributed and recycled through compartments which can synthesize nitrite over a period of time in a manner related to dose, dose rate, rate of urine formation, and rate of saliva secretion. Some examples of patterns of excretion and oral nitrite formation are shown below. The first case shows the result of a single oral dose, and clearance of nitrate into urine and saliva are approximately parallel, yielding concurrent parallel nitrate formation:

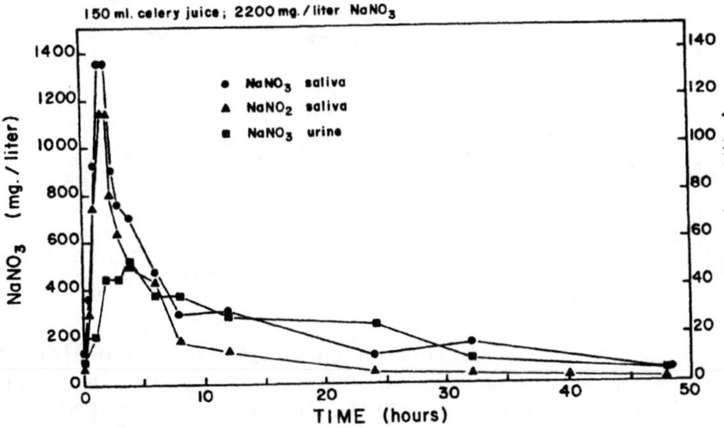

In a second example designed to investigate multiple nitrate intakes (23 mg $NaNO_3$ at each arrow) over a period of several days, a diurnal rhythm is established, especially of salivary nitrite formation:

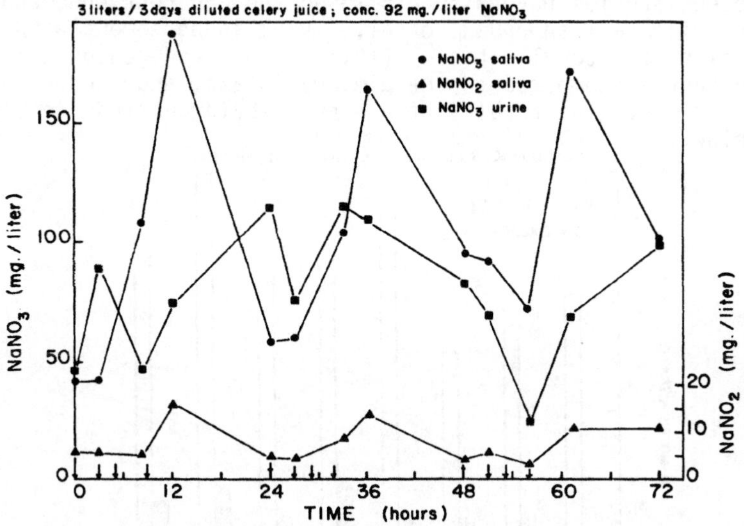

This pattern is simply the result of adding the components of successive distribution curves from consecutive doses. One important result of altering the dose patterns is shown in the cumulative urine excretion curves below:

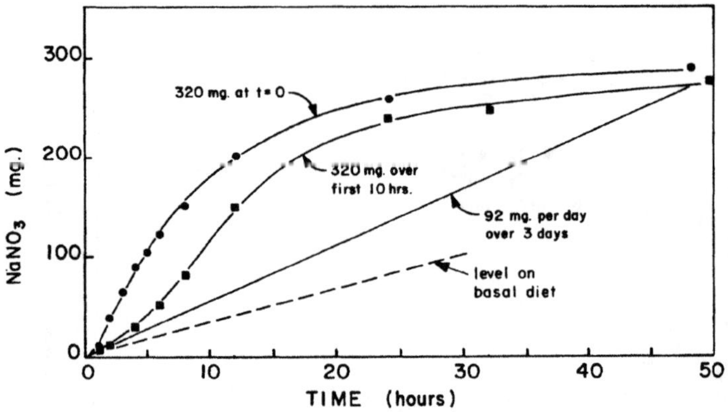

In these curves, the effect of modifying the dose rate to prolong the distribution and elimination phases can be seen to result in enhanced loss of nitrate (through increased nitrite formation). Stated in another manner, the extent of nitrite formation is directly related to the number of times nitrate is recycled through a nitrite-forming compartment.

Our recent studies have also disclosed the apparent endogenous synthesis of nitrate (Tannenbaum et al., 1978) in agreement with the earlier studies of Mitchell et al. (1916). This endogenous contribution is substantial compared to the dietary intake, and occurs in the rat as well as man. The results of a nitrate-balance study in the rat is shown below:

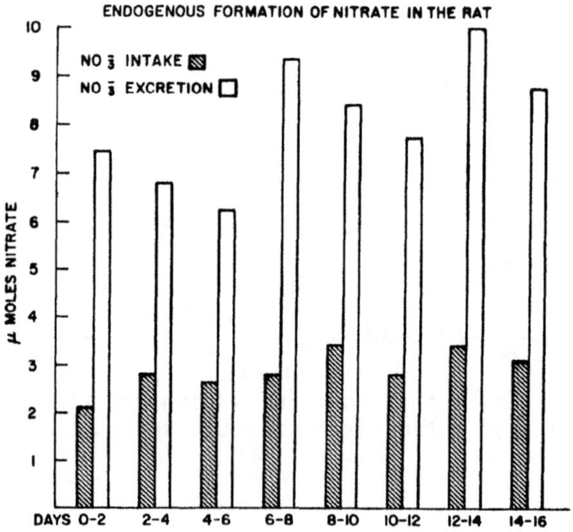

Studies in young men show a similar pattern of excess excretion over intake over periods of time as long as 84 days. There is great variability between individuals, with daily excretion as low as 0.25 and as high as 2 mmoles. Although our original hypothesis stated that the most likely source of this endogenous nitrate was intestinal microbial nitrification of ammonia, we have now demonstrated positive nitrate balance in the germ-free rat. We are currently exploring alternative mechanisms.

Preliminary experiments to characterize nitrate pharmacokinetics in man have provided valuable information on clearance, elimination rate, and volume of distribution.

The rate of elimination of nitrate following a single oral dose (10 ml) of 300 mg $Na^{15}NO_3$ is shown below in Table 2.

TABLE 2. RATE OF EXCRETION OF ^{14}N AND ^{15}N NITRATE FOLLOWING A SINGLE ORAL DOSE OF 300 MG N^{15} NO_3

	EXCRETION, MG/HR		
TIME PERIOD, HR	^{14}N + ^{15}N	^{14}N	^{15}N
-24 to 0	4.9	4.9	0
0 to 3	25.5	10.0	15.5
3 to 6	22.1	9.0	13.1
6 to 9	15.5	7.6	7.9
9 to 12	9.3	5.3	4.0
12 to 24	6.4	4.8	1.6
24 to 36	5.8	5.4	0.4
36 to 48	3.3	3.3	0
48 to 72	5.4	5.4	0

The ^{14}N values represent the initial body pool and the apparent endogenous synthesis during the phase of clearance of ^{15}N. The true rate of nitrate elimination is given by the N values, and it is first order in ^{15}N with a half-life ($t_{1/2}$) of 4.8 hrs. The data also suggest that the rate of endogenous synthesis is fairly constant.

Graphical estimation of nitrate clearance for one subject from the plot of plasma concentration vs. rate of excretion yields a value of 47.6 ml/min. The pharmacokinetic parameters for this individual may be summarized as follows:

Clearance = 47.6 ml/min

$t_{1/2}$ = 288 min; $k_e = 2.4 \times 10^{-3}$ min^{-1}

$$\text{Volume of Distribution } (V_d) = \frac{47.6}{2.4 \times 10^{-3}} = 19.8 \text{ liters}$$

Body Weight = 84 kg

$$V_d = \frac{19.8}{84} = 23.6\% \text{ of Body Weight}$$

This value of V_d corresponds well to that previously estimated for the dog (Greene and Hiatt, 1954). It is slightly greater than extracellular water, and probably includes the additional volume of the GI tract. The range of basal nitrate values, i.e. on a low nitrate diet, is as follows for a series of individual subjects studied as above.

excretion rate: 0.03 - 0.06 mmoles/hr
plasma concentration: 0.02 - 0.05 mM
total nitrate pool: 0.4 - 1 mmole

The relative significance of the various contributions of endogenous and exogenous nitrate will vary for different individuals and for different diets. To fully understand the relationship of these results to potential carcinogenic processes, we would also require a framework for the etiology of the disease. An illustration of one possible paradigm, that for gastric cancer, will be briefly expounded below.

NITRATE AND GASTRIC CANCER

The epidemiological association between nitrate and gastric cancer was mentioned in the introduction to this paper. We have developed a model for the etiology of this disease which has been fully described elsewhere (Correa *et al.*, 1975). Variations of this model might equally apply to other situations where normal or abnormal microbial growth leads to nitrite formation and concomitant formation of carcinogenic N-nitroso compounds. Of particular interest in this regard is the urinary bladder. The essential components of the gastric cancer model are shown below:

Model for Nitrate and Gastric Cancer

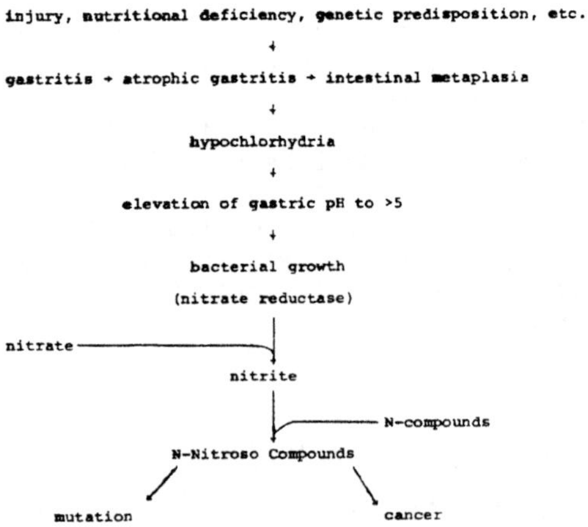

The key factor from the etiological point of view is that the processes leading to formation of the carcinogen are endogenous, but influenced by external variables. The essential questions that remain for future investigation and assessment are as follows:

1. What is the relative significance of endogenous to exogenous nitrate exposure?
2. What factors influence the extent of endogenous nitrite formation?
3. What are the important N-precursors for N-nitrosation and what is their origin?
4. How can we block the processes we have described and thereby reduce the incidence of cancer?

ACKNOWLEDGEMENT

Research on nitrate metabolism in the authors laboratory has been supported by the following: USPHS Contract NO1-CP-33315, NCI Grant 1RO1-CA26156-01, NCI Grant 1-PO1CA26731-01.

REFERENCES

1. Zaldivar, R. and Robinson, H.: 1973, Z. Krebsforsch. 80, pp. 289-295.
2. Hill, M.J., Hawksworth, G. and Tattersall, G.: 1973, Br. J. Cancer 28, pp. 562-567.
3. Cuello, C., Correa, P., Haenszel, W., Gordillo, G., Brown, C., Archer, M. and Tannenbaum, S.R.: 1976, J. Natl. Cancer Inst. 57, pp. 1015-1020.
4. Preussmann, R., this volume.
5. Tannenbaum, S.R.: 1980, Oncology, in press.
6. Magee, P.N., Montesano, R. and Preussmann, R.: 1976, In: "Chemical Carcinogens". American Chemical Society Monograph 173, pp. 491-625.
7. Tannenbaum, S.R., Sinskey, A.J., Weisman, M. and Bishop, W.: 1976, Food Cosmet. Toxicol. 14, pp. 79-84.
8. Spiegelhalder, B., Eisenbrand, G. and Preussman, R.: 1976, Food Cosmet. Toxicol. 14, pp. 549-552.
9. Tannenbaum, S.R., Weisman, M. and Bishop, W.: 1976, Food Cosmet. Toxicol. 14, pp. 549-552.
10. Kawabata, T., Ohshima, H., Vibo, J., Nakamura, M., Matsui, M. and Hamano, M.: 1979, In: "Naturally Occurring Carcinogens-Mutagens and Modulators of Carcinogenesis". E.C. Miller, J. Miller, I. Hirono, T. Sugimura and S. Takayama, eds., University Park Press, pp. 195-209.
11. Cohen and Myant: 1959, J. Physiol. 145, pp. 595-601.
12. Pastan, I.: 1957, Endocrinology 61, pp. 93-97.
13. Tannenbaum, S.R., Moran, D., Rand, W., Cuello, C. and Correa, P.: 1979, J. Natl. Cancer Inst. 62, pp. 9-12.

14. Klein, D., Gaconnet, N. Poullain, B. and Debry, G.: 1978, Food Cosmet. Toxicol. 16, pp. 111-115.
15. Tannenbaum, S.R., Fett, D., Young, V.R., Lama, P.D. and Bruce, W.R.: 1978, Science 200, pp. 1487-1489.
16. Mitchell, H.H., Shoule, H.A. and Grindley, H.S.: 1916, J. Biol. Chem. 24, pp. 461-484.
17. Greene, I. and Hiatt, E.: 1954, Am. J. Physiol. 176, pp. 463-467.
18. Correa, P., Haenszel, Cuello, C., Tannenbaum, S.R. and Archer, M.: 1975, Lancet, pp. 58-60.

NEOPLASTIC TRANSFORMATION, SOMATIC MUTATION, AND DIFFERENTIATION

Paul O. P. Ts'o, Division of Biophysics, School of Hygiene and Public Health, The Johns Hopkins University, Baltimore, Maryland 21205, U.S.A.

One of the most dreaded diseases of our time is cancer, the topic of this symposium. This disease originates from a single cell, or groups of cells, which become neoplastic or tumorigenic. Also, as indicated from the topics of this symposium, this neoplastic transformation often has a large input from external/environmental factors, particularly man-made. This dread disease then becomes closely related to our personal and economic lives and has a large impact on health care and the economic activities of our society.

One of the most important problems in biology today is the mechanism of differentiation/development, i.e. the programming of gene expression. On the other hand, one of the most significant advances in science is the molecular biology/genetics of prokaryotes. This success has now pushed forward toward the study of eukaryote and mammalian genetic apparatus, from structure to function to control.

Thus, cancer, economic/environmental problems, differentiation/control, and finally, the molecular biology of the mammalian genetic apparatus, are all wrapped up together as a major challenge to the intellectual activities of man.

In the brief space here, a panoramic view is presented of a model system for research which can provide a basis for answers to the above challenge. This experimental system is based on the careful investigation of a mammalian cellular system through which somatic mutation, neoplastic transformation, differentiation, and even senescence can be investigated concomitantly. In addition, this system allows for the possibility of investigating the correlation between cells in culture vis à vis cells in animals, in order to extrapolate from an in vitro study to an in vivo situation.

THE SYSTEM

It is a great honor to describe the use of Syrian hamster embryonic

B. Pullman, P. O. P. Ts'o and H. Gelboin (eds.), Carcinogenesis: Fundamental Mechanisms and Environmental Effects, 297–310.

cells here. This system was first described by Professor Leo Sachs and Dr. Berwald in Israel in 1963-65 (1,2). This system has the following characteristics: short life span and small animals; 44 chromosomes/cell, similar to human cells; about 10% more DNA/cell than human cells; a very low spontaneous transformation frequency upon serial passage, similar to human fibroblasts; a high propensity to enter the senescent stage in cell culture; and constant karyotypic patterns of the cells in culture that can be analyzed carefully. The animals are relatively tame in captivity, and being of highly inbred stock, they can be induced to form tumors upon injection of transformed cells into the newborn animals. In addition, Syrian hamsters have been used extensively as model animal systems for cancer study.

The embryonic fibroblasts are obtained from the primary culture of 11- to 13-day gestation fetuses. We are now working to extend this investigation to embryos from an earlier stage of development (8-10 days gestation) as well as to tissue fibroblast (skin, etc.) prepared from young and adult animals. Hopefully, the events of induced carcinogenesis can be related to the natural developmental process and aging of the animal.

PERTURBATION

Two types of perturbations have been employed in this study: (1) conventional chemical carcinogens, such as benzo(a)pyrene (B(a)P) and N-methyl-N'-nitro-N-nitrosoguanidine (MNNG), which were used to calibrate the neoplastic transformation system (3-5); and (2) perturbation that delineates the critical target molecules inside the cell for mechanistic studies. Since the common carcinogens are so reactive that they attack all kinds of molecules and organelles within the cell, it is not easy to limit the perturbation to any one type of target. Therefore, it is important to develop procedures that can systematically perturb specific target molecules or specific cellular organelles within the living cell. Initially we will study the effect of the specific attack on DNA to determine whether perturbation to DNA alone can lead to neoplasia (6,7).

BIOLOGICAL RESPONSES

As indicated in the introduction, the purpose of establishing this model system is that several biological processes can all be investigated simultaneously. Emphasis, however, is being placed here on the description of neoplastic transformation.

In the study of somatic mutation, only the two most common genetic markers are investigated here (4,8): the HGPRT locus, which depends on the selections of the 6-thioguanine-resistant (TG^r) or 8-azaguanine-resistant (AG^r) cells; and the Na^+/K^+ membrane adenosine 5-triphosphatase (ATPase) locus, which depends on the selection of the ouabain-resistant

(Oua^r) cells. The mutation at the HGPRT locus is a sex-linked, recessive mutation, while the mutation at the Na^+/K^+ ATPase locus is an autosomal dominant or codominant mutation. Colonies resistant to TG, AG, and ouabain were isolated and characterized extensively. These characterizations indicate that the AG^r and TG^r cells have a mutation in the structural or regulatory gene for HGPRT and that the Na^+/K^+ ATPase in Oua^r cells is changed.

In the study of neoplastic transformation, the ultimate endpoint is the tumorigenic potential of the transformed cells upon injection into newborn animals (9). Considerable effort has been made to put this investigation onto a quantitative basis, i.e. the number of injected cells required to induce 50% of the animal to yield histologically identifiable tumors. These tumors can often be excised for further characterization and identification of the origin. To one extreme, we can inject 10^7 cells from normal Syrian hamster embryo without causing any tumor, and on the other extreme, we have one or two cell lines in which injection of 10 cells will cause all animals or 50% of the animals to yield tumors (9,10). The next effort is to correlate other alterations in *in vitro* growth which can be associated with neoplasia. An example of this study is given in Table 1 (9). This Table is based on the studies on a series of spontaneous and benzo(a)pyrene-induced Syrian hamster clonal cell lines which differed in their tumorigenicity and *in vitro* growth properties. The correlation between various alterations of *in vitro* growth properties and tumorigenicity was analyzed statistically. It can be shown in this model study that cloning efficiency in semisolid agar, which is a property of anchorage-independent growth, is highly correlated with tumorigenicity, with a statistical confidence level of 99%. It appears also that enhancement of fibrinolytic activity and growth in 1% serum are also closely related to tumorigenicity, with a confidence level of 95%. This published study (9) has established a paradigm which demonstrates how various *in vitro* alterations in growth can be correlated with tumorigenicity. Additional effort is being made in the laboratory to extend this investigation to other phenotypic alterations and to additional neoplastic cell lines.

TABLE 1. Correlation Among *In Vitro* Growth Properties and Tumorigenicity of Syrian Hamster Cell Lines

Growth Parameter	Coefficient of rank correlation, τ
Cloning efficiency, semisolid agar	0.986[b]
Fibrinolytic activity	0.754[c]
Generation time, 1% serum	0.754[c]
Cloning efficiency, liquid medium	0.638[c]
Organization of intracellular actin	0.570[c]

TABLE 1 (Continued)

Saturation density, 10% serum	0.246
Generation time, 10% serum	0.062

[a]Computation of the Kendall coefficient of rank correlation, τ, between the TD_{50} of the cell lines and their respective growth properties was performed as described in text.
[b]Significantly correlated at the 99% confidence level ($p \leq 0.01$).
[c]Significantly correlated at the 95% confidence level ($p \leq 0.05$).

The second major problem in the in vitro neoplastic transformation is the temporal relationship of the appearance of these altered growth characteristics after the initial treatment with carcinogens. The results of this investigation show that the neoplastic transformation process appears to be progressive in nature, and is not a single-step phenomenon, as is the case with single-gene somatic mutation (5). The progression of neoplastic transformation was recognized during in vivo tumor formation by the pathologists a long time ago. However, this phenomenon has been recognized only recently in in vitro carcinogenesis.

First we established that untreated SHE cells generally senesced and did not exhibit enhanced fibrinolytic activity. Approximately one in ten untreated cultures escaped senescence and evolved as a continuous cell line; such cultures frequently exhibited enhanced fibrinolytic activity. These results suggest that the acquisition of enhanced fibrinolytic activity, while perhaps not a cause of neoplastic transformation, may reflect a loss of control of the normal function of the cellular genetic apparatus during the process of transformation. Upon further culturing to over 100 passages, we were able to develop a few spontaneously transformed tumor cell lines. This study has clearly demonstrated that the transformation event of untreated SHE cells is exceedingly rare. Most of them senesce in culture, and a few escape senescence to become permanent lines. Only after extensive, prolonged periods of culturing do some of these permanent lines evolve to tumorigenic cell lines.

The temporal acquisition of in vitro phenotypes associated with neoplasia was examined following exposure of SHE cells to a chemical carcinogen (5). Quantitative assays measuring morphological changes, enhanced fibrinolytic activity, and anchorage-independent growth were used to detect the development of transformed cells within a population of normal hamster embryo cells. Morphological transformation and enhanced fibrinolytic activity were early changes observed after treatment with B(a)P, whereas the ability to grow in semisolid agar was delayed 32-75 population doublings after carcinogen exposure. This delay was not due to selection of a small number of cells present early after treatment, but at a level below detection, since a large percent of cells isolated at early passage (10^3 total above the level of detection) was demonstrated to have developed the potential for anchorage-independent growth at later passages. These observations indicate that neoplastic transformation in vitro is a progressive process through qualitatively different stages. Thus, an analogy can be drawn to the progressive nature of in vivo carcinogenesis. These results strongly justify the study of oncogenesis in cell culture as a model for neoplastic transformation in vivo. At present, six alterations of in vitro growth properties were established which can serve as markers for neoplastic transformation with varying degrees of reliability. The sensitivities of these alterations related to the identification or selection of an altered cell among a population of normal cells as well as their approximate time requirement after the carcinogenic treatment are briefly indicated below:

1.	Alteration in clonal morphology (10^{-3});	2 weeks
2.	Enhancement in fibrinolytic activity (10^{-4});	2-4 weeks
3.	Growth on confluent cell mat (10^{-5});	4-8 weeks
4.	Growth in low serum (1-2% serum, 10^{-5}-10^{-6});	4-12 weeks
5.	Growth in low calcium (0.1 mM Ca^{++}, 10^{-5}-10^{-6});	4-12 weeks
6.	Growth in soft agar or agarose (10^{-6});	8-16 weeks

DNA AS A CRITICAL TARGET IN THE CELL FOR NEOPLASTIC TRANSFORMATION

Three different approaches were adopted in this study.

(1) Direct perturbation of DNA was accomplished by treatment of cells in culture with BrdU and near-UV light (6,7). BrdU is incorporated only into the DNA of cells, in place of its analog, thymidine. The UV absorption spectrum of DNA-containing BrdU is shifted toward a longer wavelength; thus, irradiation with light of wavelengths greater than 300 nm produces a significantly higher number of photochemical lesions in BrdU-substituted DNA than in nonsubstituted DNA. Current findings suggest that the major photochemical lesion in BrdU-substituted DNA is single-strand breaks. This breakage is caused by an initial photodissociation of the bromine atom, producing a uracilyl radical and is followed by the subsequent decomposition of this radical.

The experiment with unsynchronous Syrian hamster fibroblast indicates that both morphological transformation and neoplastic transformation, based on measurements of growth in soft agar, and tumorigenicity in newborn animals, took place (6). Concomitant with transformation, somatic mutation and DNA strand breaks were found. These results were supported by the study on synchronous Syrian hamster embryo cell cultures (7). In the synchronous culture (80% synchrony) cells in different stages of the S phase were treated with a 1-hour pulse of BrdU followed by irradiation with near-UV light. Chromosomal aberrations at the chromatid level and DNA damage, as measured by alkaline sedimentation, were induced by this treatment. No specific period during the S phase was markedly more sensitive to treatment with respect to cell survival, chromosomal aberrations, and DNA damage or its repair. The induction of morphological transformation was cell-cycle dependent, occurring only in cells synthesizing DNA. The highest incidence of morphological transformation occurred during pulse treatment in middle S phase, particularly in the second hour of S phase, while no transformation was observed in late S phase, G_1-S boundary, and G_2 phase. BrdU treatment or irradiation alone induced no changes. However, in order to test for growth in soft agar and tumorigenicity, these treated cells were required to grow continuously for 100-150 days. During this period most of the control cultures senesced and died. Generally, the results from soft agar assay and tumorigenicity support and confirm the morphological transformation.

(2) Specific DNA nuclear damage can also be introduced by the incorporation of methyl [^{3}H]-thymidine into the normal Syrian hamster fibroblast (11). In addition to cell killing as anticipated, mutagenic effects were observed for the 6-thioguanine resistance but not for ouabain resistance. This data suggest that the methyl [^{3}H]-thymidine-induced mutation acts via a frameshift or deletion mechanism predominantly. Transformation studies indicate the occurrence of subpopulations with morphological alteration at post-treatment population doubling 3-4; subpopulations with enhancement in fibrinolytic activity at post-treatment population doubling 11-16; subpopulations capable of cloning in medium containing 1% serum at post-treatment population doubling 18-32; and finally, subpopulations capable of growing in soft agar at post-treatment population doubling 25-32. All treated cells displayed tumorigenicity when examined at post-treatment population doubling of about 50, while the control cultures did not. This data again suggest the notion that breakage of DNA due to beta decay is sufficient to initiate neoplastic transformation.

(3) Somatic mutation and neoplastic transformation in diploid Syrian hamster embryo cells was also observed when treated by liposomes which contained DNase I (12). We have used phosphotidylserine liposomes to deliver active DNase I inside the cell, thus providing a direct and specific perturbation to the genetic apparatus. We showed: a) the entrance of DNase I inside the cell by the dose-dependent cytotoxicity of this treatment; and b) the entrance of DNase I into the nucleus by the induction of somatic mutation at the HGPRT locus (2). Finally, in two experiments, we have observed neoplastic transformation. In one experiment, for instance, at passage 36 or population doubling 63, the treated culture was found to be tumorigenic in nine newborn hamsters (2×10^6 cells subcutaneous injection, 3 weeks latent period), while the four control cultures were not tumorigenic at passage 38 (population doubling 85). This data suggests that damage to DNA alone, including single strand breaks, can initiate neoplastic transformation.

The results of the above three sets of experiments reinforce each other in pointing to the conclusion that specific perturbation to DNA alone, possibly through strand breaks, is sufficient to initiate neoplastic transformation. However, through such a perturbation to DNA, cell killing can be observed within a few cell divisions, and somatic mutation can also be observed within 5-10 population doublings, the required time for expression. But neoplastic transformation as indicated by anchorage independent growth or tumorigenicity cannot be observed in this culture until 50-150 population doublings after the initial insult. What kind of cellular process requires an expression time of so many cell divisions? Apparently, this is one of the most challenging puzzles, and it becomes one of the crucial differences between neoplastic transformation and somatic mutation. Even now, both processes can be initiated by specific perturbation to DNA alone.

SOMATIC MUTATION AND NEOPLASTIC TRANSFORMATION

With this Syrian hamster embryo cell system, it is now possible to make a quantitative evaluation of the relationship between somatic mutation and neoplastic transformation (8,4). Somatic mutation and neoplastic transformation of diploid SHE cells were examined concomitantly. Mutations, induced by B(a)P and MNNG, were quantitated at the HGPRT and the Na^+/K^+ATPase loci and compared to phenotypic transformations measured by changes in cellular morphology and colony formation in agar. Both cellular transformations had characteristics distinct from the somatic mutations observed at the two loci. Morphological transformation was observed after a time comparable to somatic mutation but at a frequency 25-540-fold higher. Transformants capable of colony formation in agar were detected at 10^{-5}-10^{-6} frequency, but not until 32-75 population doublings after carcinogen treatment. While this frequency of transformation is comparable to that of somatic mutations, the detection time required is much longer than the optimal expression time of conventionally studied somatic mutations. The results are summarized in Table 2.

TABLE 2. Comparison of Phenotypic Changes of SHE cells

	Somatic Mutation	Morphological Transformation	Anchorage Independent Growth[a]
Observed frequency (spontaneous)	$<10^{-6}$	$\sim10^{-4}$[b]	$<1.4 \times 10^{-8}$
Observed frequency (carcinogen treated)	10^{-5}-10^{-4}	10^{-3}-10^{-2}	10^{-5}-10-6
Expression/detection time[c]	6-8	≤ 8	32-75

[a]As measured by colony formation in soft agar.
[b]Six spontaneous, morphologically transformed colonies were observed per ~62,000 control colonies examined.
[c]Population doublings.

Neoplastic transformation of hamster embryo cells has been described as a multistep, progressive process. Various phenotypic transformations of cells following carcinogenic treatment may represent different stages in this progressive transformation. Thus, a mutational change may initiate neoplastic transformation, but neoplastic transformation may not be delineated by a single-gene mutational event involving a conventional dominant, codominant, of X-linked recessive locus. Neoplastic transformation induced by chemical carcinogens is more complex than a single-gene mutational process. Thus, this comparative study does not give experimental support to predictions of the carcinogenic potential of chemicals based on a simple extrapolation of the results obtained from conventional somatic mutation assays.

The above experiment does not clearly indicate whether somatic mutation can cause neoplastic transformation. nor does it exclude such a possibility. Neoplastic transformation of normal diploid cells most likely requires more than one mutation of a single gene to become highly tumorigenic. It is also possible that if a mutation of one gene is sufficient, then it will be a recessive autosomal mutation requiring two steps for its full expression. This hypothesis requires two types of further experimentation: (1) Characterization of neoplastic transformation into different stages during the process of progression and isolate clonally the variants obtained at these stages; and (2) To study the progression and stability of these variant clones upon further growth or upon second mutagenic perturbation.

CURRENT FINDINGS IN THE STUDY OF THE MECHANISM OF PROGRESSION

As described in the above sections, the ability of cells to grow in soft agar is a transformed phenotype (Aga^+) well-correlated with tumorigenicity. Also, spontaneously transformed cell lines can be established.

To study the role of specific alterations in cellular phenotype with ultimate tumorigenicity in such spontaneously established cell lines, we isolated clonal strains, Fol^+, derived from colonies exhibiting zonal lysis in fibrin-agarose overlay plates (13). Examination of four clonal strains, Fol^+1,2,3, and 4, derived from one such established cell culture revealed the following cellular alterations: increased cloning efficiency (14-19%), stably elevated fibrinolysis, and heteroploidy (modal range 74-85). Initially after isolation, all four strains were nontumorigenic upon subcutaneous inoculation in newborn hamsters and were incapable of growth in semisolid agar. However, in contrast to normal diploid SHE cell cultures, Fol^+ 1-4 clonal strains do exhibit anchorage-independent variants capable of growth in soft agar at a low but detectable frequency (10^{-6}/plated cell). We have further demonstrated that these Aga^+ variant clones are tumorigenic. Several results suggest a somatic mutational basis for the Aga^+ phenotype:

1. Fol^+ Aga^+ strains are stably anchorage independent, indicating a hereditable nature of the Aga^+ phenotype.

2. A Luria-Delbrück fluctuation analysis demonstrates that Aga^+ variants arise spontaneously and at a rate (5.7×10^{-8} variants/cell/generation, P_0 calculation) comparable to the rate of single-locus somatic mutation in these cells (7.8×10^{-8} Oua^r mutants/cell/generation). These rates are also in close agreement with the measured rate of mutation at the HGPRT locus (1.2×10^{-8} $HGPRT^-$ mutants/cell/generation) in BP6T cells, a subdiploid cell line derived from the same SHE cell preparation following B(a)P treatment.

3. The Aga^+ phenotype can be suppressed in cell hybrids. When diploid SHE cells are fused with a highly tumorigenic, subdiploid Aga^+ line with polyethylene glycol, hybrid cells are initially incapable of growing in semisolid agar. However, when these ioslated hybrid clones were tested after 25 cell doublings, Aga^+ cells are evident. Moreover, a wide range of growth (Aga^+ frequency 10^{-5}-10^{-1}) is observed among these isolated clonal strains this time. These results suggest a recessive nature of the Aga^+ phenotype (anchorage independence), which reexpresses in cell hybrids following chromosome segregation or rearrangement.

Several attempts to increase the Aga^+ frequency above spontaneous background by treating Fol^+ cells with MNNG and ICR-10 (frameshift mutagens) have been unsuccessful. In addition, the same mutagenic treatment protocol provided a highly significant, linear 10-200-fold induction of Na^+/K^+ ATPase mutants (Oua^r, codominant phenotype) over a dose range of 1-10 μM MNNG, demonstrating that MNNG was effective in this experiment in inducing point mutations. As described above, however, the subtetraploid Fol^+ cells exhibit Aga^+ variants at a spontaneous rate comparable to their rate of mutation at the Na^+/K^+ ATPase locus (Oua^r, codominant phenotype). Thus, it is difficult to reconcile this result with both an apparent recessive nature of the Aga^+ phenotype and solely a point mutational mechanism of the spontaneous $Aga^- \rightarrow Aga^+$ conversion in these cells, because these cells are subtetraploid. In summary, together with our observations with diploid SHE cells, these data indicate that the Aga^+ phenotype can be induced neither by single-gene mutations, which result in autosomal dominant or X-linked recessive phenotypes in diploid SHE cells, nor by point mutation alone in tetraploid Fol^+ cells.

Preliminary results from a comprehensive analysis of the karyotypes of these Fol^+ cells reveal a very high frequency of chromosomal nondisjunction in the population. It is a tempting hypothesis that the original carcinogenic insults may have caused a mutation which is recessive and autosomally-linked in nature, and which cannot be expressed or detected until the heterozygotic state of that mutation is converted to the homozygotic state through chromosomal nondisjunction. The progression may represent the period required for this conversion through karyotypic changes. There are two lines of supportive evidence to this hypothesis: 1) the measured rate of the frequency of chromosomal nondisjunction is sufficiently high to account for the proposed conversion from heterozygotes to homozygotes during the progression period; 2) the preliminary results of the study on the ploidy dependence (2N versus 4N) of neoplastic transformation indicates that a crucial (a rate-limiting) event in the process is likely to be a recessive gene mutation (13,14). So far, the cells from the 4N clone are not responsive to the treatment of mutagen/carcinogen in contrast to the ready response of the cells from the 2N clones (the positive control) in the neoplastic transformation experiment after receiving the same treatment. The observation is in agreement with the results about cell hybrids, indicating the Aga^{+} phenotype can be suppressed initially and can be expressed following chromosomal segregation or rearrangement.

It is possible, however, that the initial carcinogenic attack unleashes a stepwise, interlocking, and cascading process which leads to chromosomal abnormalities and nondisjunction exerting a profound effect on the genetic apparatus as well as disrupting the programmed activities of gene expression in differentiation. This as yet undefined process could well not be adequately atttributable to single gene mutation, the conventional type studied in somatic mutation experiments.

EARLY SYRIAN HAMSTER EMBRYO CELLS AS A MODEL FOR THE STUDY OF CELLULAR DIFFERENTIATION (15)

In the early passages of Syrian hamster embryo (SHE) cell cultures prepared from 10-day old and 13-day old embryos, we have found subpopulations which possess growth properties of contact-insensitivity (CI^{-}) and/or anchorage independence (AD^{-}). These properties are measured respectively by the cloning efficiency on lethally irradiated confluent monolayers of contact-inhibited cells (cell mat) and by the cloning efficiency in semi-solid agar. The CI^{-} and AI^{+} growth properties are invariably exhibited by the neoplastic/tumorigenic fibroblasts, but are usually not possessed by the normal diploid fibroblasts at late passage. The CI^{-}/AI^{+} subpopulations were found to disappear gradually from the cultures with an increase in population doubling upon growth on a plastic surface. To determine whether the loss of CI^{-} subpopulation in cultures is due to cellular differentiation or negative selection, SHE cells were grown on confluent cell mat, thereby excluding the possibility of negative selection. Even when cultivated on cell mat, CI^{-} populations decreased with an increase in population doubling. Clonally isolated CI^{-} colonies, when subsequently subcultured, showed drastic loss of CI^{-} properties upon further growth. However, the cultures enriched with CI^{-} cells have a higher proliferative capacity by showing longer life span and larger colonies than the normal cultures. These observations indicate that CI^{-} cells differentiate to become CI^{+} (contact insensitive) cells. AD^{-} subpopulations decreased in parallel with CI^{-} subpopulation and a small portion of the CI^{-} subpopulation exhibits AD^{-} growth properties. The above results indicate that both embryonic development <u>in vivo</u> and cell replication <u>in vitro</u> have similar influence in diminishing these subpopulations, presumably through differentiation. For the first time, neoplastic transformation, somatic mutation, and cellular differentiation can all be investigated concomitantly using this early SHE system, which could serve as a useful and unique model for studying the genetic and epigenetic control of normal differentiation versus tumorigenic transformation *in vitro*.

This question which concerns the relationship between neoplastic transformation process and the differentiation-developmental process of the animal (a man), immediately leads to another practical and challenging question: Would the neoplastic transformation of cells from a young embryo be both qualitatively and quantitatively different from cells of aged animals? To provide a cellular model to study the basic mechanism relating carcinogenesis to development, and finally, to aging, we have divided this developmental process of the cells into three stages of consideration:

1. Cells that have full potential to replicate and have little constraint of replication, such as cells from embryos that replicate rapidly and seem to be not contact-inhibited (S. Nakano *et al*., unpublished data).
2. Cells that have replicating potential but are constrained by cell-cell contact, as is expected of a differentiating tissue.
3. Cells that now have lost their reproductive capability and cease to divide, as in a terminally differentiated tissue.

It is now apparent that we can develop from the fibroblast system these three cell types in different stages of development in culture. We can ask whether the neoplastic transformation of these three different cell types will involve similar or different processes or mechanisms. Apparently we have to block the differentiation pathway for the embryonic cells to maintain them in a state of continuing replication without constraint, to achieve a state of neoplasia. For senescent cells, obviously we have to reactivate their dormant reproductive capabilities to allow a continuation of cell division and an escape of senescence. Experiments are now in progress to answer the above questions.

FUTURE PROSPECTS

The heritable biological responses, in addition to cell death, to the direct and specific DNA damage (most likely strand breaks) can be classified into two major categories: 1) immediate and direct responses, such as single-gene mutation at the HGPRT locus or Na^+/K^+ ATPase locus. This type of response is commonly described in somatic mutation studies. The altered gene products (proteins or enzymes) can, or have been identified, and the biological effects usually are not further amplified; 2) protracted, interlocking and cascading responses such as the progression phenomenon of the neoplastic transformation described here. After receiving the initial perturbation, subpopulations within the treated cultures emerge with various heritable, altered growth properties with some regular temporal relationship among these subpopulations. These subpopulations, which posses various types of altered growth properties, may or may not be heritably related to each other, though they usually have a close association. For instance, 70% of the isolated clones having enhanced fibrinolytic activity did progress to become clones capable of growing in soft agar. There is another type (Type III) of heritable and interlocking response which may exert a continuous impact on the function and structure of the genetic apparatus, such as a continuing enhancement in the propensity for chromosomal segregation, nondisjunction and rearrangement. It should be noted that so far we have not yet obtained data to indicate that this third type of response can be induced by damage to DNA, though this type of heritable property has been observed in a transformed line. Presently, the relationship between these three types of heritable responses is not yet clear, though Type I and Type II (and likely, Type III) can both be initiated by perturbation only to DNA. Furthermore, the data indicate that there is a clear difference between Type I and Type II heritable responses.

Type II responses, in many respects, are similar but not identical to the programmed process of normal differentiation and development. Therefore, it is very interesting that both neoplastic transformation and normal differentiation can now be studied simultaneously in the early Syrian hamster embryo system.

One major contribution from our previous research is the establishment of a strategy to specifically perturb one target macromolecule inside the living cell at a time. So far, the target is DNA. Obviously, there are other important targets, such as RNA and proteins or gene products. It would be of great interest to know whether specific perturbation to RNA or proteins can lead to heritable biological responses, either Type I, II, or III. Another important perturbation is to affect the karyotype of the cell, leading to aneuploidy. The knowledge of the types of heritable biological responses after the cell lines have become aneuploid, or when cells are grown continuously at a low but effective dose of antimitotic agent, will be of great importance.

The concept of sensitive and insensitive targets should be noted when considering the various types of perturbation. Obviously, the single-copy target is likely to be more sensitive to perturbation than the multiple-copy target. Target molecules involved in regulatory function are likely to be more sensitive than target molecules invovled in salvage function. It is interesting, for instance, that ^{3}H-thymidine is much more toxic than ^{3}H-uridine; however, RNase encapsulated in liposomes is quite cytotoxic when compared to DNase I encapsulated in liposomes.

Finally, regarding the very nature of the heritable biological responses, particularly those pertaining to differentiation and neoplastic transformation, we pose two concepts that may serve as a guide for the study of this problem. These concepts are based on the fact that development and differentiation are specific characteristics of multicellular organisms. Therefore, neoplastic transformation, or carcinogenesis, is also a unique pathological event of multicellular organisms, representing an abnormal pathway for development and differentiation. First, in the genome of every individual cell in a multicellular organism, there exists genetic information specifying a number of pathways, or directions, for the cell to undertake one at a time, reversibly or irreversibly. The decision to take this different pathway or direction, which is already predetermined, however, will be greatly influenced by the environment of the cell, including signals from neighboring cells or hormonal control from the central system. The interplay and the balance between the genetically determined pathway and the decision-making influence of the environmental conditions, or signals, provide the real picture of the normal developmental process of a cell or abnormal cell growth, such as cancer. Our job is to describe these genetic pathways and environmental influences in molecular terms.

Second, in describing the molecular machinery necessary to cope with the problem posed in the description of the first concept, we would like to advance the concept of genetic apparatus. In this concept, the genetic apparatus contains both DNA and the regulatory machinery that controls the expression and replication of DNA. The regulatory machinery can contain both proteins, RNA, and probably membranes in an intricate interaction with DNA. Processes affecting the regulators, such as protein and RNA, are seemingly epigenetic processes; in fact, they affect the DNA and therefore can be characterized also as genetic events. Equally important is the effect on DNA, which is generally described as a genetic process but can only be detected through the interaction with the regulatory machinery, and so may have been considered as an epigenetic process. In essence, the dynamic interactions of the entire genetic apparatus follow the present genetic information, as influenced by the environmental signal. This interaction decides the heritable changes of the cells through which both the developmental process and neoplastic event emerge. Such knowledge will not only provide us with the understanding of differentiation and neoplasia, but will also provide us the capability of controlling these cells.

REFERENCES

1. Berwald, Y. and Sachs, L.: 1963, Nature (London) 200, pp. 1182.
2. Berwald, Y. and Sachs, L.: 1965, J. Natl. Cancer Inst. 35, pp. 641.
3. Barrett, J.C., Crawford, B.D., Grady, D.L., Hester, L.D., Jones, P.A., Benedict, W.F. and Ts'o, P.O.P.: 1977, Cancer Res. 37, pp. 3815.
4. Barrett, J.C. and Ts'o, P.O.P.: 1978, Proc. Natl. Acad. Sci. USA 75, pp. 3297.
5. Barrett, J.C. and Ts'o, P.O.P.: 1978, Proc. Natl. Acad. Sci. USA 75, pp. 3761.

6. Barrett, J.C., Tsutsui, T. and Ts'o, P.O.P.: 1978, Nature 274, pp. 229.
7. Tsutsui, T., Barrett, J.C., and Ts'o, P.O.P.: 1979, Cancer Res. 39, pp. 2356.
8. Barrett, J.C., Bias, N.E. and Ts'o, P.O.P.: 1978, Mutation Res. 50, pp. 121.
9. Barrett, J.C., Crawford, B.D., Mixter, L.O., Schechtman, L.M., Ts'o, P.O.P., and Pollack, R.: 1979, Cancer Res. 39, pp. 1504.
10. Moyzis, R.K., Grady, D.L., Li, D.W., Mirvis, S.E. and Ts'o, P.O.P.: 1980, Biochemistry 19, pp. 821.
11. Lin, S.L., Klein, L. and Takii, M.: 1979, Proc. Amer. Assoc. Cancer Res., Vol. 20, pp. 118; Lin, S.L., Takii, M., and Ts'o, P.O.P.: 1980, Cancer Res., submitted.
12. Zajac, M.: 1980, Proc. Amer. Assoc. Cancer Res., Vol. 21, pp. 127; Zajac, M. and Ts'o, P.O.P.: 1980, Proc. Cell Biology Congress, Berlin, in press.
13. Crawford, B.D., Klein, L., Melville, M., Morry, D. and Ts'o, P.O.P.: 1980, Proc. Amer. Assoc. Cancer Res., Vol. 21, pp. 127; Crawford, B.D., Barrett, J.C., and Ts'o, P.O.P.: 1979, Proc. Amer. Assoc. Cancer Res., Vol. 20, pp. 135.
14. Morry, D.: 1980, Ph.D. Thesis, The Johns Hopkins University.
15. Nakano, S.: 1980, Proc. Amer. Assoc. Cancer Res., Vol. 21, pp. 127; Nakano, S. and Ts'o, P.O.P.: 1980, Proc. Natl. Acad. Sci., submitted.

ONCOGENIC TRANSFORMATION, INITIATION, PROMOTION AND MUTAGENESIS IN C3H/10T1/2 CELLS

Charles Heidelberger
University of Southern California Cancer Center, Los Angeles, CA 90033 USA

ABSTRACT

In our laboratory we have developed the C3H/10T1/2 cloned line of mouse embryo fibroblasts that is highly contact inhibited, has a very low rate of spontaneous transformation, and does not produce tumors in immunosuppressed C3H mice. These cells undergo a dose-dependent morpho logical and oncogenic transformation by polycyclic aromatic hydrocarbons, X-rays, ultraviolet light, neutrons, α-particles, and a variety of other chemical carcinogens. These cells metabolically activate benzo(a)pyrene by the cytochrome P450 monooxygenase system, have individual tumor associated transplantation antigens, and their transformation does not involve activation of a retrovirus. In this system we have reproduced two-stage carcinogenesis and can measure chemical mutagenesis to ouabain resistance. Thus, in these cells we can study mechanisms of and assay for oncogenic transformation and its initiation and promotion, and mutagenesis.

INTRODUCTION

In 1967, at the First Jerusalem Symposium on Quantum Chemistry and Biology, "Physico-Chemical Mechanisms of Carcinogenesis," I described the beginning of our research on chemical oncogenesis in organ and cell cultures (1), and I have subsequently reviewed this field (2,3). In contrast to the pioneering work of Leo Sachs and his colleagues, who were the first to report chemical transformation of primary cultures of Syrian hamster embryo cells (4), we have chosen to work with permanent cell lines. This report highlights some of the progress we have made in this field since that important symposium held 13 years ago. Our aims have been to use cell culture systems to work towards the elucidation of basic cellular and molecular mechanisms of chemical carcinogenesis and to provide potentially useful short-term tests for environmental carcinogens, initiators, tumor promoters, and mutagens.

B. Pullman, P. O. P. Ts'o and H. Gelboin (eds.), Carcinogenesis: Fundamental Mechanisms and Environmental Effects, 311–318.

ONCOGENIC TRANSFORMATION

Our first successful endeavor in this field was the development of a permanent cloned line of fibroblasts derived from the ventral prostates of C3H mice. This line was contact inhibited and nontumorigenic in immunosuppressed C3H mice, and following treatment with polycyclic aromatic hydrocarbons (PAH) the fibroblasts underwent simultaneous morphological and oncogenic transformation to cells that piled-up in distinct foci over a confluent monolayer. In this system there was an excellent quantitative correlation between the activity of a series of PAH at producing transformed foci and their carcinogenic activities in mice (5). With these cells we were able to demonstrate that 3-methylcholanthrene (MCA) directly transformed the cells and did not select for pre-exisiting malignant cells (6). We were the first to demonstrate that arene oxides were more active than the parent hydrocarbons and other derivatives at transforming these cells (7), and that arene oxides were formed in these cells by inducible mixed-function monooxygenases, whose activity was required for transformation to occur (8). We also found that chemically transformed clones had individual, non cross-reacting tumor associated transplantation antigens that were detected both by humoral and cell-mediated immunological techniques (9). The fact that multiple antigens were formed in a series of transformants from the cloned parental line demonstrated that these antigens are formed during the process of transformation, and were not present on the surfaces of individual nontransformed cells.

In 1973 we developed another cloned cell line derived from C3H mouse embryos that is more satisfactory and reproducible in its behavior than the prostate cells, and which is now being used widely throughout the world. These C3H/10T1/2 cells are highly contact inhibited, have a very low incidence of spontaneous transformation, and are not tumorigenic (10). These cells are also transformed to malignancy by a series of PAH in proportion to their in vivo carcinogenic activities (11). These cells are transformed by the short-acting alkylating agent, MNNG, in a cell cycle phase-specific manner (12). They also simultaneously exhibit on their cell surfaces individual tumor associated transplantation antigens and common embryonic antigens (13). Transformation of these cells by MCA does not involve the activation of an endogenous retrovirus, its core, or transcription of its proviral DNA (14). In these cells, as well, PAH are obligatorily activated by the inducible cytochrome P450 monooxygenases (15). The metabolic activation of benzo(a)pyrene (BaP) in microsomes from C3H/10T1/2 cells has been studied in detail by measurement of the carbon monoxide difference spectra and of arylhydrocarbon hydroxylase, whose activity is highly dependent on the growth state of the cells, and by high pressure liquid chromatography (16). In these cells, as in many other cells and tissues, BaP is activated via the 7,8-dihydrodiol-9,10-epoxide.

C3H/10T1/2 cells are transformed by X-rays (17-20), neutrons (20), ultraviolet light (uv) (21), and by α-particles (22). They are also transformed by some cancer chemotherapeutic agents (23, 24), tobacco smoke condensates (25), and hair dyes (26).

On the other hand, C3H/10T1/2 cells are not killed or transformed by aflatoxin B1, 2-acetylaminofluorene, or dimethylnitrosamine; thus, they lack the enzymatic capability to metabolically activate all chemical classes of carcinogens. In order to provide a system for cell-mediated metabolic activation and to study transformation of epithelial cells in culture, we recently developed 21 cloned lines of rapidly dividing epithelial cells derived from regenerating C3H mouse liver (27). These liver cells as lethally-irradiated "feeder layers" activate all of the above carcinogens so that they kill the target C3H/-10T1/2 cells (27); we have also obtained in this way transformation with aflatoxin B1 (28). Moreover, we have succeeded in transforming these liver cells with hepatocarcinogens to cells that produce adenocarcinomas on inoculation into syngeneic mice (unpublished data). We are now busily characterizing the biological and biochemical behavior of these cells.

We were the first to apply liver homogenate-mediated metabolic activation of chemical carcinogens to mutate Chinese hamster V79 cells at the HGPRT locus (29), based on the pioneeer work of Ames et al (30) in activating chemicals to mutagenize Salmonella typhimurium. We are now comparing the activation of a series of chemical carcinogens to transform C3H/10T1/2 cells by liver homogenates, by primary rat liver hepatocytes (31,32), and by our rapidly growing mouse regenerating liver cells.

INITIATION AND PROMOTION OF ONCOGENIC TRANSFORMATION

The two-stage model of carcinogenesis, as originally demonstrated with mouse skin by Berenblum and Shubik (33), has important implications in the causation of human cancer. The first stage, initiation, is irreversible, whereas promotion is reversible. Tumor promoters, of which 12-0-tetradecanoyl phorbol-13-acetate (TPA) is the prototype, are not mutagenic and do not require metabolic activation. In C3H/10T1/2 cells we have reproduced two-stage transformation in formal analogy with mouse skin. When subeffective concentrations of PAH were used, followed by TPA, which does not transform the cells, there was a large enhancement of transformation (34). The activity of a series of phorbol esters in this system, in general parallelled in vivo promoting activity, and we found that promotion, in this system, involves more than a stimulation of cell division (34). We also found that uv light is a pure initiator in this system (35), since in our laboratory we have never transformed cells by uv light without TPA. By contrast, Chan and Little (21) obtained transformation of C3H/10T1/2 cells by uv light alone. We also found, in collaboration with Little's group, that x-rays also act as initiators in this system (36). In a study commissioned by the United States Senate, we found that saccharin has weak promoting activity in our system following initiation with MCA (37).

Although TPA and related promoters produce hyperplasia on mouse skin, we found no significant stimulation of DNA synthesis (38). Hence, we conclude that stimulation of cell proliferation is not essential for promotion, at least in this validated model system. What, then, is the mechanism of tumor promotion? As described elsewhere in this Symposium, TPA exerts a myriad of effects in a variety of systems. The problem is to determine which effect(s) is involved in the mechanism of promotion. We have chosen to investigate these phenomena in the C3H/-10T1/2 cell system, in which promotion has been demonstrated, and to devote attention also to tumor promoters that are chemically unrelated to TPA, such as anthralin, saccharin, iodoacetic acid, and others, looking for common effects. Since TPA either stimulates or inhibits differentiation in various cells, we have investigated the effects of promoters on the 5-azacytidine-induced differentiation of C3H/10T1/2 cells into muscle cells, as originally demonstrated by Peter Jones and colleagues (39). We found, in fact, that TPA inhibits this differentiative process, as did tumor promoting phorbol esters, and that non-promoting phorbol and phorbol diacetate did not (40). Other chemical classes of tumor promoters, including those mentioned above, also inhibited 5-azacytidine-induced differentiation. However, some substances that inhibit promotion on mouse skin and in C3H/10T1/2 cells, including antipain, fluocinolone acetonide, retinoic acid, and epidermal growth factor, did not reverse the TPA-induced inhibition of differentiation. Hence, we concluded that the effects on differentiation can be dissociated from promoting activity, and consequently that effects on the induction of muscle cell differentiation cannot be used to screen for promoters or delineate their mechanism of action. This is further supported by studies on the effects of promoters on adipocyte differentiation in C3H/10T1/2 cells (S. Mondal and C. Heidelberger, manuscript in preparation). Thus even though we, as others, have not yet critically defined the mechanism of promotion, we nevertheless believe that C3H/10T1/2 cells are a valid model system to test for tumor promoters in the environment.

MUTAGENESIS IN C3H/10T1/2 CELLS

We have had considerable experience in our laboratory in studying chemical mutagenesis in Chinese hamster V79 cells at the HGPRT locus (29, 41-43). There would be obvious advantages to study mutagenesis and oncogenic transformation in the same cells at the same time, as had been done in the Syrian hamster embryo cell system by Huberman et al (44) and by Barrett and Ts'o (45), where the finite lifespan of the cells makes difficult any unequivocal characterization of drug-resistant cells as mutants. Since the HGPRT gene is an X-linked recessive, and since C3H/-10T1/2 cells appear to have three X-chromosomes, we failed to induce a single 8-azaguanine resistant colony. Hence, we turned our attention to ouabain resistance, which is a dominant mutation in which ouabain fails to bind to its receptor on the membrane Na^+/K^+ ATPase. We have succeeded in obtaining quantitative mutagenesis to ouabain resistance with several chemical carcinogens in C3H/10T1/2 cells (46), and Chan and Little (47) independently demonstrated in the same cells that uv light produces mutations to ouabain resistance. We have devoted considerable attention to the characterization of our ouabain-resistant clones as mutants, and have obtained evidence that this is so by demonstrating that various mutagens produce the mutant phenotype, that ouabain resistance is maintained when the mutant clones are passaged for prolonged periods in the absence of the selective agent, and that ouabain inhibits ^{86}rubidium uptake one thousandfold less in the mutant than in the wild-type cells (48).

In the hamster embryo cell system it had previously been reported that the ratio of the transformation to the mutation frequency is from 50-500:1 (44,45). In C3H/10T1/2 cells we found this ratio to vary from 12-52:1 (46). In fact, this ratio is now being used to calculate relative target sizes for transformation and for mutation. I strongly caution against such calculations on the grounds that the absolute transformation frequency has never been determined in any system, and that the mutation frequencies in the same cells with the same marker can vary over two orders of magnitudes in different laboratories. I venture to predict that this ratio will vary from cell to cell, from marker to marker, and from carcinogen to carcinogen, and we already have evidence that this is so. On the basis of these experiments, as yet incomplete, we can conclude that in our system mutagenesis is necessary, but not sufficient for transformation. Finally, we are comparing in C3H/10T1/2 cells, the transformation, mutagenesis, and sister chromatid exchanges produced by BaP and a series of derivatives, and have found some interesting non-parallels (49).

In conclusion, the ability to investigate quantitatively in C3H/-10T1/2 cells the processes of oncogenic transformation and its initiation and promotion, as well as chemical mutagenesis, provides a unique opportunity to critically study and compare the mechanisms of these phenomena and to validate short-term tests for substances with these properties found in the environment. In risk assessment extrapolations to humans, it would probably be more relevant to use human, rather than mouse cells. However, the difficulties of producing oncogenic transformation of human cells with different classes of chemical carcinogens, and the present inability to quantitate such transformation, justifies the continued use of mouse cells for carcinogen testing until the development and validation of quantitative systems with human cells.

REFERENCES

1. Heidelberger, C.: in, Physico-Chemical Mechanisms of Carcinogenesis, Jerusalem Symposia on Quantum Chemistry and Biochemistry: 1968, Vol. 1, pp. 45-58.

2. Heidelberger, C.: 1973, Adv. Cancer Res. 18, pp. 317-366.

3. Heidelberger, C.: 1975, Ann. Rev. Biochem. 44, pp. 79-121.

4. Berwald, Y., and Sachs, L.: 1963, Nature 200, pp. 1182-1184.

5. Chen, T.T., Heidelberger, C.: 1969, Internat. J. Cancer 4, pp. 166-178.

6. Mondal, S., Heidelberger C.: 1970, Proc. Natl. Acad. Sci. USA 65, pp. 219-225.

7. Grover, P.L., Sims, P., Huberman, E., Marquardt, H., Kuroki, T., and Heidelberger, C.: 1971, Proc. Natl. Acad. Sci. USA 68, pp. 1089-1101.

8. Marquardt, H., and Heidelberger, C.: 1972, Chem. Biol. Interactions 5, pp. 69-72.

9. Embleton, M.J., and Heidelberger, C.: 1972, Intern. J. Cancer 9, pp. 8-18.

10. Reznikoff, C.A., Brankow, D.W., and Heidelberger, C.: 1973, Cancer Research 33, pp. 3231-3238.

11. Reznikoff, C.A., Bertram, J.S., Brankow, D.W., and Heidelberger, C.: 1973, Cancer Research 33, pp. 3239-3249.

12. Bertram, J.S., and Heidelberger, C.: 1974, Cancer Research 34, pp. 526-537.

13. Embleton, M.J., and Heidelberger, C.: 1975, Cancer Research 35, pp. 2049-2055.

14. Rapp, U.R., Nowinski, R.C., Reznikoff, C.A., and Heidelberger, C.: 1975, Virology 65, pp. 392-409.

15. Nesnow, S., and Heidelberger, C.: 1976, Cancer Research 36, pp. 1801-1808.

16. Gehly, E.B., Fahl, W.E., Jefcoate,C.R., and Heidelberger, C.: 1979, J. Biol. Chem. 254, pp. 5041-5048.

17. Terzaghi, M., and Little, J.B.: 1975, Nature 235, pp. 548-549.

18. Terzaghi, M., and Little, J.B.: 1976, Cancer Res. 36, 1367-1374.

19. Miller, R., and Hall, E.J.: 1978, Nature 272, pp. 58-60.

20. Han, A., and Elkind, M.M.: 1979, Cancer Res 39, pp. 123-130.

21. Chan, G.C., and Little, J.B.: 1976, Nature 264, pp. 442-444.

22. Lloyd,E.L., Gemmell, M.A., Henning, C.B., Gemmell, B.S., and Zabransky, B.J.: 1979, Int. J. Radiat. Biol. 36, pp. 467-478.

23. Jones, P.A., Benedict, W.F., Baker, M.S, Mondal, S., Rapp, U., and Heidelberger, C.: 1976, Cancer Research 36, pp. 101-107.

24. Jones, P.A., Baker, M.S., Bertram, J.S., and Benedict, W.F.: 1977, Cancer Res. 37, pp. 2214-2217.

25. Benedict, W.F., Rucker, N., Faust, J., and Kouri, R.E.: 1976, Cancer Res. 36, pp. 857-860.

26. Benedict, W.F.: 1976, Nature 260, pp. 368-369.

27. Lillehaug, J.R., Mondal, S., and Heidelberger, C.: 1979, In Vitro 15, pp. 910-916.

28. Mondal, S., Lillehaug, J.R., and Heidelberger, C.: 1979, Proc. Am. Assoc. Cancer Res. 20, p. 62.

29. Krahn,D.F., and Heidelberger, C.: 1977, Mutation Res. 46, pp. 27-44.

30. Ames, B.N., Durston, W.E., Yamasaki, E., and Lee, F.D.: 1973, Proc. Natl. Acad. Sci. USA 10, pp. 2281-2285.

31. Langenbach, R., Freed, H., and Huberman, E.,: 1978, Proc. Natl. Acad. Sci. USA 75, pp. 2864-2868.

32. Jones, C., and Huberman, E.: 1980, Cancer Res. 40, pp. 406-411.

33. Berenblum, I., and Shubik, P.: 1949, Int. J. Cancer 3, pp. 384-386.

34. Mondal, S., Brankow, D.W., and Heidelberger, C.: 1976, Cancer Res. 36, 2254-2260.

35. Mondal, S., and Heidelberger, C.: 1976, Nature 260, pp. 710-711.

36. Kennedy, A.R., Mondal, S., Heidelberger, C., and Little, J.B.: 1978, Cancer Res. 38, pp. 439-443.

37. Mondal, S., Brankow, D.W., and Heidelberger, C.: 1978, Science 201, pp. 1141-1142.

38. Peterson, A.R., Mondal, S., Brankow, D.W., Thon, W., and Heidelberger, C.: 1977, Cancer Res. 37, pp. 3223-3227.

39. Constantinides, P.G., Taylor, S.M., and Jones, P.A.: 1978, Developmental Biol. 66, pp. 57-71.

40. Mondal, S., and Heidelberger, C.: 1980, Cancer Res 40, pp. 334-338.

41. Peterson, A.R., Peterson, H., and Heidelberger, C.: 1974, Mutation Res. 24, pp. 25-33.

42. Peterson, A.R., Peterson, H., and Heidelberger, C.: 1975, Mutation Res. 29, pp. 127-137.

43. Peterson, A.R., Peterson, H., and Heidelberger, C.: 1979, Cancer Res. 39, pp. 131-138.

44. Huberman, E., Mager, R., and Sachs, L.: 1976, Nature 264, pp. 360-361.

45. Barrett, J.C., and Ts'o, P.O.P.: 1978, Proc. Natl. Acad. Sci. USA 75, pp. 3297-3301.

46. Landolph, J., and Heidelberger, C.: 1979, Proc. Natl. Acad. Sci. USA 76, pp. 930-934.

47. Chan, G.L., and Little, J.B.: 1978, Proc. Natl. Acad. Sci. USA 1978, pp. 3363-3366.

48. Landolph, J.R., Telfer, H., and Heidelberger, C.: 1980, Mutation Res. in press.

49. Gehly, E.B., Landolph, J.R., Heidelberger, C., Nagasawa, H., and Little, J.B. manuscript in preparation.

DISSECTION OF THE EARLY MOLECULAR EVENTS IN THE ACTIVATION OF LYMPHOCYTES BY 12-0-TETRADECANOYLPHORBOL-13-ACETATE

Gerald C. Mueller, Philip W. Wertz, Cheung H. Kwong, Kristin Anderson and Steven A. Wrighton
McArdle Laboratory for Cancer Research, University of Wisconsin, Madison, Wis. 53706

In this report, normal lymphocytes from bovine lymph nodes have been used to study the metabolic interactions of tumor promoting phorbol esters and antitumor-promoting retinoids. As shown earlier with this model system, 12-0-tetradecanoylphorbol-13-acetate (TPA) and related tumor-promoting phorbol esters act as co-mitogens in cells treated with a suboptimal level of phytohemagglutinin (PHA) or concanavalin A (con A) (1,2). This mitogenic response, as measured by an increase in the level of DNA synthesis in cultures 48 hours after the initiation of TPA-PHA treatment, is prevented when the cultures are treated simultaneously with retinoic acid (RA) or a related retinoid with anti-tumor promoting activity in mouse skin. This antagonism by RA was also seen in the case of the co-induction of ornithine decarboxylase (ODC) activity by the combination of PHA and TPA during the first 18 hours in such cultures (3).

An interesting aspect of the TPA co-mitogenic response and its antagonism by RA was the finding that both agents largely exert their effects during the first 2 hours following the initiation of the mitogenic activation process by PHA. The existence of this early action window with respect to the initial membrane activation by PHA has prompted us to test for TPA and RA effects on several fundamental membrane processes: membrane movement, glucose transport, amino acid transport, and choline phospholipid synthesis. The chronology of these responses, as reflected in Figure 1, separates them into two trigger-like reactions (i.e. the activation of capping and glucose transport) and two secondary responses which ensue after a measurable lag period (i.e. the activation of aminoisobutyric acid (AIB) transport and the stimulation of phosphatidyl choline synthesis). In contrast to later events in the co-mitogenic action of phorbol esters none of these 4 early responses depend on the synthesis of new RNA or protein. The two secondary responses to TPA however differ from the trigger-like reactions in being inhibitable by retinoic acid. The action of the retinoic acid in each case is directed to the blockade

B. Pullman, P. O. P. Ts'o and H. Gelboin (eds.), Carcinogenesis: Fundamental Mechanisms and Environmental Effects, 319–333.

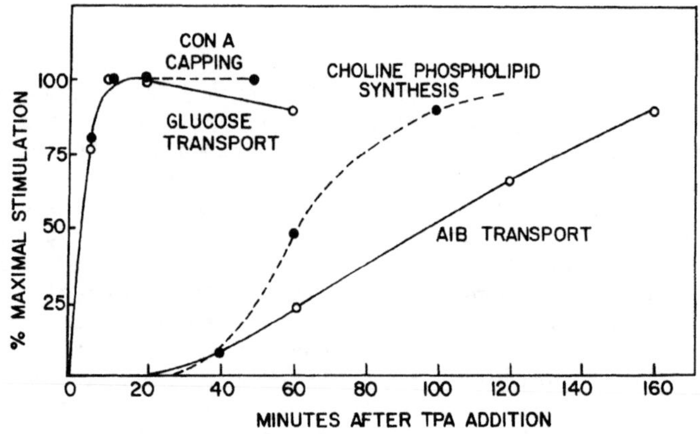

Fig. 1. Time Course of Early TPA Responses. The activation of capping, glucose transport, AIB transport, and phosphatidyl choline synthesis following treatment of bovine lymphocytes with 10 nM TPA are shown. Data are expressed as % of maximal response.

of the activation of these systems rather than the inhibition of of the established AIB transport or the on-going synthesis of phosphatidyl choline. An example of this aspect of RA action is documented in Figure 2 where it can be seen that the delayed addition of RA to TPA-treated lymphocytes prevents further activation of AIB transport without influencing the attained level of activity (4). A very similar situation prevails in the case of the TPA activation of phosphatidyl choline synthesis (5).

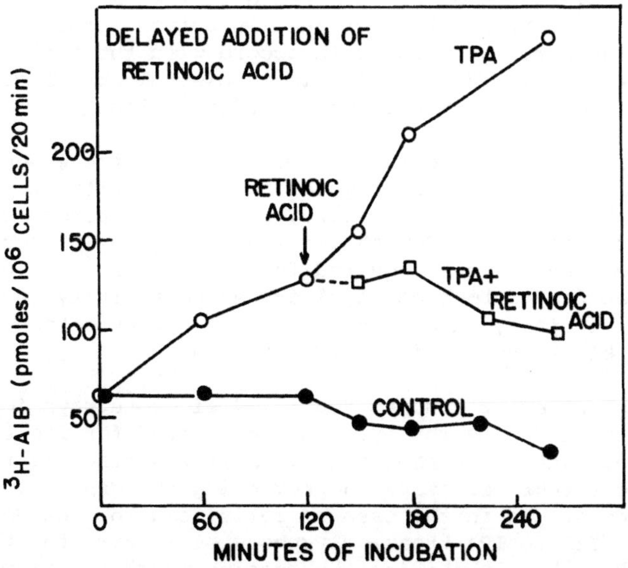

Fig. 2. Effect of the Delayed Addition of Retinoic Acid on the Activation of AIB Transport in Lymphocytes by TPA. Lymphocytes were cultured at 37° in the presence and absence of 10 nM TPA. Retinoic acid (30 μM) was added to TPA treated cells after 120 minutes. The rate of AIB transport was measured over 20 minute intervals at the indicated times using α-[^{3}H]AIB (4). The results are expressed as the uptake of AIB per 10^6 cells in a 20 minute interval at the indicated times.

Since it had been shown earlier (6) that the 5,6 epoxide of retinoic acid is a more effective inhibitor of the activation of phosphatidyl choline synthesis than retinoic acid, it was suspected that RA might be converted oxidatively in living cells to a more effective compound. To test this concept, TPA treated lymphocytes were incubated in a nitrogen atmosphere in an attempt to prevent the formation of the hypothetically active retinoic acid metabolite. Rather surprisingly, it was found that the absence of oxygen strikingly restricted the phorbol ester activation of both phosphatidyl choline synthesis and AIB transport (Table I). While this result prevented the testing of the concept that retinoic acid might have to be oxidatively activated in order to antagonize the secondary TPA responses in lymphocytes; it identified instead a requirement for oxygen in the propagation of the TPA action itself. This observation appears to be quite in accord with the earlier studies of DeChatelet et al. (8) that TPA is a powerful activator of superoxide anion production in polymorphonuclear cells.

In search of possible targets of oxidative processes which might in turn play a role in regulating the activity of phosphatidyl choline synthesis and AIB transport, several inhibitors of oxidative pathways were tested. Indomethacin inhibited the TPA response, but the levels which were required for inhibition (1.0 mM) exceeded that needed to block cyclooxygenase activity in most systems (7); in fact, indomethacin at this concentration has been reported to block phospholipase activity (9).

In this series of studies 5,8,11,14 eicosotetraynoic acid (ETYA), a lipoxidase inhibitor, was also tested for its effect on the TPA activation of phospatidyl choline synthesis; this agent proved to be rather selective in preventing the TPA activation of this process. As in the case of retinoic acid, the delayed addition of ETYA to TPA treated lymphocytes interrupted the activation process rather than inhibiting the ongoing synthesis of phosphatidyl choline (Figure 3) (7). An interesting aspect of these studies

Table I

OXYGEN REQUIREMENT FOR TPA RESPONSES

	TPA Effect (% of Control)	
Atmosphere	Phosphatidyl Choline Synthesis (7)	AIB Transport (4)
Air	235	684
Nitrogen	83	244

Lymphocytes were pre-equilibated with an atmosphere of air or nitrogen at 37°. At zero time, 10 nM TPA plus $[^3H]$choline or $[^3H]$AIB were added to replicate cultures. The incorporation of $[^3H]$choline into the phospholipid fraction was measured over a 1 hr period. AIB uptake into the cells was measured over a 2 hr period. Data are presented as % of the activity of untreated control cells in the respective atmospheres.

was the observation that arachidonic acid could oppose the inhibitory effects of ETYA on the TPA response when added concomittantly with ETYA, but was ineffective as an antagonist when added after the ETYA inhibited state had been established. This finding raised the possibility that an oxidation product of arachidonic acid might mediate the activation of phosphatidyl choline synthesis following TPA treatment of the lymphocytes. It also suggested that ETYA, in accord with its established role as an inhibitor of lipoxidase, may irreversibly titrate some component in the system when it substitutes for arachidonic acid in the activation sequence. In addition, it follows quite naturally that RA may also block TPA action at this same level through reversible or irreversible interactions with a critical component of the system -- either with the oxygenase or with the molecular target of the activation process.

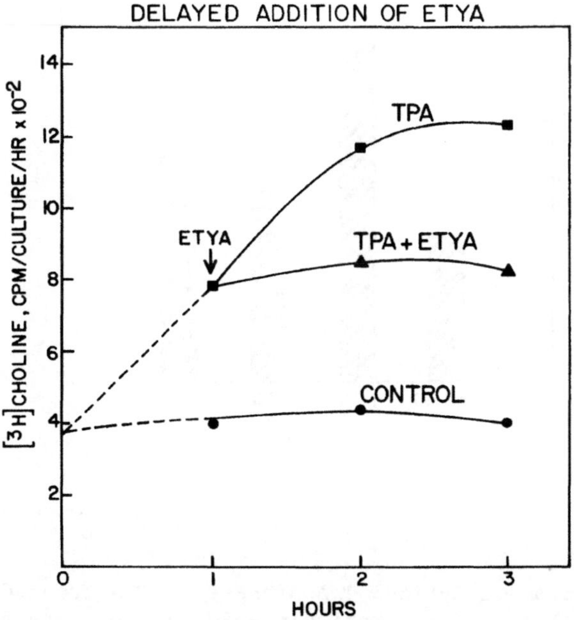

Fig. 3. Effect of ETYA on the Acceleration of Choline Phospholipid Synthesis (7). ETYA (100 μM) was added to lymphocytes after 1 hr of activation with 10 nM TPA. The rate of [^{3}H]choline incorporation into the phospholipid fraction was measured over 1 hr intervals at the indicated times and compared to TPA and control cultures. Points indicate the means of duplicate rate measurements, which did not differ by more than 10%.

To explore this area further, the influence of TPA, RA, and ETYA on the level of cytidyl tranferase activity has been measured. As shown in the studies of Vance (10,11), Schneider (12,13) and Kennedy (14) this enzyme, which catalyzes the formation of cytidine-diphosphocholine from cytidine triphosphate and phosphocholine, is the rate limiting reaction in the synthesis of choline containing phospholipids. In agreement with their studies in other tissues, we found that this enzyme in lymphocytes resides largely in the cell membrane fraction where the active complexes are multimeric and are dissociable to soluble inactive subunits by neutral salts. As shown in Figure 4, TPA treatment of bovine lymphocytes for

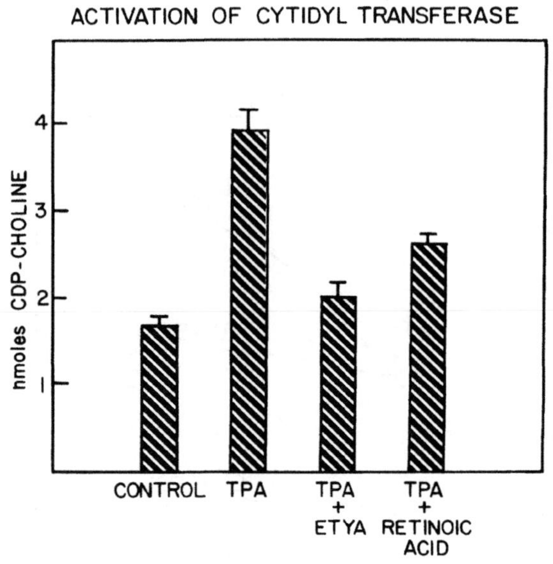

Fig. 4. Activation of Cytidyl Transferase in TPA Treated Lymphocytes. TPA (10 nM), ETYA (5 μM) and retinoic acid (30 μM) were added as indicated to replicate cultures of lymphocytes. After 2 hr of incubation at 37°, the cells were harvested, sonicated, and centrifuged at 100,000 x g for 1 hr to yield a particulate fraction. The activity of cytidyl transferase was measured in the particulate fractions using [^{14}C]choline phosphate and the charcoal assay procedure of Fiscus and Schneider (13). Results are expressed as nmoles CDP-choline formed by 16x10^6 cell equivalents in 20 min.

2 hr yields cell particulate fractions with strikingly increased cytidyl transferase activity. These increases in activity were largely prevented when the cells receiving TPA were treated simultaneously with either ETYA or RA. The results strongly support the view that TPA mediates changes in phosphatidyl choline synthesis by activating cytidyl transferase and that some step in the activation process is opposed directly or indirectly by ETYA and RA.

One approach to dissecting this problem is to reconstitute the activation of cytidyl transferase in a subcellular system. In this connection it should be acknowledged that partially fractionated preparations of cytidyl transferase have been shown earlier (11,13) to be activated by the addition of phospholipids. Following up on this report, lipid fractions were prepared from control and TPA-treated lymphocytes. The addition of these two lipid fractions to a delipidated preparation of cytidyl tranferase from control lymphocytes resulted in a 1 to 2 fold increase in cytidyl transferase activity; the lipid fraction from TPA treated lymphocytes yielded nearly twice as much activation as did the lipid fraction from control lymphocytes.

This result prompted a direct test of arachidonic acid as an activator of the cytidyl transferase in lymphocyte sonicates. The results of such an experiment are presented in Figure 5. In this case the time course of CDP-choline formation was followed in sonicates of lymphocytes pretreated for 2 hours with TPA. As shown the direct addition of arachidonic acid to the sonicates produced a nonlinear activation -- characterized by a slight induction phase, followed by a rapid activation and then a plateau of the response.

This result is in accord with the concept that arachidonic acid itself might be converted to a metabolite which in turn activates the cytidyl transferase. The dependence of the TPA response in living cells on the availability of oxygen and the sensitivity of the in vivo process to the lipoxidase inhibitor, ETYA, prompted the direct test of a lipoxidase in the activation of soluble cytidyl transferase preparations from lymphocyte sonicates. In this study arachidonic acid was first exposed to crystalline soy bean lipoxidase to form some oxidized products. Increasing concentrations of the arachidonic acid, lipoxidase mixture, or untreated arachidonic acid were then added to the soluble cytidyl transferase preparation, from control lymphocytes. As shown in Figure 6, the addition of 30 μg of arachidonic acid which had been pretreated with lipoxidase resulted in an immediate activation of the cytidyl transferase as contrasted to a slight inhibition by a comparable level of untreated arachidonic acid (i.e., unoxidized). Increasing the level of lipoxidase-treated arachidonic acid, however, resulted in a return of the activity to the base level. It is clear from this result that the overall cytidyl transferase activity is a summation of both activation and inactivation by the oxidized arachidonic acid preparation.

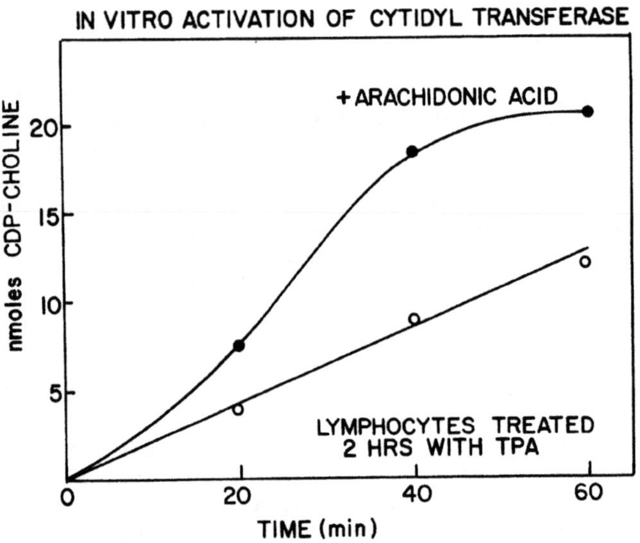

Fig. 5. Activation of Cytidyl Transferase in Lymphocyte Sonciates Supplemented with Arachidonic Acid. Lymphocytes were treated with 10 nM TPA for 2 hr prior to being collected and disrupted by sonication. Cytidyl transferase activity in the sonicate was determined in the absence or presence of added arachidonic acid (100 μg/ml). Results are expresed as nmoles of CDP-choline formed by 7.5 x 10^7 cell-equivalents at the indicated times.

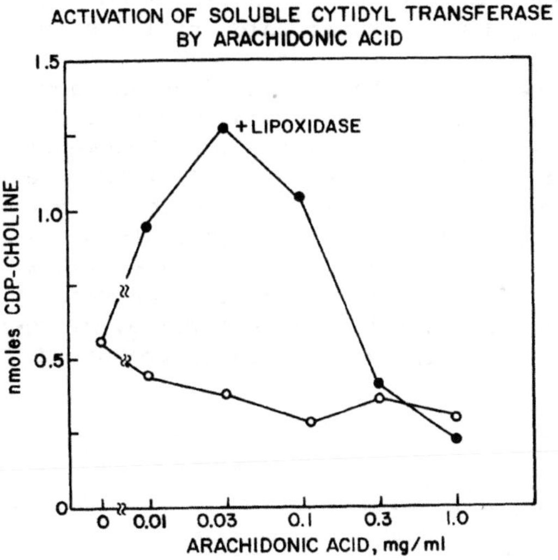

Fig. 6. Activation of Soluble Cytidyl Transferase by Lipoxidase-Treated Arachidonic Acid. Control lymphocytes were disrupted by sonication, and the sonicate was centrifuged at 100,000 x g for 1 hr. Cytidyl transferase in the resulting supernatant fraction was assayed after addition of arachidonic acid (100 μg/ml) or an equivalent amount of arachidonic acid which was preoxidized with soybean lipoxidase. Results are expressed as nmoles CDP-choline formed by 18×10^6 cell-equivalents in 20 min in the presence of the indicated level of arachidonic acid preparation.

A possible explanation of this situation is that a site specific oxidation or derivatization of the enzyme is required for activation and that over oxidation or wrong site derivatization leads to enzyme inactivation. Current investigations are directed to ascertaining the nature of the active metabolite in the lipoxidase treated arachidonic acid samples and the possible production of similar metabolites in the sonicate of TPA-activated lymphocytes. A prime consideration is the manner in which RA and ETYA interfere in this activtion process <u>in situ</u>. With the limited data available at this time, it appears most probable that they compete with the arachidonate metabolite in some terminal aspect of the activation process; however, it should be recognized that the activation process may be more complicated than anticipated and that arachidonic acid may be a substitute for a yet unidentified natural activator.

While the above studies have focused on the activation of cytidyl transferase, the requirement for oxygen and the sensitivity to both ETYA and RA suggest that a similar phenomenon may be involved in the TPA activation of AIB transport. In this case, however, the identity of the target protein remains to be established, as well as the identity of any endogenous activators. In this vane, it is anticipated that proteins, modified by a similar mechanism may play a role in the nuclear responses which underly the retinoid sensitive induction of ornithine decarboxylase, mitogenesis, and the yet unidentified signal event in tumor promotion.

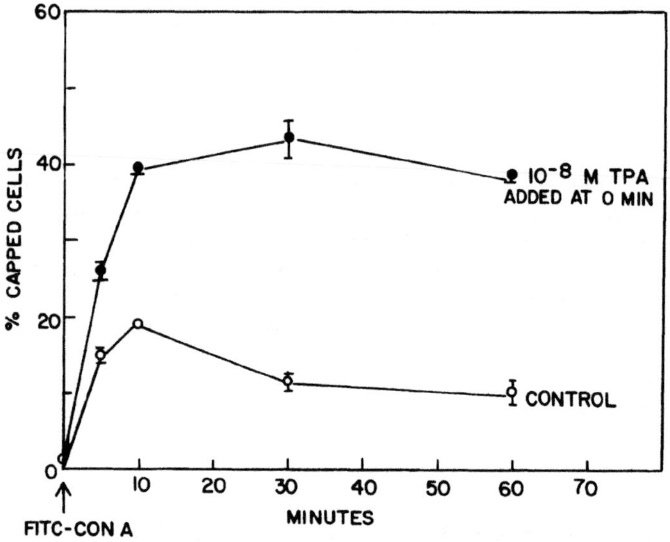

Fig. 7. Effect of TPA Treatment on the Capping of Lymphocytes with Fluorescein-Labeled concanavalin A. Cells were fixed with formalin at the indicated times and number of capped cells were counted microscopically.

An important outcome of the studies with lymphocytes is the additional evidence that TPA acts very rapidly to trigger a number of processes in the cell membrane (or possibly at other subcellular sites) which lead in turn to secondary events -- some of which require oxygen and/or are sensitive to the presence of retinoids. Evidence for a TPA receptor has been provided by the studies of Blumberg et al (15) and Estenson et al (16). The manner in which such a receptor triggers the chain of responses described above is a major interest in our program. To this end, we have attempted to probe the molecular processes that are involved in the activation of membrane movement (i.e., the concanavalin A capping (17)). As shown in Figure 7, 10 nM TPA greatly enhances capping in lymphocytes as measured by the number of cells recruited into this response. Since the extra capping is already apparent after 3 minutes and is neither sensitive to the absence of oxygen or the presence of ETYA or retinoids, it would appear to be a separate event from the ensuing responses described above. It appears, however, to be a necessary antecedent event since a number of amino acid derivatives (Table II) selectively block the TPA effect on con A capping -

Table II

INHIBITORS OF TPA INDUCED CAPPING

Compound	ID_{50} Dose
Benzoyl tyrosine ethyl ester	100 μM
Benzoyl arginine ethyl ester	250 μM
Tosyl tyrosine benzyl ester	25 μM
Tosyl phenylalanine benzyl ester	25 μM
Tosyl arginine methyl ester	1 mM
Tosyl phenylalanine chloromethyl ketone	2 μM
Tosyl lysine chloromethyl ketone	20 μM

Lymphocytes were pretreated for 30 min with 10 nM TPA at 37°. Five min prior to the initiation of capping, the indicated amino acid derivatives were added. The number of capped cells were determined 40 min after the addition of fluorescein-labeled concanavalin A. The data are expressed as the level of amino acid derivative required to depress capping to 50% of the response with TPA alone.

- and in so doing, certain of the agents also prevent the subsequent activation of phosphatidyl choline synthesis, the acceleration of AIB transport, and the mitogenic response of appropriately stimulated lymphocytes.

Among this group of amino acid derivatives which are known in general as inhibitors of proteases, benzoyl tyrosine ethyl ester (BTEE) has been studied most extensively. An unusual feature of its effect on capping is its ability to block both the forward capping when administered simultaneously with TPA and con A and to cause a temperature dependent dispersal of caps already formed in cells treated antecedantly with TPA + con A (Figure 8).

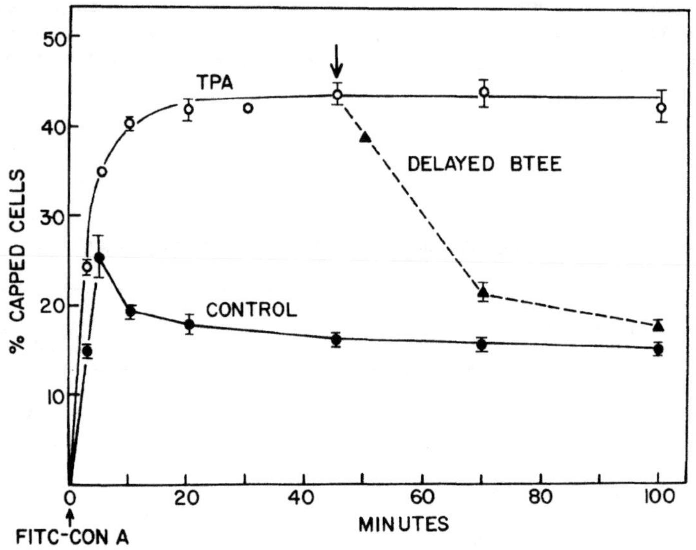

Fig. 8. Effect of Delayed Addition of Benzoyl Tyrosine Ethyl Ester (BTEE) on TPA-Promoted Concanavalin A Capping of Lymphocytes. Lymphocytes were treated with fluorescein labeled concanavalin A in the presence and absence of 10 nM TPA. After 45 minutes BTEE (200 μM) was added to the TPA treated cultures and the decrease in the number of capped cells was followed over the indicated intervals. Data are expressed as % cells capped. Addition of BTEE to control cultures had no effect on the level of capped cells.

A study of the time dependence of the cap dispersal reaction, relative to the kinetics of cap initiation with con A and TPA, supports the view that BTEE acts subsequent to the actual TPA triggering event to counteract a process which is required for the amplification of the triggering event. The possibility that the target for BTEE action is a proteolytic or esteratic process which has been triggered by TPA is suggested of course by the known protease inhibitory properties of BTEE and the other active compounds. However, the observation that BTEE can act after the establishment of the caps, when a presumed esteratic event should have already taken place, suggests that either an ongoing esteratic process may be important or BTEE and related agents may act in a way to structurally influence some protein assembly mechanism in the membrane. In any event, since BTEE can act after a TPA sensitive process has been established, its action is most likely directed to a continuing aspect of the triggered process rather than the TPA triggering mechanism itself. According to this view it is considered most probable that BTEE acts on a target of the TPA receptor mechanism rather than on the receptor itself. In agreement with the concept that BTEE acts remotely, it is has been found that the acceleration of glucose transport, another very early retinoic insensitive response to TPA, is insensitive to BTEE. This situation is understandable when one considers that a TPA receptor mechanism may influence several different types of targets -- one type which utilizes a BTEE-sensitive pathway for its amplification and another type which proceeds independently.

These relationships are summed up in our current working hypothesis of TPA action (Figure 9). In this hypothesis, TPA is visualized as first interacting with a receptor which has stereochemical restraints similar to those exhibited by steroid hormone receptors. In our current concept of receptor biology (18,19), this interaction is perceived to cause conformational changes in the receptor which are transduced to a spectrum of secondary targets which, in turn, are either rate-limiting or probability determining components in metabolic or genetic expression pathways leading to the ultimate TPA response. It is of particular interest in the present study of TPA responses that one such pathway should be inhibited by retinoids, since this raises the fascinating possibility that this pathway may be involved in the retinoid sensitive tumor promoting action of TPA. The mechanism studies to date support the view that the retinoid sensitive pathway uses oxygen to activate at least 2 secondary processes. In the case of the activation of the cytidyl transferase, which is rate-limiting for phosphatidyl choline synthesis, it appears that the activation is actually mediated by an oxidation product of arachidonic acid. In the case of the acceleration of AIB transport, however, the activator remains to be identified. Since retinoids (and ETYA) block the activation process in both cases, it is reasonable to anticipate from their hydrophobic and multiply unsaturated character that they either inhibit the overall process by blocking the oxidative formation of an endogenous lipoidal activator or compete with such an entity at the affinity sites in the respective target proteins. In this view, the important feature of the TPA is its unique ability to trigger the activation of the specific oxygen utilizing mechanisms which mediate these changes in target proteins. The specific targets and the co-factors which cooperate in their activation may range rather broadly, however, it is proposed that the oxidative mechanism underlying the activation of cytidyl transferase may constitute a fundamental mechanism for converting proteins from a non-functional state to forms which are active in specific pathways. Extending this concept, it appears reasonable to propose that a similar modification or activation of a membrane protein may promote its interaction with the cell nucleus and lead to the induction of specific genes expressions -- such as the induction of ornithine decarboxylase activity, mitogenesis, and even the signal genetic event underlying tumor promotion.

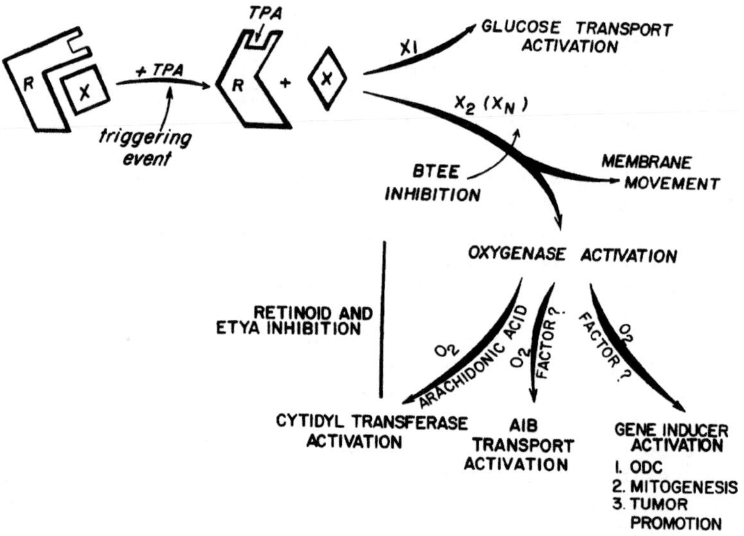

Fig. 9. Hypothesis of Phorbol Ester Action

Bibliography

1. Mastro, A. M., and Mueller, G. C.: 1974, Exp. Cell Res. 88, pp. 40-46.
2. Kensler, T. W., and Mueller, G. C.: 1978, Cancer Res. 38, pp. 771-775.
3. Kensler, T. W., Verma, A. K., Boutwell, R. K., and Mueller, G. C.: 1978, Cancer Res. 38, pp. 2896-2899.
4. Kensler, T. W., Wertz, P. W., and Mueller, G. C.: 1979, Biochim. Biophys. Acta 585, p. 43-52.
5. Wertz, P. W., and Mueller, G. C.: 1978, Cancer Res. 38, pp. 2900-2904.
6. Wertz, P. W., Kensler, T. W., Mueller, G. C., Verma, A. K., and Boutwell, R. K.: 1979, Nature 277, pp. 227-229.
7. Wertz, P. W., Mueller, G. C.: 1980, Cancer Res. 40, pp. 776-781.
8. DeChatelet, L. R., Shirley, P. S., and Johnson, R. B.: 1976, Blood 47, pp. 545-554.
9. Kaplan, L., Weiss, J., and Elsbach, P.: 1978, Proc. Natl. Acad. Sci. U.S.A. 75, pp. 2955-2958.
10. Vance, D. E., and Choy, P. C.: 1979, Trends in Biol. Sci. 4, pp. 145-148.
11. Choy, P. C., and Vance, D. E.: 1978, J. Biol. Chem. 253, pp. 5163-5167.
12. Schneider, W. C., and Behki, R. M.: 1963, J. Biol. Chem. 238, pp. 3565-3571.
13. Fiscus, W. G., and Schneider, W. C.: 1966, J. Biol. Chem. 241, pp. 3324-3330.
14. Kennedy, E. P.: 1962, The Harvey Lectures 57, pp. 143-171.
15. Driedger, P. E., and Blumberg, P. M.: 1980, Proc. Natl. Acad. Sci. U.S.A. 77, pp. 567-571.
16. Estensen, R. D., DeHoogh, D. K., and Cole, C. F.: 1980, Cancer Res. 40, pp. 1119-1124.
17. Kwong, C. H., and Mueller, G. C.: 1979, Biochim. Biophys. Acta 586, pp. 501-511.
18. Mueller, G. C.: 1980, in "Estrogens in the Environment" (ed.) John A. McLachlan, Elsevier North Holland, Inc. New York, in press.
19. Mueller, G. C., Kajiwara, K., Kim, U. H., and Graham, J.: 1978, Cancer Res. 38, pp. 4041-4045.

MODULATION OF ADIPOSE CONVERSION OF BALB/c 3T3 CELLS BY TUMOR PROMOTERS

Leila Diamond and Thomas G. O'Brien
The Wistar Institute of Anatomy and Biology
36th Street at Spruce
Philadelphia, Pennsylvania 19104

ABSTRACT

The conversion of BALB/c 3T3 preadipose cells to mature adipocytes is inhibited by the tumor promoter 12-O-tetradecanoylphorbol-13-acetate (TPA). The pH of the culture medium has a strong influence on the regulation of conversion. The inhibitory effect of TPA is dependent on the ability of the promoter to stimulate glycolysis under conditions in which the lactate produced can lower the pH of the culture medium. Thus, the mechanism by which TPA inhibits the adipose conversion of these cells appears to be stimulation of hexose transport activity and lactate production which creates an environment unfavorable for triglyceride synthesis and differentiation.

INTRODUCTION

Phorbol diester tumor promoters induce a number of diverse effects on cells in culture, including both stimulation and inhibition of terminal cell differentiation (reviewed in 1, 2). It is not yet clear which, if any, of the effects so far observed plays a role in the actual mechanism of tumor promotion _in vivo_. Nevertheless, in the past few years phorbol diesters have proven to be valuable tools in the study of various aspects of cell biology such as the process of cell differentiation. At the same time, cell systems in which terminal differentiation is affected by phorbol diesters have become models for studying the mechanisms by which these tumor promoters affect the expression of various biochemical and biological cell characteristics. Towards this end, we have been using a mouse preadipose cell system in which tumor promoters inhibit terminal differentiation into adipocytes. We have obtained evidence that the promoters act by disrupting the regulation of specific biochemical pathways that are critical for the differentiation of these cells.

Howard Green and collaborators (3) were the first to show that

B. Pullman, P. O. P. Ts'o and H. Gelboin (eds.), Carcinogenesis: Fundamental Mechanisms and Environmental Effects, 335–346.

some clones of Swiss mouse 3T3 cells were cell culture models for the development of mature tissue adipose cells in vivo. When cells of these clones are maintained as confluent monolayers or in suspension, they gradually accumulate fat droplets in the cytoplasm and acquire many of the morphological and biochemical characteristics of adipocytes (4, 5). It has been shown that the only lipid which accumulates during adipose conversion is triglyceride and that accumulation is stimulated by the addition of insulin to the medium. During conversion, there are dramatic increases in the levels of key enzymes involved in fatty acid synthesis, glycerol acylation and extracellular lipid utilization. Sensitivity to lipolytic agents also develops.

THE BALB/c 3T3 PREADIPOSE CELL SYSTEM

We have been studying clones of preadipose cells derived in our laboratory from BALB/c 3T3 clone A31 cells (6). These cells undergo adipose conversion in a manner similar to that described for Swiss 3T3 cells. Triglyceride accumulation occurs in confluent cultures with a greatly decreased growth rate, is enhanced by insulin, and results in fully differentiated adipose cells with no proliferative potential (Fig. 1a). Triglyceride accumulation can be detected by microscopic observation of unstained cultures or by staining fixed

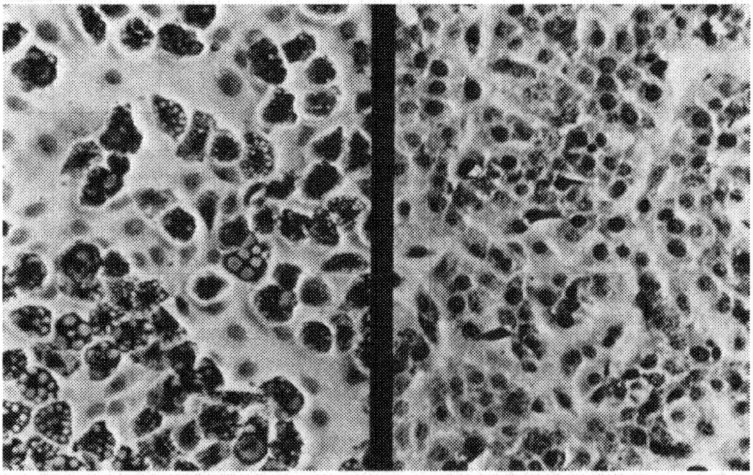

Figure 1. Adipose conversion in BALB/c 3T3 cells. Confluent cultures were refed every 2-3 days with control (left) or TPA (1.6 x 10^{-7} M)-containing (right) Eagle's minimum essential medium. After 32 days the monolayers were fixed with 10% formalin and stained with oil red O. More than 70% of the cells in the control contained many large lipid droplets. (X 300).

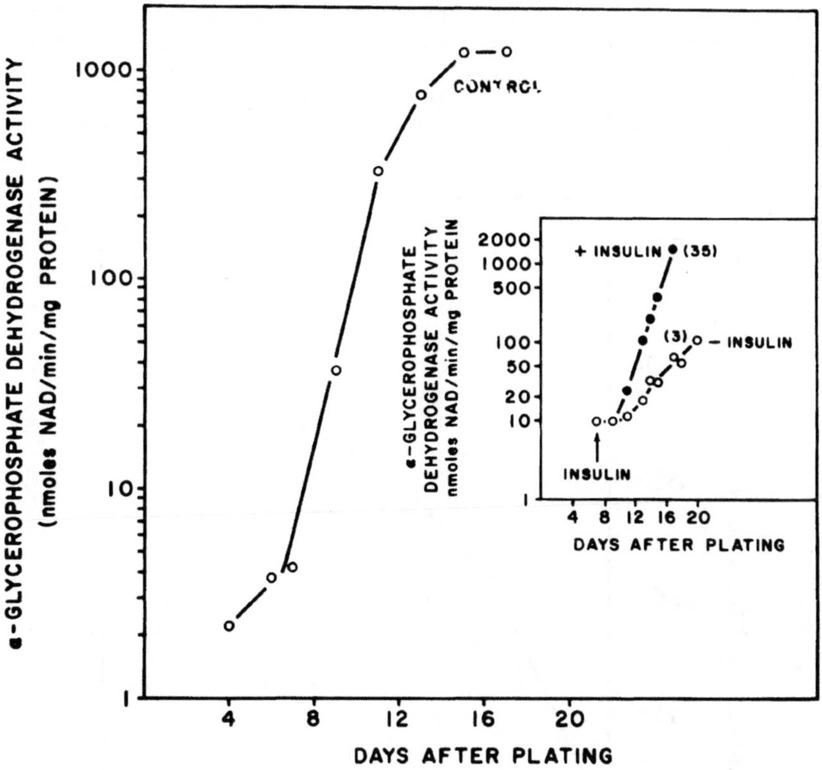

Figure 2. Development of NAD^+-linked α-glycerophosphate dehydrogenase activity in BALB/c 3T3 cells. Beginning 2 days after cells were plated in 60-mm dishes (2.4 x 10^4 cells/cm^2), the cultures were refed with insulin (1 μg/ml)-containing medium. At the indicated times, cells from duplicate dishes were harvested for determination of enzyme activity using the procedure of Kozak and Jensen (7). The percentage of differentiated cells on days 6 and 12 was 0 and 25, respectively. Inset: In another experiment, cultures grown in the absence of insulin for 7 days were refed with control or insulin-containing medium and enzyme activity was measured at the indicated times. The numbers in parentheses indicate the percentage of differentiated cells at day 17.

cultures with the fat-soluble dye, oil red O. Under certain conditions to be described below, the cultures can reach a point in which the cytoplasms of more than 85% of the cells are completely filled with lipid droplets. As adipose conversion proceeds, the cells develop the enzymes characteristic of adipocytes. For example, the activity of NAD^+-linked α-glycerophosphate dehydrogenase is increased several-hundred-fold (Fig. 2). This enzyme plays a key role in the pathway of triglyceride synthesis; it acts at the branch point from the glycolytic pathway and provides the only source of the glycerol-3-phosphate needed for the greatly increased rate of triglyceride synthesis during adipose conversion. Its development is stimulated by insulin (Fig. 2, inset), in accord with the hormone's enhancement of differentiation.

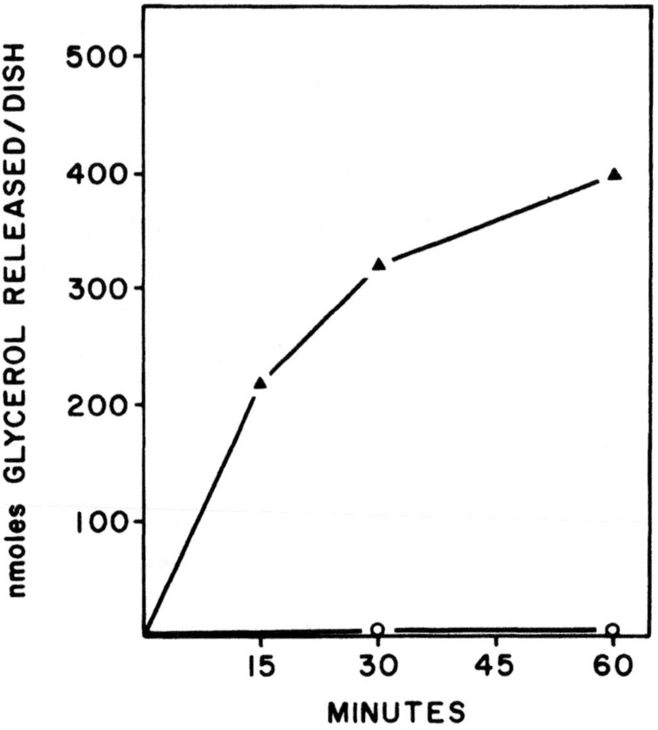

Figure 3. Effect of DL-isoproterenol on lipolysis in differentiated BALB/c 3T3 cells. Fully differentiated cultures (> 99% adipose cells with large amounts of visible triglyceride) were obtained by growing cells in Dulbecco's modified Eagle's medium containing insulin (1 μg/ml). The cultures were then washed thoroughly, incubated for 2 days in insulin-free medium and, at 0 time, washed with Krebs-Ringer phosphate buffer and incubated in this buffer containing 2% bovine serum albumin with (▲) or without (o) DL-isoproterenol (500 μM). At the indicated times, the buffer was removed and the released glycerol determined by a standard fluorometric assay (8).

Lipoprotein lipase activity is also increased during adipose conversion of BALB/c 3T3 cells. For example, by the time 70-80% of the cells have undergone conversion, the enzyme specific activity may have increased from a basal level of 1-2 mU/mg protein to

35-40 mU. Lipoprotein lipase is used by adipose cells in vivo to obtain fatty acids from circulating lipoproteins, but is probably not required for triglyceride synthesis by BALB/c 3T3 cells under the usual culture conditions. After the preadipose cells have differentiated, they become responsive to lipolytic agents and, as shown in Fig. 3, can be stimulated by the β-adrenergic agonist, isoproterenol, to hydrolyze triglyceride to free glycerol.

EFFECTS OF TPA ON BALB/c 3T3 CELLS

When preadipose BALB/c 3T3 cells are treated with the potent tumor promoter, 12-O-tetradecanoylphorbol-13-acetate (TPA), or with other phorbol diester tumor promoters, a number of biochemical and biological changes occur. Among the more rapid effects of TPA treatment are the induction of ornithine decarboxylase (ODC), which is the rate-limiting enzyme in polyamine synthesis, and stimulation of DNA synthesis (Fig. 4) (9). In the experiment shown in Fig. 4a, there was an 8-fold increase in ODC activity in confluent cultures 4 hr after treatment with TPA (1.6×10^{-7}M). It can also be seen

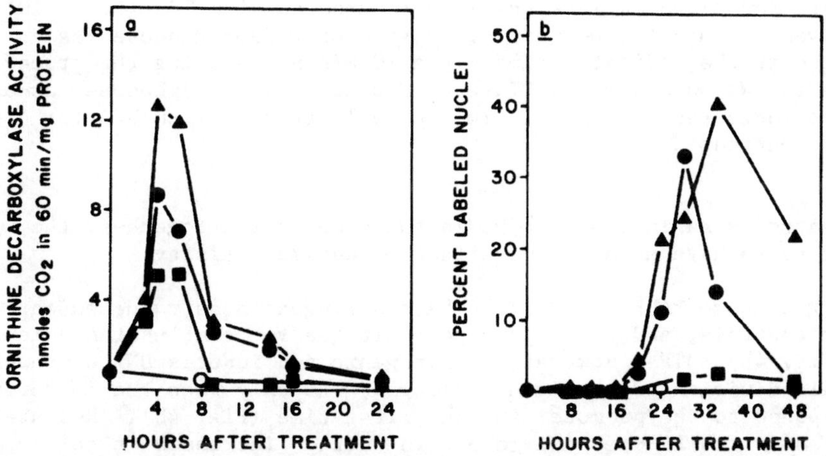

Figure 4. The effects of TPA on ODC activity and DNA synthesis in BALB/c 3T3 cells. Confluent cultures (7 days after plating) were treated in one of the following ways: TPA (dissolved in conditioned medium) was added to a final concentration of 1.6×10^{-7} M (●); the medium was replaced with fresh serum-containing medium with (▲) or without (■) TPA; or the cells were left untreated (o). At the indicated times, cells from duplicate dishes were harvested and ODC activity determined by the procedure of O'Brien et al. (10) (panel a) or coverslip preparations were harvested from cultures to which [^{3}H]thymidine had been added 1 hr earlier, and processed for autoradiography (panel b).

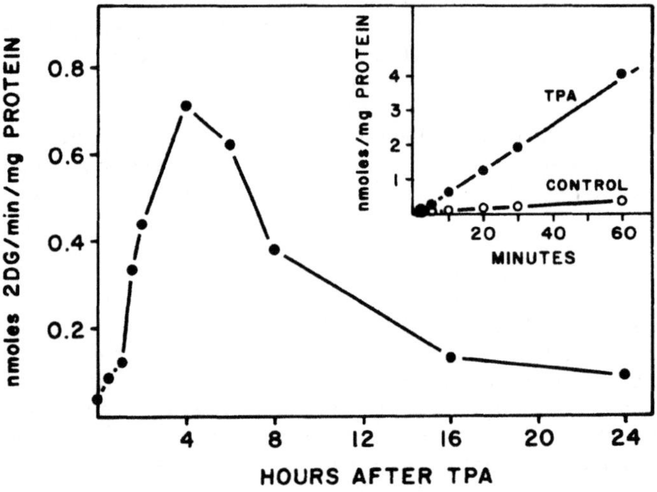

Figure 5. Time course of TPA-stimulated 2-deoxyglucose uptake in preadipose BALB/c 3T3 cells. Post-confluent cultures containing very few differentiated cells were treated with TPA (1.6 x 10^{-7} M) dissolved in conditioned medium. Uptake of 2-deoxyglucose was measured at the indicated times in a 10-min assay using the procedure of Lee and Weinstein (11). Inset: Uptake of 2-deoxyglucose was assayed for the indicated lengths of time in cells treated with TPA for 4 hr or left untreated.

that serum is an inducer of ODC in these cells and together, the two inducers have an additive effect on enzyme activity.

Preadipose BALB/c 3T3 cells have stringent density-dependent growth controls, and serum is a poor mitogen for confluent monolayers (see Fig. 4b). TPA, however, is mitogenic and induces DNA synthesis and cell division in confluent cultures, but only when the cultures are exposed to the promoter for the first time (Fig. 4b). Repeated treatment with TPA does not result in further increases in cell number, and cell density plateaus at a level approximately twice that of control cultures (6).

In preadipose BALB/c 3T3 cells, TPA stimulates hexose transport activity. It stimulates 2-deoxyglucose uptake to a greater extent than serum does, which is the opposite of what has been observed with other cells (11). The optimal effect on uptake of 2-deoxyglucose occurs at 4 hr after addition of TPA to confluent cultures, at which time there may be a 10- to 20-fold increase in transport activity (Fig. 5); the predominant change is in the V_{max}, although the K_m may also be affected. TPA-stimulated uptake of 2-deoxyglucose is inhibited completely by cytochalasin B (4 μM), a specific inhibitor of carrier-

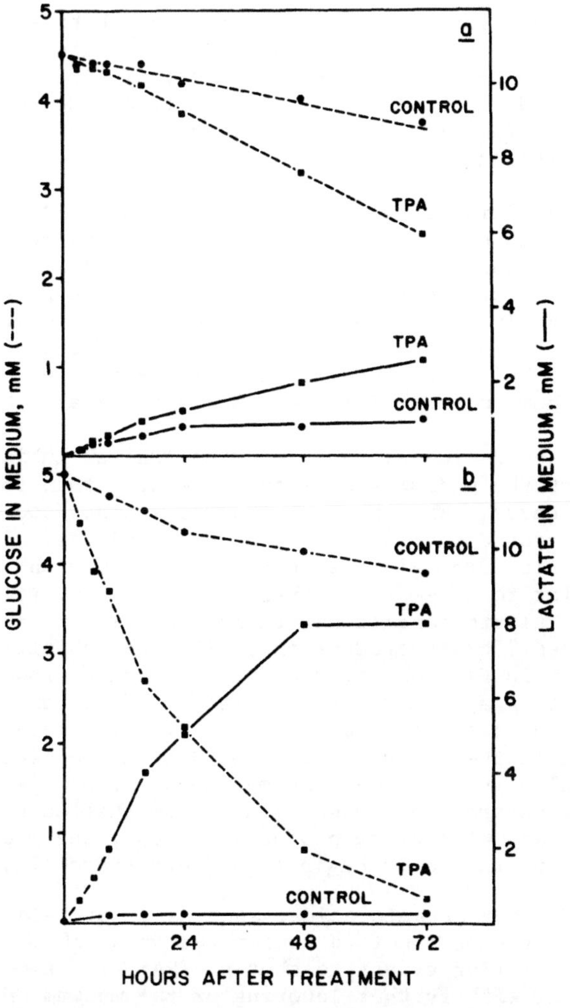

Figure 6. Glucose utilization and lactate production in preadipose BALB/c 3T3 cells treated with TPA for the first (a) or fourth (b) time. Cells were plated in 60-mm dishes at a density of 2 x 10^4 cells/cm^2 and grown to confluence. At 0 time for panel a, half of the dishes were washed twice with phosphate-buffered saline and refed with medium containing dialyzed serum (to eliminate the contribution of lactic acid from serum) with or without TPA (1.6 x 10^{-7} M); the medium was harvested for analysis at the indicated times. The other half of the dishes (panel b) were refed every 2-3 days with TPA or control medium, and after 8 days (0 time), treated as panel a. Glucose and lactic acid were measured by the methods of Slein (12) and O'Brien *et al.* (13), respectively.

mediated glucose transport, and is partially inhibited by cycloheximide. TPA also stimulates transport of the non-metabolizable glucose analog, 3-0-methylglucose; this stimulation, too, is blocked by cytochalasin B. Insulin does not stimulate glucose in undifferentiated BALB/c 3T3 cells, but does so in differentiated cells, as one would expect with adipocytes.

The most striking long-term effect of TPA on BALB/c 3T3 cells is inhibition of adipose conversion (6). In cultures refed every 2-3 days with medium containing TPA ($\geq 10^{-8}$ M), the gradual accumulation of triglyceride that occurs in confluent control cultures does not take place (Fig. 1b), and there is no increase in α-glycerophosphate dehydrogenase or lipoprotein lipase activity. The inhibition of differentiation by TPA is reversible by refeeding the cells with, or subculturing them into, medium that does not contain the tumor promoter.

We had observed in cultures in which TPA was inhibiting adipose conversion that the medium was always more acid than the medium of control cultures. When we measured the lactate concentrations in the medium, we found that there was a remarkable stimulation of lactate production from glucose in cells treated with TPA at concentrations similar to those that inhibit adipose conversion. Also, under various culture conditions, there was always a correlation between enhanced lactate production by TPA and inhibition of conversion (13). As shown in Fig. 6, stimulation of lactate production by TPA is not immediately apparent but requires 4-7 days for maximum effect. At this time, the cells are capable of converting more than 80% of the glucose in the medium (5.0 mM at 0 time) to lactate in 48 hr, which results in a decrease in medium pH from 7.4-7.5 to 7.1-7.2. If these cultures are then refed with medium that does not contain TPA, the high rate of lactate production stops, the medium pH does not fall and the cells begin to differentiate (13).

The addition of organic acids, such as D- or L-lactic acid or acetic acid, to the medium to a concentration of 10 mM (medium pH 7.1-7.2) has a similar effect to that of TPA: it stimulates L-lactic acid production, with further lowering of the medium pH, and inhibits adipose conversion (Table 1); the addition of "neutralized" lactic acid (sodium lactate) to the medium does not have these effects (14).

The differentiation process in BALB/c 3T3 cells can also be modulated with media adjusted to different pH levels by varying the $NaHCO_3$ concentration (Table 2) (14). In control cultures not treated with TPA, differentiation proceeds normally at medium pH levels greater than 7.5, is delayed slightly at pH 7.35 and is inhibited at pH 7.15. In TPA-treated cultures at pH 7.35, differentiation is completely inhibited and lactate production stimulated. On the other hand, differentiation is slightly delayed in TPA-treated cultures at the higher pH levels and then proceeds rapidly. TPA does not stimulate lactate production nor does it lower the pH of the medium at these pH levels.

Table 1. Effect of organic acids on adipose conversion of BALB/c 3T3 cells. Cells were seeded in Eagle's minimum essential medium (MEM) + 10% fetal calf serum. Beginning 4 days later, when the cultures were confluent, they were refed at 3-4 day intervals with medium containing insulin (1 μg/ml) and the indicated additions at 10 mM final concentrations. The numbers in parentheses indicate the L-lactate concentrations (mM) in the spent medium.

	Percent Adipose Cells		
TREATMENT	Day 12	Day 14	Day 21
Control	2.5 (0.8)	17 (0.6)	60 (0.3)
L-Lactic acid	0 (12.7)	0 (14.0)	0 (19.3)
D-Lactic acid	0 (3.5)	0 (4.5)	0 (4.5)
Acetic acid	0 (3.8)	0 (6.2)	0 (9.6)

Table 2. The effect of pH on adipose conversion of BALB/c 3T3 cells. Cells were seeded and grown to confluence in Eagle's MEM + 10% serum at pH 7.5. They were refed 4 days later with control or TPA (1.6×10^{-7} M)-containing medium adjusted to the indicated pH levels with $NaHCO_3$ and with added insulin (1 μg/ml). Cells treated with TPA in medium at pH 7.15 eventually degenerated because of the low pH (6.8) that the medium attained.

	Percent Adipose Cells			
CONTROL	Day 12	Day 14	Day 18	Day 20
pH 7.15	0	0	1-5	5-10
pH 7.35	< 1	1-5	70-80	65-75
pH 7.52	1-5	20-30	55-65	60-70
pH 7.90	1-5	25-35	55-65	60-70
TPA	Day 12	Day 14	Day 18	Day 20
pH 7.35	0	0	0	0
pH 7.52	0	0	65-75	75-85
pH 7.90	0	0	70-80	75-85

Thus, the medium pH has a strong influence on the regulation of adipose differentiation in these cells. The inhibitory effect of TPA is dependent on the ability of the promoter to stimulate glycolysis under conditions in which the lactate produced can lower the pH of the culture medium. If the medium pH is prevented from decreasing, for example, because of a high $NaHCO_3$ concentration, then TPA neither stimulates lactate production nor inhibits differentiation.

We have found that mouse interferon is also a reversible inhibitor of adipose conversion of BALB/c 3T3 cells at concentrations that do not affect cell survival or growth (15). When combined with low concentrations of TPA, interferon has a synergistic effect on the inhibition of differentiation. It does not, however, stimulate 2-deoxyglucose uptake or lactate production, nor does it affect the stimulation of these activities by TPA. Thus, it probably inhibits conversion through a mechanism different from that of the promoter.

DISCUSSION

The conversion of 3T3 preadipose cells to adipocytes requires "rearrangement" of the intracellular metabolism program from one geared to a relatively high rate of lactic acid production (glycolysis) and little triglyceride synthesis to one in which there is enhanced glucose utilization for fatty acid and triglyceride synthesis and no lactic acid production (see 4, 5). Our observations indicate that environmental (medium) pH is one of the determinants of this "rearrangement." TPA, through its stimulation of hexose transport activity and lactate production, may act to "lock in" the intracellular metabolism program of the preadipocyte phenotype by creating an environment (low pH) unfavorable for differentiation. Such an explanation for the mechanism of TPA's inhibition of differentiation is consistent with the fact that conversion can be inhibited by lowering the medium pH by means other than TPA, and that conversion is not inhibited when the TPA-induced decreases in pH are prevented.

If the mechanism of TPA action on the differentiation of these cells is through stimulation of glycolysis in undifferentiated cells, a number of questions still remain. First, how does TPA stimulate glycolysis? Its striking effect on hexose transport activity may play a major role, but since this effect occurs within 4 hr of TPA treatment, and maximal effects on glycolysis require 4-7 days, the promoter could also be enhancing glycolysis by acting at steps subsequent to hexose entry. It is paradoxical that, although most of the carbon atoms comprising the triglyceride that accumulates in these cells come from glucose, the stimulation of glucose transport by TPA has an inhibitory rather than a stimulatory effect on triglyceride synthesis. Second, how does the low environmental pH that develops as a result of TPA treatment prevent differentiation?

The adverse environment could be affecting any of the many steps in the triglyceride biosynthetic pathway either by blocking the development of some key enzyme(s), for example, α-glycerophosphate dehydrogenase, or by inhibiting the activity of such key enzymes after they develop. Because of the number of biosynthetic steps involved, the identification of the critical, pH-sensitive enzymes will be difficult. Finally, is the mechanism by which TPA affects the differentiation of these preadipose cells unique, or is it relevant to the effects of the promoter on other cell systems? It may seem unlikely that a common biochemical mechanism will explain all the observed effects of TPA on differentiating cells in view of the diversity of phenotypic programs that can be affected (1, 2). However, many differentiating systems, including mouse skin, presumably would be affected by altered regulation of such a critical metabolic pathway as glycolysis. Consequently, an understanding of the mechanism by which TPA and the phorbol diesters enhance glycolysis in BALB/c 3T3 preadipose cells should provide clues as to how these compounds might affect other differentiation programs and other cellular processes including tumor promotion.

ACKNOWLEDGMENTS. This work was supported, in part, by grants CA-23413, CA-21778 and CA-10815 from the National Cancer Institute, DHEW.

REFERENCES

1. Diamond, L., O'Brien, T. G., and Rovera, G.: 1978, Life Sci. 23, pp. 1979-1988.
2. Diamond, L., O'Brien, T. G., and Baird, W. M.: 1980, Adv. Cancer Res. 32, pp. 1-74.
3. Green, H., and Kehinde, O.: 1974, Cell 1, pp. 113-116.
4. Green, H.: 1978, In "10th Miami Symp. on Differentiation and Development" (F. Ahmad, ed.), pp. 13-36. Academic Press, New York.
5. Ailhaud, G.: (ed.) 1979, INSERM Symposia Series, Vol. 87. Institut National de la Sante et de la Recherche Medicale, Paris.
6. Diamond, L., O'Brien, T. G., and Rovera, G.: 1977, Nature 269, pp. 247-249.
7. Kozak, L. P., and Jensen, J. T.: 1974, J. Biol. Chem. 249, pp. 7775-7781.
8. Chernick, S. S.: 1969, Methods in Enzymology 14, pp. 627-630.
9. O'Brien, T. G., Lewis, M. A., and Diamond, L.: 1979, Cancer Res. 39, pp. 4477-4480.
10. O'Brien, T. G., and Diamond, L.: 1977, Cancer Res. 37, pp. 3895-3900.
11. Lee, L.-S., and Weinstein, I. B.: 1979, J. Cell. Physiol. 99, 451-460.
12. Slein, M. W.: 1963, In "Methods of Enzymatic Analysis" (H.-U. Bergmeyer, ed.), pp. 117-123. Academic Press, New York.

13. O'Brien, T. G., Saladik, D., and Diamond, L.: 1979, Biochem. Biophys. Res. Commun. 88, pp. 103-110.
14. O'Brien, T. G., and Saladik, D.: 1980, J. Cell. Physiol., in press.
15. Cioé, L., O'Brien, T. G., and Diamond, L.: 1980, Cell Biol. Int. Rep. 4, pp. 255-264.

ESTABLISHED CELL CULTURES AS MODEL SYSTEMS FOR CARCINOGEN METABOLISM

F.J. Wiebel[+], M. Lambiotte[‡], K.-H. Summer[+] and T. Wolff[+]

[+]Dept. Toxicol., Inst. Toxicol. Biochem., Gesellschaft für Strahlen- u. Umweltforschung, 8042 Neuherberg/München, Germany
[‡]Dept. Cell. Genetics, Inst. Molec. Biol., C.N.R.S., Univ. Paris VII, Paris, France

Abstract - Cells in permanent culture were examined for their in vitro activities of enzymes involved in the metabolism of carcinogenic chemicals. The various cell lines were found to contain different forms of monooxygenases. They were capable not only of expressing different forms of aromatic hydrocarbon-inducible cytochrome "P-448" but also of a cytochrome "P-450"-dependent form previously found to be lacking. There was a wide variation in the absolute and relative activities of conjugation reactions with sulfate, glucuronic acid or glutathione in different cell lines indicating some degree of species and tissue specificity. The findings are discussed in terms of the usefulness of established cell cultures in studies concerning carcinogen metabolism.

Cells in culture are models of choice for studying many aspects of chemical carcinogenesis e.g., the binding of reactive chemicals to nucleic acids, the repair of DNA damage or the progression of cells through various stages of malignant transformation. However, it is open to question whether cultured cells offer suitable systems for analysing the earlier steps of tumor initiation, the activation and inactivation of the carcinogenic chemicals. A growing number of observations give reason to doubt the competency of many cell lines for the metabolism of chemical carcinogens (8,11,18,27,36). In the present studies we reexamined the capacity of cells in culture to metabolize chemicals. This was done by surveying the activity of several key enzymes involved in the activation and inactivation of chemicals in various established cell cultures.

In vivo, the expression of drug metabolizing enzymes greatly differs between animal species and tissues, age and sex, as well as the hormonal and nutritional status, all factors that eventually determine the potency and tissue specificity of a carcinogen. Freshly isolated cells in culture may be capable of mimicking the drug metabolism of

B. Pullman, P. O. P. Ts'o and H. Gelboin (eds.), Carcinogenesis: Fundamental Mechanisms and Environmental Effects, 347–361.

their tissue of origin - at least for an initial short time period. However, in view of their limited life span, these cultures may be unsuitable for studying long term effects such as malignant transformation. In contrast, cells in permanent culture are amenable to long term studies but they are less likely to reflect the in vivo conditions of carcinogen metabolism since they tend to lose their differentiated functions during the process of adaptation to the culture conditions. This becomes a crucial issue, for example, in examining the activation of whole classes of procarcinogens which depends on microsomal monooxygenases that are differentiated functions of a few tissues, notably the liver.

The first part of this presentation is concerned with the expression of different monooxygenase forms in established cells in culture. Then, attention will be turned to three of the major 'inactivating' enzyme systems, the sulfo-, glucuronosyl- and glutathione-S-transferases. Finally, the findings will be discussed in regard to the applicability of established cell cultures in the study of carcinogen metabolism.

1. MONOOXYGENASE ACTIVITIES

Microsomal monooxygenases are known to comprise a family of enzymes which differ in their functional and physicochemical properties (15,25). Two major groups of monooxygenase forms can be distinguished by their substrate specificity, response to inducers and in vitro modifiers or their tissue distribution. In the following the group of monooxygenases which are inducible by aromatic hydrocarbons will be referred to as "cytochrome P-448" forms. Monooxygenases of this type occur in virtually all mammalian tissues. The second group of monooxygenases which are associated with the induction by phenobarbital will be called "cytochrome P-450" forms. These enzymes are predominantly found in the liver and to a minor degree in some extrahepatic tissues.

Most established cells in culture have been found to contain monooxygenase activities which may range from barely detectable levels to the high levels found in liver (8,11,21,33,43). These activities were almost exclusively monitored by the oxygenation of benzo(a)pyrene (BP). Closer examinations of the reaction in established cells including those of hepatic origin (34,46) showed that it is mediated by cytochrome P-448-dependent monooxygenases, possibly by a single form of this group. Monooxygenases of the cytochrome P-450 type appeared to be missing. To date, this deficiency was the major limiting factor for the applicability of cell culture systems in general studies of carcinogen metabolism.

In the following we present data to show that permanent cell cultures may express a greater variety of monooxygenases, i.e., different cytochrome P-448 forms as well as a monooxygenase of the cytochrome P-450 type thought to be lacking in cell cultures.

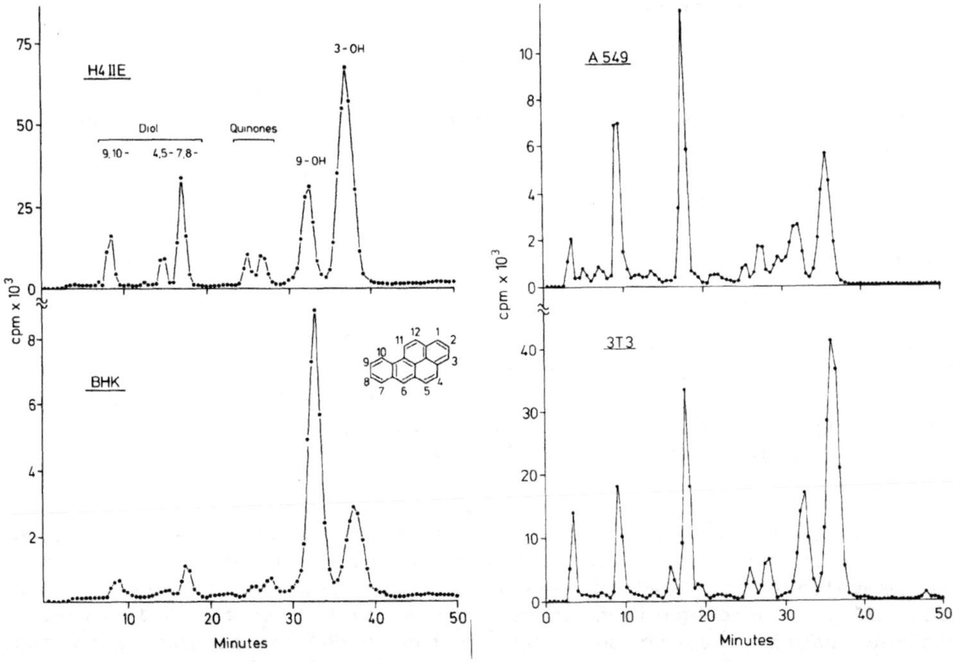

Fig. 1 BP-metabolism in cultures of H-4-II-E, BHK/21, 3T3 and A549 cells. - Cell cultures were exposed to 13 uM benz(a)anthracene for 18 hrs. Cellular preparations were incubated with the standard reaction mixture of the BP hydroxylase assay (43) including 100 nmoles of (^{3}H)-BP. After incubation the reaction mixtures were extracted as described earlier (38). The substrate, (^{3}H)-BP was separated from the products by column chromatography (50), and an aliquot of the metabolite fraction was analyzed by high pressure liquid chromatography (37). The BP-product peaks were identified by co-chromatography with non-radioactive authentic metabolites. Numbers on the ordinate represent the radioactivity per fraction (0.5 ml). Other conditions as previously described (43).

1.2. Expression of different cytochrome P-448 forms.

Monooxygenases are distinguishable by their position-specific attack of a variety of compounds (9,24,38,51) including BP (16,45). Thus the pattern of BP metabolites formed by cultured cells _in vitro_ may be taken as fingerprints of different monooxygenase forms.

The profile of BP metabolites formed in 4 cell lines, H-4-II-E (rat, hepatoma), BHK/21 (Syrian hamster, kidney), 3T3/BALB (mouse, fibroblasts) and A549 (human pulmonary carcinoma) which had been exposed to benz(a)anthracene is shown in Fig. 1. It is apparent that the ratio of metabolites differed considerably between the various cell lines.

Table 1. (^{3}H)-BP-Metabolism in Cell Cultures and Liver Microsomes

	Percent of Total Metabolites				
Metabolite Fraction	A549	3T3	BHK/21	H-4-II-E	Hepatic Microsomes
9,10-Diol	17	9	3	5	15
4,5-Diol	0.5	2	1	3	6
7,8-Diol	44	17	5	12	11
Quinones	6	8	5	9	12
9-OH	10	15	62	20	8
3-OH	21	44	23	50	47

Conditions as given in legend to Fig. 1. Microsomes were prepared from livers of male Wistar rats (150 g b.w.) which had been treated with a single dose of 3-methylcholanthrene for (40 mg/kg bw) 18 hrs. *In vitro* incubation time with (^{3}H)-BP was 10 min.

The data are summarized in Table 1: Most striking is the predominant formation of 9-hydroxy-benzo(a)pyrene (9-OH-BP) in BHK/21 cells versus the prevalence of 3-hydroxy-benzo(a)pyrene (3-OH-BP) in H-4-II-E and 3T3 cells. The comparison takes into account that the 9,10-dihydro-diol-BP fraction has to be added to the 9-OH-BP fraction since they arise from the same epoxide intermediate. It should be noted that the ratio of the group of dihydrodiols to that of the phenols varies markedly between the 4 cell lines. This has to be attributed to differences in the relative activities of hydratases and monooxygenases. Table 1 furthermore compares the metabolite pattern in the hepatoma derived H-4-II-E cells with that formed in hepatic microsomes prepared from 3-methylcholanthrene-treated rats. Except for a somewhat varying ratio of the 9,10-dihydrodiol and 9-OH-BP fractions, the pattern of metabolites show a close similarity. Thus, under the common condition of prior exposure to inducers of cytochrome P-448 type monooxygenases, established cells in culture may with some accuracy reflect the metabolism of BP in liver microsomes. That this is not a specific function of liver-derived cells is shown by the profile of BP metabolites formed in the fibroblastoid cell line, 3T3, which does not appreciably differ from that in the hepatoma cell line H-4-II-E (Table 1).

The results indicate that established cell cultures may express different monooxygenases forms which according to their response to inducers of the aromatic hydrocarbon type belong to the group of cytochrome P-448-dependent monooxygenases. The *in vivo* correlates of these monooxygenase forms are not known. Different cytochrome P-448 forms have also been observed in liver microsomes which vary in their relative amounts with animal species and developmental stage (1,13). Further exploration of the substrate specificity and distribution of these cytochrome P-448 forms in cultured cells are needed to establish their identity.

1.3. Expression of a cytochrome P-450 form in hepatoma cells

Recent observations (22) on bile acid hydroxylation and its inducibility in Reuber hepatoma cells in culture suggested that established cells may also retain the activity of cytochrome P-450 forms. The following studies more closely examine these cells for the expression of cytochrome P-450- and P-448-dependent monooxygenases by looking at the substrate specificity and the response to inducers and *in vitro* modifiers.

As shown by Deschatrette *et al.* (7) the hepatoma cells express to different degrees liver specific functions, such as secretion of albumin or the activities of liver specific isozymes of alcohol dehydrogenase and aldolase. A "differentiated" clone FAZA 967 and its subclone FaO, as well as an independently obtained "undifferentiated" clone H5, were derived from FU-5 cells, descendents of the H-4-II-EC3 line (7). From FaO and H5 a hybrid cell clone HF1 was formed which does not express the liver specific enzymes; however, a subclone HF1-4, obtained by selection in glucose-free medium, reexpresses the functions observed in the differentiated parental cell (7).

To distinguish between the various monooxygenase forms three reactions were monitored: a) the hydroxylation of BP (= aryl hydrocarbon hydroxylase, AHH) (31,42), which is preferentially mediated by cytochrome P-448 forms (26); b) the epoxidation of aldrin which is highly specific for cytochrome P-450 forms (48,49); and c) the hydroxylation of chenodeoxycholic acid which has also been found to depend on cytochrome P-450 (3).

As shown in Table 2, the differentiated lines, FAZA 967 and FaO, contained sizeable amounts of aldrin epoxidase activity, whereas AHH was barely detectable. In contrast, in the undifferentiated cell lines, H5 and HF1, AHH activity was somewhat higher but aldrin epoxidase was at the level of detectability. The hybrid cell line, HF1-4, differed from the other cell lines in that it expressed both enzymes at a relatively high level of activity.

The differentiated and undifferentiated cell lines exhibited a specific response to exposure of monooxygenase inducers. Treatment with benz(a)anthracene, an inducer of cytochrome P-448-dependent reactions, greatly increased the level of AHH activity in the undifferentiated lines H5 and HF1 and had little effect in the differentiated lines except in HF1-4, which was highly inducible for AHH. Aldrin epoxidase, however, was not induced by benz(a)anthracene in any of the cell lines nor did bile acid hydroxylation rise above the level of detectability. Exposure of cells to dexamethasone produced virtually opposite induction patterns. Dexamethasone treatment a) strongly increased chenodeoxycholate 6ß-hydroxylation in FAZA 967, FaO and HF1-4 cells; b) induced aldrin epoxidase in FAZA 967 and HF1-4 cells (but not in FaO cells); and c) induced AHH to a greater extent only in HF1-4 cells.

Table 2 Oxidative metabolism of benzo(a)pyrene, aldrin and chenodeoxycholic acid in various hepatoma cell lines exposed to dexamethasone or benz(a)anthracene.

	Benzo(a)pyrene Hydroxylation (AHH) (pmol/min/mg prot.)			Aldrin Epoxidation (pmol/min/mg prot.)			Chenodeoxycholic Acid Hydroxylation (nmol/24 hrs/mg prot.)		
Cell Line	Control	DEX	BA	Control	DEX	BA	Control	DEX	BA
FAZA 967	0.06	0.24	0.15	14.1	22.6	13.3	< 0.2	9.4	< 0.2
FaO	0.19	0.94	0.14	23.0	17.5	18.3	< 0.2	10.2	< 0.2
H5	0.50	0.75	18.00	1.7	1.5	0.7	< 0.2	< 0.2	< 0.2
HF1	0.70	1.40	48.00	1.3	1.0	0.7	< 0.2	1.5	< 0.2
HF1-4	2.80	21.00	81.90	32.0	139.0	38.0	< 0.2	14.8	< 0.2

Monolayer cultures of the cell lines were grown as previously described (23). Cultures were exposed to fresh growth medium ("Control") or to medium containing either 1 ug/ml dexamethasone (DEX) or 5 ug/ml benz(a)anthracene (BA). For the determination of BP hydroxylation (AHH) and aldrin epoxidation following an 18 hr exposure period, monolayers were washed with phosphate buffered solution, and the cells were collected and stored at -80°C as described previously (43). _In vitro_ BP hydroxylation was determined spectrofluorometrically (43); _in vitro_ aldrin epoxidation by gas chromatography (48). For the determination of chenodeoxycholic hydroxylation, cultures were exposed to BA or DEX for 48 hrs. C^{14}-Chenodeoxycholate was added from the 24th to the 48th hr. The major product, α-muricholic acid, was extracted from the medium and separated by TLC as previously described (22).

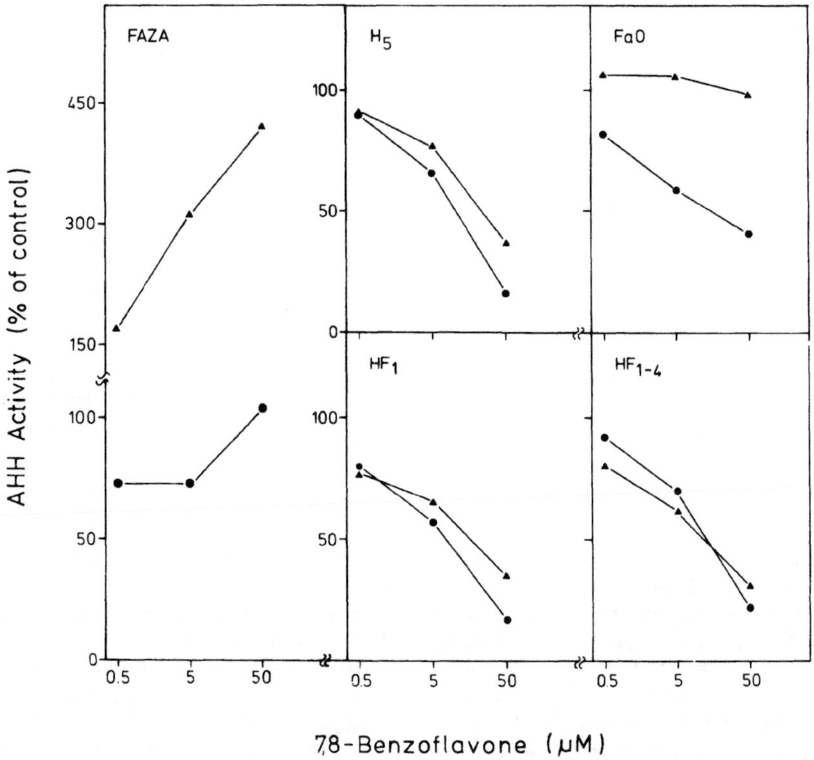

Fig. 2. In vitro effect of 7,8-benzoflavone on AHH activity in hepatoma cell lines. - Cells were exposed to dexamethasone (▲——▲) or benz(a)anthracene (●——●) and assayed for in vitro AHH activity as described in Table 2. 7,8-Benzoflavone was added at appropriate concentrations together with the substrate, BP, in 50 ul methanol/ml reaction mixture. Enzyme activities are expressed in % of control i.e., activities in the absence of 7,8-benzoflavone (cf. Table 2).

As mentioned above, monooxygenases possess some overlap for the hydroxylation of BP (26,42,44). To differentiate between the monooxygenases involved in BP hydroxylation, we determined their response to the in vitro modifiers, 7,8-benzoflavone and 2-diethylaminoethyl-2,2-diphenyl valerate (SKF-525A). The 7,8-benzoflavone has previously been shown to inhibit cytochrome P-448-dependent reactions strongly and to be without effect or to stimulate cytochrome P-450-dependent reactions (42,44). SKF-525A, on the other hand, inhibits cytochrome P-450's more strongly than cytochrome P-448's (12). In support of the data on the substrate specificity, the effects of 7,8-benzoflavone and SKF 525A on AHH activity indicate the presence of two monooxygenase forms in the hepatoma cell lines. 7,8-Benzoflavone inhibited AHH activity in benz(a)anthracene treated and dexamethasone (or untreated) cells of the undifferentiated cell lines, H5 and HF1, to a similar degree (Fig. 2).

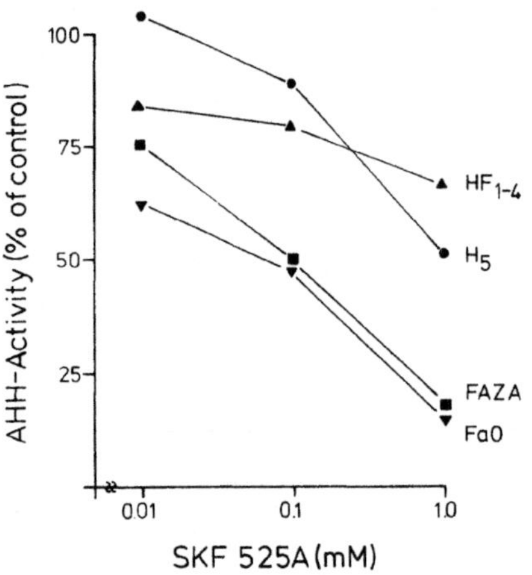

Fig. 3. *In vitro* effect of SKF-525A on AHH activity in hepatoma cell lines. - Cells were pretreated with dexamethasone as described in Table 2. Enzyme activities are expressed as % of control, i.e., the activities in the absence of the inhibitor (cf. Table 2).

In contrast, the flavone increased the level of AHH in the undifferentiated cell lines FAZA 967 and FaO, which were either untreated (data not shown) or exposed to dexamethasone. Furthermore, AHH activities in FAZA 967 and FaO cells treated with benz(a)anthracene were considerably less sensitive to 7,8-benzoflavone inhibition than those of the undifferentiated lines, which is in keeping with their relatively low inducibility by aromatic hydrocarbon-type inducers (Table 2). As expected, SKF-525A inhibits the monooxygenase in the differentiated cell lines FAZA and FaO much more efficiently than in the undifferentiated line H5 (Fig. 3). The hybrid HF1-4 again responds similarly to the undifferentiated lines.

The results indicate that the differentiated hepatoma cell lines express monooxygenase functions which differ from those of the undifferentiated lines. Following the criteria of substrate specificity, response to inducers and to *in vitro* modifiers, the monooxygenase activity in the undifferentiated hepatoma cells are of the cytochrome P-448 type commonly found in established cell cultures, whereas the monooxygenase expressed in the differentiated cells clearly belongs to the cytochrome P-450-dependent forms.

2. CONJUGATION REACTIONS

Conjugation with glucuronic acid, sulfate or glutathione (GSH) in general leads to the detoxification of reactive chemicals although these reactions may occasionally play a role in the activation of carcinogens (29,30,35). As for the monooxygenases, several forms of transferases are distinguishable of which only a few will be considered in the following. In contrast to the oxygenation by the monooxygenase systems, the conjugation reaction in cultured cells as well as *in vivo* may be limited by the availability of their co-substrates. Thus, *in*

Table 3 Monooxygenase, UDP-glucuronyltransferase and sulfotransferase activities in established cell cultures.

Cell line[a]	picomole/min/mg protein			
	3-OH-BP[b] Glucuronosyl-transferase	3-OH-BP[c] Sulfo-transferase	Benzo(a)pyrene[d] Monooxygenase Untreated	BA[e]
H-4-II-E, hepatoma, rat	750	39	6.00	80.0
HFl-4, hepatoma, rat	N.D.[f]	230	3.00	82.0
RAG, renal adenoma, BABL/C mouse	680	< 5	< 0.01	< 0.1
BHK21(13), kidney, Syrian hamster	320	50	0.10	1.65
A549, pulmonary carcinoma, human	199	<10	< 0.01	0.30
L-A9, fibroblastoid, C3H/An mouse	182	< 5	< 0.01	< 0.01
3T3, fibroblastoid, BALB/C mouse	133	< 5	0.08	0.66
V79, lung, Chinese hamster	< 5	< 5	< 0.01	< 0.01
C81, fibroblastoid, cat	< 5	N.D.	0.02	0.79
NC37, BaEV, lymphoblastoid, human	< 5	N.D.	< 0.01	< 0.01

[a] For source of cell lines see (43) and Table 2.

[b] UDP-glucuronosyltransferase directed toward 3-OH-BP was determined as previously described (37).

[c] Phenol sulfotransferase activity was determined in the 100 000 xg supernatants of sonified cells. The reaction mixture contained 50 mM Tris HCl, pH 7.4; 5 mM $MgCl_2$; 10 mM adenosine 3′:5′diphosphate (PAP); 5 mM p-nitrophenyl sulfate; 50 uM 3-OH-BP and 100-500 ug of cytosolic protein. The reaction was terminated after 10 min incubation at 37°. The fluorescence of the aqueous soluble conjugate was measured at 382 nm excitation/415 nm emission.

[d] BP monooxygenase ("aryl hydrocarbon hydroxylase") was determined according to Nebert and Gelboin (31) with modifications (43).

[e] Cells were exposed to 13 uM benz(a)anthracene (BA) for 24 hrs.

[f] Not Determined.

vitro activities which are obtained in the abundance of co-factors can only be taken as relative measures of capacity for the conjugation reactions and not as a direct reflection of the activities in intact cells.

2.1. Glucuronosyltransferase activity

In the present studies 3-OH-BP served as model substrate for the conjugation with glucuronic acid (40). The activity toward this substrate is representative for a glucuronosyltransferase form that shares many properties with cytochrome P-448-dependent monooxygenases. For example, the enzyme is inducible by aromatic hydrocarbons (4) and has been found in all mammalian tissues examined so far (5).

Table 3 shows the in vitro activities of UDP-glucuronosyltransferase in various cell lines in comparison with those of BP monooxygenase (AHH) and sulfotransferase. In agreement with the wide distribution in vivo, the majority of the cell lines contained glucuronosyltransferase activity. It is apparent that the activities of the conjugase, whenever expressed, are at least by 2 to 4 orders of magnitude greater than those of the monooxygenase either in untreated or inducer-treated cultures except for the hepatoma cell line H-4-II-E. There was no clear correlation between the relative activities of the 2 enzymes. At the extreme, either enzyme may be expressed in the absence of the other. Thus, RAG cells which lack the monooxygenase contain a high level of glucuronosyltransferase activity. On the other hand, C81 fibroblastoid cells derived from cat express monooxygenase but not the transferase activity. The latter case apparently reflects the general deficiency of cats in the activity of this glucuronosyltransferase (17). It should be noted that in two cell lines, the human lymphoblastoid cells NC37 BaEV and the Chinese hamster cells V79, neither monooxygenase nor the transferase activity was detectable.

2.2. Phenol sulfotransferase activity.

Phenol sulfotransferases compete with UDP-glucuronosyltransferases for many of their substrates including the monohydroxybenzo(a)pyrenes (6,19,28,32). Table 2 shows the in vitro sulfotransferase activity directed toward 3-OH-BP in the cytosol from various cell cultures. In most of the cell lines, sulfotransferase activity is below or near the level of detectability. Low activities are found in H-4-II-E and BHK/21 cells. Only the hepatoma hybrid cell, HF1-4, contained an appreciable level of the transferase. Under our assay conditions, sulfotransferase activity in the cytosol of rat liver is about 10 times that in the HF1-4 (unpublished observation). Previously, intact HF1-4 cells have also been shown to convert deoxycholic acid to its sulfate ester (22). The relevance of the in vitro activity of 3-OH-BP-sulfotransferase or of its absence in cell cultures is difficult to evaluate. Presently, there is not enough information on tissue and species distribution of the enzyme available to make a valid comparison between in vivo and our in vitro activities. Conjugation of monohydroxybenzo(a)pyrene (and of other BP metabolites) has been shown to be a major pathway in short-term organ cultures of human lung (28) and in freshly isolated rat hepatocytes (19) but not in short-term cultures of rat lung (28). Low activities were detectable in cultured human bronchus (2) and in hamster and rat trachea (6). Studies using various substrates other than BP or its metabolites suggest that cells of rat kidney tubule (10) and intestinal mucosa (39) contain phenol sulfotransferase activities which greatly exceed the activities of the monooxygenases. Clearly, more has to be learned to understand the lack of sulfotransferase activity in the majority of the cell lines tested and to predict sulfate conjugation in the intact cell from the *in vitro* activities, when the enzyme is expressed.

2.3. Glutathione S-transferase activities

The activities of glutathione (GSH) S-transferases in cultured cells were determined using 3 substrates, 1-chloro-2,4-dinitrobenzene (CDNB), 1,2-dichloro-4-nitrobenzene (DCNB) and 1,2-epoxy-3-(p-nitrophenoxy)-propane (PO). In contrast to the glucuronosyl- and sulfotransferases, the GSH S-transferase activities are found in all the cultured cells (Table 4). There is a rough correlation between the activities directed toward DCNB and CDNB, with the notable exception of the human cell lines. The expression of the activity directed toward PO appears to follow a different pattern in the various cell lines. The activities of this form range from 1.7 to 9.7 nmol/min/mg protein; only in one cell line, BHK/21 is the activity at the level of detectability, i.e., well below the activities in the other cell lines.

Table 4 Activities of glutathione S-transferases in cultured cells.

Cell line		nmol/min/mg protein		
		CDNB	DCNB	PO
V79	hamster	227	2.8	2.6
3T3	mouse	110	1.0	3.1
BHK/21	hamster	84	0.3	0.1
L-A9	mouse	114	1.9	4.1
RAG	mouse	181	0.9	9.5
H-4-II-E	rat	174	3.0	6.6
A549	human	73	0.1	3.3
Chang	human	77	0.1	1.7
HEp-2	human	38	0.1	1.7

GSH S-transferase activities toward the substrates 1,2-dichloro-4-nitrobenzene (DCNB), 1-chloro-2,4-dinitrobenzene (CDNB) and 1,2-epoxy-3-(p-nitrophenoxy)propane (PO) were determined in sonified homogenates of cells according to the method of Habig *et al*. (15) with minor modifications (21).

The apparent lack of the transferase activity directed toward DCNB observed in the three human cell lines, A549 (pulmonary adenocarcinoma), Chang (liver) and HEp-2 (larynx carcinoma), has also been found in other human cell lines (41). Since the assay involved is relatively insensitive, i.e., about two orders of magnitude less sensitive than those for glucuronosyl- and sulfotransferase, low activities of the GSH conjugation reaction might readily escape detection. It is interesting to note that in human liver the activity directed toward DCNB is very low in comparison to the activities directed toward the other substrates (20), and likewise amounts to only a fraction of the activity in rat liver (14). Thus, in this point cells in culture reflect some of the species specificity of drug metabolism.

CONCLUSIONS AND OUTLOOK

The data presented shed light only on a small section of the large spectrum of carcinogen-metabolizing enzymes. However, some general conclusions may be drawn: Established cells in culture differ widely in their absolute and relative levels of activating and inactivating enzyme. Nearly any constellation in the expression of the various enzymes seems to be possible. In most cell lines the conjugases are by orders of magnitude more active than the monooxygenases which is reminiscent of the conditions in extrahepatic tissues. On the other hand, hepatoma cells have become available which bear some resemblance to liver tissue in their monooxygenase specificity and overall high level of drug-metabolizing enzymes. Since any or all of the three enzymes, UDP-glucuronosyltransferase, phenol-sulfotransferase and monooxygenase may be lacking in established cell cultures, they appear not to be essential for the survival of cells under 'normal' culture conditions. In contrast, the ubiquitous presence of at least one form of GSH S-transferase in the cell cultures underlines the vital function of these enzymes and dims the prospects of finding or developing cells deficient in these activities.

It is apparent that the possibilities and limitations of using established cell cultures in studies involving the carcinogen metabolism largely depend on the specific field of application. Three major areas may be envisioned:

1) Cell cultures may be employed to dissect the metabolic pathways of carcinogens and to determine their biological significance. Here, the lack of some of the drug-metabolizing enzymes that we observe in most cell lines is an advantage rather than a shortcoming. Cell lines may be taken as carriers of specific combinations of carcinogen activating and inactivating enzymes with built-in indicators for the biological effects of the chemicals. Presently, from the point of carcinogen metabolism, this is the area in which permanent cultures show their greatest strength.

2) Cell cultures, furthermore, could serve as systems for the screening of potentially mutagenic or carcinogenic chemicals. In this

case, cells should be fully competent in the 'activating' functions and, optimally, be devoid of 'inactivating' functions. As shown in this presentation, no single cell line is available which meets these requirements. However, although the survey of the drug-metabolizing enzymes in cell cultures is still spotty, the data suggest that we might come close to the goal by establishing a set of cell lines that is capable of detecting a wide range of genotoxic substances.

3) Finally, cells in culture may be taken as models for species and organ specific carcinogen metabolism. This is the greatest challenge to the culture systems and at present also their weakest point. As mentioned above, cells adapted to the growth in culture are generally not likely to mirror the metabolic conditions of their tissue of origin. However, our results have shown that cells in culture may retain essential differentiated functions of drug metabolism and display some degree of species and organ specificity. With further development of culture systems selective for the expression of differentiated functions, cell lines may conceivably be established that more closely resemble their tissue of origin. Once their capacity for carcinogen metabolism is defined and the metabolic pathways of their *in vivo* correlates, the human tissues, are known, cell cultures may become a valuable tool to probe not only for the carcinogenic potential of chemicals but also for their potency, i.e., providing instruments for the assessment of risk in chemical carcinogenesis.

ACKNOWLEDGMENTS

We are indebted to Dr. M. Weiss, Gif-sur-Yvette, France, for generously supplying the hepatoma cell lines, FAZA 967, FaO, H5, HF1 and HF1-4. We thank Dr. J. Singh for his valuable contributions to the studies of UDP-glucuronosyltransferase activities. The excellent technical assistance of Mrs. E. Schindler and the expert secretarial help of Ms. J. Byers are gratefully acknowledged.

REFERENCES

1. Atlas, S.A., Boobis, A.R., Felton, J.S., Thorgeirsson, S.S., and Nebert, D.W.: 1977, J. Biol. Chem. 252, pp. 4712-4721.
2. Autrup, H., Curtis, C.H., Stoner, G.D., Selkirk, J.K., Schafer, P.W., and Trump, B.G.: 1978, Lab. Invest. 38, pp. 217-224.
3. Bjorkhem, I., Danielsson, H. and Wiawall, K.: 1974, J. Biol. Chem. 249, pp. 6439-6445.
4. Bock, K.W., and Lilienblum, W.: 1979, Biochem. Pharmacol. 28, pp. 695-700.
5. Bock, K.W., v. Clausbruch, U.C., Kaufmann, R., Lilienblum, W., Oesch, F., Pfeil, H., and Platt, K.L.: 1980, Biochem. Pharmacol. 29, pp. 495-500.
6. Cohen, G.M., Haws, S.M., Moore, B.P., and Bridges, J.W.: 1976, Biochem. Pharmacol. 25, pp. 2561-2570.
7. Deschatrette, J. and Weiss, M.C.: 1968, J. Biol. Chem. 243, pp. 6242-6249.

8. Diamond, L.: 1971, Int. J. Cancer 8, pp. 451-462.
9. Frommer, U., Ullrich, V., and Orrenius, S.: 1974, FEBS Lett. 41, pp. 14-16.
10. Fry, J.R., Wiebkin, P., Kao, J., Jones, C.A. Gwynn, J., and Bridges, J.W.: 1978, Xenobiot. 8, pp. 113-120.
11. Gelboin, H.V. and Wiebel, F.J.: 1971, Ann. N.Y. Acad. Sci. 179, pp. 529-547.
12. Goujon, F.M., Nebert, D.W., and Gielen, J.E.: 1972, Molec. Pharmacol. 8, pp. 667-680.
13. Guenthner, T.M. and Nebert, D.W.: 1978, Eur. J. Biochem. 91, pp. 449-456.
14. Habig, W.H., Pabst, M.J. and Jakoby, W.B.: 1974, J. Biol. Chem. 249, pp. 7130-7139.
15. Haugen, D.A., van der Hoeven, T.A., and Coon, M.C.: 1975, J. Biol. Chem. 250, pp. 3567-3570.
16. Holder, G., Yagi, H., Dansette, P., Jerina, D.M., Levin, W., Lu. A.Y.H., and Conney, A.H.: 1974, Proc. Natl. Acad. Sci. USA 71, pp. 4356-4360.
17. Hirom, P.C., Idle, J.R., and Millburn, P.: 1977, in, Drug Metabolism--From Microbe to Man (D.V. Parke and R.L. Smith, eds.), Taylor and Francis Ltd., London, pp. 299-330.
18. Huberman, E. and Sachs, L.: 1974, Int. J. Cancer 13, pp. 326-333.
19. Jones, C.A., Moore, B.P., Cohen, G.M., Fry, J.R., and Bridges, J.W.: 1978, Biochem. Pharmacol. 27, pp. 693-702.
20. Kamisaka, K., Habig, W.H., Ketley, J.N., Arias, I.M. and Jakoby, W.B.: 1975, Eur. J. Biochem. 60, pp. 153-161.
21. Kouri, R.E., Kiefer, R., and Zimmerman, E.M.: 1974, In Vitro 10, pp. 18-25.
22. Lambiotte, M. and Sjövall, J.: 1979, Biochem. Biophys. Res. Comm. 86, pp. 1089-1095.
23. Lambiotte, M. and Thierry, N.: 1979, Biochem. Biophys. Res. Comm. 89, pp. 933-942.
24. Lotlikar, P.D., Hong, S.Y., and Baldy, W.J. Jr.: 1978, Toxicol. Lett. 2, pp. 135-139.
25. Lu, A.Y.H.: 1976, Fed. Proc. 25, pp. 2460-2463.
26. Lu, A.Y.H., Kuntzman, R., West, S., Jacobson, M., and Conney, A.H.: 1972, J. Biol. Chem. 247, pp. 1727-1734.
27. Marquardt, H. and Heidelberger, C.: 1972, Cancer Res. 32, pp. 721-725.
28. Mehta, R. and Cohen, G.M.: 1979, Biochem. Pharmacol. 28, pp. 2479-2484.
29. Miller, J.: 1970, Cancer Res. 30, pp. 559-576.
30. Mulder, G.J., Hinson, J.A., and Gillette, J.R.: 1977, Biochem. Pharmacol. 26, pp. 189-196.
31. Nebert, D.W. and Gelboin, H.V.: 1968, J. Biol. Chem 243, p. 6242-6249.
32. Nemoto, N., Hirakawa, T., and Takayama, S.: 1978, Chem.-Biol. Interact. 22, pp. 1- .
33. Niwa, A., Kumaki, K., and Nebert, D.W.: 1975, Mol. Pharmacol. 11, pp. 399-408.
34. Owens, I.S. and Nebert, D.W.: 1975, Mol. Pharmacol. 11, pp. 94-104.

35. Rannug, U., Sundvall, A., and Ramel, C.: 1978, Chem. Biol. Interact. 20, pp. 1-16.
36. San, R.H.C. and Stich, H.F.: 1975, Int. J. Cancer 16, pp. 284-291.
37. Selkirk, J.K., Croy, R.G., and Gelboin, H.V.: 1974, Science 184, pp. 169-171.
38. Selkirk, J.K., Croy, R.G., Wiebel, F.J., Gelboin, H.V.: 1976, Cancer Res. 36, pp. 4476-4479.
39. Shirkey, R.J., Kao, J., Fry, J.R., and Bridges, J.W.: 1979, Biochem. Pharmacol. 28, pp. 1461-1466.
40. Singh, J. and Wiebel, F.J.: 1979, Analyt. Biochem. 98, pp. 394-401.
41. Smith, G.J., Huebner, K., and Litwack, G.: 1977, Biochem. Biophys. Res. Comm. 76, pp. 1174-1180.
42. Wiebel, F.J.: 1980, in, Carcinogenesis, Vol. 5, (T.J. Slaga, ed.), Raven Press, New York, pp. 57-84.
43. Wiebel, F.J., Brown, S., Waters, H.L., and Selkirk, J.K.: 1977, Arch. Toxicol. 39, pp. 133-148.
44. Wiebel, F.J., Leutz, J.C., Diamond, L., and Gelboin, H.V.: 1971, Arch. Biochem. Biophys. 144, pp. 78-86.
45. Wiebel, F.J., Selkirk, J.K., Gelboin, H.V., Haugen, D.A., van der Hoeven, T.A., and Coon, M.J.: 1975, Proc. Natl. Acad. Sci. (Wash.) 72, pp. 3917-3920.
46. Wiebel, F.J., Schwarz, L.R., and Goto, T.: 1980, in: Short-term Mutagenicity Test Systems in Detecting Carcinogens, eds. Norpoth, K. and Garner, R.C., pp. 209-225.
47. Wiebel, F.J. and Singh, J.: 1980, Arch. Toxicol. 44, pp. 85-97.
48. Wolff, T., Deml, W., and Wanders, H.: 1979, Drug Metab. Dispos. 7, pp. 301-305.
49. Wolff, T., Greim, H., Huang, M.T., Miwa, G.T., and Lu, A.Y.H.: 1980, Eur. J. Biochem., in press.
50. Yang, S.K., Gelboin, H.V., Trump, B.F., Autrup, H., and Harris, C.C.: 1977, Cancer Res. 37, pp. 1210-1215.
51. Zampaglione, N., Jollow, D.J., Mitchell, J.R., Stripp, B., Hamrick, M., and Gillette, J.R.: 1973, J. Pharmacol. Exp. Therap. 187, pp. 281-227.

REGULATION OF THE EXPRESSION OF SV40 T ANTIGEN IN STEM VERSUS DIFFERENTIATED CELLS

Alban Linnenbach, Kay Huebner and Carlo M. Croce
The Wistar Institute of Anatomy and Biology
36th Street at Spruce
Philadelphia, Pa. 19104
United States of America

ABSTRACT

We have inserted the entire SV40 genome into a plasmid carrying the herpes simplex viral thymidine kinase gene and we have transformed thymidine kinase deficient F9 teratocarcinoma stem cells with this recombinant DNA. The SV40 containing transformed F9 cells did not express SV40 T antigen. Following exposure to the inducer retinoic acid the transformed stem cells differentiated and expressed SV40 T antigen.

It has been shown that the SV40 T antigen is not expressed in mouse teratocarcinoma cells that have been infected with SV40 (1,2). On the contrary SV40 infected differentiated cells have been found to express SV40 T antigen (1,2). In order to study the regulation of the expression of SV40 T antigen in the stem versus differentiated cells we constructed a recombinant DNA vector that contained the entire genome of SV40 in the plasmid pHSV106 which carried a herpes simplex virus type 1 (HSV-1) thymidine kinase gene at the BamH1 site of pBR322 [Fig.1] (3,4). This recombinant plasmid designated C6 (4), was used to transform thymidine kinase deficient F9 (5) mouse teratocarcinoma cells by the calcium phosphate precipitation method described by Graham and van der Eb (6). HAT medium (7) selected colonies of teratocarcinoma stem cells were grown continuously in HAT medium and studied for the expression of HSV-1 thymidine kinase (4). As shown in figure 2, the HAT selected clones expressed HSV-1 thymidine kinase activity. DNA derived from two of these clones (12-1 and 13-1) was extracted according to the Hirt method (8) and divided in Hirt pellet (high molecular weight DNA) and Hirt supernatant (low molecular weight DNA) fractions (8). The DNA of each of the fractions was digested with Xbal, Kpnl and BamH1 restriction enzymes that cut C6 DNA 0, 1 and 3 times respectively. The restricted DNA fragments were separated by agarose gel electrophoresis, blotted on nitrocellulose filters by the Southern method (9) and hybridized with SV40 DNA labeled with ^{32}P by the nick translation method (10).

B. Pullman, P. O. P. Ts'o and H. Gelboin (eds.), Carcinogenesis: Fundamental Mechanisms and Environmental Effects, 363–369.

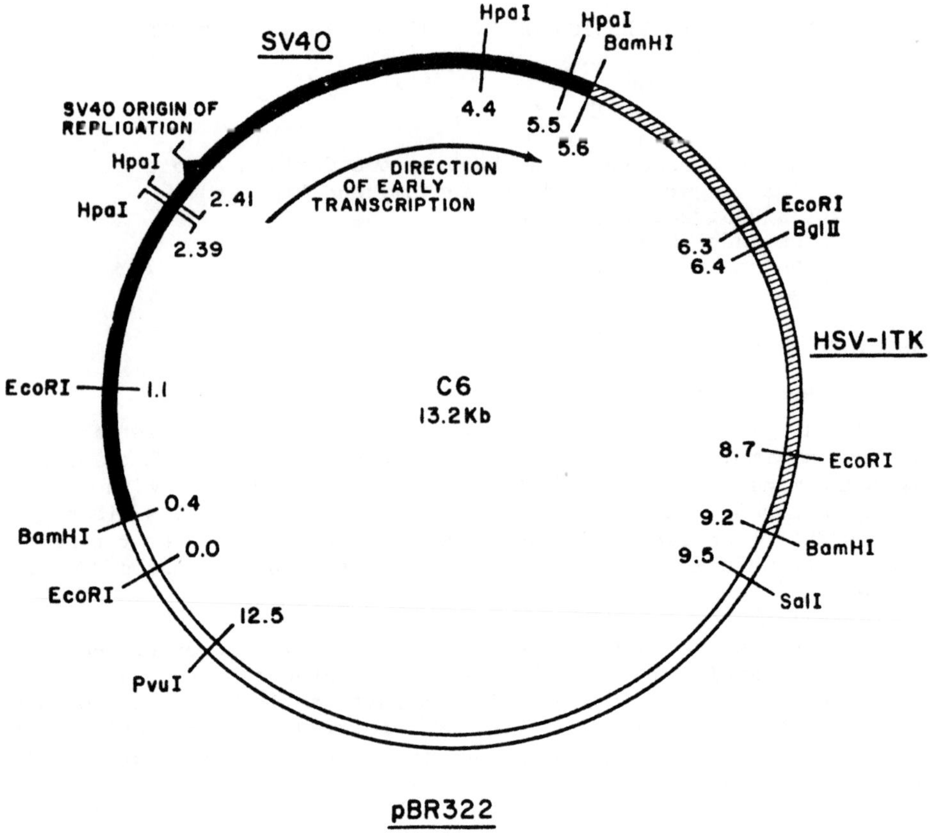

Figure 1. Restriction map of plasmid C6. One copy of BamH1 digested SV40 DNA is inserted into the pBR322- HSV-1 TK vector, pHSV-106.

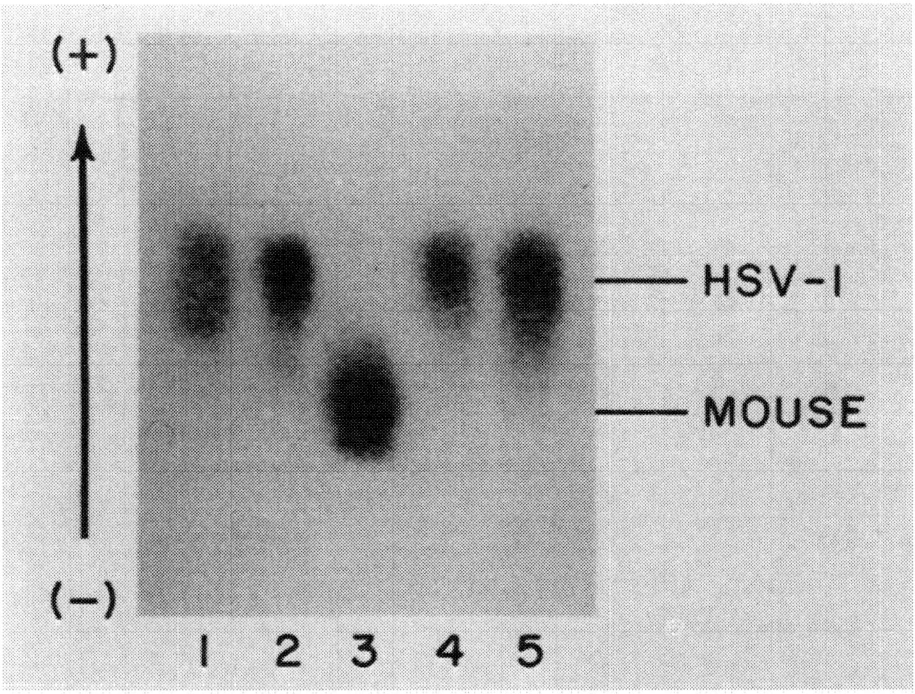

Figure 2. Starch gel electrophoretic separation of thymidine kinase gene products. Enzyme migration is visualized by the binding of titrated thymidine monophosphate product to DE-81 paper overlaid on electrophoretically separated thymidine kinase enzymes. Enzymes in lanes 2 and 5 are from F9 stem cell transformants 13-1 and 12-1; lane 3 contains murine thymidine kinase derived from MC57G cells; lane 4 contains HSV-1 thymidine kinase derived from 41-1-1 cells. Lane 1 contains HSV-1 thymidine kinase derived from an F9 stem cell transformant not described in this report.

As shown in figure 3, only the Hirt pellet of clone 12-1 contained SV40 DNA, that migrated as a 15 Kb single band following digestion of the cellular DNA with XbaI, which does not cut C6 DNA. This observation indicates that the SV40 genome is integrated into a single site in 12-1 cells. On the contrary, the DNA derived from clone 13-1 did not hybridize with the SV40 probe indicating that it had lost the SV40 genome, while retaining the TK gene.

Restriction of the 12-1 DNA Hirt pellet with Kpnl resulted in the appearance of two bands as expected for a C6 integrated DNA molecule. Restriction of the 12-1 DNA Hirt pellet with BamHl resulted in the appearance of a single band of 5.2 Kb in size that represents the intact SV40 DNA linear molecule [Fig. 4].

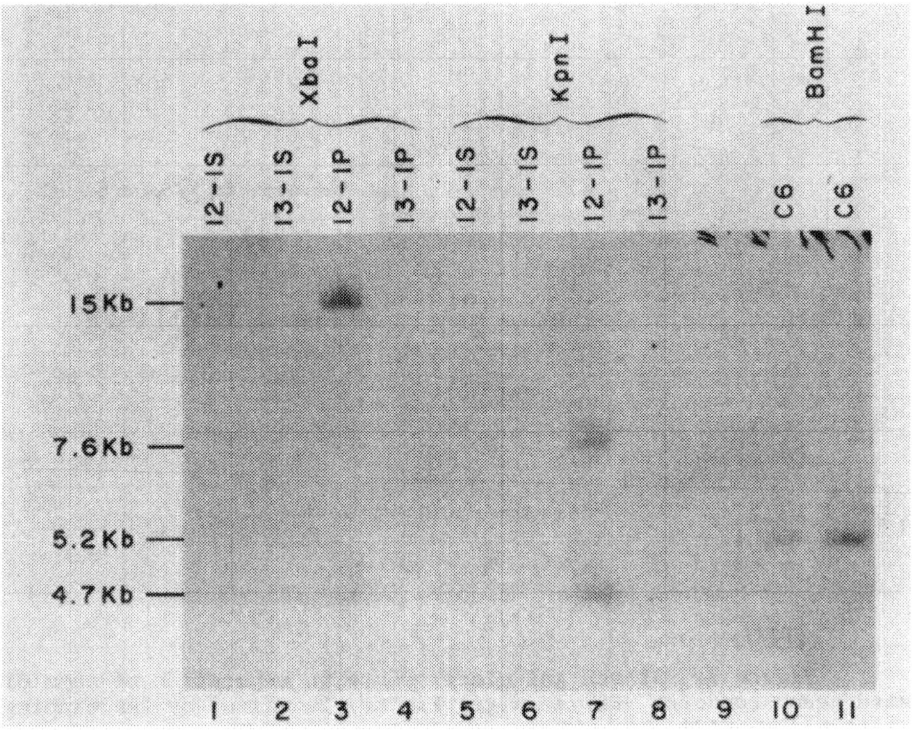

Figure 3. Hybridization of ^{32}P-labeled SV40 DNA to 13-1 and 12-1 cellular DNA after restriction endonuclease digestion and transfer from agarose gel. Hirt Supernatant (S) and pellet (P) DNA (10 µg/lane) from 12-1 and 13-1 stem cells were cleaved with Xbal (lanes 1-4) and Kpnl (lanes 5-8); electrophoresed in a 0.7% agarose slab gel; denatured; transferred to a nitrocellulose sheet, and hybridized to ^{32}P-labeled SV40 DNA. BamHl cleaved and similarly treated C6 DNA (lane 10, 12.5 pg; lane 11, 6.25 pg) was included as a marker.

We have also restricted the 12-1 cell DNA and the C6 DNA with Xbal and BamHl and hybridized the DNA fragments that were separated by agarose gel electrophoresis and blotted on a nitrocellulose filter with three different ^{32}P probes; pBR322 (a), HSV TK (b) and SV40 (c).

As shown in figure 4 in all three cases, the digestion of the 12-1 cell DNA with Xbal resulted in the appearance of a 15 Kb DNA fragment that hybridized with the three probes. This experiment indicates that all three DNA molecules that formed the C6 DNA are contained within the same cellular DNA fragment in the transformant.

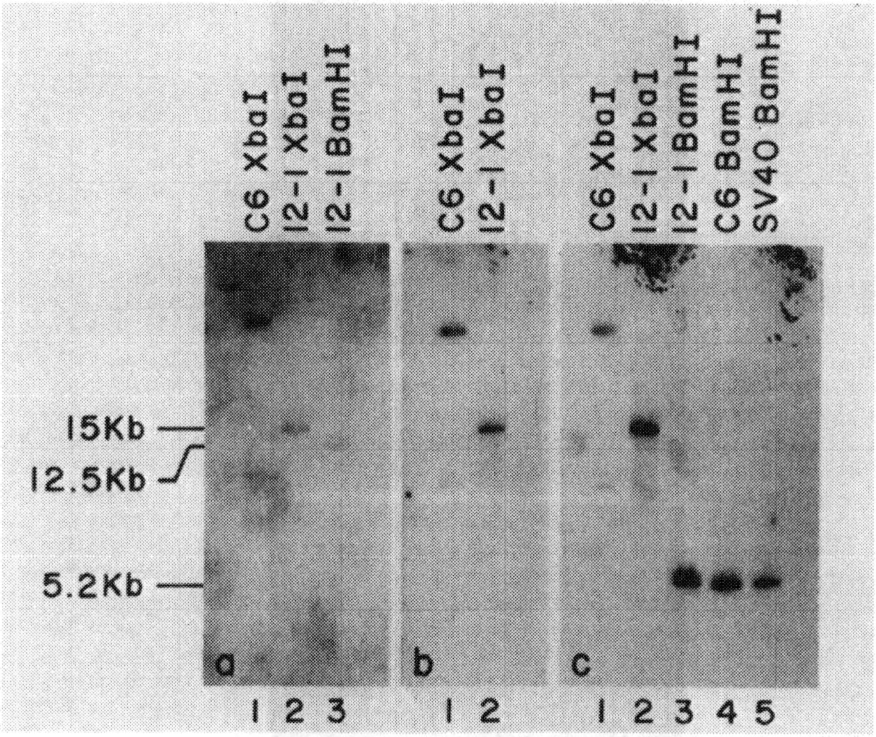

Figure 4. Hybridization of ^{32}P-labeled pBR322 DNA (a), HSV-1 TK DNA (b) and SV40 DNA (c) to Xbal and BamHl cleaved 12-1 cellular DNA. Hirt pellet DNA (10 µg/lane) from 12-1 cells was cleaved with Xbal and BamHl restriction endonucleases and applied to an agarose slab gel as indicated above each lane in this figure. After electrophoresis, the gel was cut into three parts and the DNA in each gel was denatured, transfered to nitrocellulose sheets and each of the three nitrocellulose sheets was hybridized to a different ^{32}P-labeled probe: a) hybridized to ^{32}P-labeled pBR322 DNA; b) hybridized to ^{32}P-labeled HSV-1 TK DNA; c) hybridized to ^{32}P-labeled SV40 DNA. C6 DNA, BamHl cleaved C6 DNA, and BamHl cleaved SV40 DNA are included as markers.

We then tested the 12-1 transformed mouse teratocarcinoma cells for the expression of SV40 T antigen and found them negative (Fig. 5a).

We then induced the cells to differentiate by exposing them to retinoic acid (11). As shown in figure 5b, the differentiated cells expressed SV40 T antigen. We conclude, therefore, that we have constructed a useful cellular system to study the molecular basis of the regulation of gene expression in stem versus differentiated cells.

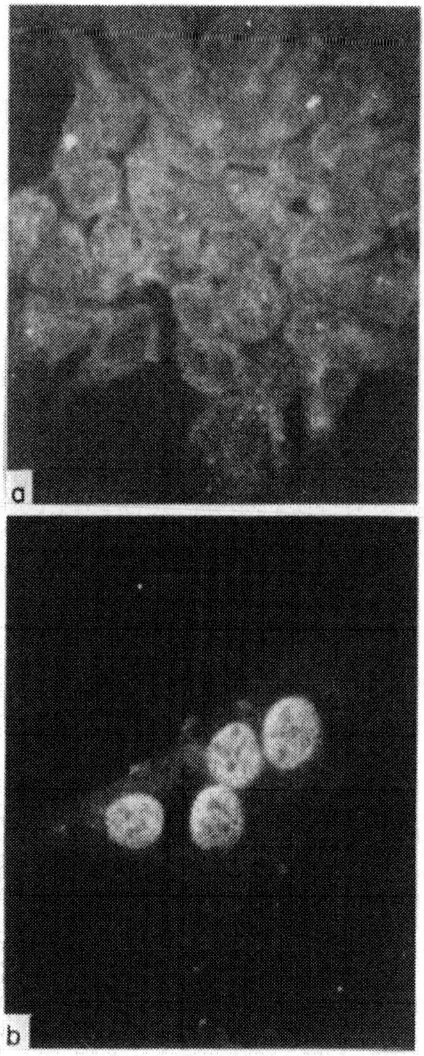

Figure 5 (a,b). Indirect immunofluorescence of SV40 T-antigen in transformant 12-1 cells before (a) and after (b) induction of differentiation with 0.1 μM retinoic acid.

ACKNOWLEDGEMENTS. We wish to thank Drs. M. Shander, T. Dolby, P. Botchan, K. B. Tan, and R. Weinmann of the Wistar Institute, who have each given us advice in setting up many of the techniques involved in this research. C. M. Croce thanks Dr. S. McKnight and Dr. D. Brown with whom he began learning recombinant DNA technology during sabbatical leave at the Carnegie Institute, Baltimore, Maryland.

We thank P. Tan for ^{32}P-labeled HSV-1 TK DNA and K. Edelberg for preparation of plasmid DNA.

This work was supported by United States Public Health Service Research grants CA10815, CA16885, CA20741, CA21069, CA21124 from the National Cancer Institute, GM20700 from the Institute of General Medical Services and by grant 1-522 from the National Foundation - March of Dimes.

C. M. Croce is a recipient of a Research Career Development Award CA00143 from the National Cancer Institute.

REFERENCES

1. Swartzendruber, D. E., and Lehman, J. M. : 1975, J. Cell. Physiol. 85, pp 179-188.
2. Swartzendruber, D. E., Friedrich, T. D., and Lehman, J. M. : 1977. J. Cell. Physiol. 93, pp 25-30.
3. McKnight, S. L., and Croce, C. M. : 1979, Carnegie Institute Yearbook. 98, pp 56-61.
4. Linnenbach, A., Huebner, K., and Croce, C. M. : 1980, Proc. Natl. Acad. Sci. USA. In press.
5. Gmür, R., Solter, D., and Knowles, B. B. : 1980, J. Exp. Med., In press.
6. Graham, F. L., and van der Eb, A. J. : 1973, Virology 52, pp 456-467.
7. Littlefield, J. W. : 1966, Exp. Cell. Res. 41, pp 190-196.
8. Hirt, B. : 1967, J. Mol. Biol. 26, pp 365-369.
9. Southern, E. : 1975, J. Mol. Biol. 98, pp 503-515.
10. Maniatis, T., Kee, S. G., Efstratiadis, A., and Kafatos, F. C. : 1976, Cell 8, pp 163-182.
11. Strickland, S., and Mahdavi, V. : 1978, Cell 15, pp 393-403.

THE EFFECT OF 12-O-TETRADECANOYL-PHORBOL-13-ACETATE (TPA) ON CELL TRANSFORMATION BY SIMIAN VIRUS 40 MUTANTS

R. MONIER, L. DAYA-GROSJEAN and A. SARASIN
Institut de recherches scientifiques sur le cancer (Villejuif)

The effect of TPA on the transformation of Swiss 3T3 cells to anchorage-independent growth by the tsA 58 mutant of SV40 has been investigated. At the restrictive temperature of 39°C, the combined action of the tsA 58 mutant and of TPA allows the formation of soft-agar colonies from Swiss 3T3 cells. No colonies were observed when either one or the other agent was omitted. Antipaïn had no inhibitory effect on this phenomenon. TPA also stimulates the anchorage-independent growth of cells previously transformed by either the tsA 58 mutant or the dl 2112 mutant.

1. INTRODUCTION

The enhancement by tumor promoters of *in vitro* transformation of fibroblast cell cultures previously exposed to chemical carcinogens, ultraviolet light or X rays has already been demonstrated (Lasne, Gentil and Chouroulinkov, 1974 ; Mondal, Brankow and Heidelberger, 1976 ; Mondal and Heidelberger, 1976 ; Kennedy *et al.*, 1978).

It has also been shown that the potent tumor promoter 12-O-tetradecanoyl-phorbol-13-acetate (TPA) increases the yield of transformed foci from adenovirus-infected secondary rat embryo cells (Fisher *et al.*, 1978). TPA also has an effect on the anchorage-independent growth of cells previously transformed by adenovirus 5 (Wigler and Weinstein, 1976 ; Fisher, Bozzone and Weinstein, 1979 ; Fisher *et al.*, 1979), and on the behavior of SV40-transformed cells (Sivak and van Duuren, 1967).

The availability of SV40 mutants, in which the transforming early region of the genome has been mutated, provides the opportunity of analysing in greater detail the interaction between a transforming DNA virus and promoters. In the present report, we describe experiments performed using two types of SV40 early mutants : tsA 58 and dl 2112. Our results show that, in the presence of TPA, transformed cells can be obtained at the restrictive temperature of 39°C from tsA 58-infected

B. Pullman, P. O. P. Ts'o and H. Gelboin (eds.), Carcinogenesis: Fundamental Mechanisms and Environmental Effects, 371–378.

mouse 3T3 cells. TPA also stimulates the anchorage-independent growth of cells previously transformed with either mutant.

2. MATERIALS AND METHODS

2.1. Cells and viruses

Swiss 3T3 cells, obtained from ATCC, were grown in Dulbecco-modified Eagle medium containing 10% new-born calf-serum. The mouse cell line ME Δ 121 was obtained by transformation of secondary mouse embryo cultures by the dl 2112 mutant. The rat cell line tsA-F was obtained by transformation of secondary rat embryo cultures by the tsA 58 mutant. Both were a gift of C.A. Petit and J. Feunteun. Wild type SV40-strain LP (Suarez et al., 1974) and the tsA 58 mutant (Tegtmeyer, 1972) were grown on monkey CV1 cells as previously described (L. Daya-Grosjean and R. Monier, 1978).

2.2. Transformation procedure

Swiss 3T3 cells were plated at 33°C in 2.5 cm plastic Petri dishes (2×10^5 cells per dish). They were infected 24 hrs later with virus (10 p.f.u. per cell) and incubated at 33°C for 24 hrs. The cells were trypsinized and plated in soft-agar (10^4 cells per 6 cm Petri dish) according to MacPherson and Montagnier (1964). TPA (100 ng/ml) and/or antipaïn (50 μg/ml) were incubated in the soft-agar as required. Further incubation was carried out at either 33°C or 39°C. Cell cultures were refed weekly with 1 ml per dish of soft-agar medium eventually containing TPA and/or antipaïn. Colonies were scored after 3 weeks at 33°C and after 2 weeks at 39°C.

2.3. Kinetics of T antigen appearance in tsA 58-infected cells

Swiss 3T3 cells were infected at 33°C as described above. 24 hrs after infection cells were incubated at 33°C or 39°C with or without TPA (100 ng/ml). Cells were examined for T antigen immunofluorescence (Daya-Grosjean and Monier, 1978) at various times. The results are expressed as per cent of T antigen-positive nuclei.

2.4. Cloning efficiency of SV40-transformed cells in soft-agar

Cells previously transformed by SV40 mutants were plated in soft-agar with or without TPA (100 ng/ml) (10^4 cells per 6 cm dish) and incubated at the appropriate temperature. Macroscopic colonies were scored after 3 weeks at 33°C and 2 weeks at 39°C.

2.5. Products

TPA was obtained from Dr P. Borchert (University of Minnesota). Antipaïn was a gift from Dr T. Sugimura.

3. RESULTS

3.1. Transformation of Swiss 3T3 cells to anchorage-independent growth by TPA and SV40 tsA 58 at 39°C

It is well known that the SV40 tsA 58 mutant cannot produce macroscopic colonies after infection of rodent cells at the restrictive temperature of 39°C (for a discussion, see Fluck and Benjamin, 1979). As shown in Table I, the inclusion of TPA in the soft-agar allows the appearance of macroscopic colonies at 39°C in cells previously infected with tsA 58. At 33°C, colony formation occurs after tsA 58 infection in the absence of TPA but the efficiency of transformation is increased 10-fold in the presence of TPA. Under our experimental conditions, no colonies were observed when virus infection was omitted.

In parallel experiments (see Table II), TPA had no significant effect on the efficiency of transformation by the wild type virus at 39°C.

Table I. Effect of TPA on the transformation efficiency of Swiss mouse 3T3 cells by SV40 tsA 58 at 33°C and 39°C.

TPA addition (ng/ml)	uninfected cells		tsA 58-infected cells (10 p.f.u./cell)	
	33°C	39°C	33°C	39°C
0	10^{-5}	10^{-5}	3×10^{-4}	10^{-5}
100	10^{-5}	10^{-5}	4.6×10^{-3}	1.8×10^{-3}

3.2. Kinetics of T antigen production in tsA 58-infected cells in the presence of TPA

In order to test for the possible effect of TPA on the expression of early viral functions in tsA 58-infected Swiss 3T3 cells, the percent of T antigen-specific immunofluorescent nuclei was measured as a function of time after infection. As shown in fig. 1, TPA had no effect whatsoever on the production of T antigen as estimated by this procedure. At 33°C, neither the time course of T antigen production nor the final percentage of positive nuclei were affected by the presence of TPA. No significant immunofluorescence was observed at 39°C with or without TPA. This observation makes it unlikely that the effect of TPA is simply due to an enhancement of early viral expression, which would increase the leakiness of the tsA mutant.

3.3. Effect of antipaïn on TPA-induced tsA 58-transformation

The protease inhibitor, antipaïn, has been shown to inhibit the promotion phase of mouse skin tumorigenesis (Troll, Klassen and Janoff, 1970 ; Hozumi et al., 1972). It also inhibits some of the effects of

Effect of TPA on the synthesis of T antigen in SV40 tsA58-infected Swiss 3T3 cells

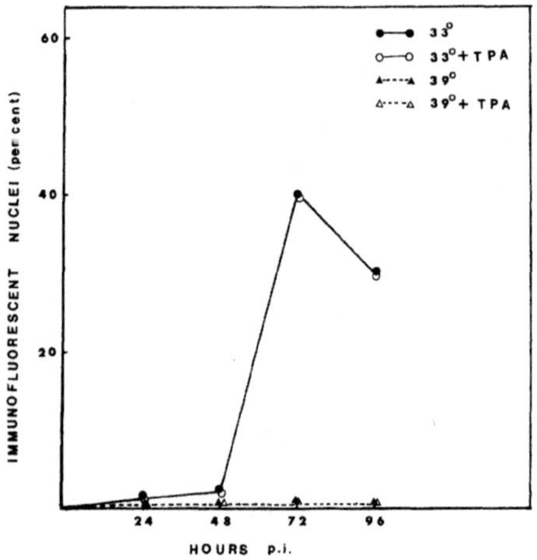

Figure 1

TPA on cultured cells, such as the stimulation of sister chromatid exchanges (SCE) (Kinsella and Radman, 1978 ; Nagasawa and Little, 1979). On the other hand it does not impair the enhancement of anchorage-independent growth of adenovirus-5-transformed cells (Fisher, Bozzone and Weinstein, 1979). It was therefore of interest to investigate the effect of antipaïn on TPA-induced transformation of Swiss 3T3 cells by tsA 58 at 39°C. The inclusion in the soft-agar of a dose of antipaïn which is known to inhibit the TPA-induced SCE did not reduce the number of colonies formed at 39°C after tsA 58-infection and TPA treatment.

Table II. Effect of TPA and antipaïn on the transformation efficiency of Swiss mouse 3T3 cells by SV40 WT and tsA 58 at 39°C.

TPA (ng/ml)	WT infected cells		tsA 58-infected	
	- antipaïn	+ antipaïn+)	- antipaïn	+ antipaïn
0	7×10^{-4}	2.7×10^{-4}	10^{-5}	-
100	4.8×10^{-4}	4×10^{-4}	2.2×10^{-3}	2.0×10^{-3}

+) 50 μg/ml

In the absence of virus, any combination of treatments gave less than 10^{-5} colonies.

In the same experiment, we investigated the effect of TPA and/or antipaïn on the formation of agar colonies at 39°C after infection by the wild type virus. Both compounds had no significant effect on the frequency of transformation under these conditions.

3.4. Effect of TPA on anchorage-independent growth of cells previously transformed by SV40 mutants

The viable deletion SV40 mutant dl 2112 (Feunteun et al., 1978) is missing one of the borders used in the splicing of small t antigen mRNA (Volckaert et al., 1979). As a consequence, dl 2112-infected cells do not produce any small t antigen while the production of large T antigen is normal (May, Kress and May, 1978). After infection of secondary mouse embryo cell cultures by the dl 2112 mutant, transformed cells were selected on the basis of their ability to form colonies on plastic when seeded at low density. Such transformed cells display growth properties similar to continuous 3T3 cell lines. In particular, they do not form macrocolonies (> 0.1 mm diameter) when seeded in soft-agar (Petit, C.A. and Feunteun, J., unpublished data). One of these transformed cell lines, ME Δ 121 was suspended in soft-agar including or not TPA. In the absence of TPA, no colonies larger than 0.1 mm diameter were observed, whereas, in the presence of TPA, macrocolonies (> 0.5 mm diameter) were observed after 21 days with an efficiency of 6×10^{-3}.

Similar experiments were carried out on a secondary rat embryo cell-derived tsA 58-transformed cell line, tsA-F. This cell line derived from a colony formed on plastic at 33°C as above. It is fully thermosensitive for growth on plastic at 39°C and does not exhibit anchorage-independence at either 33°C or 39°C (Petit, C.A. and Feunteun, J., unpublished data). tsA-F cells were plated in soft-agar at 39°C plus or minus TPA. Colonies only appeared in TPA-containing plates.

4. DISCUSSION

In the present report, we have described the results of experiments in which two types of SV40 mutants have been used in an attempt to further analyze the interactions between an oncogenic DNA virus and a promoter at the level of in vitro cell transformation.

The tsA class of early SV40 mutants are known to affect the thermal stability of one of the two tumor antigens produced in SV40-infected cells, namely the 90 K large T tumor antigen. At the restrictive temperature, such mutants are blocked in viral DNA replication at the level of initiation (Tegtmeyer, 1972). They cannot induce permanent in vitro transformation at the restrictive temperature, probably because their genome cannot integrate in the cellular genome (Fluck and Benjamin, 1979).

The SV40 dl 54-59 viable deletions occur in the part of the early region which specifically codes for the C-terminal portion of the second tumor antigen, the 17 K small t antigen (Crawford et al., 1978). dl 54-59-infected cells therefore produce either a shortened small t antigen or, as is the case with the dl 2112 mutant, no small t antigen at all (May, Kress and May, 1978). dl 54-59 mutants are still tumorigenic in new-born hamsters, although with an increased latent period as compared to wild type virus (Lewis and Martin, 1979). Depending upon the experimental conditions, they can transform secondary rodent embryo cells into the equivalent of continuous cell lines, unable of anchorage-independent growth (Sleigh et al., 1978 ; Petit, C.A. and Feunteun, J., unpublished data). It has been suggested that the important factor in the outcome of in vitro transformation experiments with dl 54-59 mutants is the growth state of the cells at the time of transformation : actively growing cells usually lead to anchorage-independent transformants (Seif and Martin, 1979). On the basis of these observations on tumorigenicity and in vitro transformation, Seif and Martin (1979) have proposed that the small t polypeptide could act as a promoter.

We have first investigated the effect of TPA treatment on the transformation of mouse cells by the tsA 58 mutant at the restrictive temperature. Our results show that the presence of TPA can formally "complement" the tsA 58 mutation as regards in vitro transformation of mouse cells. A trivial explanation for this observation might be that TPA produces an enhancement of the early gene expression, leading to an increased leakiness of the tsA mutant, sufficient to allow normal integration to occur. Although our observation on the kinetics of appearance of tumor antigen-specific immunofluorescence in tsA 58-infected cells does not support this interpretation, the lack of sensitivity of the immunofluorescence technique does not allow a definite conclusion. Further experiments, using the more sensitive immunoprecipitation method, are in progress.

A more interesting hypothesis would be that TPA treatment of infected cells provides an integration pathway independent of the large T antigen. Our observation on the effect of antipaïn nevertheless indicates that the effect of TPA on the tsA 58-cell interaction may be dissociated from its effect on induced SCE. It might be of interest to mention in this regard that antipaïn does not inhibit the TPA-induced enhancement of anchorage-independence of adenovirus-5-transformed cells (Fisher, Bozzone and Weinstein, 1979).

Besides its effect on transformation, TPA is also known to affect the properties of already transformed cells (Sivak and van Duuren, 1967 ; Wigler and Weinstein, 1976 ; Fisher, Bozzone and Weinstein, 1979 ; Fisher et al., 1979). We have confirmed these observations in a preliminary way in the case of both tsA 58- and dl 54-59-transformed cells, the anchorage-independence of which is stimulated by TPA. This effect is nevertheless much less spectacular than the effect of TPA on tsA 58-transformation under restrictive conditions, which probably involves different mechanisms. Further work is in progress to understand what these mechanisms may be.

ACKNOWLEDGMENTS

This work was supported in part by D.G.R.S.T. grant 76.7.1667.

REFERENCES

Crawford, L.V., Cole, C.N., Smith, A.E., Paucha, E., Tegtmeyer, P., Rundell, K., and Berg, P.: 1978, Proc.Natl.Acad.Sci., USA 75, pp. 117-121.

Daya-Grosjean, L., and Monier, R.: 1978, J.Virol. 27, pp. 307-312.

Feunteun, J., Kress, M., Gardes, M., and Monier, R.: 1978, Proc.Natl. Acad.Sci., USA 75, pp. 4455-4459.

Fisher, P.B., Weinstein, I.B., Eisenberg, D., and Ginsberg, H.S.: 1978, Proc.Natl.Acad.Sci., USA 75, pp. 2311-2314.

Fisher, P.B., Dorsch-Häsler, K., Weinstein, I.B., and Ginsberg, H.S.: 1979, Nature 281, pp. 591-594.

Fisher, P.B., Bozzone, J.H., and Weinstein, I.B.: 1979, Cell 18, pp. 695-705.

Fluck, M.M., and Benjamin, T.: 1979, Virology 96, pp. 205-228.

Hozumi, M., Ogawa, M., Sugimura, T., Takeuchi, T., and Umezawa, H.: 1972, Cancer Res. 32, pp. 1725-1728.

Kennedy, A.R., Mondal, S., Heidelberger, C., and Little, J.B.: 1978, Cancer Res. 38, pp. 439-443.

Kinsella, A., and Radman, M.: 1978, Proc.Natl.Acad.Sci., USA 75, pp. 6149-6153.

Lasne, C., Gentil, A., and Chouroulinkov, I.: 1974, Nature 247, pp. 490-491.

Lewis, A.M., Jr, and Martin, R.G.: 1979, Proc.Natl.Acad.Sci., USA 76, pp. 4299-4302.

MacPherson, I., and Montagnier, L.: 1964, Virology 23, pp. 291-294.

May, E., Kress, M., and May, P.: 1978, Nucl.Acids Res. 5, pp. 3083-3100.

Mondal, S., and Heidelberger, C.: 1976, Nature 260, pp. 710-711.

Mondal, S., Branrow, D.W., and Heidelberger, C.: 1976, Cancer Res. 36, pp. 2254-2260.

Nagawa, H., and Little, J.B.: 1979, Proc.Natl.Acad.Sci., USA 76, pp. 1943-1947.

Seif, R., and Martin, R.G.: 1979, J.Virol. 32, pp. 979-988.

Sivak, A., and van Duuren, B.L.: 1967, Science 157, pp. 1443-1444.

Sleigh, M.J., Topp, W.C., Hanich, R., and Sambrook, J.F.: 1978, Cell 14, pp. 79-88.

Suarez, H.G., Cassingena, R., Estrade, S., Wicker, R., Lavialle, C., and Lazar, P.: 1974, Arch.gesamte Virusforsch. 46, pp. 93-104.

Tegtmeyer, P.: 1972, J.Virol. 10, 591-598.

Troll, W., Klassen, A., and Janoff, A.: 1970, Science 169, pp. 1211-1213.

Volckaert, G., Feunteun, J., Crawford, L.V., Berg, P., and Fiers, W.: 1979, J.Virol. 30, pp. 674-682.

THE USE OF CELL CULTURES TO ASSAY THE EFFECTS OF CHEMICALS ON BONE MARROW

Eugene A. Arnold, Wen-shing Liaw, and Paul O.P. Ts'o, Department of Biochemical and Biophysical Sciences (Division of Biophysics), School of Hygiene and Public Health, and Department of Pathology, School of Medicine, The Johns Hopkins University, Baltimore, Maryland, U.S.A.

ABSTRACT. Fresh bone marrow and liquid cultures of marrow were exposed to MNNG, 4NQO, MAMA, or PDD. MNNG was toxic in all systems; 4NQO was either toxic or stimulatory depending upon concentration and type of cell system; MAMA and PDD stimulated cell proliferation. Differentiation was directed to either myeloid or myelomonocytic pathways.

INTRODUCTION

Results from several laboratories (1-6) indicate that cell lines derived from hematopoietic or lymphoid tissues may be used to study the effects of chemicals on cellular differentiation or proliferation. The human myeloid leukemia cell line, HL-60, differentiates to granulocytes when exposed to dimethyl sulfoxide or actinomycin D or to macrophages when treated with tumor-promoting phorbol esters (1). Mouse erythroleukemic cells are also susceptible to induced differentiation (2) and Lotem and Sachs (3) have shown that cloned mouse myeloid leukemic cells may be induced to differentiate following incubation with 12-0-tetradecanoylphorbal-13-acetate. Choy and Littlefield (4) generated clones of human lymphoblastoid cells resistant to ouabain by the use of several mutagens while Penman, Wong, and Thilly (5,6) produced mutant human lymphoblasts resistant to 6-thioguanine by exposure of these cells to 5-halodeoxyuridines or methylnitronitrosoguanidine (MNNG).

The use of normal bone marrow cells for in vitro studies of chemical or drug interactions is much less than that of leukemic or lymphoblastoid cell lines. In several studies, Byron has used short term incubations of mouse marrow with drugs, hormones, or carcinogens to determine the effects and mechanisms by which these substances produce cell cycling in hematopoietic stem cells (HSC)(for review, see 7). The effects were assayed by determining losses of HSC in spleen colony assays following hydroxyurea or thymidine suicide. Recently, it has been demonstrated that tumor promoting phorbal esters stimulate hematopoietic colony formation in soft agar cultures (3,8). Phorbal and non-promoting esters were ineffective.

B. Pullman, P. O. P. Ts'o and H. Gelboin (eds.), Carcinogenesis: Fundamental Mechanisms and Environmental Effects, 379–388.

The recent development of liquid culture techniques for bone marrow (9) provides a new system for the study of the effects of different agents on cellular differentiation and proliferation. To the best of our knowledge these types of cultures have not been utilized for experiments with chemicals although Greenberger et al. (10) have reported the first induction of an acute promyelocytic leukemia in vitro using either Friend or Abelson murine leukemia virus.

In this paper, we wish to present some data from initial studies of the interaction of MNNG, 4-nitroquinolineoxide (4NQO), and methyl azoxymethanol acetate (MAMA) on normal marrow cells in vitro.

MATERIALS AND METHODS

Adult inbred Syrian hamsters (Bioline FID) were obtained from Trenton Labs, Bar Harbor, Me. All animals were 6-8 weeks of age.

Following ether anesthetization, the femurs were removed under sterile conditions, and the marrow contents both flushed and scraped out using a sterile needle and syringe containing either Fisher's or RPMI-1640 medium containing 20% horse serum, penicillin (200 units/ml) and streptomycin (200 μg/ml). The contents of the bones were pooled and inoculated at a density of 10^6 cells per ml into plastic tissue culture flasks. The cultures were incubated in a humid atmosphere of 5.5% CO_2 in air at 37°C. Feeding was accomplished by removing half the supernatant medium and suspension cells and replacing the removed medium with an equal volume of fresh medium.

Supernatant cells were removed from continuous cultures by the demi-depopulation procedure after the establishment of the adherent layer. These cells were then inoculated into fresh culture flasks at a density of 10^6 cells per ml. Fresh and conditioned medium were used for the initial inoculation. Thereafter, the cultures were fed in the same manner as continuous adherent layer cultures.

Fresh marrow cells, adherent layer culture cells, and supernatant culture cells were exposed to varying concentrations of 4NQO, MNNG, and MAMA for 2 hrs in serum-free medium in culture flasks at 37°C in the culture incubation. The chemicals were removed by washing with fresh culture medium and returned to the culture flasks or used in soft agar cultures. PDD was not removed from the cultures or marrow cells prior to culture.

Colony assays in 0.3% agar were performed using the method of Johnson and Metcalf (11). Cells were plated in 35 mm plastic petri dishes at $2x10^4$ or $4x10^4$ cells per dish and then exposed to hamster spleen conditioned medium (12) (50 microliters per dish). After 7 days incubation, the colonies were evaluated in situ following benzidine staining. Colonies (> 40 cells) were counted and evaluated using an inverted phase microscope.

Conventional light microscopy of cell suspensions was carried out on Wright's stained cytospin preparations. Cells were enumerated using a TOA automatic cell counter.

RESULTS

Fresh marrow cells were incubated with MNNG, MAMA, and PDD and then evaluated for colony formation in soft agar using spleen-conditioned medium as the source of growth factors. Since variability of growth promoting activity may occur with different batches of conditioned medium, a single preparation of spleen conditioned medium was used which had been tested with normal marrow cells for linearity of colony stimulation. Colony stimulation was linear for cell inputs between $2x10^4$ to 10^5 cells per dish and for additions of conditioned medium from 20 to 200 microliters. Non-mixed colony stimulation (committed precursors) increased linearly with either increased cell density or medium concentration. Mixed colony formation (pluripotent stem cells) increased linearly with increasing cell density but was constant for all concentrations of added medium indicating maximal stimulation by the preparation used.

Colony growth was decreased for all concentrations of MNNG tested (Fig. 1). The loss of colony-forming cells followed a dose-response relationship with total absence of colonies at concentrations of MNNG exceeding $2x10^{-6}$M. Mixed colonies were reduced to 1-4 percent of controls at $5x10^{-7}$M and surviving colonies at higher concentrations were either myeloid or monocytic. Benzidine-positive colonies (erythroid) or megakaryocytic colonies were not observed. At $2x10^{-6}$M surviving colonies were markedly reduced in cell number with some growth being in the cluster category (< 40 cells). Toxicity was not observed in cultures treated with either MAMA or PDD (Fig. 1). At concentrations between 10^{-7} to 10^{-5}M, no significant stimulatory effect was noted in cells exposed to MAMA. At 10^{-4}M MAMA there was an observed stimulatory effect with the increased colonies being both mixed and myeloid in type. Colony size was also increased. PDD produced a marked enhancement of the stimulatory effect of the spleen conditioned medium at concentrations of 0.01 micrograms per ml or higher. The colonies formed were primarily monocytic with some increase in myeloid forms also. Colony size was increased as in the observed MAMA effect.

Treatment of continuous cultures with 4NQO produced two separate observed effects. Following exposure of the culture at 28 days *in vitro* there was a sustained or increased cell density for the period of observation to 60 days culture (Fig. 2). This response of the culture to 4NQO did not appear to be dose-related as no significant differences were found at concentrations between 10^{-4} to 10^{-7}M. Examination of cytospin preparations of the treated cells revealed a significant difference in the type of cells in the culture. Normal controls in these cultures displayed a marked predominance of monocytic forms whereas the 4NQO treated cells were as much as 68% myeloid even at 60 days

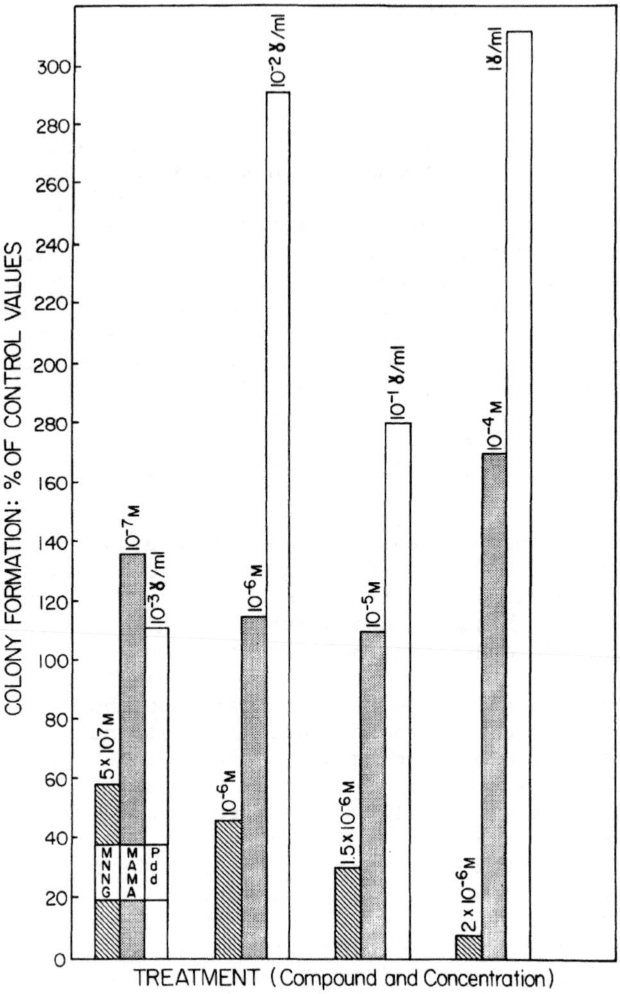

Fig. 1. Colony growth in the presence of spleen conditioned medium after exposure of bone marrow to various chemicals. One hundred percent equals control number. Concentrations are given at the top of the bar.

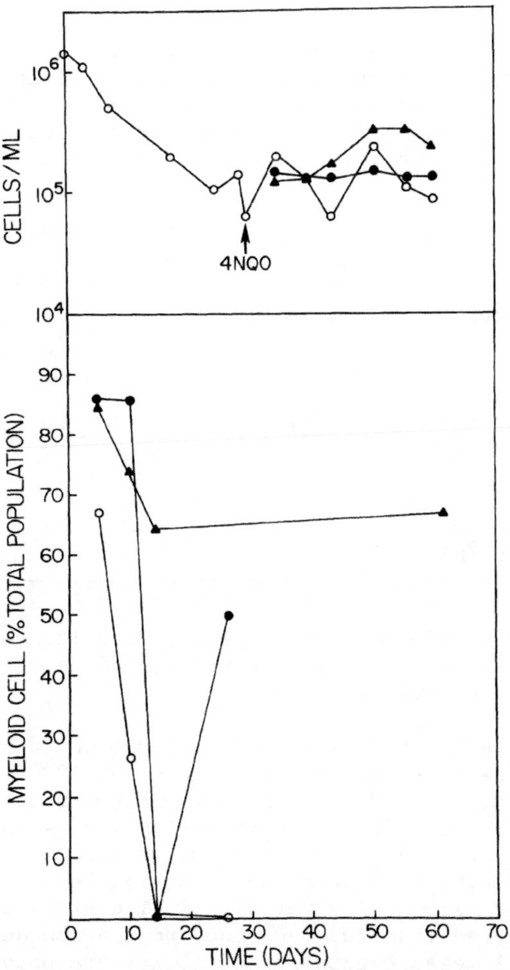

Fig. 2. Properties of adherent cell cultures after exposure to 4NQO. Top panel is cell density at various periods after exposure. Arrow indicates exposure. Bottom panel presents the differential morphology in percent of total cell population. Open circles equal control, solid circles equal 10^{-5}M, solid triangles equal 10^{-7}M 4NQO.

post-treatment. The sustained cell density and the differential percentage indicate a marked absolute increase in myeloid cells produced by 4NQO.

In cultures tested for colony formation, the control cultures displayed a loss of precursors at 17 days to a figure of 2 colonies per 10^5 cells and 0 colonies at 33 days (Table 1). 4NQO treated cultures, however, maintained a significant precursor population at either time period. The colonies observed were primarily myeloid.

TABLE 1. Colony Formation by Cultured Marrow Cells After Exposure to 4NQO

Treatment Group	Days After Treatment	Colony Number (per 10^5 cells)
Control		2
4NQO(10^{-6}M)	17	198
4NQO(10^{-7}M)		206
Control		0
4NQO(10^{-5}M)	33	108
4NQO(10^{-7}M)		43

The effects of treatment of supernatant cultures produced results that were different than those observed for either fresh marrow cells or adherent layer cultures.

Cultures were treated with 4NQO at three concentrations and then observed for 11 days (Table 2). At a level of 10^{-4}M, 4NQO produced death of all cultures by day 6. This is in contrast to adherent layer cultures where cell densities and viability were maintained to 60 days of observation. At 10^{-5}M 4NQO there was a steady decline of cell density to day 6 with proliferation occurring between days 6 and 11 to a cell density of $2x10^5$ cells per ml. Cell density changes in this table are calculated from the decreased population density occurring at feeding since all measurements and cytology were obtained from the cells obtained during the demi-depopulation procedure. The cell increase over the 5 day period was $1.5x10^5$ cells per ml. Cytologically, however, the cell population was predominantly monocytic. Colony formation was also markedly suppressed in contrast to the situation in adherent layer cultures. At the lowest concentration tested (10^{-6}M) decline in cell population and density paralleled the control cultures except that no loss in cells occurred during the three day period following the demi-depopulation at day 3. Proliferation led to an increase in cell density of $3.0x10^5$ cells per ml between days 6 and 11. Colony formation was maintained and mixed colonies were twice control levels at day 6. The morphology of the cells paralleled the control cultures with myeloid, monocytic, megakaryocytic, and blast cells being present. No myeloid predominance was seen.

TABLE 2. 4NQO Effect on Supernatant Cultures

Days After Exposure	Treatment Group	Cell Density (Cells/Mlx10^{-5})	Change in Cell Density	Total Colonies %	Mixed Colonies %
	Control	10	-	100	100
0	10^{-4}M	10	-	1	1
	10^{-5}M	10	-	8	0
	10^{-6}M	10	-	117	133
	Control	4.5	-5.5×10^{5} (per ml)	100	100
	10^{-4}M	4.0	-6.0×10^{5}	0	0
3	10^{-5}M	3.0	-7.0×10^{5}	9	0
	10^{-6}M	4.0	-6.0×10^{5}	112	74
	Control	2.0	-2.5×10^{4}	100	100
	10^{-4}M	99% non-viable		0	0
6	10^{-5}M	1.0	-5.0×10^{4}	4	0
	10^{-6}M	2.0	0	109	200
	Control	4.0	$+2.0 \times 10^{5}$		
11	10^{-5}M	2.0	$+1.5 \times 10^{5}$	-	-
	10^{-6}M	4.0	$+3.0 \times 10^{5}$		

TABLE 3. MAMA Effect on Supernatant Cultures

Days After Exposure	Treatment Group	Cell Density (Cells/Ml$\times 10^{-5}$)	Change in Cell Density	Total Colonies %	Mixed Colonies %
	Control	10	-	100	100
0	10^{-3}M	10	-	64	50
	10^{-4}M	10	-	108	215
	Control	4.0	-6.0×10^{5} (per ml)	100	100
3	10^{-3}M	3.0	-7.0×10^{5}	93	125
	10^{-4}M	5.2	-4.8×10^{5}	105	100
	Control	4.0	$+2.0\times 10^{5}$	100	100
6	10^{-3}M	1.0	-5.0×10^{4}	127	400
	10^{-4}M	3.0	$+4.0\times 10^{4}$	139	100
	Control	3.0	$+1.0\times 10^{5}$		
11	10^{-3}M	4.0	$+4.5\times 10^{5}$	-	-
	10^{-4}M	10.4	$+12.5\times 10^{5}$		

Two concentrations of MAMA were used in supernatant cultures and the results were markedly different than seen in the 4NQO experiment (Table 3). MAMA at 10^{-3}M produced an initial decline in colony formation followed by a rise to a level where four times the number of mixed colonies (compared to controls) was observed on day 6. At 10^{-4}M, MAMA sustained colony formation and produced a 12-fold greater proliferation rate between days 6 and 11 when compared to controls. At 10^{-3}M, the proliferation rate during this period was 4-5 times the control level. At all periods, the morphology of the treated cell populations was the same as control cultures with no predominant type of cell.

Treatment of supernatant cultures with PDD produced no significant differences when compared to controls (data not shown).

DISCUSSION

The effects of a single short term incubation of bone marrow cells with the chemicals used in these experiments is perhaps best illustrated by the 4NQO data. Although the nominal incubation period was two hours, the actual period of effective interaction was probably much less since 4NQO is stable for only 30 minutes at 37°C (13). In the adherent layer cultures, this short exposure resulted in a sustained maintenance of cell density with an absolute stimulation to myeloid differentiation and the enhancement of colony formation for 33 days post-treatment. No dose response was evident. The effects on supernatant cultures was different in that toxicity was observed at 10^{-4}M and the patterns of differentiation and proliferation at lower concentrations varied from the adherent layer cultures. Although we have no data on 4NQO mechanisms in marrow cells, it should be noted that adherent cultures have large numbers of medullary stromal elements which are not present in supernatant cultures. Since these stromal elements may have regulatory function (9) it may be that the differences observed are due to 4NQO interaction with stromal cells.

MAMA produced proliferative effects in the systems examined with no evidence of preferential differentiation. Byron (7) has observed that MAMA triggers the HSC from G_o into S phase and found that the deacetylation of the parent compound could be blocked by cholinesterase inhibitors with a loss of triggering capacity. This suggests that the carcinogenic activity of MAMA is mediated by interaction of the compound with a drug receptor site. Studies are currently underway to determine if MAMA is either mutagenic or carcinogenic in our bone marrow test systems.

MNNG displayed toxic effects on bone marrow with a dose response similar to that seen in other systems. We are currently investigating the use of this mutagen to produce mutants resistant to ouabain and 6-thioguanine.

The response of bone marrow cells was similar to that observed by others (3,8). PDD increased the response to colony stimulating factors in spleen conditioned medium in a fashion like the responses of mouse marrow cells to macrophage-granulocyte inducer (3) or to colony stimulating factor (8). PDD produced no significant effects in supernatant only cultures which suggests that PDD acts by stimulating medullary stromal cells to produce growth factors or by modifying receptors on cellular elements retained in the adherent cultures but absent in supernatant cultures. A third possibility is that supernatant culture cells are maximally stimulated and show only slight responses to phorbal ester action (Eastment et al., submitted for publication).

REFERENCES

1. Collins, S.H., Ruscetti, R.W., Gallagher, R.F., and Gallo, R.C.: 1978, Proc. Natl. Acad. Sci. USA 75, pp. 2458-2462.
2. Miao, R.N., Fieldsteel, A.H., and Fodge, D.W.:1978, Nature 247, pp. 271-272.
3. Lotem, J., and Sachs, L.:1978, Proc. Natl. Acad. Sci. USA 76, pp. 5158-5162.
4. Choy, W.N., and Littlefield, J.W.:1980, Proc. Natl. Acad. Sci. USA 77, pp. 1101-1105.
5. Penman, B.W., Wong, M., and Thilly, W.G.:1976, Life Sci. 19, pp. 563-568.
6. Penman, B.W., and Thilly, W.G.:1976, Somat. Cell. Genet. 2, pp. 325-330.
7. Byron, J.W.:1975, Exp. Hemat. 3, pp. 44-53.
8. Stuart, R.K., and Hamilton, J.A.:1980, Science 208, pp. 402-404.
9. Dexter, T.M., Allen, T.D., and Lajtha, L.G.:1977, J. Cell Physiol. 91, pp. 335-344.
10. Greenberger, J.S., Davisson, P.B., Gans, P.J., and Maloney, W.C.: 1979, Blood 53, pp. 987-1001.
11. Metcalf, D., and Johnson, G.R.:1979, J. Cell Physiol. 99, pp. 159-174.
12. Metcalf, D., and Johnson, G.R.:1978, J. Cell Physiol. 96, pp. 31-42.
13. Endo, H.:1969, RRCR. 34, pp. 32-52.

STRUCTURAL MODIFICATIONS AND THEIR EFFECTS ON THE GENETIC FUNCTIONS OF DNA GENERATED BY INTERACTIONS WITH BENZO(A)PYRENE METABOLITES

Tsuyoshi Kakefuda, Hiroshi Mizusawa, Che-Hung Robert Lee, Patricia Madigan, and Richard J. Feldman*
Laboratory of Molecular Carcinogenesis, National Cancer Institute, N.I.H., Bethesda, Md.,U.S.A.
*Division of Computer Research and Technology, National Institutes of Health, Bethesda, Md. U.S.A.

Abstract: The structural modification and its effects on the genetic functions of DNA induced by a benzo(a)pyrene metabolite, diol epoxide (DEBP) was studied. Covalent binding to DNA of less than 0.2% did not significantly alter the superhelical density of circular DNA as detected by the mobility of the modified DNA in gel electrophoresis. Molecular coordination models constructed by computer graphics technology and CPK models showed that DEBP covalently bound to the 2 amino group of guanine residue oriented diagonally, about 30 degrees to the axis of the DNA helix and did not necessarily generate a drastic torsional effect. An increased amount of covalent binding, however, may cause a cumulative microenvironmental change in the minor groove of the DNA helix, which becomes suitable for other forms of interaction with DEBP, such as binding with adenine residues. Strand scission and spontaneous adduct elimination occurred only when the winding angle of DNA was forced to change. These findings led us to propose a multiphasic, synergistic mechanism of modification generated by DEBP interaction with different structural components of DNA. The BP metabolite with hydroxylation at the 7,8 position was found to be capable of intercalating DNA. Intercalation of DEBP covalently bound to the guanine residue is, however, unlikely to occur because of highly restricted molecular configuration. A single molecule of DEBP covalently bound to pBR322 DNA is enough to block the elongation of the polynucleotide chain during replication in vitro. Our new method of plasmid mediated mutation in E.coli allowed us to analyze the effect of a small number of DEBP covalently bound to DNA on mutation, with virtually no toxic effect on the host cells. The large quantity of mutated DNA which were cloned from the transformed cells will be useful for the analysis at a single molecular level of the mechanism of mutation caused by chemical carcinogens.

I. INTRODUCTION

Benzo(a)pyrene (BP) is a major chemical environmental pollutant produced as a pyrolysis product of fossil fuels, cigarette

B. Pullman, P. O. P. Ts'o and H. Gelboin (eds.), Carcinogenesis: Fundamental Mechanisms and Environmental Effects, 389–407.

smoking and industrial byproducts. It remains in the atmosphere, soil, water and common food chains with a high degree of stability. Statistical studies indicate that the recent increase in the incidence of lung cancer may be attributed primarily to this polyaromatic hydrocarbon through bodily uptake.

BP is converted to a number of metabolites by a microsomal enzyme system; the mixed function oxidases containing cytochrome P450 and epoxide hydratase. Among the variety of BP metabolites produced enzymatically, those metabolites with epoxidation at the 2,3-, 4,5-, 7,8-, and 9,10-positions have drawn considerable attention because of their potent chemical reactivity with biological macromolecules. Some epoxides such as 2,3-epoxide are highly unstable and open nucleophilic sites directly, other epoxides are further converted by the epoxide hydratase to form either dihydrodiols by addition of water or to produce nucleophilic sites.

The (±) r-7,t-8-dihydroxy-t-9,10-oxy-7,8,9,10-tetrahydrobenzo-(a)pyrene (DEBP)[24], which will be discussed in this chapter is a BP metabolite which is formed enzymatically and stereospecifically by subsequent oxygenation of trans-r-7,t-8-dihydroxy-7,8-dihydro BP (7,8 diol BP) at the 9,10-positions.[24,26,32,33] DEBP has been shown to be a potent mutagen and is also believed to be an ultimate carcinogen form of BP.[7,22,24]

The development of synthetic isomers of diol-epoxides allowed us to investigate the interaction of isomeric DEBP interacting with DNA <u>in vitro</u> as well as <u>in vivo</u>. Major covalent binding occurs between the carbon 10 position of DEBP and the 2-amine group of guanosine residues.[10,25,30] Binding to the N-7 position of guanosine[23] and to the N-6 position of adenosine[9,25], and to the phosphate groups[4] of DNA have also been identified. Binding to the highly polymerized DNA double helix and the subsequent conformational changes were investigated extensively. The following phenomena have been associated with covalent binding of DEBP: strand breakage[4,12,13], reduction in superhelical densities of circular DNA[4,5,12,13], local denaturation[12,13], and covalent binding of DEBP with[5,17] or without intercalation.[15] However, information available in the literature is inadequate for clarification of the interrelationship between these various types of structural modification and their effects on the genetic functions of DNA. The molecular mechanisms involved in generating mutation and carcinogenesis are not fully understood.

This chapter describes several different forms of DNA modifications induced by covalent and non-covalent interactions of DEBP and other closely related BP metabolites. <u>In vitro</u> model systems which have allowed us to study the effects of DNA binding on DNA replication and mutation will also be described.

II. MODIFICATION OF DNA CONFORMATION INDUCED BY COVALENT BINDING WITH DIOL-EPOXIDE I: A HYPOTHESIS OF TWO STEP MODIFICATION

The major target site of covalent binding of DEBP to double stranded DNA is the 2-amino group of guanine residues (about 90% of total binding).[10,25,30,31] This is mainly due to the strong chemical reactivity of DEBP with the exocyclic amino group which can rotate freely in the minor groove of the DNA helix. Although the adenine residue also possesses an exocyclic amino group at the N6 position, the covalent binding with this DEBP is only one tenth of that with the 2-amino group of guanine.[9,12,13,25,30,31] The lesser degree of binding with 6-amino adenine has never been adequately explained.

In the present study, we found no detectable changes in superhelical configuration of pK0482 DNA (3.15×10^6 daltons, 4500 bp) at a low level of covalent binding, as noted by the degree of retardation in gel electrophoresis of DNA which are covalently bound to less than 10 molecules of DEBP (Fig. 1).

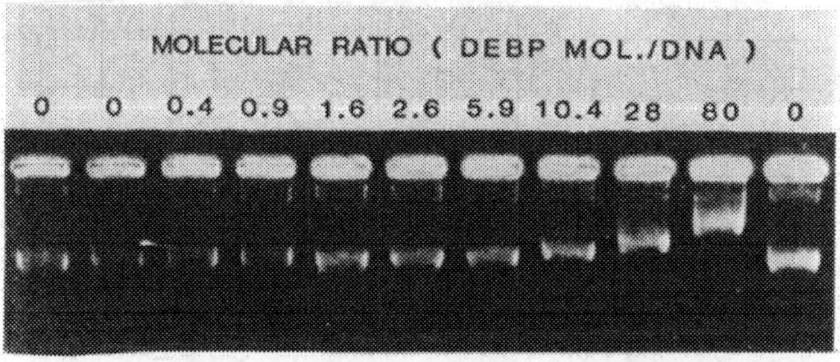

Fig. 1 Agarose gel electrophoresis (1%, 3.5 v/cm, room temperature) of plasmid pK0482 DNA covalently bound to different numbers of ^{14}C-DEBP molecules. The molecular ratio was determined by the radioactivity of ^{3}H-DEBP covalently bound to DNA. The concentration of DNA was determined by fluorometry (excitation 355 nm; emission 455 nm) after staining with bisbenzimidazole (Hoechst H 33258)[1] and uv absorption (254 nm).

Under this level of DNA modification, covalent binding to residues other than guanine is negligible. The rate of retardation in gel electrophoresis was linearly increased when the total number of DEBP molecules bound to DNA exceeded ten.

Space filling models, as well as the molecular coordination of the DNA helix and DEBP constructed by computer graphics technology were studied (Figs. 2 and 3). These observations demonstrated that the C-10 position of DEBP molecules when covalently bound to the 2-amino group of guanine will give rise to two possible alternative positions of the DEBP molecules. The pyrene moiety of DEBP in the minor groove of the DNA helix was held diagonally, about 30 to 40 degrees to the long axis of DNA, with a highly restricted movement in the minor groove (Fig. 2). This molecular orientation was consistent with the previous observation by Weinstein et al, which was based on linear and circular dichromism studies.[30,31]

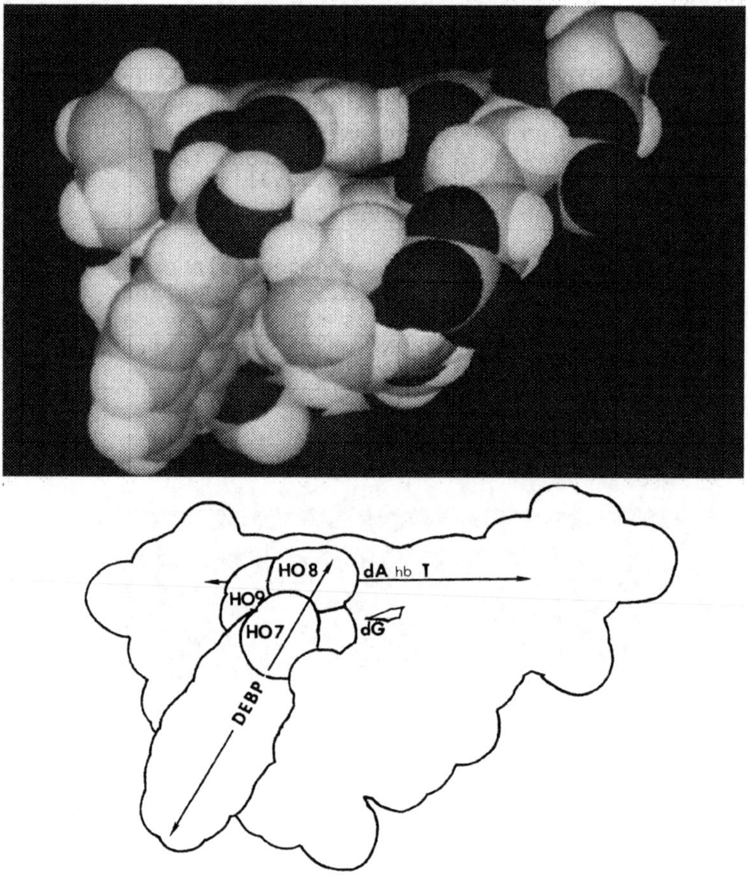

Fig. 2 Three base pairs of DNA linked with the 2-amino group of guanine residue constructed by computer graphic technology. The horizontal line at the top indicates the plane of adenine-thymine base pair that is perpendicular to the long axis of the DNA helix. The carbon-10 position of DEBP (invisible from this front view) is linked with the guanine residue that is located in the middle of this model. The plane of the flat DEBP moiety is about 30 degrees to the long axis of the DNA helix. "hb" is a hydrogen bond between deoxyadenosine and thymine residues.

Torsional stress generated on the DNA helix by the covalent binding of a DEBP molecule to the 2 amino group of the guanine residue was found to be minor and did not generate a drastic change in the winding angle of the helix. However, an increasing amount of

DEBP binding may result in a moderate reduction of winding angle of the helix because of the cumulative effect of microenvironmental changes caused from the bulky, aromatic pyrene rings residing in the minor groove. We assume that binding of up to 10 molecules of DEBP per SV40 DNA (5100 base pairs) or pK0482 generates only a moderate degree of modification.

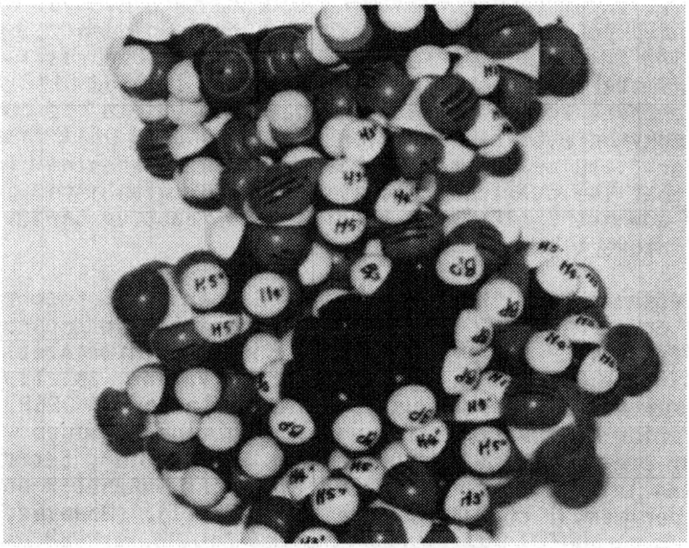

Fig. 3 A CPK atomic model of double stranded DNA covalently bound with a DEBP molecule similar to that of Fig. 2.

As a result of a slight change in the winding angle (in the direction which decreases the negative superhelical turn), the 6-amino group of adenine which normally resides deep in the major major groove becomes more exposed to the exterior. The accessability of the terminal ring. (C-10 position) of DEBP to the 6-amino group of the adenine residue is then greatly increased. The covalent linkage to the adenine residue, which is involved in base pairing with the thymine residues would presumably be associated with the disruption of the hydrogen bond. This would thus generate a torsional effect on the DNA strand which would cause a more drastic conformational change. Our previous study indicated that the bindings to adenine residues were preferentially hydrolyzed by endonuclease S1 which hydrolyzes the single stranded region of DNA.[12,13]

The key element of our proposal of two step DNA modification by covalent binding with DEBP is that the 6-amino group of the adenine residue whose stereoconfiguration in the double stranded DNA helix is not as favorable as the 2 amino group of guanine in terms of accessibility and interreactivity with DEBP which is also stereospecific. A sufficient amount of binding with the guanine residues will result in dehydration as well as electrostatic change generated by an increasing number of DEBP molecules residing only in the minor groove of the DNA helix. Such change in the microenvironment would thus open secondary target sites in the helix. This hypothetical concept is partially supported by the fact that there is a significantly increased proportion of covalent binding of DEBP with deoxyadenosine in single stranded polydeoxynucleotides. The ratio of deoxyguanosine to deoxyadenosine in the single stranded calf thymus DNA reacted with DEBP was approximately 1:1, and the proportion of labeled DEBP found as the deoxyadenosine adduct was 40%, compared to 10% in the double-stranded polydeoxynucleotide. Reaction with denatured DNA "randomizes" adduct formation (loss of stereoselectivity and more adenine binding).[18] This suggests that the chemical reactivity of DEBP with amino groups of guanine and adenine residues is equal if the reactive target sites were open for interaction.

During preparation of this manuscript, Gamper et al reported a similar result concerning DNA conformational change, but interpreted the result differently.[5] They suggested a possible intercalation of the BP moiety in the DNA helix. A similar observation was also reported by Drinkwater et al previously.[3] Intercalation of DEBP in the DNA helix prior to covalent binding with guanine residues was proposed.[18] Our present study indicated that 7,8-diol BP, tetrol BP, and triol BP can intercalate non-covalently in the DNA helix under an appropriate experimental condition (See Section III). However, we found that the carbon-10 position of the 7,8-diol BP or DEBP was a few Å away from the 2-amino group of guanine when the pyrene moiety was inserted in the position parallel to the plane of the DNA base pairs. (Note the distance between the carbon 10 position of tetraol DEBP and the 2 amino group of the guanine after intercalation in Fig. 7).

From these studies we concluded that DEBP might intercalate the DNA helix in transition. However, conversion to the chemically inactive tetrol BP rather than covalent binding with a DNA base may be a significantly greater chemical consequence which follows after intercalation. This hypothesis is supported by our previous experiments, in which ^{14}C-trans 7,8-diol BP was incubated with SV40 DNA for intercalation and subsequently treated with purified cytochrome P450. The covalent binding of enzymatically converted ^{14}C-DEBP was significantly less than that without preincubation, suggesting that the intercalation of DEBP, if it occurs, reduces the frequency of covalent binding with the target site of DNA (in preparation for publication).

The strand breakage which occurred in DEBP bound DNA is another important phenomenon noted.[4,12,13] It was recognized as an increasing number of nicked open circles and linear molecules of heavily modified (more than 0.1% of DNA bases) superhelical DNA in gel electrophoresis. The strand breakage was attributed to the formation of phosphotriester bonds or depurination followed by β-elimination.[4]

Although it is most unlikely that such heavy modification may take place leading to strand breakage in the cells, our recent study indicated that strand breakage can occur at a low level of DEBP modification when additional torsional stress is added to the DNA helix. Fig.4 shows SV40 DNA treated with ^{14}C-labeled trans-7,8-diol BP (7,8-diol BP), and cytochrome P450 and P450 reductase purified from rabbit liver.[1*] In Fig. 4B, the amount of ^{14}C-labelled 7,8-diol BP in the reaction mixture was twice as much as that of (A). The DNA extracted by phenol after the incubation was analyzed by ethidium bromide-CsCl density gradient centrifugation. The two peaks in the (A) represent closed circular (cc) and the nicked open circle (oc), plus linear DNA. The binding occurred about 3.5 molecules of DEBP per DNA in (B), which is about twice as much as that on (A), and shows only open circles and linear DNA molecules. The specific activity of ^{14}C-7,8-diol BP enzymatically activated and subsequently bound to ^{3}H-DNA is 10-fold higher in closed circles than nicked DNA.

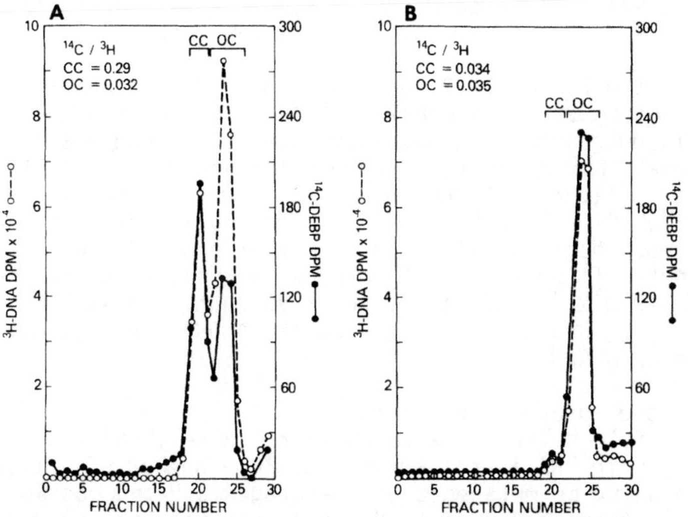

Fig. 4 CsCl-ethidium bromide density gradient centrifugation of superhelical SV40 DNA with ^{14}C-trans-7,8-diol BP and cytochrome P450. ^{3}H-SV40 DNA (90 μg) was incubated with the reaction mixutre (1ml) by a method similar to that reported by Deutsch et al.[2] [100 uM Tris-HCl pH 7.7, 15 umol MgCl, 1.13 umol NADPH, 0.6 nmol P450 LM4, 2.8 units NADPH cytochrome reductase, 60 μg sonicated dilauroylglyceryl-1,3-phosphorylcholine, and 10 nmol ^{14}C-trans-7,8-diol BP (11.9 Ci/mol) for (A)]. (B) received 20 nmol of ^{14}C-trans-7,8-diol BP. The DNA was extracted by phenol twice and analyzed in a CsCl-ethidium bromide density gradient centrifugation.

These findings indicated that the DNA strand is not only nicked, but the alkylated adducts were eliminated from the DNA stand as the binding number increased. Since the strand breakage did not occur at a low level of modification (Fig. 1), the unwinding force of the DNA strand from intercalation of ethidium bromide presumably generated the strand scission and elimination of DEBP adducts from DNA strands. Considering the dynamic changes constantly occurring on functioning DNA during transcription, replication, and chromosomal organization in the cell, the non-enzymatic nicking and spontaneous elimination of adducts may play an important role, particularly in the repair of alkylated DNA.

III. NON-COVALENT INTERACTION OF BENZO(A)PYRENE METABOLITES WITH DNA

In the simple model of metabolic activation of a drug, the parent compound converted to a relatively stable and non-reactive intermediate will accumulate in the cells more than reactive and unstable intermediates. A BP metabolic intermediate, 7,8-diol BP is a relatively stable compound as compared with DEBP; the half life of DEBP is known to be a matter of a few minutes.

In fact, the 7,8-diol BP has been isolated as a major metabolic intermediate after incubation of BP with hepatic microsomes or with cells possessing the necessary activating enzymes,[7,33] suggesting that this metabolic intermediate accumulated in the cells in a significantly large quantity. However, the biological activity of 7,8-diol BP without further metabolic activation is not well known. Nagao et al have shown that trans-7,8-diols was mutagenic to the TA100 strain of Salmonella typhimurium in the absenece of S-9 mix.[22] Hsu et al reported that trans-7,8-diol preferentially binded to single strand ϕX174 DNA, but not with double stranded DNA.[8]

Whether the mutagenic activity found is a result of a further metabolism in Salmonella cells or whether trans-7,8-diol BP interacts directly with DNA is not known. Since a number of intercalating drugs are known to be mutagenic or co-mutagenic,[6] the possibility of intercalation of trans-7,8-diol BP was our particular interest.

An elegant study has been carried out by Keller using DNA relaxing enzyme to determine the number of superhelical turns of circular DNA in the presence of ethidium bromide.[14] The function of DNA relaxing enzyme is to convert superhelical DNA to the relaxed form by a repetitive nicking and sealing process. The presence of an inceasing concentration of intercalating drug reduces the average rotation angle of the helix, which, in turn results in aquiring negative superhelical turns after removal of the intercalated molecules. The number of acquired turns is determined by the number of molecules initially intercalated during relaxation.

	BP METABOLITES	COVALENT BINDING	NON-COVALENT INTERACTION
PHENOL	1-OH	–	–
	2-OH	–	–
	3-OH	–	–
	5-OH	–	–
	6-OH	–	–
	8-OH	–	–
DIOL	t 4, 5 DIOL	–	–
	c 4, 5 DIOL	–	–
	t 7, 8 DIOL	–	+
	t 9, 10 DIOL	–	–
TRIOL	t 7, 8, 9 TRIOL	–	+
TETROL	t 7, 8, 9, 10 TETROL	–	+
	c 7, 8, 9, 10 TETROL	–	+
DIOL-EPOXIDE	t 7, 8 DIOL 9, 10 EPOXIDE	++	*
	c 7, 8 DIOL 9, 10 EPOXIDE	++	*
EPOXIDE	4, 5 EPOXIDE	+	–
	7, 8 EPOXIDE	+	–

Table 1. Intercalation of BP metabolites to SV40 DNA assayed by DNA relaxing enzyme system (see text and ref. 14). *DEBP had potential intercalating activity similar to 7,8-diol BP if any portion of the molecule was not covalently bound to DNA. Covalent binding prevents the free insertion of pyrene moiety between base pairs. Torsional effect on the DNA strand made the assay for intercalation difficult in this experimental system. Parent BP and other metabolites were not sufficiently soluble in aqueous enzyme assay mixture.

A

B

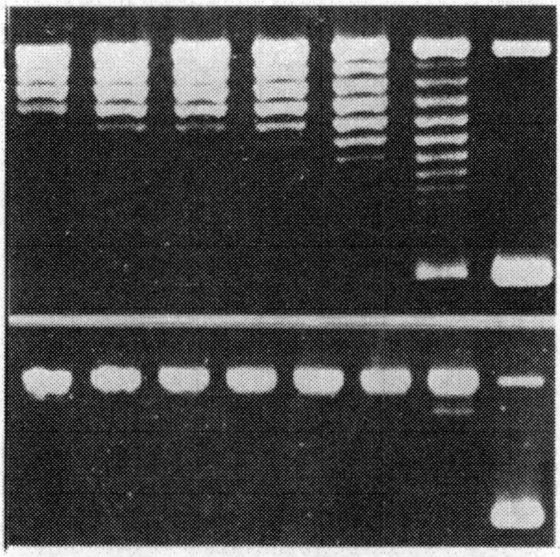

Fig. 5 Agarose gel electrophoresis of SV40 DNA incubated with DNA relaxing enzyme in the presence of either 7,8-diol BP (A) 7,8-epoxide (B). The reaction mixture (27 μl) contained 20 mM Tris-HCl pH7,8, 2 mM $MgCl_2$, 2.3 units enzyme, 2.3 μg DNA. The reaction as quenched by addition of SDS or phenol extraction of DNA with no difference in result. Concentraitons of metabolites from right to left are: 306 μM (no enzyme), 306 μM, 153 μM, 77 μM, 38 μM, 19 μM, 10 μM.

We examined BP metabolites using this method and the results are shown in Figs 5A and B and Table 1. All of the phenols examined showed no DNA interaction. BP with epoxidation at the 4,5-, 7,8-, (Fig. 5A) and 9,10- positions and the 9,10-dihydrodiol also showed no intercalation. When trans-7,8-diol BP was added in the concentration range from 10 μM to 300 μM in the enzyme reaction mixture, the agarose gel electrophoresis of the DNA showed a ladder formation as shown in Fig.5B. Since the nicked ends can be joined and sealed by the enzyme only after an integral number of 360 degree turns around the opposite strand, the difference in the topological winding number appears as the rungs of ladder formation rather than smearing. This is a clear difference from a DNA covalently linked to DEBP (Fig. 1 MR 28 and 80), in which the change in winding angle of DNA was not determined by intercalating activity. This will be discussed in a later section. The average values of topographic winding numbers are proportional to the number of intercalated molecules in the DNA helix. This value should be in a narrow range if the number, size, and topological orientation of the intercalating molecule are the same. Minor differences in the number of superhelical turns appear as a Gaussian like distribution (Fig. 6). The position of the Gaussian center shifted from the positions of the relaxed molecule (OC) to that of the superhelical (CC) DNA (from top to bottom of the gel) as an increasing amount of trans-7,8-diol BP is added in the reaction mixture. Interestingly, similar experiments with trans- and cis-7,8,9,10-tetrahydroxy-tetrahydro BP (tetrol) and trans- and cis -7,8,9-trihydroxypentahydro BP (triol BP) with the same range of concentration showed an identical ladder formation as with 7,8-diol BP. These results suggest that dihydroxylation of the terminal ring of BP, particularly at the 7,8-position play an important role in intercalation. Additional hydroxylation at the 9 and 10 positions are optional, but not essential for intercalating activity. Hydroxylation or epoxidation only at the 9,10-position failed to provide the intercalating activity.

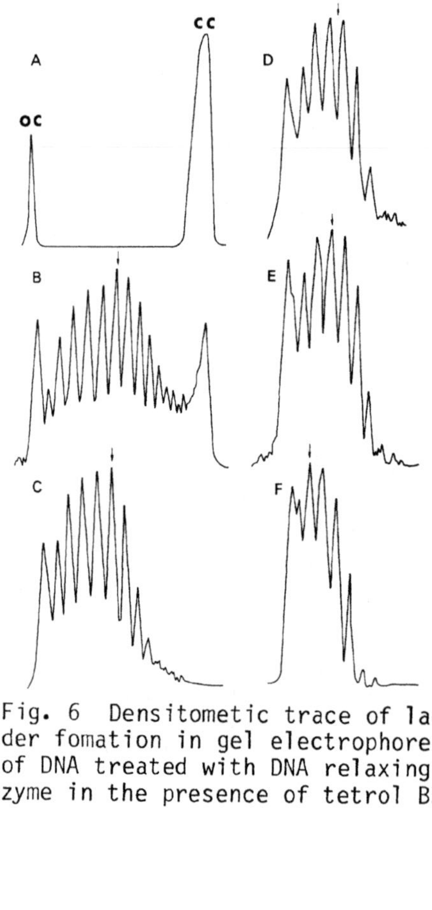

Fig. 6 Densitometic trace of ladder fomation in gel electrophoresis of DNA treated with DNA relaxing enzyme in the presence of tetrol BP.

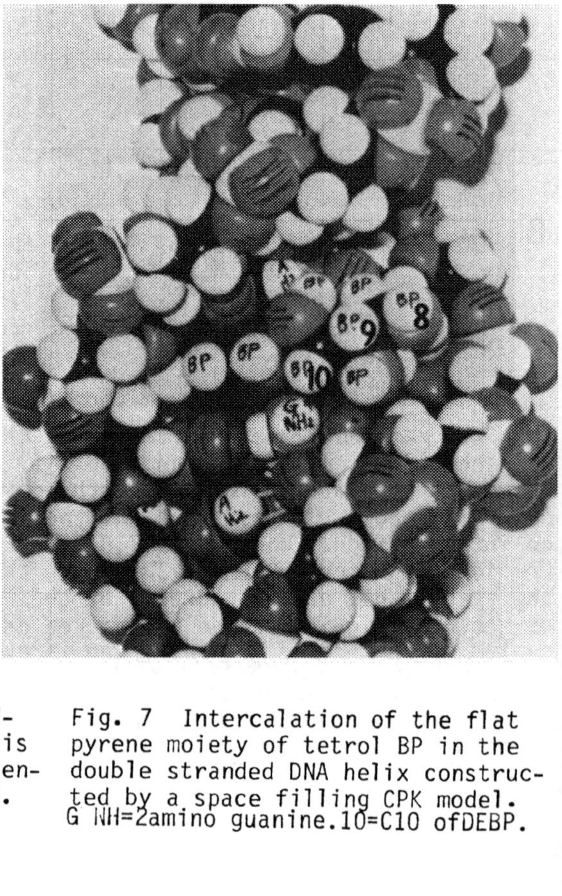

Fig. 7 Intercalation of the flat pyrene moiety of tetrol BP in the double stranded DNA helix constructed by a space filling CPK model. G NH=2amino guanine.10=C10 ofDEBP.

These results suggest that the portions which are actually inserted into the DNA helix are aromatic benzorings of pyrene because this portion of 7,8-diol BP, tetrol BP, and triol BP shares a common angular shape, size, and hydrophobic nature, providing an identical degree of rotational change (winding angle) in the DNA helix after intercalation. The hydroxylated terminal ring presumably stays outside and is closely oriented to the sugar-phosphate backbone of the helix (Fig. 7). A space-filling model produced by computer graphics corrdination techniques supported the feasibility of this postulated model of intercalation.

Insertion of the pyrene moiety of DEBP in the DNA helix is also likely to occur[5,18] because it shares the identical molecular configuration as 7,8-diol BP. However, covalent binding with the 2-amino groups of the guanine residue (with C-10 position of DEBP) requires almost complete repositioning of the pyrene moiety as described in Section II.

The intercalation of trans-7,8-diol BP to DNA, however, does not explain why tetrol BP, which also intercalates is not a mutagen.[22] The intracellular concentration, turnover rates and enzymatic activation, and detoxification of these metabolic intermediates may be involved in the complex biological processes which lead to mutation and malignant transformation of the cells.

IV. INHIBITION OF DNA SYNTHESIS IN VITRO BY BINDING OF DIOL-EPOXIDE I BENZO(A)PYRENE

Available evidences suggest that the synthesis of DNA is inhibited at the sites of DNA modification induced by UV irradiation and mutacarcinogen bindings such as AAF,[31] BP [8,19] and other alkylating agents. Such effect may be lethal to living cells as a result of the loss of template activity of DNA.

In order to understand the mechanism of the inhibition of DNA synthesis by DEBP binding, we established an in vitro assay method in which the inhibitory effect of the binding of DEBP on DNA replication could be analyzed quantitatively and qualitatively. pBR322 DNA is a double stranded superhelical DNA having the origin of replication and gene of B lactamase derived from Col El plasmid. It has 4362 base pairs and the molecular weight is 2.6 X 10^6 daltons. The pBR322 DNA replicates in vitro semiconservatively and completely using proper origins for replication.[19] The mode of replication and gene products required for replication are identical to that of living cells.[19,27,28,29] When unmodified DNA was incubated with crude cell extract of E.coli. with deoxy- and ribonucleoside triphosphates, ^{32}P-TMP was incorporated into pBR322 DNA linearly for at least 90 minutes. When DNA was covalently bound to DEBP in a molecular ratio (MR) of 1.4 per DNA and subjected to replication in vitro, the incorporation was reduced by about 30%. From MR=0 to

MR=5.9, the rate of reduction was linearly related to the numbers of molecules of DEBP bound to each pBR322 DNA. When MR=5.9, the initial incorporation (up to 20 minutes) was normal, but the total ^{32}P-TMP incorporation after 100 minutes was almost completely inhibited (Fig. 8).

Fig. 9 shows an alkaline sucrose density gradient analysis of the newly synthesized daughter strand under conditions where different numbers of DEBP were covalently bound to the template DNA strand. Under normal conditions (MR=0), the closed circular DNA appeared as a single peak with a skewed portion toward higher molecular weight in a neutral sucrose gradient (data not shown). The skewed portion represents closed circular DNA associated with newly synthesized DNA fragments. Unit sized single linear and single circular molecules banded at the middle of the alkali sucrose density gradient (fractions 30, in Fig. 9 15S). A peak which appeared at the top of the gradient represents the 6S initiation fragments dissociated from the origin of replication. The size classes of DNA synthesized during the first 30 minutes of incubation are identical and unrelated to the number of DEBP molecules bound to template DNA as observed in an identical pattern in Fig. 9 a, b, and c.

After 60 minutes of incubation, about 4 times more closed circular molecules had been synthesized from unmodified DNA than after 30 minutes (Fig. 9 a,d). When DEBP was bound to DNA at MR=1.4, the closed circular DNA synthesized was only doubled after an additional 30 minutes incubation (Fig. 9 b, e). When MR=5.8, the amount of closed circular DNA did not increase after an additional 30 minutes incubation (Fig. 9c,f). The number of unit-sized single-stranded linear or circular molecules (15S) markedly increased when intact DNA was subjected to replication. This class of DNA also increased as much as that in the control during 30 to 60 minutes incubation with DNA of MR=1.4. However, the peak at fraction 26-32 in Fig. 9e is markedly skewed toward the top. Moreover, there was no distinctive peak in regions of fraction 26-32 when MR=5.8. In contrast, the relative proportion of 6S initiation fragment synthesized did not change significantly in either control or modified DNA during the first 30 minutes incubation. These results indicate that DEBP binding effectively blocked the chain elongation with little effect on the initiation of DNA replication.

Note that pBR322 DNA, which carries the same base sequence as Col El DNA at the initiation site, contains a high proportion of adenine and thymine residues, which have only less than 10% of the chemical reactivity with DEBP in comparison to more than 90% of binding which usually occurs with guanine residues under these conditions. The chances of DEBP binding occurring at sites other than the guanine residue in the present experiment are very low. It is a reasonable assumption, therefore, that the initiation site of DNA replication is free from the binding of DEBP and capable of generating a normal initiation. The elongation of the

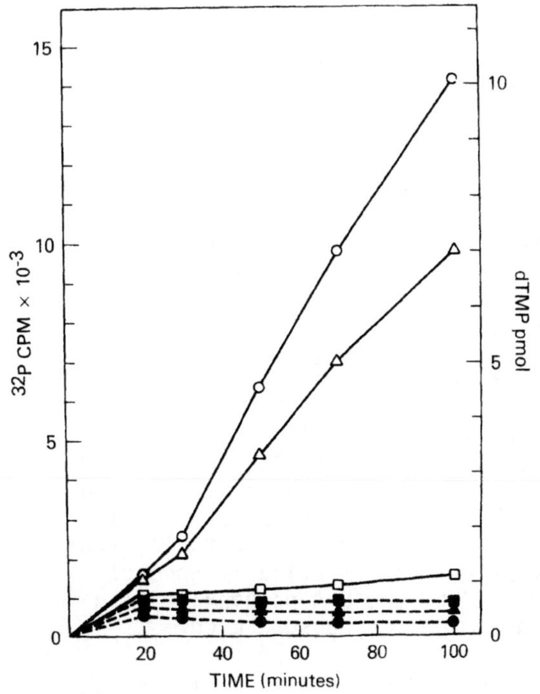

Fig. 8 Assay of incorporation of ^{32}P-TMP into pBR322 DNA. The DNA was treated with different concentrations of ^{14}C-DEBP. Number of DEBP molecules covalently bound was determined by ^{14}C counts and uv absorption (254 nm) and designated as molar ratio (DEBP mol./DNA). Assay mixture contained: [four 25 μM dNTP, four 200 μM rNTP, 25 mM potassium phosphate, 67 mM KCl, 7.5 mM $MgCl_2$, 2 mM spermidine, 2 μg pBR322 DNA, 0.2 μCi ^{32}P-TTP, and crude cell extract][19,27,28,29] in 200 μl. 30 ul aliquots were taken at time shown. Radioactivity of acid precipitate on filter papers was counted. (o-o MR=0), (△MR=1.4), (□-□MR=5.8). Discontinuous lines represent experiments carried out in presence of rifampicin, an inhibitor of initiation of DNA synthesis in this system.[27]

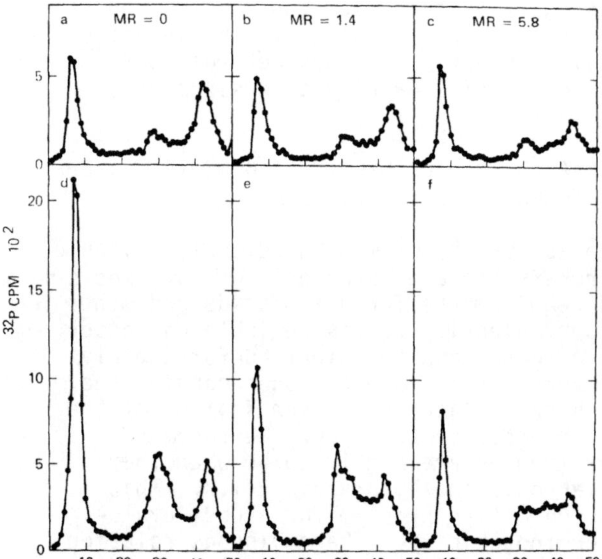

Fig. 9 Analysis of newly synthesized DNA products in alkaline sucrose density gradient. Assays were performed as described in Fig. 8 legend. Reaction was stopped by addition of 0.1 M EDTA and 1% SDS and chloroform-isoamyl alcohol mixture was added (24:1 V/V). After centrifugation (500 rpm, 30 min, 4 C) supernatant was dialyzed and added to 0.25 M NaOH. Samples were then applied to alkaline sucrose density gradient (5-20%). [a,d: control MR=0]; [b,e: MR= 1.4]; [c,f: MR=5.8]; a,b,c, = 30 minutes; d,e,f = 60 minutes of incubation for <u>in vitro</u> replication of DNA.

deoxypolynucleotide chain was effectively blocked at the binding site, presumably by the DEBP modified guanine residue, encountered by the moving DNA polymerase. Hsu et al have reported similar results regarding inhibition of the conversion of single stranded ϕX174 DNA to the double stranded replicative intermediate, which correlated well with inhibition of viral infectivity.[8] Moore also observed frequent blockage of chain elongation at binding sites with guanine as well as adenine in DEBP modified ϕX174.[21]

Some apparent questions raised in these experiments are: whether or not DNA polymerase stops at every binding site or can DNA polymerase "read through" the DNA sequence by bypassing some of the regions; or are some binding binding sites removed by an excision reair mechanism independently or coupled with DNA replication. In a series of experiments carried out in our laboratory, we found no detectable DEBP alkylated regions in the pBR322 DNA, which have completed a round of replication in vitro. This was assayed by observing the conversion of superhelical pBR322 DNA to nicked open circles after treatment with heat-alkali. The binding sites which are heat-alkali labile and present before replication in the parent strand (MR=1.4) were no longer detectable after a round of replication.[19] This implies that practically none of the binding sites were "read through" by the DNA polymerase with the frequency detected by the biochemical methods we used.

We also found that nicotinamide mononucleotide, a ligase inhibitor, has no effect on either intact or DEBP bound DNA when added to the cell extract, suggesting that excision and subsequent ligation of DEBP binding region were not involved in the in vitro replication (data not shown). However, this does not rule out the possibility of excision repair of DEBP alkylated sites occurring in living cells.

V. PLASMID DNA MUTATION GENERATED BY DIOL-EPOXIDE I BENZO(A)-PYRENE BINDING AND EXPRESSED IN TRANSFECTED E.COLI CELLS

Somatic mutation can be assayed by examining several heritable phenotypic alterations in prokaryotic and eukaryotic cells. Resistance to certain drugs, and requirements for amino acids and other nutrients are commonly and conveniently used as reliable characteristic markers for detection of genetic alteration. Unfortunately, these assay methods provide very little information regarding the precise molecular event which take place in a given segment of the gene assigned for a specific phenotypic function. Furthermore, many potent mutagens and carcinogens are highly toxic when they are applied directly to prokaryotic and eukaryotic cells. This potent toxicity has been a prohibitive problem for proper evaluation of mutagenicity and carcinogenicity. The ambiguous relationship between the rate of cell survival and mutation frequency at a critical dose has been of major concern when the conventional back mutation assay system was used.

In addition to this, the extraordinarily unstable nature of DEBP in aqueous solution contributes to the complexity of dose-effect relationships since it rapidly converts to the chemically inert tetraol BP.

In order to overcome these problems, we have been examining a plasmid mediated expression in recipient E.coli.[20] This system is entirely free from direct toxic effects of DEBP to the assayed cell. Plasmid pKO482 DNA[2] (a kind gift from Drs. K. McKenney, H. Shimatake, and M. Rosenberg), which has only two dominant genes, β-lactamase (ampicillin resistancy; amp-r) and galactokinase (gal K) with its own promoter (482 promoter) (Fig. 10) was modified with a preselected number of DEBP molecules prior to transfection into E.coli. Transformants were selected by ampicillin resistance and mutations were analyzed by the altered expression of the gal K gene.

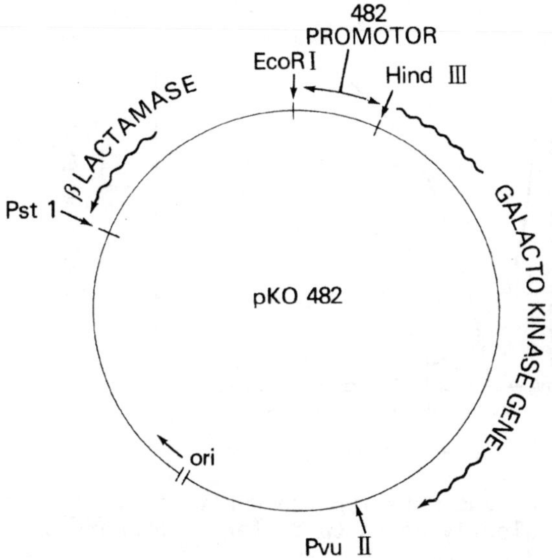

Fig. 10 Genetic map of pKO482 DNA.

In the present experiment, plasmid pKO482 DNA was incubated with different amounts of ^{3}H-labeled DEBP (obtained through the National Cancer Institute Carcinogenesis Research Program; 400 mCi/mmol). pKO482 DNA covalently bound with 0 to 20 molecules of DEBP were transfected to competent cells which were prepared by the method of Mandel and Higa.[16] The recipient E.coli cells used were N99 (gal K^-) and were sensitive to ampicillin. This strain grew as "white" colonies on MacConkey galactose agar plates without ampicillin because of a deficiency in galactose metabolism (gal K⁻). Because the pK0482 DNA contains a galactokinase structural gene and its own promoter, most of the cells grew as "red" conlonies when transfected by the plasmid. Only transfected cells can grow on the MacConkey galactose agar plate in the presence of ampicillin. Transfection efficiency in these experiments was 1×10^6 colonies per microgram of DNA when unmodified pK0482 DNA was used. The transformation by DEBP bound pK0482 DNA was determined by counting the number of "red" and "white" colonies.

As shown in Fig. 11, pK0482 DNA covalently bound to DEBP resulted in the marked reduction in transformation efficiency; about three molecules of DEBP per molecule of pK0482 DNA resulted in 37% transformation efficiency.

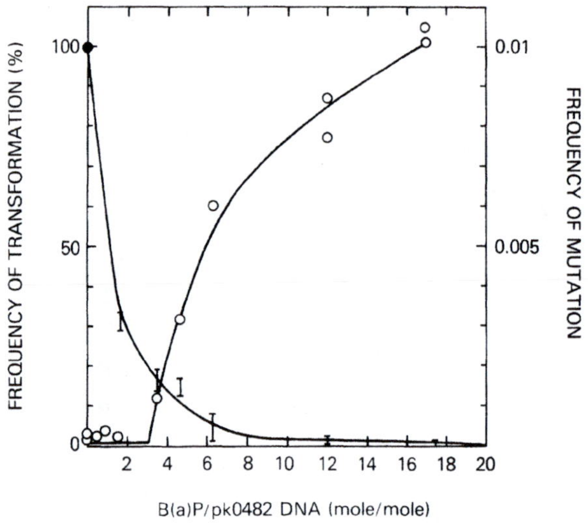

Fig. 11 Transformation and mutation frequency of E.coli strain N99 by pK0482 DNA covalently bound to different numbers of DEBP molecules. ●—● transformation: ○—○ mutation.

A mutation anywhere on the gal K region of the pK0482 will cause the resultant colonies to grow as "white" because of the lack of formation of the gal K gene in both the recipient E.coli and the plasmid DNA.

Since pK0482 transformants were selected by attained resistance to ampicillin (50 ug/ml), there was no selection for the expression of the galactokinase gene. The mutation frequencies were determined by counting the white colonies over the total number of red and white colonies (Fig. 11).

To insure that these mutants had occurred in the gal K gene of pK0482 DNA, closed circular plasmid DNAs were extracted from the mutated white colonies. These DNAs had the same molecular weight and restriction endonuclease Hpa II digestion pattern as the original pK0482 in an agarose gel electrophoresis, except for two exceptional cases (out of more than 30 colonies) in which several hundred base pairs were deleted from the gal K region of DNA.[20]

Retransformation by the mutated DNA isolated from the white colonies in N99 cells resulted in the formation of "white" colonies repeatedly, indicating that the mutation had been in the gal K gene of plasmid pK0482 DNA.

The maximum mutation frequency was about 2% with DEBP molecules bound to DNA. This 2% value is lower than that expected judging by the size of the gal K gene (about 40%) in pK0482 DNA and assuming that the binding of DEBP occurred randomly throughout the DNA strand. This discrepancy may be explained by the fact that most of the DEBP which bind to DNA lose their biological activity by the effective inhibition of replication, while unmodified DNA which was presumably present in the transfecting DNA material might replicate more quickly, resulting in a lower value of mutation frequency (white/red colony counts).

Potential excision repair of binding sites and more than two plasmid DNA in a recipient cell among which a normal plasmid DNA were present (including multimeric DNA; colonies will be "red" in these cases) can also contribute to a lower value of mutation frequency. Nevertheless, the rate of 2% mutation frequency is significantly higher than most of the back mutation assay systems available at the present time.

The plasmid mediated mutation assay system developed in the present study has several advantages over conventional back mutation assays: (1) the recipient cells are not exposed to toxic mutagens directly, therefore the cells are entirely free from toxic effects of mutagens. Thus, the survival rate of the cells is totally dependent upon the functional gene of β-lactamase introduced into the recipient cells by the plasmid; (2) The number and location of mutagen molecules covalently bound can be determined by the radioactivity of the mutagen and functional markers of plasmid DNA, facilitating the interpretation of the result to becoming more simple and straightforward; (3) The selection of transformed cells is carried out under conditions where genetic expression of the mutated gene (gal K) is not cross related; (4) The mutated plasmid DNA can be isolated in a large quantity, allowing us to examine the location and physical nature of DNA modification at a single base level; (5) The stability of mutation was able to be examined by repeated retransfection of the mutated DNA to the same recipient cells. The different rate of transformation and mutation frequency obtained in uvr A^- and rec A^- strains of E.coli will be described elsewhere.[20] This plasmid DNA mediated mutation assay can be extended not only to ultimate foms of muta-carcinogens, but can also be applied to parent compounds with purified P450 enzyme activation systems which are available in many laboratories.

ACKNOWLEDGEMENTS

*1. Purified cytochrome P450 and P450 reductase isolated from rabbit liver was a kind gift from Dr. M.J. Coon and his associates of the University of Michigan through Dr. H.V. Gelboin.

*2. We thank Drs. K. McKenney, H. Shimatake and M. Rosenberg for their kind gift of E.coli strain SA1943 carrying the pK0482 plasmid.

*3. We also thank Dr. H.V. Gelboin and his colleagues for helpful discussions and support.

REFERENCES

1. Brunk, C.F., Jones, K.C., & James, T.W.: 1979, Analy. Biochem. 92, pp. 497-500.
2. Deutsch, J., Leutz, J.C., Yang, S.K., Gelboin, H.V., & Chiang, Y.L.: 1978, Proc. Natl. Acad. Sci. U.S.A. 75, pp. 3123-3127.
3. Drinkwater, N.R., Miller, J.A., Miller, E.C. & Yang, N.C.: 1978, Cancer Res. 38, pp. 3247-3255.
4. Gamper, H.B., Tung, A.S-C., Straub, K., Bartholomew, J.C. & Calvin, M.: 1977 Science 197, pp. 671-674.
5. Gamper, H.B., Straub, K., Calvin, M. & Bartholomew, J.C.: 1980, Proc. Natl. Acad. Sci. USA 77, pp. 2000-2004.
6. Hayashi, K., Nagao, M. & Sugimura,T.: 1977, Nucleic Acid Research 4, pp. 3679-3685.
7. Huberman, E., Sachs, L., Yang, S.Y. & Gelboin, H.V.: 1976, Proc. Natl Acad. Sci. USA 73, pp. 607-611.
8. Hsu, W-T., Lin, J.S., Harvey, R.G. & Weis, S.B.: 1977, Proc. Natl. Acad. Sci. USA 74, pp. 3335-3339.
9. Jeffrey, A.M., Grzeskowiak, K., Weinstein, I.B., Nakamishi, K., Roller, P. & Harvey, R.G.: 1979, Science 206, pp. 1309-1311.
10. Jeffrey, A.M., Weinstein, I.B., Jennette, K.W., Grzeskowiak, K., Harvey, R.G. Autrup, H. & Harrix, C.: 1977, Nature 269, pp. 348-350.
11. Jennette, K.W., Jeffrey, S.H., Blobstein, F.A., Harvey, R.G. & Weinstein, I.B.: 1977, Biochem. 16, pp. 932-938.
12. Kakefuda, T. & Yamamoto, H.: 1978, Proc. Natl. Acad. Sci. USA 75, pp. 415-419.
13. Kakefuda, T. & Yamamoto, H.: 1978, Acad. Press, N.Y., San Francisco and London, pp. 63-74.
14. Keller, W.: 1975, Proc. Natl. Acad. Sci. USA 72, pp. 4876-4880.
15. Lefkowitz, S.M., Brenner, H.C., Astorian, D.G. & Clarke, R.H.: 1979, FEBS Letters 105, pp. 77-79.
16. Mandel, M. & Higa, A.: 1970, J. Mol. Biol. 53, pp. 159-162.
17. Meehan, T., Straub, K. & Calvin, M.: 1977, Nature 269, pp. 725-727.
18. Meehan, T. & Straub, K.: 1979, Nature 277, pp. 410-412.

19. Mizusawa, H. & Kakefuda, T.: 1979, Nature 279, pp. 75-78.
20. Mizusawa, H. & Kakefuda, T.: 1980, submitted for publication.
21. Moore, P. & Strauss, B.S.: 1979, Nature 278, pp. 664-666.
22. Nagao, M., Sugimura, T., Yang, S.K. & Gelboin, H.V.: 1978, Mutation Res. 58, pp. 361-365.
23. Osborn, M.R., Harvey, R.G., & Brookes, P.: 1978, Chem.-Biol. Interac. 20, pp. 23-
24. Sims, P., Grover, P.L., Swaisland, A., Pal, K. & Hewer, A.: 1974, Nature 252, pp. 326-327.
25. Straub, K.M., Meehan, T., Burlingame, A.L. & Calvin, M.: 1977, Proc. Natl. Acad. Sci. USA 74, pp. 5285-5289.
26. Thakker, D.R., Yagi, H., Lu, A.Y., Levin, W. Conney, A.H. & Jerina, D.M.: 1976, Proc. Natl. Acad. Sci. USA 73, pp. 3381-3385.
27. Tomizawa, J., Sakakibara, Y. & Kakefuda, T.: 1974, Proc. Natl. Acad. Sci. USA 71, pp. 2260-2264.
28. Tomizawa, J., Sakakibara, Y. & Kakefuda, T.: 1975, Proc. Natl. Acad. Sci. USA 72, pp. 1050-1054.
29. Tomizawa, J. Ohmori, H. & Bird, R.E.: 1977, Proc. Natl. Acad. Sci. USA 74, pp. 1865-1869.
30. Weinstein, I.B., Jeffrey, A.M., Jennette, K.W., Bolstein, S.H., Harvey, R.G., Harris, C., Autrup, H., Kasai, H. & Nakanishi, K.: 1976, Science 193, pp. 592-595.
31. Weinstein, I.B., Jeffrey, A.M., Leffler, S., Pulkrabek, P., Yamasaki, H. & Grunberger, D.: 1978, Academic Press, N.Y., San Francisco and London, pp. 4-30.
32. Yagi, H., Hernandez, O. & Jerina, D.M.: 1975, J. Am. Chem. Soc. 97, pp. 6881-6883.
33. Yang, S.K., McCourt, D.W., Roller, P.P. & Gelboin, H.V.: 1976, Proc. Natl. Acad. Sci. USA 73, pp. 2594-2598.

NEOPLASTIC TRANSFORMATION OF HUMAN MUTANT CELLS BY A TUMOR PROMOTER

Levy Kopelovich
Sloan-Kettering Institute for Cancer Research
1275 York Avenue
New York, New York 10021, USA

ABSTRACT

Neoplastic transformation is a multi-phase process apparently caused by carcinogens and subject to the influence of promoters. The naturally occurring phorbol esters such as 12-0-tetradecanoyl phorbol-13-acetate (TPA) are potent tumor promoting agents. We have studied the effects of TPA on human mutant fibroblast cell strains derived from individuals with hereditary adenomatosis of the colon and rectum (ACR), an autosomal dominant trait. We have previously demonstrated in these fibroblasts abnormal phenotypic expressions which often appear in transformed cells. In these studies, we have assumed that the ACR cell exists in an 'initiated state' due to a dominant mutation and that expression of the malignant state might only require treatment with a promoting agent. This simple experimental protocol provided a novel system for the study of cancer promotion _in vitro_. We have now demonstrated the growth _in vivo_ and growth properties _in vitro_ of human mutant cells exposed to TPA alone.

INTRODUCTION

It is generally believed that all forms of cancer are due to heritable and permanent changes in the cell genome (1-3). A view that considers tumor cells as an expression of a particular state of differentiation rather than a genetic variant has also been considered (4). Presumably, genetic and epigenetic mechanisms might be associated with both the initiation and maintenance of the malignant state (3). Malignant transformation is a multiphase process apparently caused by carcinogens and subject to the influence of promoters (2,3,5,6). A potent class of tumor promoting agents are the naturally occurring phorbol esters (7,8), such as TPA. Through the use of phorbol esters, a two-stage process of malignant transformation has been demonstrated in the mouse skin model (5-8) and more recently in cell culture systems (9-13). Studies _in vitro_ suggest that TPA reversibly affects terminal differentiation in certain model systems, and that its function is presumably

B. Pullman, P. O. P. Ts'o and H. Gelboin (eds.), Carcinogenesis: Fundamental Mechanisms and Environmental Effects, 409–416.

to increase the probability of expression of the malignant phenotype (14).

THE MODEL SYSTEM: EXPERIMENTAL APPROACHES

Our approach toward the elucidation of mechanisms associated with initiation and promotion in human cancer has been to study in detail an inherited form of cancer, adenomatosis of the colon and rectum (ACR). ACR is a disease in which numerous adenomatous polyps develop from the mucosa of the large intestine. The polyps vary in number and are distributed throughout the large bowel with high densities in the rectum (15-17). The close association of ACR with malignancy has been established in a large number of cases in which frank tumors presumably develop from the polyps. At present, it is believed that the disease is carried by an autosomal dominant gene (15-17). However, it seems probabl that additional genes may pleitropically modify its expression (15,17). The Gardner syndrome is considered a special variant of ACR, although no sharp distinction between the two has been made (16). ACR may be regarded as a disease in which there is a general tendency to develop both benign and malignant growths from normal flat epidermoid tissue of the colon and, in some instances, from extracolonic tissues. The list of tumors referred to as extracolonic is extensive and has been described elsewhere (17). Based on our experience and that of others (18), we think that autosomal dominant traits which predispose individuals to cancer truly reflect genetic information directly related to this disease An autosomal dominant pattern has also been recently recognized in famili aggregates predisposed to cancer. These comprise a large segment of all cancers reported (19,20).

As part of a study of human mutant cells in vitro, we have found that, while cutaneous biopsies of ACR patients and their progeny are apparently normal, the cultured fibroblasts were abnormal in several aspects of in vitro growth control, increased proteolytic activity, cell architecture, anchorage insensitivity, and susceptibility to further transformation by viral probes (Table I). These findings suggested a systemic disorder of stromal cells in ACR patients that might provide clues to cancer initiation and cancer promotion in man (21-24).

To date, no specific biological feature has been identified with the process of cancer initiation (14). Based on our results, we have assumed that the ACR cell exists in an initiated state due to a dominant mutation and expression of the malignant phenotype might presumably only require treatment with a tumor promoting agent (24). This provided a unique and uncomplicated system for the study of certain aspects of cance promotion in a human cell system in vitro. In this connection, an associ ation of the transformed state with a dominant mutation has been previous suggested (25,26).

STUDIES ON THE EFFECTS OF TPA ON HUMAN SKIN FIBROBLASTS

We have shown that treatment of sparse ACR cell cultures (about 0.25×10^2) with TPA (0 - 100 ng/ml) produced a biphasic (concave upward) dose-response curve with maximum inhibition in the range of 2 - 10 ng/ml of TPA (24). Experiments on thymidine labeling indices suggest that while the fraction of cells undergoing mitosis in exponentially growing cultures is not appreciably affected by TPA, the absolute number of cells and possibly the rate at which DNA synthesis proceeds might be responsible primarily for the effect seen with this compound. However, we have not been able to induce ornithine decarboxylase in this cell system (T. O'Brien and L. Kopelovich, unpublished). This effect was also demonstrated in cultures of normal human cell strain, but they were considerably more sensitive to the toxic effects of TPA. However, this effect was not seen in established normal rodent fibroblastic cell lines. It is of interest that the greatest toxicity due to TPA in our cell system coincided almost exactly with the maximum of SCE induction by TPA in rodent cell lines (John B. Little, personal communication). In contrast, addition of TPA to high-density ACR cell cultures (about 1×10^5 cells/cm^2) resulted in a dose-dependent stimulation of cell proliferation (E. Gansler and L. Kopelovich, unpublished).

The apparent discrepancy in the dose-response pattern of TPA may suggest a functional difference between sparse and confluent human cell cultures. It may also indicate that TPA affects at least two distinct processes of cell proliferation: one which is inhibitory to cell growth, but can be saturated at relatively low concentrations, and another at the higher concentration range which stimulates cell proliferation. Whether these results suggest the existence of at least two cell populations, each of which displays a distinct type of receptor for TPA, or a single cell population with at least two types of receptors for TPA, remains to be established. These seemingly diametrically opposed results may possibly be related to some, but not all, of the tumor promoting effects of TPA. The non-tumor promoter analogues of TPA, phorbol, and 4-O-methyl-12-O-tetradecanoyl-phorbol-13-acetate (24), gave a usual dose-response pattern characteristic of a drug with a single mode of action. It would be of interest to extend these findings to study the effects of TPA on metabolic cooperation (27) and differentiation (28,29) of human fibroblasts. If, as has been suggested, the dominant cancer trait occurs in a class of tissue differentiating genes (32), the elucidation of their mechanism of action in response to tumor promoters vis a vis differentiation would be important not only for the problem of cancer but also for the understanding of normal development.

The chronic application of TPA (100 ng/ml) to ACR cell cultures effected a change in cell morphology from the smooth fibroblastic-like to a dentritic-like cell. This effect was most pronounced at the early periods of exposure to TPA and largely disappeared in later passages. This amount of TPA has also effected an increase in growth rate, saturation density, plasminogen-dependent activity occurring in very dense foci, and

TABLE I

SYSTEMIC MANIFESTATIONS ASSOCIATED WITH LOSS OF REGULATORY CONTROL MECHANISMS AND BIOCHEMICAL ALTERATIONS IN HEREDITARY ADENOMATOSIS OF THE COLON AND RECTUM

Human Phenotypes	Phenotypic expressions in cultured human skin fibroblasts											
	Growth in nutrients deprived environment	Loss of contact inhibition	Formation of cell aggregates	Increased proteolytic activity	Deformed actin cables	Increased agglutination by lectins	Increased 2-deoxyglucose uptake	Decreased toxicity to TPA	Susceptibility to transformation by an oncogenic virus	Anchorage independence	Embryo specific proteins	Ability to form palpable nodules
The preneoplastic phase (multiple steps presumably due to a single mutation)												
The neoplastic phase (multiple steps)												
Clinically symptomatic	+	+	+	+	+	+	+	+	+	+	+	+
Clinically asymptomatic progeny												
a) pos.	+	+	+	+	+	ND	ND	ND	+	+	+	+
b) neg.	-	-	-	-	-	ND	ND	ND	-	(+)	(+)	(+)

TABLE I (continued)

	Growth in nutrients deprived environment	Loss of contact inhibition	Formation of cell aggregates	Increased proteolytic activity	Deformed actin cables	Increased agglutination by lectins	Increased 2-deoxyglucose uptake	Decreased toxicity to TPA	Susceptibility to transformation by an oncogenic virus	Anchorage independence	Embryo specific proteins	Ability to form palpable nodules
Normals	-	-	-	-	-	-	-	-	-	(+)	(+)	(+)

Vertical arrow indicates the transforming event; in this system, the transforming agents were KiMSV and SV40, as possibly TPA as well. The efficacy of virus transformation of cells from normal individuals (designated as (+) was considerably less than that from ACR individuals, but all transformed cells gave rise to the same phenotypic expressions. The clinically asymptomatic progeny has been subdivided into positive and negative according to our experimental findings.

and anchorage insensitive colonies. However, these effects by TPA did not necessarily occur concurrently; they were transient during consecutive passages, and they were variable for a given human cell strain during different periods of TPA application (23). Furthermore, the sensitivity to TPA following chronic exposure was both qualitatively and quantitatively similar to that observed in cells acutely exposed to this compound. Although agar growing cells showed a tighter clonal morphology, increased proliferation in monolayer cultures and in agar and a dose response to TPA considerably less concaved than shown by non agar growers, they could not be sustained beyond two passages in agar, whereupon they eventually senesced (23). This is in contrast to a TPA-induced permanent loss of anchorage dependence reported with stable initiated mouse epidermal cell lines in culture (31). Our results, put together from a large number of experiments, suggest a high degree of heterogeneity, possibly coupled with an adaptive response (30), and/or reversibility during selection of human cells in the continuous presence of TPA.

Previous attempts to inoculate TPA-treated ACR cells sc in the nude mouse have failed to yield any tumors. Recently, the inoculation of these cells from 2 different individuals into the anterior chamber of the eye of a nude mouse apparently gave rise to a tumor which is characterized by uniformly appearing, highly basophilic, fibroblast-like cells (23). The higher consistency in TPA-induced tumors may suggest that selection of human cells *in vivo* is more rigorous than selection in culture, or through agar. However, the number of different individuals tested to date and the number of animals examined do not provide us with any good measure about the probability of such an event occurring upon repeated exposure of cells from the same individuals and cells taken from different gene-carriers.

TPA has been shown to enhance the stable transformation of murine, and more recently, of human foreskin fibroblasts previously exposed to a carcinogen (9-13). Thus, our results may indicate that the ACR mutation is a complete one for malignancy, representing an initiated state (23,24), and that the chronic application of TPA, in support of the two-stage 'Berenblum-hypothesis" (33) can precipitate the final oncogenic event. Alternatively, this cell mass growing *in vivo* may represent an intermediary state, similar perhaps to the TPA-induced papillomas in the mouse skin model (5-8) or to the clinical appearance of polyps in the colon; these may or may not regress spontaneously upon withdrawal of the stimulating factor. The latter would suggest that an additional mutation(s) is necessary for the malignant transformation of ACR cells with certain promoters acting during all phases of oncogenesis to increase the probability of expression of the malignant phenotype. In this regard, our ability to understand reversibility and adaptation (acquired resistance) in relation to inflammation, hyperplasia (transient and sustained), and promotion in TPA-sensitive (mouse) and TPA-refractile animals (hamster, rat, guinea pig) (30) would provide insight about the effects of TPA in human cells *in vitro*.

The way in which the chronic exposure of ACR cells to TPA might effect a transition from the initiated state to the neoplastic state is now being investigated. We have recently demonstrated that TPA-treated ACR cells which grew in athymic mice showed about 33% increase in cell aneuploidy (L.K. and R. Moon, unpublished). The same proportion of cells has also been shown to be positive with respect to a first trimester human fetal antigen in the perinuclear region (L. Kopelovich and P. Higgins, unpublished). However, these cells did not necessarily grow in agar, nor did they acquire an infinite life-span in vitro (23). Indeed, not all cells obtained from spontaneously occurring human tumors appear to grow indefinitely in culture (34). We speculate that both initiators and promoters, in that order (5-14), can be mutagenic. Carcinogens are mutagenic due to their direct interaction with DNA, whereas TPA is mutagenic, presumably because it might destabilize DNA and cause certain deletions of inserted sequences and chromosomal alterations (35,36). These chromosomal alterations, if stable, may conceivably be consistent with a somatic mutation(s). Clearly, the proper monitoring of mutations and mutation frequencies in human cells due to TPA will be of great significance.

ACKNOWLEDGEMENTS

We thank Ms. P. Monaghan, Ms. R. Vuolo, and Ms. T. Shapiro for excellent technical assistance. This work was supported by grants CA-19529 and CA-21623 from the National Large Bowel Cancer Project, and NCI grant CA-08748.

REFERENCES

1. Ames, B.N., McCann, J. and Yamasaki, B.:1975, Mutation Res. 31, pp. 347-364.
2. Heidelberger, C.:1975, Ann. Rev. Biochem. 44, pp. 79-121.
3. Miller, E.C.:1978, Cancer Res. 38, pp. 1479-1496.
4. Mintz, B. and Illimensee, K.:1975, Proc. Natl. Acad. Sci. USA 72, 3583-3589.
5. Berenblum, I.:1978, J. Natl. Cancer Inst. 60, pp. 723-726.
6. Boutwell, R.K.:1974, Prog. Exp. Tumor Res. 4, pp. 207-250.
7. Hecker, E.:1971, Methods Cancer Res. 6, pp. 439-484.
8. Van Duuren:1976, in Chemical Carcinogens, Monograph 173, edited by C.E. Searle, American Chemical Society, Washington, D.C., pp. 24-51.
9. Lasne, C., Gentil, A. and Chouroulinkov, I.:1974, Nature 247, pp. 490-491.
10. Mondal, S., Brankow, D.W. and Heidelberger, C.:1976, Cancer Res. 36, pp. 2254-2260.
11. Mondal, S. and Heidelberger, C.:1976, Nature 260, pp. 710-711.
12. Kennedy, A.R., Mondal, S., Heidelberger, C. and Little, J.B.:1978, Cancer Res. 38, pp. 439-443.

13. Milo, G.F. and Dipaolo, J.A.:1978, Nature 275, pp. 130-132.
14. Weinstein, I.B., Yamasaki, H., Wigler, M., Lee, L.S., Fisher, P.B., Jeffrey, A. and Grunberger, D.:1979, in Carcinogens: Identification and Mechanisms of Action, edited by A. Clark Griffin and Charles R. Shaw, Raven Press, New York, pp. 399-418.
15. Alm, T. and Licznerski, G.:1973, Clin. Gastroenterol. 2, pp. 577-601.
16. Gardner, E. and Richards, R.:1953, Am. J. Hum. Genet. 5, pp. 139-148.
17. Morson, B. and Bussey, H.:1970, Cur. Probl. Surg., pp. 1-50.
18. Knudson, A.G.:1977, in Advances in Human Genetics, edited by H. Harris and K. Kirshhorn, Raven Press, New York, pp. 1-66.
19. Fraumeni, J.F., Jr.:1977, in Genetics of Human Cancer, edited by J.J. Mulvihill, R.W. Miller and J.F. Fraumeni, Jr., Raven Press, New York, pp. 223-235.
20. Lynch, H.T., Harris, R.E., Lynch, P.M., Guirgis, H.A., Lynch, J.P. and Bardawil, W.A.:1977, Cancer 40, pp. 1845-1849.
21. Kopelovich, L.:1977, Cancer 40, pp. 2534-2541.
22. Kopelovich, L., Conlon, S. and Pollack, R.:1977, Proc. Natl. Acad. Sci. USA 74, pp. 3019-3022.
23. Kopelovich, L., Bias, N. and Helson, L.:1979, Nature 282, pp. 619-621.
24. Kopelovich, L.:1980, in Colorectal Cancer: Prevention, Epidemiology and Screening, edited by S. Winawer, D. Schottenfeld and P. Sherlock Raven Press, New York, pp. 97-108.
25. Comings, D.E.:1973, Proc. Natl. Acad. Sci. USA 70, pp. 3324-3328.
26. Stanbridge, E.J. and Wilkinson, J.:1978, Proc. Natl. Acad. Sci. USA 75, pp. 1466-1469.
27. Yotti, L.P., Chang, C.C. and Troski, J.:1979, Science 206, pp. 1089-1091.
28. Bell, E., Marek, L.F., Levinstone, D.S., Merrill, C., Sher, S., Young, I.T. and Eden, M.:1978, Science 202, pp. 1158-1163.
29. Kontermann, K. and Bayreuther, K.:1979, Gerontology 25, pp. 261-274.
30. Pullman, B., Ts'o, P.O.P. and Gelboin, H.:1980, in 13th Jerusalem Symposium, "Carcinogenesis: Fundamental Mechanisms and Environmental Effects", Jerusalem, Israel, April 28-May 1, 1980.
31. Colburn, N.H., Former, B.F., Nelson, K.A. and Yuspa, S.H.:1979, Nature 281, pp. 589-591.
32. Knudson, A.G.:1979, J.A.M.A. 241, p. 279.
33. Berenblum, I.:1954, Cancer Res. 14, pp. 471-477.
34. Smets, L.:1980, Biochem. Biophys. Acta 605, pp. 93-111.
35. Chopan, M. and Kopelovich, L.:1980, Exp. Cell Biol., in press.
36. Kinsella, A.K. and Radman, M.:1978, Proc. Natl. Acad. Sci. USA 75: pp. 6149-6153.

USE OF HUMAN EPIDERMAL KERATINOCYTES IN STUDIES ON CHEMICAL CARCINOGENESIS

Kuroki, T., Nemoto, N.* and Kitano, Y.**
Department of Pathobiochemical Cell Research, Institute of Medical Science, University of Tokyo, Shirokanedai, Minato-ku, Tokyo 108, *Department of Experimental Pathology, Cancer Institute, Kami-ikebukuro, Toshima-ku, Tokyo 170, and **Department of Dermatology, Osaka University School of Medicine, Fukushima-ku, Osaka 553, Japan.

ABSTRACT

Epidermal keratinocytes were obtained from the skin of normal persons for use in studies on chemical carcinogenesis. The cultures consisted of tightly packed polygonal cells with a characteristic epithelial appearance. These cells were shown to metabolize BP, but the percentage of BP metabolized varied considerably in cultures from different individuals. Analysis of BP metabolites by high pressure liquid chromatography indicated that epidermal keratinocytes metabolize BP preferentially at non-K-regions (positions 7,8,9, and 10) forming a moderate amount of 7,8-diol BP. Conjugate formation was examined by treating the medium with β-glucuronidase and arylsulfatase, but no appreciable amount of conjugates was detected. The metabolic activity of human epidermal keratinocytes on BP was further demonstrated by cell-mediated assay, in which V79 Chinese hamster cells were cultured on top of sheets of keratinocytes and treated with BP for 48 hr. Other problems that can be investigated with the present culture system are listed.

INTRODUCTION

Many in vitro systems for studying chemical carcinogenesis using mesenchymal cells of rodents have been developed, because these cells are easy to cultivate and transform. However, the realization that almost 90% of all human tumors are derived from epithelial cells has prompted attempts to use epithelial cells in studies on transformation. The epithelial cells that have so far been reported to be transformed by chemicals include liver parenchymal cells of rats, epidermal keratinocytes of rats and mice, bladder epithelium of rats and mice, submandibular glands of mice and tracheal epithelium of rats (see recently published monograph 1).

B. Pullman, P. O. P. Ts'o and H. Gelboin (eds.), Carcinogenesis: Fundamental Mechanisms and Environmental Effects, 417–426.

Increasing emphasis is being placed on the use of human cells in studies on chemical carcinogenesis. Use of cultured human cells is the only way possible to test risk of chemical carcinogens in man, as in vivo tests are obviously impossible. Such cultures may be valuable in providing a link between experimental studies in animals and human cancers for assessing the validity of extrapolating carcinogenicity data on animals to humans. However, only recently have human cells such as cultured embryo cells (2, 3), blood lymphocytes (4), blood monocytes (5), bronchial cells (6, 7), peripheral lung tissues (8), pulmonary macrophages (9), colon tissues (10) and fibroblasts (11 - 14) been used for this purpose. After a long period of discussion on the difficulty of transforming human cells, chemical transformation has recently at last been achieved using human skin fibroblasts (13, 14).

Because of the importance of the use of epithelial cells and particularly human epithelial cells, as discussed above, we have recently started a project on the use of human epidermal keratinocytes in studies of chemical carcinogenesis. We chose epidermal keratinocytes for this purpose because they can be obtained relatively easily during plastic surgery and because a technique for culture of keratinocytes has been established by Kitano and Hata (15). Furthermore, epidermal keratinocytes show terminal differentiation, reorganizing to form a multilayered epidermal sheet (16, 17), and this process offers unique opportunities for investigating cellular differentiation and carcinogenesis.

STRUCTURE OF HUMAN SKIN AND CULTURE OF EPIDERMAL KERATINOCYTES

A vertical section of the human epidermis shows one layer of basal cells on the basal membrane 5 to 10 layers of spinous cells and 2 to 3 layers of granular cells from the bottom to the top. In the uppermost portion of the epidermal sheet, the keratinocytes lose their nuclei and form a horny layer. However, viewing the the skin from the surface shows another feature of the epidermis, i.e. columns of hexagonal maturing and matured cells on the top of about 10 basal cells (Figure 1). Surprisingly, this unique hexagonal structure has been realized only within the last 10 years. Potten referred to this hexagonal structure as 'the epidermal proliferation unit', by analogy to the proliferation unit seen in intestinal mucosa (18). This structural unit can be observed in the epidermis in many sites and in many animal species.

Since the epidermis is closely associated with the dermis, cell suspension obtained by enzymatic digestion of skin are usually contaminated by mesenchymal cells, which must be removed to obtain pure cultures of keratinocytes. In cultures of rodent skin, keratinocytes have usually been separated from fibroblasts in a discontinuous Ficoll gradient (19, 20). Rheinwald and Green (21) used a feeder layer of lethally irradiated 3T3 cells which suppress fibroblasts but stimulate the growth of keratinocytes obtained from human foreskin. We used the method

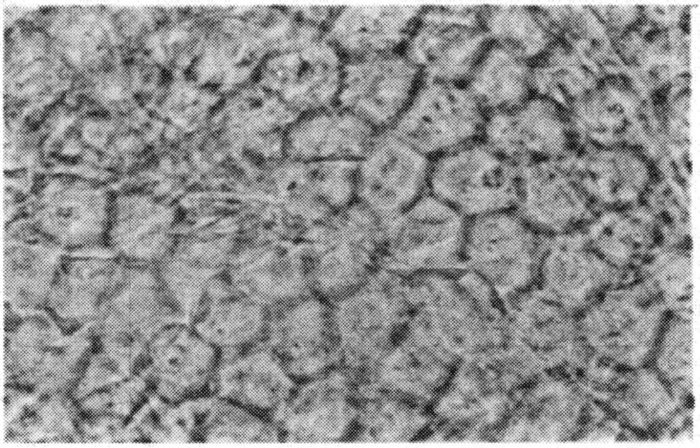

Figure 1. Surface view of a sheet of epidermis from human buttock after removing basal cells. Phase-contrast picture. Note the hexagonal structure of epidermal proliferating units.

developed by Kitano and Hata (15), in which the epidermis is peeled off with fine forceps from the dermis after overnight incubation of dermatome-sections of skin in 0.25% trypsin solution at 4°C, and then a suspension of keratinocytes is obtained by mechanical shaking. The viability of the cells obtained by this method is more than 90%. Although the cell suspension obtained by this method seems to be essentially free of fibroblasts, the cells must be plated at high density, i.e. more than 2×10^5 basal cells per ml (5 ml per 60 mm dish), to prevent growth of fibroblasts. The culture medium used in our studies was Eagle's MEM supplemented with 20% or 30% foetal calf serum (FCS). The presence of FCS at higher concentrations than 20% apparently enhances the growth of keratinocytes.

The epithelial character of these cultures was apparent from the fact that they formed tightly packed sheets of polygonal cells. Abundant filamentous structures, possibly keratin filaments, were often seen in the cytoplasm. These cells stained red with both Rhodanile blue stain and modified Mallory stain. Material staining orange was observed on cell sheets stained with Papanicolou stain, indicating the formation of keratin. Some melanocytes were seen scattered singly in the sheets of keratinocytes, but scarcely any fibroblasts were found in the primary cultures. These morphological characteristics and the electron microscopic appearance (16) of the cells indicate that they were keratinocytes.

There has long been a debate about whether mesenchymal support is necessary for growth of epithelial cells with certain differentiated functions. The epidermal keratinocytes purified as described above showed considerable organization and maturation in the absence of underlying living mesenchyme (16, 22, 23), although previous studies indicated that growth of epidermal cells was dependent on the presence of collagen gels which were conditioned by viable fibroblasts (24). Recently, Rheinwald and Green (21) used a feeder layer of lethally irradiated mouse 3T3 cells to support the growth of human epidermal keratinocytes at a small plating number for an extended period. We found that human or rat fibroblasts could be used as a feeder layer for cultures of human epidermal keratinocytes (unpublished data). These mesenchymal cells may produce type I and III collagen, which are widely distributed, while epidermal cells grow in association with the basement membrane which contains basement membrane specific type IV collagen. Type IV collagen seems to be a product of epidermal cells (unpublished data). In keeping with these findings, Murray et al. (25) recently reported that epidermal cells from guinea pig skin preferentially grew on type IV collagen (25). Mesenchymal support of epithelial cells should be investigated further in terms of the specificity of collagen gels.

Like diploid fibroblasts, human epidermal keratinocytes have a finite life time in culture (26). Under our culture conditions, most of the cultures derived from normal subjects of different ages could grow until at least the second transfer, but were hardly transferred beyond the third passage. After the second passage cultures often contained fibroblasts, which could not be seen in primary cultures. Rheinwald and Green (26) reported that the culture life time of epidermal cells could be increased to 150 generations from 50 by addition of epidermal growth factor (EGF).

METABOLIC ACTIVATION OF BENZO[A]PYRENE IN HUMAN EPIDERMAL KERATINOCYTES

Being on the surface of the body, epidermal keratinocytes are exposed to environmental carcinogens, which are thought to cause a large proportion of all human cancers. Benzo[a]pyrene (BP) and other polycyclic aromatic hydrocarbons are prevalent contaminants of air, water and soil. Therefore, we investigated the metabolism of BP in cultured human epidermal keratinocytes obtained from normal subjects using thin layer chromatography (TLC), high pressure liquid chromatography (HPLC) and cell-mediated mutagenesis assay (paper submitted). All experiments were performed on cells in primary culture.

Human epidermal keratinocytes metabolized BP significantly. The percentage of BP metabolized varied in cultures derived from different individuals: when HUSKI-1 cells were treated with 20 μM BP for 48 hr, 84.6% of the radioactivity was recovered unchanged on TLC, while when HUSKI-5 cells were treated with 5 μM BP, almost all (97%) the BP was metabolized. Figure 2 shows the profiles of BP metabolites obtained after 48-hr incubation, and analyzed by HPLC. Human epidermal keratinocytes produced almost the whole series of metabolites of

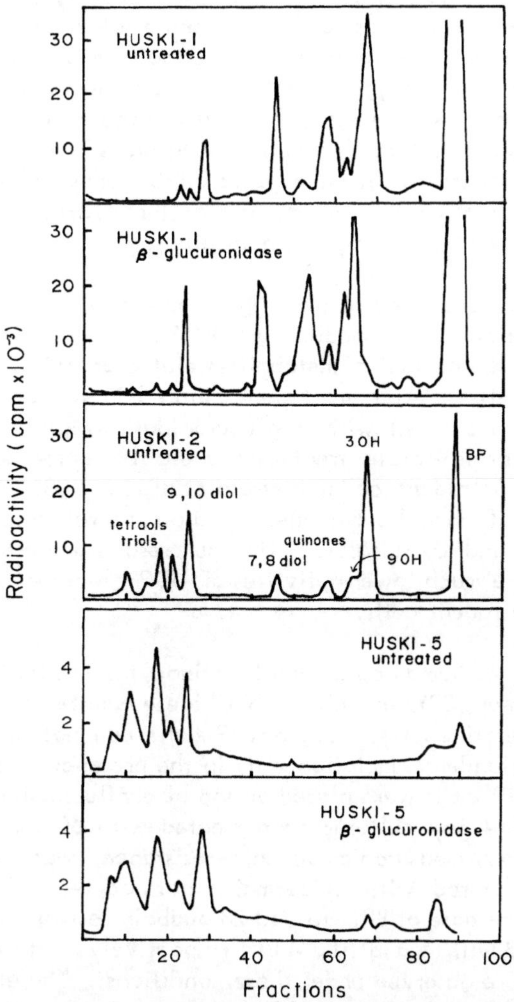

Figure 2. HPLC analysis of BP metabolites in the medium of human epidermal keratinocytes with and without β-glucuronidase treatment. The media 48 hr after treatment with [^{3}H]-BP at 20 μM (HUSKI-1) or 5 μM (HUSKI-2 and -5) were extracted 3 times with ethyl acetate and the extracts were subjected to HPLC. β-Glucuronidase treatment (1 mg/ml) was performed at 37°C for 3 h before ethyl acetate extraction. HUSKI-1, -2 and -5 cells were obtained from an 8-year-old boy, a 46-year-old woman and a 21-year-old man and had been cultured for 25, 7 and 10 days, respectively, at the time of BP treatment. BP metabolites and their corresponding peaks are shown on the profile of HUSKI-2 cells.

BP : 9,10-diol, 7,8-diol, quinones, and 9-hydroxy and 3-hydroxy BP. The peaks eluted before 9, 10-diol seem to be those of tetraols and triols at the 7, 8, 9 and 10 positions. The absence of metabolites other than tetraols, triols and the 9, 10-diol in medium of HUSKI-5 cells suggests that these cells metabolized BP very rapidly in the non-K-region (positions 7, 8, 9 and 10) of the molecule. The 7, 8-diol, a precursor of the ultimate form 7,8-diol-9,10-oxide BP, constituted 1.3 and 4.0% of the total radioactivity in the organic soluble phase of the media of HUSKI-1 and -2 cells, respectively. The amount of the 4,5-diol, which was eluted between the 9,10- and 7,8-diols, was negligible.

Conjugate formation was examined by treating the media with β-glucuronidase and arylsulfatase. Incubation of the medium of HUSKI-1 cells with β-glucuronidase did not change the amount of metabolites eluted on HPLC (Figure 2). The medium of HUSKI-5 cells showed small peaks eluted in the phenol region (fractions 66 to 75) after treatment with β-glucuronidase (Figure 2), but no metabolites released by arylsulfatase treatment could be detected by TLC (Figure 3). In the culture of rodent cells, Nemoto et al. (27, 28) demonstrated that considerable amounts of phenols, quinones and diols are released after treatment with β-glucuronidase and arylsulfatase. Thus our results suggest that human epidermal keratinocytes have much lower activities of UDP-glucuronyl-transferase and sulfate transferase than rodent cells.

The metabolic activity of human epidermal keratinocytes was further demonstrated by cell-mediated assay (29), in which V79 Chinese hamster cells were co-cultured with epidermal keratinocytes. Since V79 cells can not activate most indirect carcinogens, they undergo mutation only in the presence of a metabolic activation system. The V79 cells were plated on top of confluent sheets of non-irradiated HUSKI-2 cells (14 day cultures) and treated with BP for 48 hr. As shown in Figure 4, mutation, measured as ouabain-resistance, was induced when the V79 cells were co-cultured with epidermal keratinocytes. The mutation frequency increased with the dose of BP: 18 and 23 ouabain resistant colonies per 10^5 survivors were observed with 5 and 10 μM BP, respectively. However, few, if any, cytotoxic effects were observed under these conditions. The efficiency of mutation of V79 cells by BP with human epidermal keratinocytes was compared with that with rat embryo fibroblasts. As shown in Figure 4, the mutation frequency increased with increase in the BP concentration to 5 μM, and then reached a plateau. At concentrations of more than 5 μM BP, higher mutation frequencies were obtained with keratinocytes than with rat embryo fibroblasts.

Harris and his colleagues (30, 31) used human cells and tissues as an activating layer in cell-mediated assay. They found that both human pulmonary alveolar macrophages and human bronchial explants activated BP and its proximate form, the 7,8-diol, causing mutation of co-cultured V79 cells, but that the activity of the human cells was much lower than that of rodent cells. In the present study,

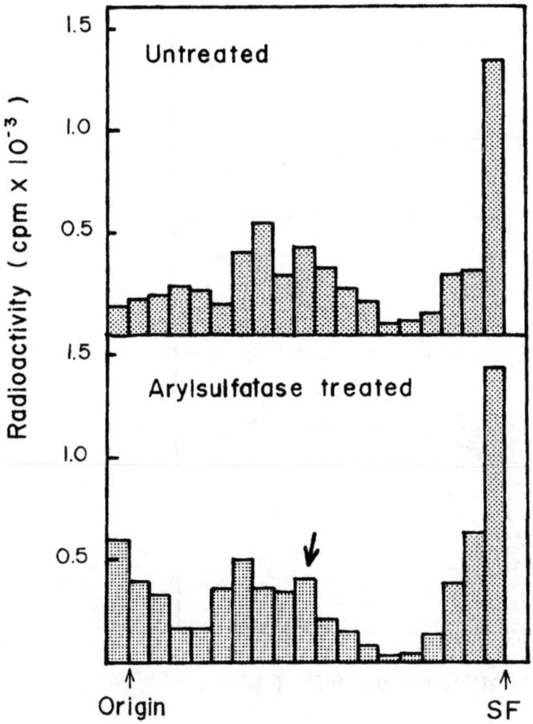

Figure 3. TLC analysis of BP metabolites in the medium of HUSKI-5 cells, with and without arylsulfatase treatment. Arylsulfatase treatment (333 µg/ml) was done at 37°C for 3 hr. The medium was subjected to silica gel TLC with a solvent mixture of ethyl acetate: methanol : H_2O : formate (100 : 25 : 20 : 1). A sulfate conjugate, if present, can be recovered in the region indicated by the arrow. SF : solvent front.

however, the extent of induced mutation obtained using human keratinocytes was higher than that observed using rat embryo fibroblasts as the activating layer, although the shapes of the dose-response curves were different.

FUTURE PROBLEMS

The use of human epidermal keratinocytes should make it possible to investigate a variety of problems in the field of chemical carcinogenesis, most of which have so far been investigated only using rodent mesenchymal cells. These problems include the following:

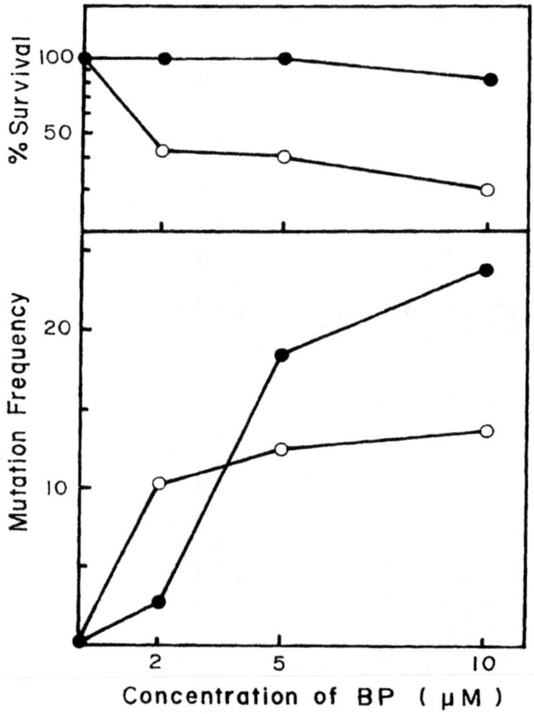

Figure 4. Cytotoxicity (top) and mutagenicity (ouabain resistance, bottom) of BP to V79 Chinese hamster cells co-cultured with non-irradiated human epidermal keratinocytes (HUSKI-2, ●) and irradiated rat embryo cells (○).

a) Metabolic activation of various chemical carcinogens by epidermal keratinocytes

b) Possible differences in metabolic activation by epidermal cells and mesenchymal cells

c) Differences between individuals in metabolic activation and the possibility of using these cells for estimating risk

d) Binding of chemical carcinogens to DNA and the correlation of binding with metabolism

e) DNA repair induced by carcinogens

f) The effects of growth factors, such as EGF and cholera toxin, on the growth and differentiation of keratinocytes and identification of their receptors

g) The effects of tumor promoters on the growth and differentiation of keratinocytes

h) Malignant transformation of human epidermal keratinocytes in vitro by chemical carcinogens.

REFERENCES

1. Franks, L.M.,and Wigley, C.B. (editors): 1979, Neoplastic Transformation in Differentiated Epithelial Cell Systems in vitro. Academic Press, N.Y., pp.314.
2. Huberman, E., and Sachs, L.: 1973, Int. J. Cancer 11, pp.412-418.
3. Brookes, P., and Duncan, M.E.: 1971, Nature 234, pp.40-43.
4. Bushee, D., Shaw, D.R., and Cantrell, D.: 1972, Science 178, pp.315-316.
5. Bast, R.C.Jr., Okuda,T., Plotkin, E., Taron, R., Rapp, H.J., and Gelboin, H.V.: 1976, Cancer Res. 36, pp.1967-1974.
6. Yang, S.K., Gelboin, H.V., Trump, B.F., Autrup, H., and Harris, C.C.: 1977, Cancer Res. 37, pp.1210-1215.
7. Stonner, G.D., Harris, C.C., Autrup, H., Trump, B.F., Kingsbury, E.W., and Myers, G.A.: 1978, Lab. Investigation 38, pp.685-692.
8. Cohen, G.M., Mehta, R., and Meredith-Brown, M.: 1979, Int. J. Cancer 24, pp.129-133.
9. Cantrell, E., Bushee, D., Warr, G., and Martin, R.: 1973, Life Sci. 13, pp.1649-1654.
10. Autrup,H., Harris, C.C., Trump, B.F., and Jeffrey, A.M.: 1978, Cancer Res., 38, pp.3689-3696.
11. Levin, W., Conney, A.H., Alvares, A.P., Merkatz, I., and Kappas, A.: 1972, Science 176, pp.419-420.
12. Rudiger, H.W., Marxen, J., Kohl, F.V., Melderis, H., and Wichert, P. V.: 1979, Cancer Res. 39, pp.1083-1088.
13. Kakunaga,T.: 1978, Proc. Natl. Acad. Sci. USA 75, pp.1334-1338.
14. Milo, G.E.,Jr., and DiPaolo, J.A.: 1978, Nature 275, pp.130-132.
15. Kitano,Y., and Hata,S.: 1972, Arch. Derm. Forsch. 245, pp.203-210.
16. Kitano,Y., and Endo, H.: 1977, in Biochemistry of Cutaneous Epidermal Differentiation, eds. Seiji,M. and Bernstein, I.A., Univ. Tokyo Press, Tokyo, pp.319-335.
17. Green, H.: 1977, Cell, 11, pp.405-416.
18. Potten, C.S., and Allen,T.D.: 1975, Differentiation 3, pp.161-165.
19. Fusenig, N.E., and Worst, P.K.M.: 1975, Exp. Cell Res. 93, pp.443-457.
20. Marcelo, C.L., Kim, Y.G., Kaine, J.L., and Voorhees, J.J. :1978, J. Cell Biol. 79, pp.356-370.
21. Rheinwald, J.G., and Green, H.: 1975, Cell 6, pp.331-344.
22. Voigt, W.H., and Fusenig, N.E.: 1979, Biologie Cellu. 34, pp.111-118.
23. Indo,K., and Wilson, R.B.: 1977, J. Natl. Cancer Inst. 59, pp.867-880.
24. Karasek, M.A., and Charlton, M.E.: 1971, J Invest. Derm. 56, pp.205-210.
25. Murray, J.C., Stingl, G., Kleinman, H.K., Martin, G.R., and Katz, S. I.: 1979, J. Cell Biol. 80, pp.197-202.

26. Rheinwald, J.G., and Green, H.: 1977, Nature 265, pp.421-424.
27. Nemoto, N., Hirakawa, T., and Takayama, S.: 1978, Chem. -Biol. Interactions 22, pp.1-14.
28. Nemoto, N., Takayama, S., and Gelboin, H.V.: 1978, Chem. -Biol. Interactions 23, pp.19-30.
29. Huberman, E., and Sachs, L.: 1974, Int. J. Cancer 13, pp.326-333.
30. Harris, C.C., Hsu, I.C., Stoner, G.D., Trump, B.F., and Selkirk, J.K.: 1978, Nature 272, pp.633-634.
31. Hsu, I.C., Stoner, G.D., Autrup, H., Trump, B.F., Selkirk, J.K., and Harris, C.C.: 1978, Proc. Natl. Acad. Sci. USA 75, pp.2003-2007.

STUDIES ON WHY 12-O-TETRADECANOYL-PHORBOL-13-ACETATE (TPA) DOES NOT PROMOTE EPIDERMAL CARCINOGENESIS OF HAMSTERS

J. Carl Barrett and Enid E. Sisskin
Environmental Carcinogenesis Group, Laboratory of Pulmonary Function and Toxicology, National Institute of Environmental Health Sciences. Research Triangle Park, North Carolina, USA

ABSTRACT

TPA, which is a potent promoter of mouse epidermal carcinogenesis, is inactive as a promoter on hamster, rat, and guinea pig skin. In order to probe the basis for this species specificity, the effects of TPA on hamster epidermis and hamster epidermal cells in culture were studied and compared to the known effects of TPA on mouse skin. We first wanted to determine whether TPA was reaching the target cells or if it was metabolized. Our studies revealed that metabolic inactivation of TPA does not account for its lack of tumor promoting activity on hamster skin. In addition, hamster epidermis treated with 81 nmol of TPA responded in a manner analogous to TPA-treated mouse epidermis. A hyperplastic response of the hamster skin was observed reaching a maximum at 48 hours after TPA treatment. Ultrastructural changes similar to those induced in mouse skin by TPA were observed in hamster skin. Hamster epidermal cells grown in culture also responded to TPA. TPA inhibited terminal differentiation of hamster epidermal cells and stimulated DNA synthesis in the cultures. These results demonstrate that TPA is very active on hamster epidermis and epidermal cells in culture. Since tumor promotion requires long-term exposure of cells to TPA and all the above experiments were acute experiments, the effects on hamster epidermis of TPA treatment for 1 to 4 weeks were studied. The most significant difference between mouse and hamster skin is in their hyperplastic response to TPA. In the mouse there is a potentiation in TPA-induced hyperplasia after two or more exposures to the promoter; while hamster skin adapts to the TPA treatment and no longer responds hyperplastically after multiple doses. The mechanism of this adaptation is unknown, but is currently under investigation. Examining differences in adapted and nonadapted hamster skin should be useful in understanding the mechanism of TPA promoting action. The role of sustained hyperplasia in tumor promotion and current models of tumor promotion are discussed.

B. Pullman, P. O. P. Ts'o and H. Gelboin (eds.), Carcinogenesis: Fundamental Mechanisms and Environmental Effects, 427–439.

INTRODUCTION

It is well documented that neoplastic development occurs by a multistage, progressive process (1-3). In mouse epidermal carcinogenesis, at least two stages, initiation and promotion, can be identified by their response to qualitatively different stimuli (3-5).

Initiating and promoting stimuli differ by a number of criteria (3-5). Initiating agents are effective with a single exposure, induce an irreversible action, are additive with increasing doses with no apparent threshold and can be activated to electrophiles which bind covalently to cellular macromolucules and are mutagenic (4,5). On the other hand, promoting agents require multiple exposures which are not additive, are reversible to some extent, and are active only above a threshold dose (4,5). Promoters are not mutagenic in conventional mutation assays (6).

Two stage carcinogenesis has been demonstrated in a number of other species and tissues (7) indicating that this is a general principle of neoplastic development. However, the importance of promoting stimuli to malignant tumor development in humans is unknown. Since most initiating agents are mutagens, simple short-term tests for detecting potential environmental carcinogens are presently available. However, such tests for promoters are not available due to the lack of our understanding of the mechanism of action of tumor promotion.

The most widely studied tumor promoter is 12-O-tetradecanoyl-phorbol-13-acetate (TPA), a diterpene phorbol ester, which is a potent promoter of epidermal carcinogenesis in mice (3,4). Studies with TPA and related phorbol derivatives have revealed pleiotypic effects of these compounds on mouse epidermis (7) and on numerous cell types in culture (5, 7-9). It is difficult, however, to discern TPA-induced changes which are critical to tumor promotion from possible secondary effects of the chemical. A genetic approach to this problem might resolve this question. Certain strains of mice do not appear to be sensitive to carcinogen and promoter-induced epidermal carcinogenesis and might be useful for such studies (10-13). Furthermore, a species specificity for tumor promotion exists; croton oil, of which TPA is the most active component, does not promote epidermal carcinogenesis in hamsters, guinea pigs, or rats under conditions where it is a very active promoter in mice (14-18).

It is unclear whether the lack of responsiveness of certain strains of animals to two stage carcinogenesis is due to insensitivity to the action of the carcinogen during the initiation phase or to the action of the promoter during the promotion phase of carcinogenesis. The former explanation is very likely for some strains of mice (11), however, it is unlikely for hamsters which are known to actively metabolize carcinogens, particularly polycyclic aromatic hydrocarbons. These animals are highly sensitive to carcinogen-induced tumorgenesis (14, 15,17,18).

Using species differences as an experimental tool is an attractive approach because it can be employed to learn which critical effects of TPA may be lacking in the nonresponsive animal, and also because it may yield information on the universality of tumor promotion.

For these reasons, we have begun a systematic study of the effects of TPA and other phorbol derivatives on hamster epidermis *in vivo* and hamster epidermal cells *in vitro*.

STUDIES ON THE EFFECTS OF TPA ON SYRIAN HAMSTER EPIDERMIS AND EPIDERMAL CELLS IN CULTURE

Shubik and coworkers (14,16,17) have reported that croton oil does not promote epidermal carcinogenesis in hamsters and Goerttler *et al.* (15) have recently confirmed this observation with TPA, although promotion of melanoma formation was observed. We attempted to duplicate in hamsters the two-stage carcinogenesis model as developed with mouse skin. We used two different strains of inbred hamsters which are sensitive to tumorigenesis (19), two doses (50μg and 400μg) of the initiator DMBA, and two different promoters, PDD and TPA at doses which we knew to induce hyperplasia after a single exposure. After a year of skin painting, we found a significant increase in the number of melanomas in the promoted animals as compared to those treated with DMBA alone, but could not demonstrate any recognizable papillomas or carcinomas of epidermal origin, again confirming the lack of promotion in hamster epidermal carcinogenesis.

O'Brien *et al.* (20) have reported that cells of Syrian hamster origin when grown in culture have the ability to deacylate TPA to its inactive monoesters and phorbol. Using the method of O'Brien *et al.* (20), we examined the metabolism of ^{3}H-TPA in hamster skin and in cultures of hamster epidermal cells and dermal fibroblasts. Although dermal fibroblasts in culture rapidly degraded TPA to phorbol-13-acetate, epidermal cells in culture degraded less than 50% of the TPA after two days, and no metabolism of TPA was observed in hamster skin *in vivo*. Thus, metabolic degradation of TPA does not appear to be the major cause of its lack of tumor promoting activity on hamster skin.

Hamster epidermis is capable of responding to TPA as shown by our studies on the ultrastructural changes and hyperplastic response of hamster skin exposed to TPA. In the first few hours after a single application of 81 nmoles of TPA, effects resembling those seen in mouse epidermis were induced. The intercellular spaces dilated, with the cell membranes forming many papillary projections. With increasing time, the cells began to hypertrophy and the epidermis became significantly hyperplastic by 48 hours. The cells also underwent a change of orientation, from being parallel to becoming perpendicular to the basement membrane. This contributed to the approximately three-fold increase in the thickness of the noncornified epidermis at 48 hours, while the number of nucleated cell layers in the interfollicular epithelium increased from 1-2 cell layers to 5-6 cell layers. This hyperplasia was maximal at 48 hours and began to decrease at 72 hours and further at 96 hours; the epidermis was undistinguishable from untreated epidermis by 168 hours.

Further evidence for the ability of hamster epidermal cells to respond to TPA is provided by our studies on the effects of TPA on a cell culture system developed in our laboratory which is suitable for studies on the growth and differentiation of hamster epidermal cells *in vitro* (21). Epidermal cells were isolated from 1 day old newborn Syrian hamsters by separating the epidermis from the dermis by cold trypsin treatment using the method of Yuspa and Harris (22). A large number of cells were isolated by this procedure without contamination with dermal fibroblasts. When grown in culture the epidermal cells divide rapidly for 2 to 3 days until the cells reach confluency at which time the cells stratify and differentiate, elaborate abundant keratin-like amorphous material, stain red with rhodanile blue (which is specific for cornification) and form cornified envelopes.

TPA markedly inhibits the differentiation process of the hamster epidermal cells in culture. When grown in the presence of 0.1μg/ml of TPA for three or more days the epidermal cells fail to stain positively for cornification with rhodanile blue. The differentiation of the epidermal cells is quantitated by measuring the percentage of cells with cornified envelopes and TPA is shown to reduce by 50% the number of these terminally differentiated cells. Phorbol didecanoate also inhibits the differentiation of hamster epidermal cells in culture while phorbol is inactive. The effect of TPA is reversible. If TPA is removed from the media, the cells rapidly differentiate to the same degree as untreated cells. TPA also stimulates DNA synthesis of the epidermal cells especially after 10 days in culture at which time the vast number of cells in cultures have ceased DNA synthesis. These results demonstrate that TPA can inhibit differentiation of hamster epidermal cells. Inhibition of differentiation has been proposed as the primary mechanism of action of TPA in tumor promotion (7-9).

It should be noted, that while TPA inhibits differentiation of hamster epidermal cells in culture, it does not promote epidermal carcinogenesis in hamster skin. There are several possible explanations for these seemingly disparate observations: (I) Inhibition of differentiation plays no role in tumor promotion. However, evidence from ours and other systems, and theoretical considerations support at least a partial role for inhibition of differentiation in tumor promotion. (II) TPA is metabolized to inactive products *in vivo* but not *in vitro*. Our results on the metabolism of TPA on hamster skin indicate that this is not the case. (III) Inhibition of differentiation of hamster epidermal cells occurs *in vitro* but not *in vivo*. Our ultrastructural studies of hamster skin exposed to TPA indicate a pronounced TPA effect on hamster epidermis, analogous to those observed by Raick *et al*. (23,24) in mouse skin exposed to TPA. These changes are consistent with an alteration of differentiation by TPA of mouse and hamster epidermis *in vivo*: (IV) The transitory effects of TPA on hamster epidermal cells *in vitro* and *in vivo* do not persist for a sufficient time to allow tumor promotion. This hypothesis suggests that TPA induces all the changes necessary for tumor promotion *in vivo* but the effects are reversed more rapidly in hamsters than in mice. (V) Tumor promotion requires an inhibition of cellular differentiation but other effects are also needed and these effects are not induced in hamster epidermis. We feel that the most probable explanation of the different effects *in vivo* and *in vitro* are a combination of the fourth and fifth hypotheses. Studies on other effects of TPA on hamster epidermis and epidermal cells in culture are currently in progress in our laboratory. The current status of these investigations is summarized in Table 1.

Table 1: COMPARISON OF EFFECTS OF TPA ON MOUSE SKIN, HAMSTER SKIN AND HAMSTER EPIDERMAL CELLS IN CULTURE

EFFECT	MOUSE SKIN	HAMSTER SKIN	HAMSTER EPIDERMAL CELLS IN CULTURE
Promotion of Neoplastic Transformation	+	-	?
Metabolism of TPA	-	-	+
Inhibition of Differentiation	+	+	+
Acute Hyperplasia	+	+	N/A
Increased DNA Synthesis	+	?	+
Increased ODC	+	?	?
Ultrastructural Changes	+	+	+
Induction of Dark Cells	+	?	N/A
Stimulation of Prostaglandins	+	?	+
Receptor for TPA	+	?	?
Sustained Hyperplasia	+	-	N/A

?, not known +, TPA induced effect
N/A, not applicable -, no effect

The effects of multiple treatments of TPA on hamster skin was examined to determine if long term effects of TPA were similar to the acute effects described for mouse and hamster epidermis. If the hamsters are treated with TPA once a week for one or two weeks, hyperplasia of the epidermis is induced at 48 hours after treatment. However, if the weekly exposures are continued for four weeks, no hyperplasia is observed. This loss of response is accelerated if the TPA exposures are closer together. After just one week of two exposures per week, there is virtually no hyperplasia measureable at 48 hours after the last exposure. Thus, while a single application of TPA induces a transient hyperplasia in hamster skin, with repeated exposures this hyperplastic response diminishes rapidly. This is in sharp contrast to the response of mouse skin to repeated TPA applications. Our results suggest that hamster skin can adapt to multiple TPA treatments.

The mechanism of this tachyphylaxis is unknown but we are currently examining the metabolism of TPA and the presence of TPA receptors (25) in adapted and nonadapted hamster skin. The study of hamster skin in these two states may prove very useful for studying the molecular mechanism of TPA action.

ROLE OF SUSTAINED HYPERPLASIA IN TUMOR PROMOTION

The results of our study show that while TPA induces hyperplasia of hamster epidermis following a single treatment, sustained hyperplasia of the epidermis elicited by multiple treatments of TPA does not occur. The question that then arises is whether the lack of sustained hyperplasia is sufficient to explain the lack of tumor promoting activity of TPA on carcinogen treated hamster epidermis. Although the answer to this question is unknown, it may be instructive to examine the evidence for and against a role of sustained hyperplasia in promotion of epidermal carcinogenesis. Numerous studies have addressed this problem and many authors have concluded that hyperplasia is either not involved in promotion (26-29) or that hyperplasia is necessary but not sufficient (30-36).

These conclusions are not supported by most studies with the phorbol ester series which have indicated a good correlation between tumor promoting activity and hyperplasia inducing ability (31,32,34). Some phorbol esters, such as phorbol dioctanoate and phorbol dibenzoate, are highly inflammatory but only weakly or moderately tumor promoting. However, their ability to induce hyperplasia was found to correlate with their tumor promoting ability (31,43), indicating a difference between inflammation and hyperplasia. Unfortunately, many authors do not make this distinction in their discussions of this topic. Some studies have indicated a different dose response for hyperplasia and promotion induced by although others do not (31). The former studies indicate that low doses of TPA which do not promote epidermal carcinogenesis are still hyperplastogenic. These results can be easily explained by either a lack of sustained hyperplasia with these doses of TPA or by the requirement for promotion of a minimum degree of hyperplasia not manifest by this treatment (37).

The relationship between hyperplasia and promotion has been discounted mainly on the basis of several studies which indicated that many hyperplasia-inducing agents are not promoters (26,30,35,40,45,46). However, Frei and Stephens (47) demonstrated a good correlation between promotion of tumor growth and induction of epidermal hyperplasia for a number of agents. Their paper has been criticized (46), because a high dose of carcinogen in a solvent that favored retention at the site of application was used for the initiation step.

The hyperplastic agents that have been investigated to the greatest extent are ethylphenylpropiolate (EPP) (37,39), acetic acid (38,40), cantharidin (32) and mezerein (42) (Table 2). Unfortunately, most of these studies compare the hyperplastic response of a single exposure of an agent to its tumor promoting activity following multiple treatments (37). As with the TPA treatment of hamster epidermis, treatment of mouse skin with some hyperplastic agents causes an adaptation to the treatment and hence multiple treatments are much less active in inducing hyperplasia than is predicted from a single exposure. This is apparently true for acetic acid (38). Although acetic acid is highly inflammatory on mouse skin, it induces only a moderate degree of hyperplasia with a single exposure (40) and this is reduced with multiple exposure (38). Its weak promoting activity (40) is therefore consistent with its weak ability to induce sustained hyperplasia.

A single exposure of EPP induces the same degree of hyperplasia as TPA, yet EPP is a much weaker tumor promoter. While tachyphylaxis does not appear to result from multiple exposures of EPP, the potentiation of hyperplasia observed with multiple treatments of TPA does not occur with multiple treatments of EPP (35,37,39). Both EPP and TPA induce hyperplasia to about 3-4 nucleated cell layers following a single exposure. This increases to 4-5 cell layers with multiple EPP treatments but to 12-13 cell layers with multiple TPA treatments. The level of sustained hyperplasia with EPP is, therefore, very consistent with its low tumor promoting activity. Frei (37) has shown that once a certain level of sustained hyperplasia is reached, a good correlation exists between the degree of hyperplasia and tumor promotion. He demonstrated that the tumor promoting activity of EPP is very close to that predicted from this correlation.

The sustained level of hyperplasia induced by multiple treatments with cantharidin and mezerein has not been reported. However, it has been noted that multiple treatments with high doses of cantharidin is toxic and this may limit the tumor promoting activity of this compound (32).

TABLE 2: COMPARISON BETWEEN TUMOR PROMOTING ACTIVITY AND HYPERPLASIA INDUCING ACTIVITY ON MOUSE SKIN OF SELECTED COMPOUNDS

Compound(s) (dose)	Promoting Activity	Hyperplasia Inducing Activity (Cell Layers)		Reference
		Acute	Sustained	
Acetone	Negative	1-2	1-2	35,37,41
TPA (16 nmol)	Strong	4-5	12-13	35,37
TPA (1.6 nmol)	Weak	3-4	4-5	35,37,39
TPA and Steroids	Negative	1-2	?	49
TPA and Retinoids	Negative	4-5	?	50
Ethylphenylpropiolate (20μmol)	Weak	2-3	4-5	37,39
Acetic Acid	Weak	3-4	(reduced)	38,40
Tween 60	Weak	5-6	4	41
Cantharidin (50μg)	Weak	4-5	?	32
Mezerein (1.7 nmol)	Negative	3-4	?	42
Mezerein (17 nmol)	Weak	?	?	42

It could be argued that hyperplasia induced by these chemicals is secondary to their tumor promoting activity. However, this argument does not seem plausible when the promoting stimulus is wound-healing (32,48). Argyris (48) has demonstrated that regenerative epidermal hyperplasia of adequate intensity is a sufficient stimulus for tumor promotion.

Further insight into the role of hyperplasia in promotion is provided by studies of inhibitors of promotion. It is well documented that steroids (49) and retinoids (50) specifically inhibit the TPA induced promotion phase of carcinogenesis. Steroids are also potent inhibitors of the hyperplasia induced by TPA in mouse epidermis (49). Retinoids are an interesting class of chemicals that play an important role in the development of mucous and surface epithelia. They have been reported to inhibit the TPA-induced ornithine decarboxylase activity in mouse epidermis (51), but not the hyperplastic response elicited by the promoter in the same tissue (52). In contrast, the retinoids do inhibit carcinogen-induced hyperplasia and metaplasia in prostate and tracheal organ cultures (50). If, indeed, retinoids can inhibit TPA induced promotion while being unable to inhibit the hyperplasia induced by the promoter, these chemicals will be important in defining the role of hyperplasia in tumor promotion.

Taken in total, the evidence for a role of hyperplasia in promotion is compelling. However, it is obvious that general hyperplasia would not a priori result in tumor promotion. Clayson (53) has defined tumor promotion as selective proliferation of initiated cells. The mechanisms controlling proliferation of initiated cells are probably only slightly different from the control mechanism of normal epidermal cells. These mechanisms include not only positive and negative growth controls but also a balance between cells which continue to divide (stem cells) and cells which terminally differentiate (54). TPA has been demonstrated to have an effect on all of these processes. Hyperplasia of normal epidermis induced by tumor promoters is essential for tumor promotion to the extent that it reflects the proliferative response of all cells, including initiated cells. The basis for the increased proliferative response of initiated cells very likely resides in the nature of the initiated cell not in the tumor promoter. To further understand tumor promotion of epidermal carcinogenesis will require an understanding of the control mechanisms of epidermal cell differentiation and proliferation and the alterations in these properties in initiated cells.

ESSENTIAL EFFECTS OF TUMOR PROMOTERS

The reluctance of many investigators to accept a role of hyperplasia in tumor promotion probably arises from the fact that although many other hyperplastic agents are tumor promoters, none are as potent or as effective as TPA in this process. This observation suggests that TPA induces essential changes for promotion in addition to induction of hyperplasia.

Boutwell (4) has proposed two steps in the promotion phase of carcinogenesis. The first step is termed "conversion" and represents a TPA induced alteration of initiated cells to dormant tumor cells. The second step is termed "propagation" and represents the growth of the dormant tumor cell to a visible papilloma. These two steps were demonstrated by the sequential treatment of initiated epidermis first with TPA and then with a nonspecific hyperplastic agent such as turpentine. Neither a short treatment with TPA nor continuous treatment with turpentine were sufficient for promotion.

This model has recently been supported by Slaga et al. (55) who demonstrated that following from 1 to 4 exposures of initiated mouse skin to TPA, treatment with mezerein was as effective for tumor promotion as continuous TPA treatment. Mezerein by itself had no effect on tumor promotion. This was interperted as TPA induced conversion of initiated cells followed by mezerein induced propagation. A single treatment with TPA was sufficient for partial conversion of the initiated cells and four treatments yielded maximum conversion.

Recognition that TPA induces a direct alteration of initiated cells changes some of the previous concepts of tumor promotion. This stage of promotion does not require multiple treatments with TPA and may be irreversible, two features previously attributed only to initiation. This is supported by the observations that TPA also induces a direct, irreversible alteration of growth control of rat tracheal cells (56) and mouse epidermal cells (57) in culture.

If it is accepted that TPA induces conversion of some initiated cells and that the propagation step is analogous to a proliferation of these cells, then many features of promotion can be explained. Many hyperplastic agents are tumor promoters because they can induce proliferation of cells that have been converted during the initiation phase. TPA is a much more potent promoter than other hyperplastic agents because it induces additional conversion of initiated cells and hence more cells can be induced to proliferate to papillomas. It is reasonable to assume that initiators can also convert initiated cells to a stage where they can be stimulated to proliferate selectively under a hyperplastic stimulus. This would explain why Frei and Stephens (47) when using a high concentration of carcinogen observed a good correlation between promoting activity and hyperplastic activity of a number of agents.

It should be pointed out that many papillomas which have progressed through the conversion and proliferation step are still TPA dependent for maintenance and growth and will regress in the absence of TPA. Since a neoplasm is defined as a growth which persists in the absence of the stimulus that induced it (59), these papillomas may not be considered neoplastic. We can, therefore, define a third stage in promotion as the neoplastic step, which is the progression of the papillomas to independent growth in the absence of a tumor promoter.

Finally, progression of papillomas to the malignant stage (Table 3) can be considered as a fourth step in this process. Burns et al. (58) have proposed that carcinomas arise predominantly from autonomous growing papillomas in TPA induced epidermal carcinogenesis of mice.

Table 3 lists the four steps in tumor promotion we have defined and the possible role of known TPA effects on individual steps. Elucidation of the molecular mechanism of each step in promotion will be important in understanding tumor promotion. Particular emphasis of future studies should be on the nature of the crucial events which determine the progression of cells to the malignant and premalignant stages.

In conclusion, we have studied the basis for the species specificity of tumor promotion by TPA. Hamsters are not sensitive to TPA mediated promotion of epidermal carcinogenesis possibly because the hamster epidermis can adapt to multiple TPA treatments and fail to undergo sustained hyperplasia. The mouse and the hamster epidermis, thus, have different physiological responses to multiple treatments with TPA. The mouse epidermis displays a potentiation of the effects with additional treatments, while the hamster epidermis becomes insensitive to further stimulation. Studies of the physiological and pharmacological basis for species specificity of tumor promotion may provide insight into the mechanism of action of tumor promoters; the factors involved in cellular growth control processes; and the extrapolation of risk of tumor promoters from one species to another.

TABLE 3: STEPS IN TUMOR PROMOTION AND POSSIBLE ROLES OF TPA INDUCED EFFECTS

STEP	TPA INDUCED EFFECTS	REFERENCES
I Conversion Step	Dedifferentiation	23,24
	Inhibition of Differentiation	9
	Alteration of Gene Expression	4
	Induction of Dark Cells	24
	Mitotic Aneuploidy	60
II Proliferation Step (Propagation)	Comitogen	61
	Inhibition of Differentiation	9
	Loss of Chalone Responsiveness	54
	Enhancement of Neoplastic Phenotype	8
	Inhibition of Immune Surveillance	46
	Alteration of Gene Expression	4
	Loss of Metabolic Cooperation	61
	Induction of ODC, PG's, uptake of small molecules	9
III Neoplastic Step (TPA Independent Papillomas)	Mitotic Aneuploidy	60
	Alteration of Gene Expression	4
	Alteration of Differentiation	9
IV Malignant Step	?	

REFERENCES

1. Barrett, J.C., Crawford, B.D. and Ts'o,P.O.P.:1980,in: Mammalian Cell Transformation by Chemical Carcinogens, ed. by V.C. Dunkel and R.A. Mishra, Pathtox Publishing Co., in Press.
2. Foulds, L.:1969,Neoplastic Development Academic Press, London.
3. Berenblum, I.:1954, Cancer Res:14,pp 471-477.
4. Boutwell, R.K.:1974, Crit. Rev. Tox. pp. 419-443.
5. Weinstein, I.B., Wigler, M., Fisher, P.B., Sisskin, E.E. and

Pietropaolo, C.:1978,in: Carcinogenesis, Vol 2, Mechanisms of Tumor Promotion and Cocarcinogenesis, ed. by T.J. Slaga, A. Sivak, and R.K. Boutwell, Raven Press, New York, pp. 313-333.
6. Trosko, J. E., Chang C., Yotti, L. P., and Chu, E.H.Y.:1977, Cancer Res. 37, pp. 188-193.
7. Colburn, N.:1979, in Carcinogenesis, Vol 5, Modifiers of Carcinogenesis, ed. by T.J. Slaga, Raven Press, New York, pp. 33-56.
8. Weinstein, I.B. and Wigler M.:1977, Nature 270,pp. 659-660.
9. Diamond, L., O'Brien, T.G., and Rovera, G.:1978, Life Sciences 23, pp. 1979-1988.
10. Bonser, G.M.:1938, J. Path. Bact. 46, pp. 581-602.
11. Gelboin, H.V., Kinoshita, N. and Wiebel, F.J.:1972, Fed. Proc. 31, pp. 1298-1309.
12. Nebert, D.W., Benedict, W.F., and Gielen, J.E.: 1972, Molec. Pharm. 8, pp. 374-379.
13. Kreyberg, L.:1934, Acta Path Microbiol Scand 11, pp. 174-182.
14. Della Porter, G., Rappaport, H., Saffiotti, U. and Shubik, P.: 1956, Arch. Path 61, pp. 305-313.
15. Goerttler, K., Loehrke, H., Schweizer, J. and Hesse, B.:1980, Cancer Res. 40, pp. 155-161.
16. Shubik, P.:1950, Cancer Res. 10, pp. 13-17.
17. Shubik, P., Pietra, G. and Della Porta, G.:1960, Cancer Res. 20, pp. 100-105.
18. Saffiotti, U. and Shubik, P.:1963, J. Nat. Cancer Inst. Monograph No. 10, pp. 489-507.
19. Homburger, F., Hsueh, S.-S., Kerr, C.S. and Russfield, A.B.:1972, Cancer Res. 32, pp. 360-366.
20. O'Brien, T.G. and Diamond, L.:1978, Cancer Res. 38, pp. 2562-2566.
21. Sisskin, E.E. and Barrett, J.C.:1979, Proc. Am. Assoc. Cancer Res. 20, p. 797.
22. Yuspa, S.H. and Harris, C.C.:1974, Exptl. Cell Res. 86, pp. 95-105.
23. Raick, A.N.:1974, Cancer Res. 34, pp.2915-2925.
24. Raick, A.N.:1973, Cancer Res. 33, pp. 269-278.
25. Driedger, P.E. and Blumberg, P.:1980, PNAS, pp. 567-571.
26. Raick, A.N. and Burdzy, K.:1973, Cancer Res. 33, pp. 2221-2230.
27. Van Duuven, B.L., Sivak, A., Segal, A., Seidman, I. and Katz, C.: 1973, Cancer Res. 33, pp. 2166-2172.
28. Van Duuven, B.L. and Sivak, A.:1968, Cancer Res. 28, pp. 2349-2356.
29. Raick, A.N., Thumm, K. and Chivers, B.R.:1972, Cancer Res. 32, pp. 1562-1568.
30. Setala, K., Merenmies, L., Stjernvall, L., Aho, Y. and Kajanne, P.: 1959, J. Nat. Cancer Inst. 23, pp. 925-951.
31. Slaga, T.J., Scribner, J.D., Thompson, S. and Viaje, A.:1976, J. Nat. Cancer Inst. 57, pp. 1145-1149.
32. Hennings, H. and Boutwell, R.K.:1970, Cancer Res. 30, pp. 312-320.
33. Baird, W.M. and Boutwell, R.K.:1971, Cancer Res. 31, pp. 1074-1079.
34. Baird, W.M., Sedgwick, J.A. and Boutwell, R.K.:1971, Cancer Res. 31, pp. 1434-1439.
35. Salaman, M.H. and Roe, F.J.C.:1964, Brit. Med. Bull. 20, pp. 139-144.
36. Berenblum, I.:1978, in Carcinogenesis, Vol 2, Mechanisms of Tumor

Promotion and Cocarcinogenesis, ed. by T.J. Slaga, A. Sivak, and R.K. Boutwell, Raven Press, New York, pp. 1-10.
37. Frei, J.V.:1977, J. Nat. Cancer Inst. 59, p. 299.
38. Slaga, T.J.:1977, J. Nat. Cancer Inst. 59, pp. 299-300.
39. Raick, A.N.:1974, Cancer Res. 34, pp. 920-926.
40. Slaga, T.J., Bowden, G.T. and Boutwell, R.K.:1975, J. Nat. Cancer Inst. 55, pp. 983-987.
41. Dammert, K.:1961, Acta Path. Microbiol. Scand. 53, pp. 33-49.
42. Mufson, R.A., Fisher S.M., Verna, A.K., Gleason, G.L., Slaga, T.J. and Boutwell, R.K.:1979, Cancer Res. 39, pp. 4791-4795.
43. Viaje, A., Slaga, T.J., Wigler, M. and Weinstein, I.B.:1977, Cancer Res. 37, pp. 1530-1536.
44. Berenblum, I.:1944, Arch Path 38, pp. 233-244.
45. Elgjo, E.:1968, Eur. J. Cancer 3, pp. 519-530.
46. Scribner, J.D. and Suss, R.:1978, in: International Review on Experimental Pathology, ed. by G.W. Richter and M.A. Epstein, Academic Press, New York, pp. 138-198.
47. Frei, J.V. and Stephens, P.:1968, Brit. J. Cancer, 22, pp. 83-92.
48. Argyris, T.S.:1980, Manuscript in press.
49. Slaga, T.J., Fischer, S.M., Viaje, A., Berry, D.L., Bracken, W.M., LeClerc, S. and Miller, D.R.:1978, in: Carcinogenesis, Vol 2, ed. by T.J. Slaga, A. Sivak and R.K. Boutwell, Raven Press, New York, pp. 173-195.
50. Nettesheim, P.: 1980, Can. Med. Assoc. J., in press.
51. Verna, A.K. and Boutwell, R.K.:1977, Cancer Res. 37, pp. 2196-2201.
52. Slaga, T.S., Klein-Szanto, A.J.P., Fisher, S.M., Weeks, C.E., Nelson, K. and Major, S.:1980. Proc. Natl. Acad. Sci., in press.
53. Clayson, D.B:1962, in: Chemical Carcinogenesis, Little, Brown & Co., Boston, pp. 290-314.
54. Marks, F.:1976, Cancer Res. 36, pp. 2636-2643.
55. Slaga, T.J., Fischer, S.M., Nelson, K. and Gleason, G. L:1980, Proc. Natl. Acad. Sci., in press.
56. Colburn, N.H., Former, B.F., Nelson, K.A. and Yuspa, S.H.:1979, Nature, 281, pp. 589-591.
57. Colburn, N.H., Former, B.F., Nelson, K.A. and Yuspa, S.H.:1979, Nature, 281, pp. 589-591.
58. Burns, F.J., Vanderlaan, M., Snyder, E. and Albert, R.E.:1978, in: Carcinogenesis, Vol 2, ed. by T.J. Slaga, A. Sivak, and R.K. Boutwell, Raven Press, New York, pp. 91-96.
59. Robbins, S.L. and Cotran, R.S.:1979, Pathological Basis of Disease, W.B. Saunders Co., Philadelphia, P. 141.
60. Parry, J., unpublished.
61. Dicker, P. and Rozengurt, E.:1978, Nature 276, pp. 723-726.
62. Yotti, L.P., Chang, C.C. and Trosko, J.:1979, Science, 206, pp. 1089-1091.

MODULATION OF REPAIR OF DNA DAMAGES INDUCED BY NITROSAMINES

Montesano, Ruggero* and Margison, Geoffrey P.**
*Division of Chemical and Biological Carcinogenesis, International Agency for Research on Cancer, 150 cours Albert Thomas, 69372 Lyon Cédex 2, France, ** Paterson Laboratories, Christie Hospital and Holt Radium Institute, Manchester M20 9BX, Great Britain.

ABSTRACT

The persistence of various DNA alkylation products (7-methylguanine, 3-methyladenine and O^6-methylguanine) was examined in the liver of rats following chronic administration of low doses of dimethylnitrosamine. The treatment resulted in an increased removal of O^6-methylguanine due to an inducible DNA repair process. These observations in rat liver are discussed with reference to a similar inducible repair process observed in *E. coli* exposed to low doses of the alkylating agent, *N*-methyl-*N'*-nitro-*N*-nitrosoguanidine.

INTRODUCTION

The initiation of the carcinogenesis process by nitrosamines is linked to the metabolic competence of the target organ or cell to convert these carcinogens into mutagenic metabolites (Magee *et al.*, 1976) and more specifically to the binding of the metabolites to cellular macromolecules. The methodology developed particularly by Lawley (1976a) has permitted the qualitative and quantitative examination of the relevance of various binding products to DNA or RNA to nitrosamine carcinogenesis. As discussed in this Symposium, there are various reasons for believing that mutational events are critical in the initiation of the carcinogenesis process and in the case of alkylating agents, like nitrosamines, various attempts have been made over a period of more than 10 years to identify the critical DNA adduct(s). Among the various sites of base alkylation in DNA, the N-7 and O^6 positions of guanine and the 3 position of adenine are the major alkylated sites, this being dependent upon the mechanism by which the different alkylating species react with DNA (see Lawley, 1976b; Singer, 1979). Loveless (1969) indicated that alkylation at the O^6 position of DNA guanine might be more critical in the mutagenic effect of alkylating agents, since this position, and not the N-7 position, of guanine is involved in base pairing. In keeping with this hypothesis, it has been

B. Pullman, P. O. P. Ts'o and H. Gelboin (eds.), Carcinogenesis: Fundamental Mechanisms and Environmental Effects, 441–451.

shown that the presence of O^6-methylguanine (O^6-meGua) in synthetic polymers resulted in the incorporation of non-complementary bases during polyribo- or polydeoxyribonucleotide synthesis *in vitro*, whereas other methylated bases, such as 7-methylguanine, 3-methylcytosine and 3-methylguanine, did not result in an appreciable misincorporation using *E. coli* DNA polymerase I (Gerchman and Ludlum, 1973; Abbott and Saffhill, 1979). This observation is substantiated by the findings of Coulondre and Miller (1977) that various mutagenic alkylating agents induce G-C — A-T transition in *E. coli* and by those of Vogel and Natarajan (1979) and Newbold *et al.* (1980), who reported a good correlation between the mutagenic activity of various alkylating agents in *Drosophila* and V79 Chinese hamster cells and their capacity to alkylate preferentially the oxygen atoms of the DNA molecule. The most direct evidence pointing to O^6-meGua as the 'promutagenic' lesion was derived from the parallelism observed between accumulation of mutations in *E. coli* by *N*-methyl-*N'*-nitro-*N*-nitrosoguanidine (MNNG) and the accumulation of this modified base, but not of 7-methylguanine and 3-methyladenine, in their DNA (Schendel and Robins, 1978; Jeggo, 1979). There are minor DNA adducts (O^2- and O^4-alkylthymine and O^2-alkylcytosine) that have been shown to mispair (see Singer, 1979), but it is not possible at present to assess their significance in the mutagenicity of alkylating agents.

These *in vitro* findings are consistent with studies *in vivo* aimed at correlating the organ- or species-specificity in nitrosamine carcinogenicity with the presence and persistence of various DNA adducts in the target organ. The induction of tumours in the kidney of rats treated with a single dose of various methylating or ethylating agents does not parallel the extent of formation in DNA of N-7-alkylguanine (Swann and Magee, 1968, 1971), indicating that alkylation of the N-7 position of guanine may not be relevant to the process of carcinogenesis. Following the initial observation of Goth and Rajewsky (1974), it was observed in various experimental systems in which tumours are induced by a pulse dose of nitroso compounds that the organs in which tumours occur most frequently are generally those in which the greatest persistence of the promutagenic base, O^6-alkylguanine, in DNA occurs (see Pegg, 1977). The removal of O^6-alkylguanine from DNA is an enzymic process, as indicated by various observations: its chemical stability *in vitro* under physiological conditions; its removal from DNA *in vivo* at a rate higher than that explicable by DNA turnover; its removal occurs at different rates among various organs of the same animal species or in the same organ of different animal strains and species; certain tissue extracts can catalyse specifically the loss of O^6-meGua, but not 7-methylguanine, from DNA *in vitro* (see Pegg, 1977; Margison and O'Connor, 1979; Roberts, 1980).

The persistence of this promutagenic base in the DNA presumably results in an increased probability that a miscoding event will take place during DNA synthesis and result in a permanent heritable change in the base sequence. Therefore, the rapidity of the removal of this product from DNA may be an important protective mechanism against carcinogenesis, provided that a substantial degree of removal can occur prior to DNA synthesis. The relatively rapid removal of O^6-meGua from liver DNA of rats treated with a single dose of DMN (Nicoll *et al.*, 1975) might be the reason why no tumours are induced in this organ under these conditions, but a significant indicence of liver tumours is induced if the same dose of dimethylnitrosamine (DMN) is administered 24 hrs after partial hepatectomy, that is at the time when DNA synthesis is at a peak (Craddock, 1978). Conversely, tumours are induced by a similar dose of DMN in the liver of hamsters (Margison *et al.*, 1976; Stumpf *et al.*, 1979) which have a low capacity to remove this alkylation product from their DNA.

Most of these findings were observed in experimental systems in which tumours are induced by a single dose of nitrosamine. The exposure of human beings or the induction of tumours in experimental animals mainly occur, however, through chronic exposure to one or many carcinogens, and it was thus of interest to examine the behaviour of O^6-meGua and other alkylation products in the liver DNA of rats during continuous treatment with DMN. Studies carried out so far in our laboratory are summarized here and indicate that an increase in the enzymic activity responsible for the removal of O^6-meGua occurred after prolonged exposure to low doses of DMN. These findings are discussed in relation to the adaptive response against mutagenesis by MNNG in *E. coli* (Samson and Cairns, 1977) and to carcinogenesis dose-response with DMN in rats.

EFFECTS OF CHRONIC TREATMENT OF RATS WITH DMN ON THE REMOVAL OF O^6-METHYLGUANINE FROM LIVER DNA

In a series of experiments originally aiming to investigate the possibility that chronic administration of DMN might result in an accumulation of O^6-methylguanine in liver DNA, male BDIV rats were treated with labelled DMN (2 mg/kg) on weekdays for up to 24 weeks. At various times (from 2 to 24 weeks) the amounts of 7-methylguanine and O^6-meGua were determined 72 hrs after the final DMN treatment. No O^6-meGua was detected in liver DNA even after 24 weeks of treatment with labelled DMN, indicating that at this dose schedule no accumulation had occurred (Margison *et al.*, 1977). Since, on the basis of 7-methylguanine present in the DNA, some O^6-meGua would have been expected, these findings suggested that the chronic treatment with DMN affected the removal of O^6-meGua from liver DNA. In fact, subsequent experiments (Montesano *et al.*, 1979, 1980a) have shown that pretreatment with DMN for 44 days at various dose levels affected the removal of O^6-meGua from liver DNA in a dose-related fashion. Fig. 1 shows the O^6-meGua/7-methylguanine ratios observed in liver DNA 6 or 12 hrs after a dose of 2 mg/kg [^{14}C]-DMN given to control rats or to rats pretreated with 0.2 or 2.0 mg/kg. Variation in the amount of 7-methylguanine in DNA due to chemical depurination over this time is not significant and thus changes in the ratios reflect the removal of O^6-meGua due to an enzymic process.

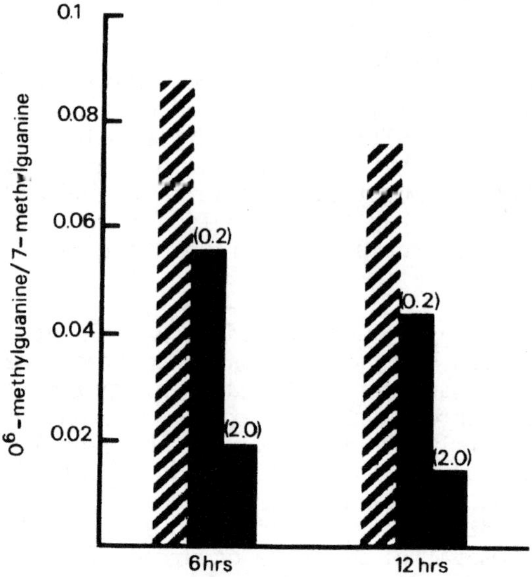

Figure 1. O^6-/7-Methylguanine ratios in liver DNA 6 and 12 hrs after 2 mg/kg [^{14}C]-DMN to BDIV rats pretreated (■) with 0.2 or 2.0 mg/kg DMN for 44 days, or not (▨).

As shown in Fig. 1, the ratios in the control rats at 6 and 12 hrs were 0.087 and 0.076, whereas in the rats pretreated with 0.2 mg/kg DMN the ratios were 0.056 and 0.044 and an even greater increase in removal of O^6-meGua was observed in rats pretreated with 2.0 mg/kg DMN. The values found in the pretreated rats may have been influenced by repair of DNA containing O^6-meGua produced by the preceding doses of unlabelled DMN; this may give a comparative indication of the magnitude of the effect. These data also indicate that the pretreatment did not alter the ability of the liver to metabolise the nitrosamine, because in such a case this should be reflected in the amounts of 7-methylguanine or 3-methyladenine in liver DNA which were, however, similar in all pretreated and control rats (for further details see Montesano *et al.*, 1980a).

Variations in the length of pretreatment show that the enhanced removal of O^6-meGua is produced within 2 weeks of pretreatment with DMN. The O^6-/7-methylguanine ratios in liver DNA 6 or 24 hrs after administration of [^{14}C]-DMN (2 mg/kg) to control rats or to rats which had received unlabelled DMN (2 mg/kg) over a period of 1, 2, 3, 4 and 6 weeks are shown in Fig. 2. The treatment with DMN resulted in a lower O^6-meGua/7-methylguanine ratio within one week of treatment. A still

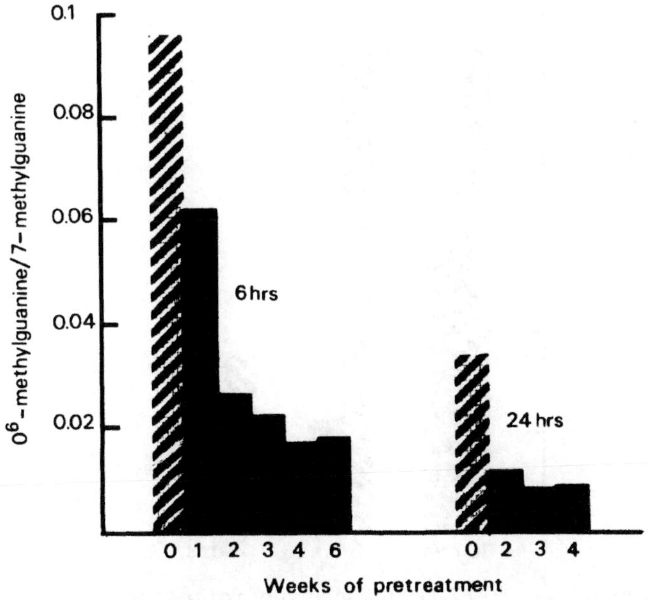

Figure 2. O^6-/7-Methylguanine ratios in liver DNA 6 and 24 hrs after 2 mg/kg [^{14}C]-DMN to BDIV rats pretreated with 2 mg/kg DMN for varying numbers of weeks.

lower amount of O^6-meGua was found by 2 to 3 weeks of pretreatment, and no further significant decline was observed by continuing the pretreatment up to 6 weeks.

The results obtained 24 hrs after administration of the [^{14}C]-DMN parallel those obtained at 6 hrs. At both times no effect on the formation and persistence in DNA of other alkylated adducts, like 7-methylguanine and 3-methyladenine, was detected, indicating that the treatment specifically affects the removal of O^6-meGua. More recently, comparable results have been reported by Swann and Mace (1980).

The lower amount of O^6-meGua found in the DNA of pretreated rats, as compared to controls, is the result of an increased rate of removal which occurs during the early period of time between administration and measurement of the amounts of alkylated adducts. This is very likely the consequence of the induction of the enzyme(s) responsible for the removal of O^6-meGua from DNA. The increased activities of cell-free preparations from liver of pretreated rats in removing O^6-meGua from methylated DNA *in vitro* as compared to those of liver extracts from control animals, is in keeping with this conclusion (Pegg, 1980; Montesano *et al.*, 1980b).

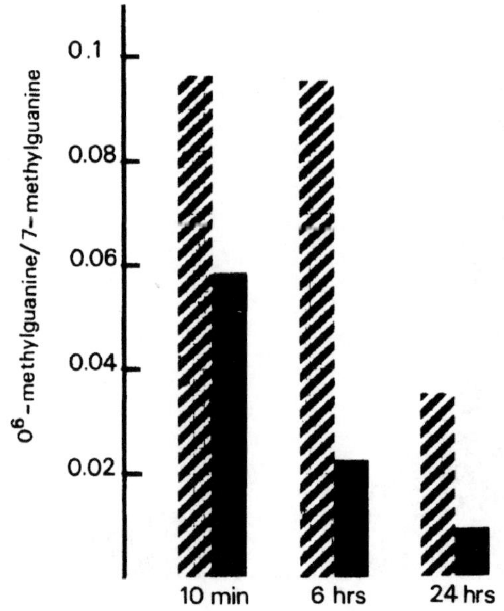

Figure 3. O^6-/7-Methylguanine ratios in liver DNA 10 min, 6 and 24 hrs after administration of 2 mg/kg [^{14}C]-DMN to BDIV rats pretreated (■) or not (▨) with 2 mg/kg DMN for 3 weeks.

The kinetics of this process indicate that the removal of O^6-meGua in liver of pretreated rats occurs very early after the dose of [^{14}C]-DMN and the enzyme is able to cope with a limited amount of this DNA adduct only (Montesano *et al.*, 1980b).

Fig. 3 shows that in liver DNA of pretreated rats at 10 min after a dose of 2 mg/kg [^{14}C]-DMN the value of the O^6-/7-methylguanine ratio is 0.058, whereas in control rats it is 0.096. When various challenge doses of [^{14}C]-DMN (0.2, 2.0 and 20 mg/kg) are administered to rats pretreated with 2 mg/kg of unlabelled DMN an increased removal of O^6-meGua is observed with the doses of 2.0 and 0.2 mg/kg, but not with the dose of 20 mg/kg (Fig. 4)

These findings indicate that the increased enzymic activity induced by pretreatment with DMN has a limited and finite capacity to cope with the removal of an increased amount of DNA damage, that is the removal of an increased number of O^6-meGua molecules; above this level no differences are detected in liver DNA of pretreated or control rats in the rate of removal of O^6-meGua by the constitutive enzyme.

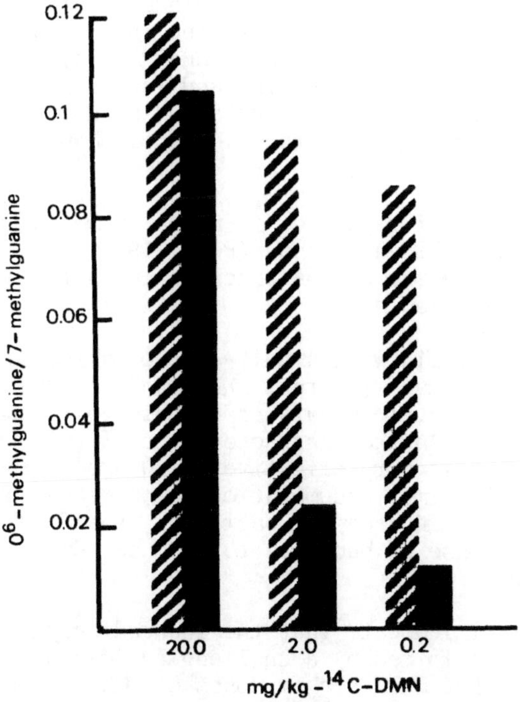

Figure 4. O^6-/7-Methylguanine ratios in liver DNA 6 hrs after various doses of [^{14}C]-DMN to BDIV rats pretreated (■) or not (▨) with 2.0 mg/kg DMN for 3 weeks

Extensive studies by Pegg and Hui (1978) on the removal of O^6-meGua from rat liver DNA after various single doses of DMN indicated that there may be two enzyme systems for this process, one of which is saturated at much lower levels of DNA damage than the other.

THE ADAPTIVE RESPONSE IN *E. COLI*

The observations of an increased removal of O^6-meGua from liver DNA of rats treated with low doses of nitrosamine parallel very closely the finding in *E. coli* of an inducible error-free repair process during exposure to low doses of the alkylating agent, MNNG. The induction of this repair process results in the development of a resistance to the mutagenic effect of the same agent (adaptive response) (Samson and Cairns, 1977; Jeggo *et al.*, 1977). Evidence was also provided that this error-free repair process is distinct from the SOS repair process and affects specifically certain DNA alkylation products. In fact, no differences were observed between adapted and nonadapted bacteria in the accumulation of 7-methylguanine or 3-methyladenine in their DNA, whereas no accumulation of O^6-meGua was detected in the adapted bacteria as compared to nonadapted bacteria (Schendel and Robins, 1978; Schendel *et al.*, 1978). This indicates that the induced repair specifically affects O^6-meGua and not the other DNA alkylation products and is responsible for the lack of or limited mutagenicity of MNNG in adapted bacteria.

Although the mechanism of repair of O^6-meGua in DNA of bacteria or mammalian cells is not well understood (Pegg, 1978; Karran *et al.*, 1979; Verly, 1980), the kinetics of the reaction in bacteria and mammalian cells show close resemblance.

The reaction is very rapid with a half-life of O^6-meGua in adapted bacteria of less than one second (Cairns, 1980a) and the reactant molecule responsible for the removal appears to be able to act only once (Robins and Cairns, 1979). Another characteristic of this repair process is that this reaction is very effective below a certain amount of alkylation, but above this limit accumulation of O^6-meGua and mutations occur at the same rate as in nonadapted bacteria. For further discussion on the adaptive response of bacteria to MNNG see Cairns (1980b) and Cairns *et al.* (1980).

The data in rat liver indicate that the induced enzyme is very rapid in removing O^6-meGua formed by a challenging dose of 2 mg/kg DMN (Fig. 3), and also show that when higher doses of DMN (20 mg/kg) are used the enzyme seems to be saturated (Fig. 4).

DISCUSSION

The modulation of repair in liver DNA of O^6-meGua following various dose schedules of chronic DMN to rats is probably reflected in the carcinogenic dose-response of DMN in this organ. Although the presence of O^6-meGua in DNA cannot be seen separately from the rate of DNA synthesis as a critical determinant in the initiation of carcinogenesis by DMN, these findings suggest that at some point the risk of cancer developing would increase more rapidly than in proportion to the dose rate of carcinogen. Recent long-term carcinogenicity studies (R. Peto, personal communication) are in keeping with this suggestion; these studies indicate that, above a certain dose level of DMN, the liver cancer risk in rats increases rapidly with the dose of DMN administered, resulting in at least a 1000-fold increase of tumour incidence with a 10-fold increase in daily dose rate.

Table 1 compiles the various animal systems so far reported in which the effect of treatment with various carcinogens on the removal of O^6-meGua was examined. A proper analysis of these data is however

Table 1. Effect of chronic administration of various carcinogens on the removal from rat liver DNA of O^6-alkylguanine

Pretreatment	Challenge	Species	Effect[a]	References
DMN	DMN	BDIV rats Sprague-Dawley rats Wistar rats	+ + +	Montesano *et al.*, 1979, 1980a Pegg, 1980 Swann and Mace, 1980
AAF	DMN	Wistar rats	+	Buckley *et al.*, 1979
NMU	DMN	Wistar rats	-	Unpublished data
MMS	DMN	Wistar rats	-	Unpublished data
DMN	DMN	Chinese hamsters	-	Margison *et al.*, 1979a
DEN	DMN	Wistar rats	+	Margison *et al.*, 1979b
DEN	DEN	Wistar rats	+	Margison *et al.*, 1979c
DMPT	DMPT	BDIX rats	+	Cooper *et al.*, 1978

[a] **(+) indicates that an increased removal from DNA of O^6-alkylguanine, but not of 7-alkylguanine or 3-alkyladenine, was observed in pretreated as compared to control rats.**

conditioned by the limited knowledge on the reaction of removal of O^6-meGua from DNA and on the mechanism responsible for the induction of this repair process.

It is not unreasonable to imagine that similar inducible repair processes must exist for other classes of chemical carcinogens.

ACKNOWLEDGEMENTS

We express our thanks to Dr L. Tomatis for the helpful criticism of the manuscript and to Miss J. Mitchell for secretarial assistance. The work reported here was supported in part by Contract NO1 CP-55630 of the National Cancer Institute of the USA.

REFERENCES

Abbott, P.J. and Saffhill, R.: 1979, Biochim. Biophys. Acta, 562, pp. 51-61
Buckley, J.D., O'Connor, P.J. and Craig, A.W.: 1979, Nature (Lond.), 281, pp. 403-404
Cairns, J.: 1980a, Nature (Lond.), in press
Cairns, J.: 1980b, Proc. Roy. Soc. Lond. B, in press
Cairns, J., Robins, P., Sedgwick, B. and Talmud, P.: 1980, Progr. Nucleic Acid Res. Molec. Biol., in press
Cooper, H.K., Hauenstein, E., Kolar, G.F. and Kleihues, P.: 1978, Acta Neuropathol. (Berl.), 43, pp. 105-109
Coulondre, C. and Miller, J.H.: 1977, J. Mol. Biol., 117, pp. 577-606
Craddock, V.M.: 1978, in Primary Liver Tumours, eds. Remmer, H., Bolt, H.M., Bannasch, P. and Popper, H., MTP Press, Lancaster, UK, pp. 377-383
Gerchman, L.L. and Ludlum, D.B.: 1973, Biochim. Biophys. Acta, 308, pp. 310-316
Goth, R. and Rajewsky, M.F.: 1974, Proc. Natl. Acad. Sci. USA, 71, pp. 639-643
Jeggo, P.: 1979, J. Bacteriol., 139, pp. 783-791
Jeggo, P., Defais, M., Samson, L. and Schendel, P.: 1977, Mol. Gen. Genet., 157, pp. 1-9
Karran, P., Lindahl, T. and Griffin, B.: 1979, Nature (Lond.), 280, pp. 76-77
Lawley, P.D.: 1976a, in Screening Tests in Chemical Carcinogenesis, IARC Scientific Publications No. 12, eds. Montesano, R., Bartsch, H. and Tomatis, L., International Agency for Research on Cancer, Lyon, pp. 181-210
Lawley, P.D.: 1976b, in Chemical Carcinogens, ACS Monograph Series 173, ed. Searle, C.E., American Chemical Society, Washington, D.C., pp. 83-244
Loveless, A.: 1969, Nature (Lond.), 223, pp. 206-207
Magee, P.N., Montesano, R. and Preussmann, R.: 1976, in Chemical Carcinogens, ACS Monograph Series 173, ed. Searle, C.E., American Chemical Society, Washington, D.C., pp. 491-625
Margison, G.P., Craig, A.W., Brésil, H., Curtin, N., Snell, K. and Montesano, R.: 1979b, Br. J. Cancer, 40, pp. 815
Margison, G.P., Curtin, N.J., Snell, K. and Craig, A.W.: 1979c, Br. J. Cancer, 40, pp. 809-813
Margison, G.P., Margison, J.M. and Montesano, R.: 1976, Biochem. J., 157, pp. 627-634
Margison, G.P., Margison, J.M. and Montesano, R.: 1977, Biochem. J., 165, pp. 463-468
Margison, G.P. and O'Connor, P.J.: 1979, in Chemical Carcinogens and DNA, ed. Grover, P.L., CRC Press, Baltimore, pp. 111-159
Margison, G.P., Swindell, J.A., Ockey, C.H. and Craig, A.W.: 1979a, Br. J. Cancer, 40, pp. 816
Montesano, R., Brésil, H. and Margison, G.P.: 1979, Cancer Res., 39, pp. 1798-1802

Montesano, R., Brésil, H., Planche-Martel, G., Margison, G.P. and Pegg, A.E.: 1980a, Cancer Res., 40, pp. 452-458
Montesano, R., Brésil, H., Planche-Martel, G. and Margison, G.P.: 1980b, Proc. Am. Assoc. Cancer Res., 21, p. 2.
Newbold, R.F., Warren, W., Medcalf, A.S.C. and Amos, J.: 1980, Nature (Lond.), 283, pp. 596-599
Nicoll, J.W., Swann, P.F. and Pegg, A.E.: 1975, Nature (Lond.), 254, pp. 261-262
Pegg, A.E.: 1977, Adv. Cancer Res., 25, pp. 195-269
Pegg, A.E.: 1978, Biochem. Biophys. Res. Commun, 84, pp. 166-173
Pegg, A.E.: 1980, Arch. Toxicol., in press
Pegg, A.E. and Hui, G.: 1978, Cancer Res., 38, pp. 2011-2017
Roberts, J.: 1980, Br. Med. Bull, 36, pp. 25-31
Robins, P. and Cairns, J.: 1979, Nature (Lond.), 280, pp. 74-76
Samson, L. and Cairns, J.: 1977, Nature (Lond.), 267, pp. 281-282
Schendel, P.F., Defais, M., Jeggo, P., Samson, L. and Cairns, J.: 1978, J. Bacteriol., 135, pp. 466-475
Schendel, P.F. and Robins, P.E.: 1978, Proc. Natl. Acad. Sci. USA, 75, pp. 6017-6020
Singer, B.: 1979, J. Natl. Cancer Inst., 62, pp. 1329-1339
Stumpf, R., Margison, G.P., Montesano, R. and Pegg, A.E.: 1979, Cancer Res., 39, pp. 50-54
Swann, P.F. and Mace, R.: 1980, Chem.-Biol. Interactions, in press
Swann, P.F. and Magee, P.N.: 1968, Biochem. J., 110, pp. 39-47
Swann, P.F. and Magee, P.N.: 1971, Biochem. J., 125, pp. 841-847
Verly, W.G.: 1980, Biochem. Pharmacol., 29, pp. 977-982
Vogel, E. and Natarajan, A.T.: 1979, Mutation Res., 62, pp. 55-100

DNA REPAIR IN HUMAN CELLS EXPOSED TO COMBINATIONS OF CARCINOGENIC AGENTS

R. B. Setlow and Farid E. Ahmed[1]
Biology Department, Brookhaven National Laboratory, Upton, New York 11973 USA

Normal human and XP[2] fibroblasts were treated with UV plus UV-mimetic chemicals. The UV dose used was sufficient to saturate the UV excision repair system. Excision repair after combined treatments was estimated by unscheduled DNA synthesis, BrdUrd photolysis, and the loss of sites sensitive to a UV specific endonuclease. Since the repair of damage from UV and its mimetics is coordinately controlled we expected that there would be similar rate-limiting steps in the repair of UV and chemical damage and that after a combined treatment the total amount of repair would be the same as from UV or the chemicals separately. The expectation was not fulfilled. In normal cells repair after a combined treatment was additive whereas in XP cells repair after a combined treatment was usually less than after either agent separately. The chemicals tested were AAAF, DMBA-epoxide, 4NQO, and ICR-170.

INTRODUCTION

Skin cancer is the most common cancer in the United States and experimental and epidemiological data indicate that it arises from the shorter UV (290-320 nm) that penetrates the stratospheric ozone layer and reaches the surface of the earth (Scott and Straf). The knowledge that cells from individuals with the disease XP are defective in one or more pathways for repairing UV damage to DNA (Cleaver and Bootsma; Setlow, 1978; Arlett and Lehmann; Friedberg et al.) reinforces the idea that damages to DNA are initiating events in skin carcinogenesis. A comparison of the dose-response curve for skin cancer incidence in the white population of the United States and in XPs indicates that DNA repair is effective in removing > 85% of UV damage from the average person and that as a result of this removal the skin cancer incidence is 10^2-10^4-fold less than in XPs (Setlow, 1980). Such estimates are very crude because of heterogeneity in both the average and in the XP populations, not only in their life-style and in the UV transmission of their skin, but in their repair capabilities. For example, there are seven different

B. Pullman, P. O. P. Ts'o and H. Gelboin (eds.), Carcinogenesis: Fundamental Mechanisms and Environmental Effects, 453–464.

complementation groups of XP and they vary widely in their abilities to repair UV damage (Friedberg et al.). Nevertheless, the calculation indicates that relatively small changes in repair might have large effects on cancer initiation.

It is important to know the times of repair compared to replication and transcription since damage to DNA will only have biological consequences if the altered template is enzymatically read. The fixation of mutations will depend on the relative time of replication compared to DNA repair (Maher, et al.) and the cytotoxic action of UV on fibroblasts could well involve transcriptional events that result in the synthesis of aberrant proteins which would not permit cells to remain attached to their substrate (Kantor and Hull). Hence, much experimental effort has gone into the measuring the kinetics of DNA repair. To do so properly it is necessary to know what alterations are being repaired and whether the alterations would have severe biological consequences if not repaired. For example, N-7 alklyguanine seems to be an innocuous damage whereas 0^6 methylguanine is much more important biologically although it is numerically inferior to the N-7 product. UV irradiation of DNA results in the formation of many products (Setlow and Setlow) but one of these--pyrimidine dimers--can be shown to be a major lethal and mutagenic lesion in prokaryotic and eukaryotic systems by virtue of the fact that many of them contain photoreactivating enzyme. The enzyme binds to DNA and in the presence of light monomerizes the dimers and reverses the biological effect of UV. There is no other known substrate for the enzyme and hence it may be used as a diagnostic reagent for the role of pyrimidine dimers in biological effects. By this test dimers account for $\sim 80\%$ of the UV killing of frog cells in culture (Rosenstein and Setlow). Such data do not indicate the molecular mechanisms by which dimers affect cells.

Most laboratory experiments are carried out with UV of wavelength 254 nm but the effective wavelengths in sunlight seem to be closer to 305 nm (Setlow, 1974). The inference that dimers are as important at the longer wavelengths as at the shorter comes not only from the constancy of the photoreactivable sector as a function of wavelength (Rosenstein and Setlow) but from the shape of the action spectrum for killing mammalian cells which parallels closely that for the production dimers in both Chinese hamster cells (Rothman and Setlow) and in normal human and XP fibroblasts (Kantor, et al.). The latter argument is strengthened by the observation that the yield of other UV photoproducts relative to dimers increases dramatically at longer wavelengths (Cerutti and Netrawali). A further indication that 254 nm is a good model for the effects of many sunlight irradiations is the equality of the ratio of sister chromatid exchanges to endonuclease-sensitive sites in Chinese hamster cells irradiated by 254 nm and by simulated sunlight (Reynolds, et al.).

EXCISION REPAIR

Despite the fact that UV-induced pyrimidine dimers are well identified, major lesions, it is important to remember that both physical and chemical carcinogens make a multitude of products in DNA, for example, DNA protein crosslinks (Fornace and Little). The damage resulting from the physical agents is more or less randomly distributed along DNA, but chemical changes seem concentrated in the spacer regions between nucleosomes (Jahn and Litman). The removal of damages by excision repair is not completely random. It seems preferentially to begin in spacer regions but the data indicate that such regions are not static ones but seem to move along the DNA duplex (Smerdon et al.).The latter observations probably account for the changing kinetics of excision repair with time (see below).

The presumed sequence of steps in excision repair has been described many times (Hanawalt, et al.). Such descriptions show that an early step in excision repair involves attack on the damaged polynucleotide by an endonuclease such as a UV-specific endonuclease. Because few single strand breaks accumulate during repair and because fewer breaks are observed in excision-defective cells than in normally repairing ones, it seems as if the endonucleolytic step is the rate-limiting one. This point of view is reasonable but it should be remembered that the only UV-specific endonucleases that have been well characterized are those from T4 phage infected E. coli and from M. luteus (Paterson). These endonuclease preparations are specific for pyrimidine dimers and make one single strand nick per dimer. However, the M. luteus activity seems to be a combination of two separate ones--a glycosylase that splits half of the dimer from the polynucleotide (Grossman et al.) and an endonuclease acting at a later time (Setlow and Grist). The UV-endonuclease activity in E. coli seems to be a complex of proteins (Seeberg) and that from calf thymus is a large unstable protein (Waldstein, et al.). The fact that there are seven complementation groups of XP implies that any one of seven mutations results in a decrease in the ability to excise dimers and that many proteins or cofactors are associated with the endonuclease step in human cells.

The observations that many chemicals mimic UV (Regan and Setlow; Setlow, 1978; Friedberg, et al.) (see Table 1) suggests that there are similar rate-limiting steps for the repair of UV damage and damage from mimetics and that the chemical damages should compete for the repair system working on dimers in vivo. We have investigated this competition.

In the work to be described three techniques were used to measure excision. They are: 1) UDS measured radioautographically by the incorporation of ^{3}HdThd for 3 hr after treatment; 2) the photolysis of BrdUrd incorporated into parental DNA for 12 hrs after treatment (Regan and Setlow); and 3) the loss during 24 hrs after treatment of sites in DNA sensitive to a UV endonuclease preparation

Table 1. Ways in which some chemical damages mimic UV damage in human cells.

1. UV- sensitive cells (XP) are more sensitive to the chemical than normal cells.
2. Chemically treated viruses show a higher survival on normal cells than on XP cells.
3. XP cells deficient in repair of UV damage are also deficient in excision of chemical damage.
4. Excision repair of UV and of chemical damage involves long patches (approx. 100 nucleotides).
5. XP complementation groups observed for repair of chemical damage are the same as those for UV damage (Zelle and Bootsma).

from M. luteus (Paterson, 1978). The first two techniques detect most types of excision repair and their quantitative values depend on the number of sites repaired, the patch size, and the concentration of thymine in the patches. The third technique measures only the loss of pyrimidine dimers and so is well suited to measure dimer repair in cells treated with combinations of UV and chemicals that mimic UV.

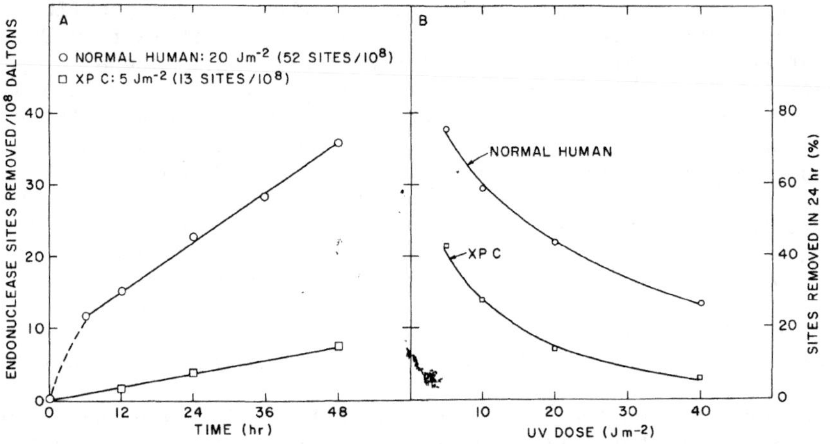

Figure 1. Excision repair as a function of time and dose for normal and XP fibroblasts (Ahmed and Setlow, 1978a).

The use of the endonuclease assay in measuring the excision of dimers is indicated in Fig. 1 which shows (A) excision versus time at a fixed dose for normal and XP cells and (B) the percentage excised at a fixed time for different initial doses. The latter figure shows not only the defect in excision in complementation group C of XP cells but indicates that the defect is not an absolute one and that because the repair system becomes saturated (see below) the fraction of dimers excised decreases as the dose increases. Hence, at high doses it is difficult to measure repair by the loss of specific products. At such doses however, it is possible to measure repair easily by the other techniques.

If such experiments are to be done properly one must work at dose levels that saturate the repair systems. Figure 2 shows data for dimer removal versus dose for a number of cell strains. Not shown is the fact that the initial number of dimers increases

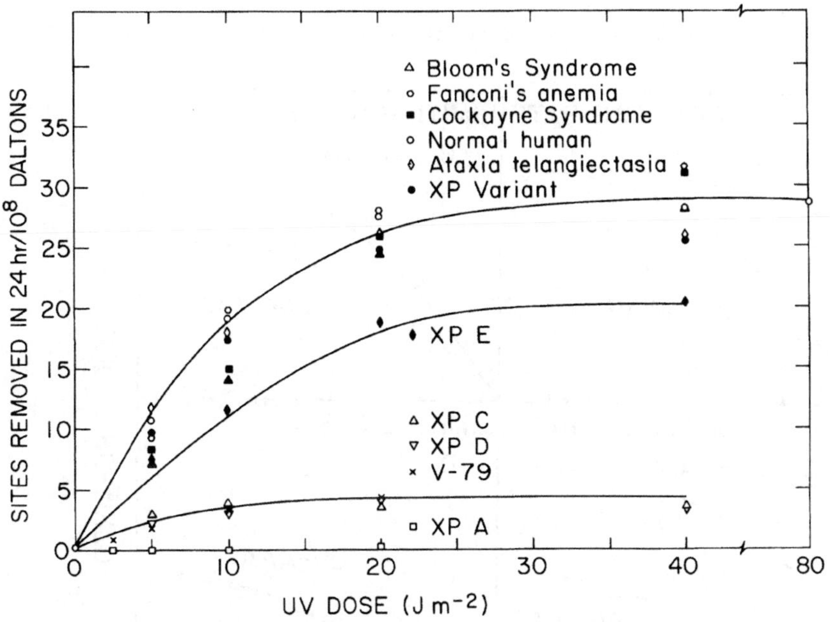

Figure 2. Endonuclease sensitive sites removed in 24 hrs as a function of initial UV dose for a number different human cell strains and Chinese hamster V79 cells (Ahmed and Setlow, 1979a).

proportionately with dose (Ahmed and Setlow, 1979b). Three conclusions are apparent from the data. 1) The repair rate in normal cells is saturated at doses of approximately 20 Jm^{-2} (approximately 50 sites per 10^8d) and at somewhat lower doses in XP and Chinese hamster cells. 2) XP cells are deficient compared to normal excising strains. 3) V79 cells are similar in their excision properties to XP C cells. Similar saturation data are observed, using the UDS and BrdUrd photolysis techniques for cells treated with UV or AAAF. The amount of UDS is consistent with the observed dimer removal and patch size (Ahmed and Setlow, 1979b). AAAF acted like UV if its concentration in μM equalled the UV dose in Jm^{-2}. We also obtained saturation data for human cells treated with an epoxide of DMBA but for both normal and XP cells the saturation level was only 10-20% of that observed after UV (Ahmed, et al.). This observation in itself indicates that although the repair of UV and DMBA damage may be coordinately controlled, the repair pathways must be different.

COMPETITION BETWEEN UV AND ITS MIMETICS

Expectation.

Most of our work has been done at doses (20 Jm^{-2}) close to the

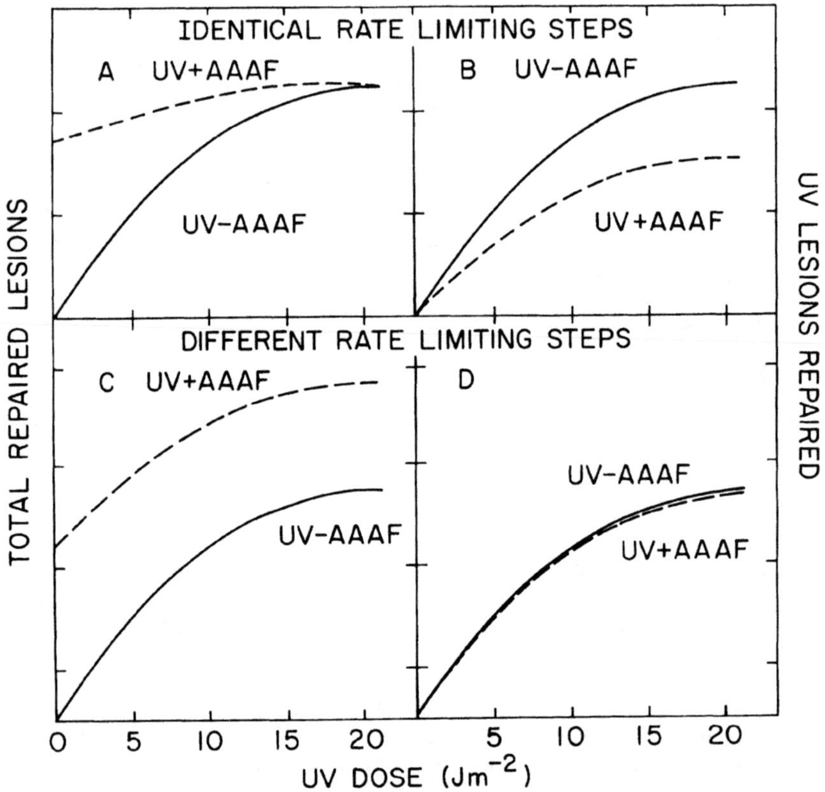

Figure 3. Possibilities for repair as a function of UV dose for cells treated ± 20 μM AAAF. (A and C): total repair; (B and D): dimers removed in 24 hr.

saturation level. Such doses are much greater than those received in sunlight. Noon sunlight in Texas would affect DNA in skin at the equivalent of a 254 nm dose rate of ~0.01Jm^{-2}/min. The possible results from a combined treatment with UV and one of its mimetics, AAAF, are shown in Fig. 3. Panels A and C give the possibilities for UDS or BrdUrd photolysis and panels B and D give the possibilities for the loss of endonuclease-sensitive sites. If the repair of AAAF damage had identical rate-limiting steps as UV repair, we would expect no increase in UDS and a decrease in dimer excision, since the repair would be shared between UV and AAAF lesions. If there were different rate-limiting steps, UDS would be additive and AAAF would not inhibit dimer excision.

UV PLUS AAAF

Typical UDS data are shown in Fig. 4. The results

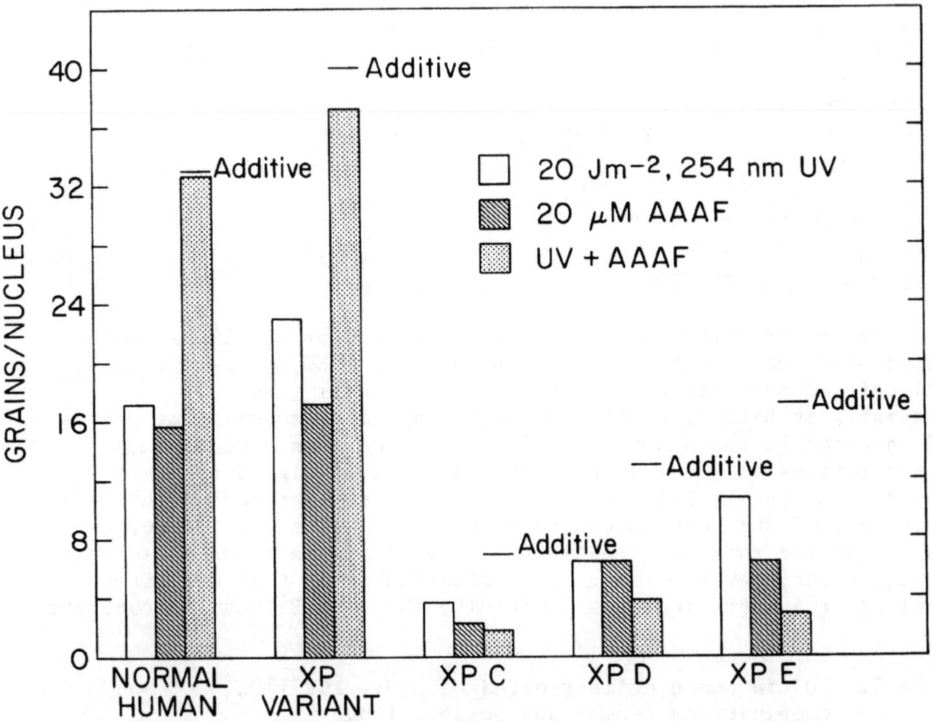

Figure 4. UDS in normal or XP cells exposed to UV or AAAF or the combination of the two (Ahmed and Setlow, 1979b).

indicate additivity of repair for normal excising strains and, not predicted in Fig. 3, an inhibitory effect for XP cells. The latter data cannot be explained by cell toxicity since the measurements were made on the plateau of the repair versus dose curve (see Fig. 2). Similar results using the other two assays and a number of other cell strains are shown in Table 2 (Ahmed and Setlow, 1978b, 1979b). The generalization is clear. AAAF damage does not compete with the repair of UV damage in normal excising strains, but in XP cells each agent inhibits the repair of the other.

TABLE 2.

Measures of excision repair in human cells treated with UV, AAAF and combinations

Cell, line	Unscheduled synthesis(1)			Endonuclease assay(2)		BrdUrd photolysis(3)		
	$20 Jm^{-2}$	$20 \mu M$	$20 Jm^{-2} + 20 \mu M$	$20 Jm^{-2}$	$20 Jm^{-2} + 20 \mu M$	$20 Jm^{-2}$	$10 \mu M$	$20 Jm^{-2} + 10 \mu M$
Normal human								
Par Bel (CRL 1191)	18.6	16.1	33	27.5	27.4	3.5	1.4	4.9
Rid Mor (CRL 1220)	17.4	16.4	32.9	23.1	23.3	3.3	1.4	4.4
Ataxia telangiectasia								
NeNo (CRL 1347)	19.7	16.6	35	24.6	24.2	3.2	2.5	5.4
Se Pan (CRL 1343)	22.3	14	33.2	24.5	24.9	3.0	1.4	3.9
AT 4BI	19.6	14.4	35	26.3	26.4	2.2	1.4	3.4
Fanconi's anemia								
Ce Rel (CRL 1196)	15.4	13.9	28.2	27.5	27.4	3.4	0.9	4.4
Cockayne syndrome								
GM 1098	14.3	19.8	33.3	26.1	26.1	3.2	2.6	6.2
GM 1629				26.9	27.0	4.1	2.1	6.6
Xeroderma pigmentosum								
Variant; Wo Mec (CRL 1162)	23.0	17.1	37.2	24.7	24.0	2.8	1.1	4.3
C; Ge Ar (CRL 1161)	3.7	2.2	1.8	3.7	1.4	1.0	0.1	0.1
D; Be Wen (CRL 1160)	6.4	6.6	3.8	3.9	0.8	1.4	0.2	0.4
E; XP2R0 (CRL 1259)	10.8	6.4	2.8	19.3	9.6	4.8	1.0	3.8

(1) Grains/Nucleus incorporated in 3 hr (8 days exposure).
(2) Sites removed in 24 hr/10^8 Daltons.
(3) $\Delta(1/M_w) \times 10^8$ at highest 313 nm dose (12 hr repair).

UV PLUS 4NQO OR ICR-170

Repair after treatment of cells with 4NQO or ICR-170 is more complicated than after UV (Ahmed and Setlow, 1980). There is no well-defined saturation dose for these chemicals (Fig. 5). Presumably at high doses the chemicals damage proteins or membranes and as a result the amount of DNA repair goes down. We used concentrations well below the peak responses in Fig. 5 so there should be no possibility of a combined treatment exceeding the equivalent of the peak concentrations shown in Fig. 5. Hence, the results of the combined treatment are not as dramatic as those shown in Fig. 4 but, nevertheless, it is clear(Table 3) that UDS after combined treatments indicate additivity for normal human fibroblasts

Table 3. UDS in human cells treated with UV, ICR-170, 4NQO and combinations (Ahmed and Setlow, 1980).

	UV	ICR-170	Combined	UV	4NQO	Combined
normal	37.3	6.4	44.6	37.3	9.3	44.8
XP C	6.1	5.1	4.1	5.9	1.9	8.8

UV: $20 Jm^{-2}$; ICR-170: 5 μM; 4NQO:0.5 μM

and an inhibitory effect for XP C fibroblasts treated with UV and ICR-170 but additivity for UV plus 4NQO. Similar results were obtained by the BrdUrd photolysis technique. A summary of our results is shown in Table 4.

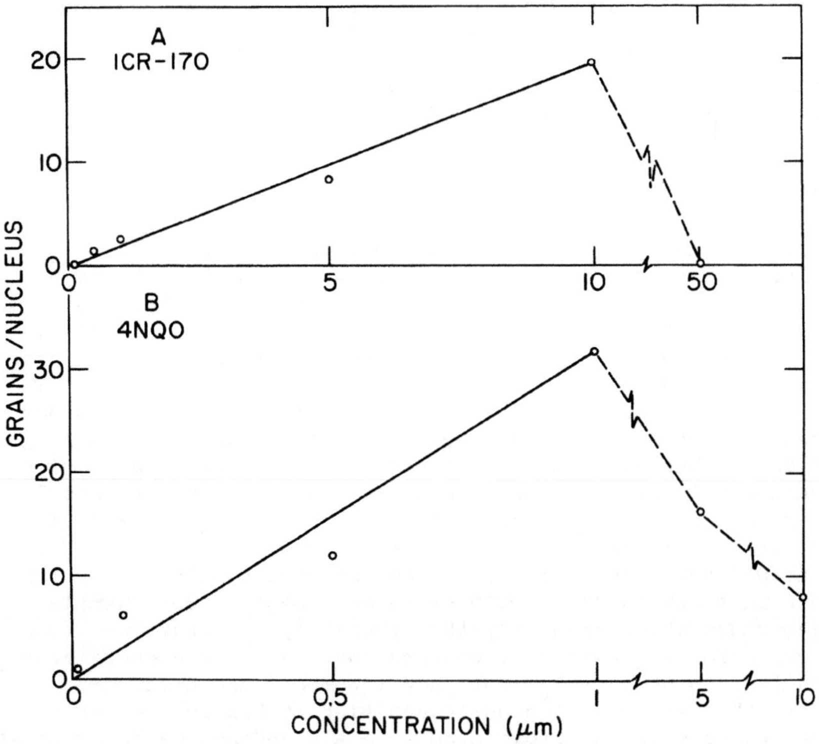

Figure 5. UDS in normal human fibroblasts versus concentration for ICR-170 and 4NQO (Ahmed and Setlow, 1980).

Table 4. DNA repair responses of human cells exposed to combinations of UV and its mimetics.

	Normal	XPC
UV + AAAF	additive	inhibitory
UV + DMBA-epoxide	additive	inhibitory
UV + ICR-170	additive	inhibitory
UV + 4NQO	additive	additive

Another group of investigators (Brown, et al.) has not observed additivity of repair in normal cells treated with UV plus AAAF or UV plus 4NQO. Since both their and our experiments seem definitive but contradictory, we suspect that they have not replicated our experiments in some as yet undefined way.

CONCLUSION

Obviously there is no simple explanation for all the observations we report. Nevertheless, we conclude that the repair pathways in XP cells are qualitatively different from those in normal human cells. If they were similar, but just had a lower level of the necessary enzymes or repair systems, we would expect that both groups of cells would either show additivity or an inhibitory effect. They do not. One could explain the results by hypothesizing that minor UV photoproducts might inhibit the repair enzymes in XP cells working on chemical damage but not inhibit those that work on UV damage and vice versa. Alternatively, one could construct rather elaborate models based on groups or complexes of proteins or cofactors to explain the observations that XP cells are defective in the repair of many damages to DNA but that normal cells do not have similar rate-limiting steps in the repair of these damages. For example Yarosh describes the possibility that there may be separate endonucleases for the different damages and that such endonucleases are normally present in relatively small numbers per cell. He further hypothesizes that the nucleases bind to DNA but do not nick it unless two or more cofactors common to all endonucleases associate with the nuclease bound to DNA. In normal cells there might be an excess of cofactors and the nucleases would be saturated with them and the repair would be additive. If the number of cofactors in XP cells was small, say 2- or 3-fold greater than the number of UV endonucleases, then after a combined treatment with UV and AAAF both UV and AAAF endonucleases would bind to DNA and the cofactors would be distributed equally between the two types of nucleases. As a result, only a small fraction of the nucleases would have two associated cofactors and the observed UV endonuclease activity would be depressed extensively as would be that of the putative AAAF endonuclease activity. These involved speculations are only presented to indicate the complexity of the problem and as a guide to further research.

ACKNOWLEDGEMENT

The research reported here was carried out at Brookhaven National Laboratory under the auspices of the U. S. Department of Energy.

REFERENCES

Ahmed, F. E., Gentil, A., Rosenstein, B. S., and Setlow, R. B.: 1980, Biochim, Biophys. Acta., in press.
Ahmed, F. E. and Setlow, R. B.: 1978a, Biophys. J. 24, pp. 665-675.
Ahmed, F. E. and Setlow, R. B.: 1978b, Biochim. Biophys. Acta 521, pp. 805-817.
Ahmed, F. E. and Setlow, R. B.: 1979a, Photochem. Photobiol. 29, pp. 983-989.
Ahmed, F. E. and Setlow, R. B.: 1979b, Cancer Res. 39, pp. 471-479.
Ahmed, F. E. and Setlow, R. B.: 1980, Chem.-Biol. Interactions, in press.
Arlett, C. F. and Lehmann, A. R.: 1978, Ann. Rev. Genet. 12, pp. 95-111.
Brown, A. J., Fickel, T. H., Cleaver, J. E., Lohman, P. H. M., Wade, M. H., and Waters, R.: 1979, Cancer Res. 39, pp. 2522-2527.
Cerutti, P. A. and Netrawali, M.: 1979, Radiation Research (Proc. 6th Int. Cong. Radiat. Res., Tokyo), pp. 423-432.
Cleaver, J. E. and Bootsma, D.: 1975, Ann. Rev. Genet. 9, pp. 19-38.
Fornace, A. J. and Little, J. B.: 1979, Cancer Res. 39, pp. 704-710.
Friedberg, E. C., Ehmann, U. K., and Williams, J. I.: 1979, Adv. Radiat. Res. 8, pp. 85-174.
Grossman, L., Riazuddin, S., Haseltine, W. A., and Lindan, C.: 1978, Cold Spr. Harb. Symp. Quant. Biol. 43, pp. 947-955.
Hanawalt, P. C., Cooper, P. R., Ganesan, A. K., and Smith, C. A.,: 1979, Ann. Rev. Biochem. 48, pp. 783-836.
Jahn, C. L. and Litman, G. W.: 1979, Biochemistry 18, pp. 1442-1449.
Kantor, G. J. and Hull, D. R.: 1979, Biophys. J. 27, pp. 359-370.
Kantor, G. J., Sutherland, J. C., and Setlow, R. B: 1980, Photochem. Photobiol., in press.
Maher, V. M., Dorney, D. J., Mendrala, A. L., Konze-Thomas, B., and McCormick, J. J.: 1979, Mutat. Res. 62, pp. 311-323.
Paterson, M. C.: 1978, Adv. Radiat. Biol. 7, pp. 1-53.
Regan, J. D. and Setlow, R. B.: 1974, Cancer Res. 34, pp. 3318-3325.
Reynolds, R. J., Natarajan, A. T., and Lohman, P. H. M: 1979, Mutat. Res. 64, pp. 353-356.
Rosenstein, B. S. and Setlow, R. B.: 1980, Photochem. Photobiol., in press.
Rothman, R. H., and Setlow, R. B.: 1979, Photochem. Photobiol. 29, pp. 57-61.
Scott, E. L. and Straf, M. L.: 1977, Origins of Human Cancer, Cold Spring Harbor Lab., pp. 529-546.
Seeberg, E.: 1978, Proc. Nat. Acad. Sc. USA 75, pp. 2569-2573.
Setlow, R. B.: 1974, Proc. Nat. Acad. Sci. USA 71, pp. 3363-3366.
Setlow, R. B.: 1978, Nature 271, pp. 713-717.
Setlow, R. B.: 1980, Arch. Toxicol.: in press.
Setlow, R. B. and Grist, E.: 1980, 8th Int. Photobiol. Cong., in press.
Setlow, R. B. and Setlow, J. K.: 1972, Ann. Rev. Biophys. Bioengineer. 1, pp. 293-346.

Smerdon, M. J., Kastan, M. B., and Lieberman, M. W.: 1979, Biochemistry 18, pp. 3732-3739.
Waldstein, E. A., Peller, S., and Setlow, R. B.: 1979, Proc. Nat. Acad. Sci. USA 76, pp. 3746-3750.
Yarosh, D.: 1979, Amer. Soc. Photobiol. Abstracts 7, p. 157.
Zelle, B. and Bootsma, D.: 1980, Mutat. Res., in press.

Footnotes

1. Present address: Pharmacopathics Research Laboratories, Inc., 9705 North Washington Blvd., Laurel, MD 20810.

2. Abbreviations used:

AAAF	- N-acetoxy-acetylaminofluorene
DMBA-epoxide	- 7, 12 dimethylbenz[a]anthracene 5, 6-oxide
ICR-170	- Acridine mustard
4NQO	- 4 nitroquinoline oxide
XP	- Xeroderma pigmentosum
UDS	- Unscheduled DNA synthesis
UV	- Ultraviolet

REACTIONS OF AFLATOXIN B_1 DAMAGED DNA IN VITRO AND IN SITU IN MAMMALIAN CELLS

P.A. Cerutti, V.T. Wang and P. Amstad
Swiss Institute for Experimental Cancer Research,
Dept. of Carcinogenesis - CH-1066 Epalinges s/Lausanne

ABSTRACT

The primary DNA adducts formed by metabolically activated aflatoxin B_1 (AFB_1), i.e. adducts released by acid hydrolysis as 2,3-dihydro-2-(N^7-guanyl)-3-hydroxyaflatoxin B_1 (AFB_1-N^7-Gua), are of low chemical stability and possess a half life time of 12 hr in free DNA at pH 7 and 37° and *in situ* in C3H 10T1/2 clone 8 mouse embryo fibroblasts. They undergo three major spontaneous reactions (1) release of 2,3-dihydro-2,3-dihydroxyaflatoxin B_1 resulting in the reconstitution of intact guanine residues; this reaction predominates under physiological conditions (2) release of free AFB_1-N^7-Gua presumably under formation of aguaninic sites (3) Formation of secondary products in DNA which possess increased chemical stability probably by opening of the 5-membered ring of guanine under formation of the putative 2,3-dihydro-2-(N^5-formyl-2',5',6'-triamino-4'-oxo-N^5-pyrimidyl)-3-hydroxyaflatoxin B_1 (AFB_1-triamino-Py).

The formation and removal of AFB_1-DNA adducts was studied in human lung cells A549 and 10T1/2 mouse embryo fibroblasts since it was found that both possess the capability to metabolize AFB_1. A549 but not 10T1/2 cells actively eliminated a portion of the AFB_1-DNA-adducts by a cellular repair process. While adducts which are released by acid hydrolysis in the form of the putative AFB_1-triamino-Py accumulated in the DNA in 10T1/2 cells at a rate which was comparable to free AFB_1-DNA these lesions were actively removed from the DNA in A549 cells. DNA was identified as the major target for the cytotoxic effect of AFB_1 towards 10T1/2 cells. The spontaneous loss of a portion of the adducts from DNA upon incubation of AFB_1-treated 10T1/2 cells in confluency before replating at low cell density for the determination of the colony forming ability was reflected in a corresponding diminution of the cytotoxic effect. Only very low transformation frequencies were obtained when confluent 10T1/2 cells were treated with 0.2 μM AFB_1 for 4 hr.

B. Pullman, P. O. P. Ts'o and H. Gelboin (eds.), Carcinogenesis: Fundamental Mechanisms and Environmental Effects, 465–477.

INTRODUCTION

Most ultimate carcinogens are electrophiles which react with multiple nucleophilic centres of DNA and other macromolecules. However, a favoured model of carcinogenesis proposes that DNA rather than the other macromolecules represents the crucial target for initiation of malignant transformation. Support for this notion derives from the high correlation between mutagenicity and carcinogenicity of a long list of agents. If DNA damage indeed represents the starting point for transformation it should be possible to establish quantitative relationships between DNA damage concentrations and transformation frequencies. Prerequisites for such correlation studies are the knowledge of the structures and chemical properties of the DNA lesions induced by a particular carcinogen as well as of the cellular pathways accomplishing their elimination and/or modification

In this report we are summarizing our recent studies of the chemical and biological properties of the DNA lesions induced by the mycotoxin aflatoxin B_1 (AFB_1) in rodent and human cells. The inherent chemical instability of the primary DNA adducts formed by AFB_1 in metabolically active tissues represented a special difficulty in this work. The contribution of enzyme catalyzed reactions to the processing of AFB_1-lesions in the intact cell only became discernible upon direct comparison with the spontaneous reactions of free AFB_1-DNA *in vitro*. It was found that human lung cells A549 but not C3H 10T1/2 clone 8 mouse embryo fibroblasts actively eliminate a fraction of AFB_1-DNA adducts by a cellular repair process. Nevertheless, a majority of the primary AFB_1-adducts disappeared from the DNA during post-treatment incubation by spontaneous hydrolysis also in 10T1/2 cells. This situation made it possible to study the effect of changes in AFB_1-adduct concentration and -composition on viability of 10T1/2 cells in the absence of excision repair. A linear relationship was obtained between the logarithm of the colony forming ability of 10T1/2 cells and the AFB_1-adduct concentration identifying DNA as the major target for AFB_1 cytotoxicity. Partial recovery of the viability of AFB_1-treated 10T1/2 cells upon post-treatment incubation in confluency before replating at low density for the determination of CFA was related to the spontaneous release from DNA of a portion of the AFB_1-adducts.

SPONTANEOUS REACTIONS OF AFLATOXIN B_1-DAMAGED DNA

Upon activation with a rat liver microsome system AFB_1 binds covalently to free DNA *in vitro* (1-3). Under these conditions more than 90% of the covalent adducts can be released by mild acid treatment in the form of 2,3-dihydro-2-(N^7-guanyl)-3-hydroxyaflatoxin B_1 (AFB_1-N^7-Gua) (2,4,5) as is evident from the HPLC profile shown in Fig. 1A. Since AFB_1-N^7-Gua is also released spontaneously at neutrality, albeit with a lower rate and lower yield, it is likely that the predominant primary adduct in DNA possesses the structure of 2,3-dihydro-2-(N^7-deoxyguanosyl)-

3-hydroxyaflatoxin B_1 (8). These primary adducts in DNA are of low chemical stability and undergo spontaneous reactions probably mostly because the 5-numbered ring of guanine carries a positive charge. The understanding of these spontaneous reactions is a prerequisite for the study of the cellular processing and the biological effects of AFB_1 induced DNA damage. Therefore, we have carried out a detailed analysis of these reactions *in vitro*. Figure 1A shows the changes occurring in the High Pressure Liquid Chromatography (HPLC) profiles of acid hydrolysates of alcohol precipitable AFB_1-DNA upon incubation at 37° in 0.01 M sodium phosphate, 0.1 M sodium chloride, 0.01 M sodium citrate (PSC) pH 7 for 24 hr. The total adduct content of the DNA has been reduced substantially during this time period mostly because of the reduction of the AFB_1-N^7-Gua peak III. It should be noted that peak I which contains the putative 2,3-dihydro-2-(N^5-formyl-2',5',6'-triamino-4'-oxo-N^5-pyrimidyl)-3-hydroxyaflatoxin B_1 (AFB_1-triamino-Py) (4) had increased in size, on the other hand. Peak IV which also increased upon incubation most likely corresponds to the peak with the same numeral of Lin et al. (4) for which these authors have tentatively proposed the structure of 2,3-dihydro-2-(8,9-dihydro-8-hydroxy-7-guanyl)-3-hydroxyaflatoxin B_1; comparison with an authentic marker (6) showed that peak II consists of 2,3-dihydro-2,3-dihydroxyaflatoxin B_1 (AFB_1-dhd) while the identity of peak V remains unknown. The following should be kept in mind for the interpretation of these data. Since mild acid treatment is routinely being used for the analysis of the adduct content of alcohol precipitable AFB_1-DNA it is conceivable that the hydrolysis products are secondary derivatives which may or may not possess the same structures as the adducts attached to DNA. As mentioned above this appears unlikely for AFB_1-N^7-Gua but may be the case for other adducts. *For simplicity it is assumed in our present discussion that the hydrolysis products AFB_1-N^7-Gua and the putative AFB_1-triamino-Py possess the structural properties of the adducts in DNA from which they derive*. Fig. 1B shows the concomittant changes occurring in the HPLC profile of the material which is being released upon incubation from AFB_1-DNA in alcohol soluble form. Most of this material chromatographs as a broad heterogenous peak with low retention time. In addition, two well defined peaks are discernible which were identified as AFB_1-dhd and AFB_1-N^7-Gua, respectively. When the samples were preincubated at pH 5 before HPLC analysis the heterogenous early peak essentially disappeared and the AFB_1-dhd peak increased in size correspondingly (not shown; cf. refs. 6-8). From these observations together with double label experiments using DNA specifically labeled with ^{14}C-guanine and reacted with ^3H-AFB_1 it is concluded that the heterogenous early peak mostly consists of labile secondary products of AFB_1-dhd (8). Therefore, the spontaneous disappearance of the primary adduct AFB_1-N^7-Gua from DNA under physiological conditions is mostly due to the release of AFB_1-dhd. The release of AFB_1-N^7-Gua and the modification on DNA to AFB_1-triamino-Py represent minor reactions. The release of AFB_1-dhd presumably results in the reconstitution of intact guanine residues in DNA while the release of AFB_1-N^7-Gua is accompanied by the formation of aguaninic sites. The modification of the primary adducts to AFB_1-triamino-Py results in lesions with increased chemical stability which remain in DNA over prolonged periods.

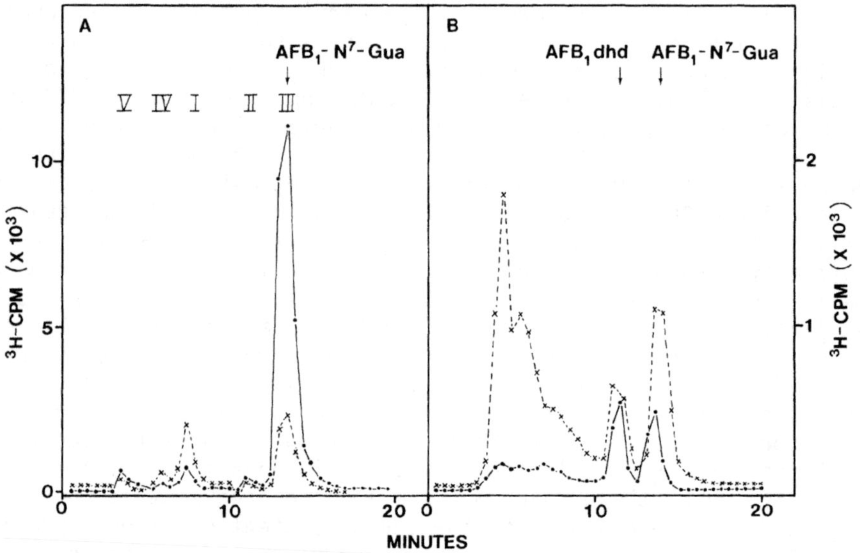

Figure 1
HPLC-analysis of acid hydrolysates of ethanol precipitable ^3H-AFB_1-DNA and of the ethanol soluble material released from ^3H-AFB_1-DNA. ^3H-AFB_1-DNA was prepared by the exposure of calf thymus DNA to ^3H-AFB_1 in the presence of rat liver microsomes. The samples were chromatographed on μBondapak C_{18} with 95% ethanol-1-propanol-water (10:5:85 v/v) at a flow rate of 1 ml/min. The retention times of authentic markers are indicated. A acid hydrolysates of ethanol precipitable material; ●—●, *no incubation* X--X, *24 hr incubation in PCS buffer pH 7 at 37° before ethanol precipitation. B ethanol soluble material;* ●—●, *material released within 2 hr incubation in PCS buffer pH 7 at 37°;* X--X, *material released within 24 hr.*

A detailed quantitative analysis was carried out of these spontaneous reactions of AFB_1-DNA in the pH range from 6.7 to 7.3 at 37°. HPLC analysis of acid hydrolysates of alcohol precipitable AFB_1-DNA and of the material released in alcohol soluble form was carried out as described above and in Fig. 1. The $[^3H]AFB_1$-DNA used in these experiments had been prepared by the rat liver microsome mediated reaction of $[^3H]AFB_1$ with calf thymus DNA and contained one covalent AFB_1 adduct per 8000 deoxynucleotides. Of the adducts, 91% could initially be released in the form of AFB_1-N^7-Gua by mild acid treatment. The following reactions were monitored : (1) loss of total [^{3}H]-AFB_1 radioactivity from ethanol-precipitable DNA (2) disappearance of AFB_1-N^7-Gua in the acid hydrolysates of the ethanol-precipitable DNA (3) appearance of the putative AFB_1-triamino-Py in the acid hydrolysates of the ethanol-precipitable DNA (4) appearance of free AFB_1-N^7-Gua in the ethanol-soluble fraction (5) appearance of AFB_1-dhd plus its degradation products in the ethanol-soluble fraction.

The following major conclusions can be drawn from these studies. The initial rates of the loss of total AFB_1-moieties and the disappearance of AFB_1-N^7-Gua from DNA increased from pH 6.7 to 7.3. The half-life times at 37° for the disappearance of AFB_1-N^7-Gua were 19 hr at pH 6.7, 12 hr at pH 7.0 and 8 hr at pH 7.3. This is mostly due to a substantial increase in the rate of release of AFB_1-dhd plus its secondary products with rising pH and to a lesser degree also to an increase in the rate of the modification of the primary adducts to AFB_1-triamino-Py which remain attached to DNA. In contrast, the rate of release of AFB_1-N^7-Gua declines slightly under these conditions (8). Our results have the following implications for the understanding of the biological effects of AFB_1. The primary adducts in AFB_1-DNA possess a short half-life time under physiological conditions. Nevertheless, they are present long enough to exert their effect on the processes of DNA metabolism in mammalian cells. At later times after exposure to AFB_1, the DNA is expected to contain mostly secondary lesions of increased chemical stability, in particular aguaninic sites and the putative AFB_1-triamino-Py. Direct evidence for the presence of aguaninic sites in AFB_1-DNA has been obtained (9). Fig. 2 gives a synopsis of the major spontaneous reactions of AFB_1-DNA under physiological conditions *in vitro*.

Figure 2
Major spontaneous reactions of AFB_1-DNA under physiological conditions in vitro. It is assumed that the major primary adduct in DNA possesses the structure of AFB_1-N^7-deoxyguanylic acid. Approximate rate constants k (in h^{-1}) are indicated (from ref. 8).

FORMATION OF AFLATOXIN B_1-DNA ADDUCTS IN MAMMALIAN CELLS

AFB_1 is a procarcinogen which has to be activated by mixed function oxidases for macromolecular binding to occur. Therefore, the level of AFB_1-DNA binding depends on the metabolic activity of a particular tissue. It is likely that the same major adduct, i.e. AFB_1-N^7-Gua, is initially formed in all tissues. However, because of the inherent chemical instability of these primary adducts discussed above, especially to small changes in pH, the conditions used for DNA preparation and analysis are expected to affect the measurements of adduct composition. Besides such experimental factors, the local cellular pH, enzymes which may catalyze the secondary reactions of AFB_1-N^7-Gua or participate in repair processes may introduce further variations in the lesion spectra between different tissues. In Table 1 the level and composition of AFB_1-DNA adducts formed in various cultured human tissues under similar conditions are compared. With the exception of the lung tumor line A549, primary fetal hepatocytes possessed the highest capacity for DNA adduct formation. An HPLC-profile of acid hydrolysates of DNA extracted from ^3H-AFB_1 treated primary fetal hepatocytes is shown in Fig. 3

TABLE 1 FORMATION OF AFB_1-DNA ADDUCTS IN HUMAN TISSUES (1)

	AFB_1 conc. μM	Total Adduct μmoles/mole DNA-P	AFB_1-N^7-Gua (2)	AFB_1-triamino-Py (2)
Fetal hepatocytes	1.0	0.34	61 %	22 %
Human lung A549	0.82	0.16-1.34	52 %	---
Bronchial mucosa (3)	1.5	0.32	33 %	38 %
Colon mucosa (3)	1.5	0.12	--- (4)	--- (4)

(1) Treatment for 24 hr

(2) HPLC of HCl hydrolysates

(3) From ref. 10

(4) A ratio of 2:1 was found for AFB_1-N^7-Gua / AFB_1-triamino-Py

Despite the low radioactivity content of the sample the characteristic peaks for AFB_1-triamino-Py and AFB_1-N^7-Gua are clearly discernible (11).

REMOVAL OF AFB_1-ADDUCTS FROM DNA IN CULTURED HUMAN AND RODENT CELLS

The removal of AFB_1-adducts from DNA during post-treatment incubation was studied in the human lung tumor line A549 (12) and in 10T1/2 mouse embryo fibroblasts (13). These particular cell lines were chosen because they have retained the capacity for oxidative drug metabolism in culture

(12, 14, 15). AFB_1 concentrations of low toxicity were chosen on the basis of survival data for these cell lines (16,19).

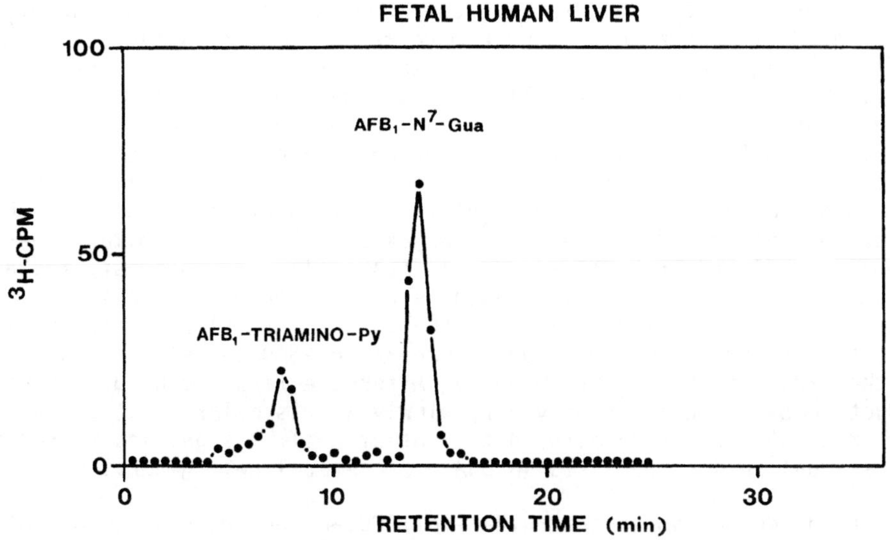

Figure 3
HPLC-analysis of acid hydrolysates of DNA isolated from primary human fetal hepatocytes which had been exposed to 1 μM ^3H-AFB_1 for 24 hr (11). The chromatography conditions were as described in the legend to Fig. 1.

Before treatment with ^3H-AFB_1 the cells were prelabeled in their DNA with ^{14}C-thymidine. This made it possible to express the adduct concentrations as ratios of the two radioactive isotopes, to correct for differences in the yields of DNA extraction between individual samples and to detect eventual cell loss. Immediately following ^3H-AFB_1 treatment a portion of the cells was harvested and the DNA extracted. This DNA was used for the measurement of the initial adduct concentration and -composition but also for the determination of the kinetics of the spontaneous reactions of the primary AFB_1-adducts under physiological conditions *in vitro*. The rest of the cell cultures were incubated for different lengths of time before DNA extraction and analysis were carried out. This experimental design allowed a direct comparison of the kinetics of the reactions of free AFB_1-DNA *in vitro* with AFB_1-DNA *in situ* in the intact cell at identical initial adduct concentrations and adduct compositions.

In the experiments with A549 cells rapidly growing cultures were treated with 0.82 μM ^{3}H-AFB$_1$ for 24 hr and then incubated up to 120 hr. In different experiments 0.18 to 1.34 μmol total adduct per mol DNA-phosphate were present immediately after treatment 42 to 64% of which corresponded to AFB$_1$-N^7-Gua. The kinetics of the disappearance of total adducts from DNA in the intact cell and from free DNA extracted immediately after treatment *in vitro* are compared in Fig. 4. It is evident that early product removal occurs much more rapidly in the intact cell. While 60% of the initial adducts had disappeared from high-molecular-weight DNA in 24 hr in the intact cell, only 15% had been lost from free DNA within the same time period. It is concluded that at least a portion of the covalent AFB$_1$-DNA adducts are removed by a cellular process in A549 cells. Determinations of the adduct composition as a function of post-treatment incubation by analysis of formic acid hydrolysates on Sephadex LH 20 columns showed that AFB$_1$-N^7-Gua disappeared with similar kinetics from free DNA and from DNA *in situ* in A549 cells. In contrast, the AFB$_1$-triamino-Py concentration increased as a function of incubation time in free DNA *in vitro* but decreased in the DNA *in situ* (16). It is concluded that the active process of adduct removal in A549 cells mostly consists in the excision of AFB$_1$-triamino-Py. After the first 30 hr of incubation adduct removal continues only very slowly with similar kinetics *in vitro* for free DNA and DNA *in situ* in the intact cell. A significant fraction of residual lesions persists in DNA over several cell generations.

As discussed above the release of AFB$_1$-N^7-Gua represents one of the spontaneous reactions of free AFB$_1$-DNA. In the cell this reaction may occur both spontaneously and catalyzed by a glycosylase. Indeed, evidence for the presence of such an enzyme activity has been obtained in extracts of E. coli (M. Olsson, T. Lindahl, V. Wang, P. Amstad and P. Cerutti, unpublished). The removal of AFB$_1$-N^7-Gua results in the formation of aguaninic sites which may be repaired by excision via the action of an apurinic site endonuclease or possibly by an insertion mechanism. Evidence for active cellular repair of AFB$_1$-induced DNA damage in human fibroblasts has been obtained by other investigators. Liver microsome activated AFB$_1$ induced unscheduled DNA synthesis (17) and repair replication (18) in normal fibroblasts but not in *Xeroderma pigmentosum* fibroblasts of complementation group A.

No evidence for active enzymatic removal of AFB$_1$-DNA adducts was obtained in studies with confluent cultures of 10T1/2 mouse embryo fibroblasts. 10T1/2 cells proved to be metabolically more active than A549 cells. After 16 hr incubation with 0.2 μM AFB$_1$ the DNA contained approximately 7 μmol total adduct per mol DNA-phosphate. Immediately after treatment 66 to 80% of the adducts were AFB$_1$-N^7-Gua and 8 to 13% AFB$_1$-triamino-Py. In contrast to A549 cells the kinetics of the disappearance of total AFB$_1$-adducts and of AFB$_1$-N^7-Gua were indistinguishable for free DNA and DNA *in situ* in the cell. Similar kinetics were also obtained for AFB_1-triamino-Py which increased in concentration in the intact cell and in free DNA as a function of post-treatment incubation. It is concluded that the disappearance from DNA of AFB_1-adducts in 10T1/2 cells mostly occurs by spontaneous hydrolysis reactions rather than enzymatic repair (19). However, it is entirely possible that aguaninic sites which are formed as a consequence of the release of AFB_1-N^7-Gua are repaired by cellular processes.

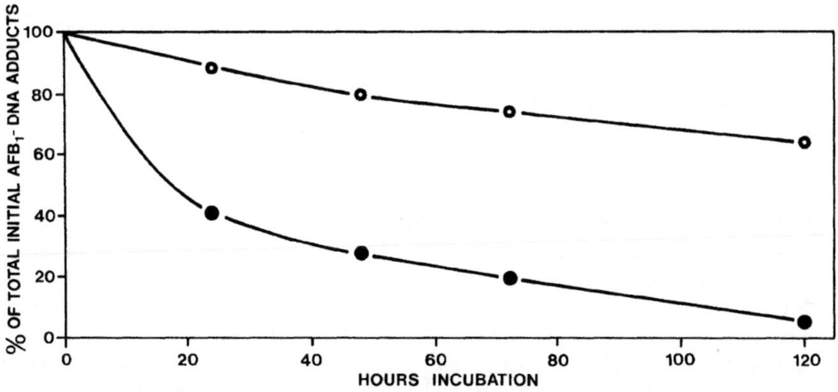

Figure 4
Kinetics of disappearance of total AFB_1 adducts from DNA in situ in human lung cells A549 and from free A549 DNA in vitro. A549 cells were treated for 24 hr with 0.82 μM ^3H-AFB_1, the DNA extracted and its AFB_1-adduct content determined after different lengths of post-treatment incubation. For comparison free DNA obtained from A549 cells at 0 hr post-treatment incubation was kept at 37° in PCS buffer pH 7.0 for the same lengths of time. The data is plotted as percentage of the initial adduct concentration. For experimental details see Ref. 16. ●, disappearance of AFB_1-adducts from DNA in situ in A549 cells; O, disappearance of AFB_1-adducts from free DNA in vitro.

MECHANISM OF AFLATOXIN B_1 CYTOTOXICITY

The relationship between the concentration of AFB_1-DNA adducts and colony forming ability (CFA) was investigated with 10T1/2 cells. Two treatment regimens were used for the introduction of increasing concentrations of AFB_1-DNA adducts in confluent cultures of 10T1/2 cells. Either the cultures were exposed to increasing concentrations of AFB_1 in the range of 0.05 to 0.228 μM for a constant duration of 4 hr or for increasing amounts of time from 1 to 16 hr at a constant concentration of 0.2 μM AFB_1. After the treatment a portion of the cells were replated at low densities for the determination of their CFA. From the remaining cells the DNA was extracted and the total adduct concentration derived from the $^3H/^{14}C$ ratios. As described in the previous section the 10T1/2 cultures had been prelabeled in their DNA with ^{14}C-thymidine before they reached confluency and ^3H-AFB_1 was used for the treatment. Survival of CFA was plotted in the usual semilogarithmic fashion as a function of the total adduct concentration in DNA immediately after termination of the drug treatment. The same linear survival curve was obtained regardless of the treatment regimen. No shoulder was discernible at low adduct concentration. The initial adduct concentration which effects a decrease of CFA to 37% (i.e. 1/e) is 1.9 μmoles adduct per mole DNA-phosphate (19). This value is approximately 3 times lower than for rapidly growing human lung cells A549. The lower cytotoxicity of AFB_1 per initial DNA adduct in A549 cells may be due to the presence of cellular excision repair processes in this cell line.

In a second series of experiments the effect of post-treatment incubation of 10T1/2 cells in confluency on CFA was investigated. As discussed in the previous section a large portion of the covalent AFB_1-DNA adducts initially present disappear from the DNA over a 40 hr post-treatment incubation period probably mostly due to spontaneous hydrolysis reactions. Fig. 5 shows that the CFA recovered rapidly within the first 20 hr of incubation. Survival was enhanced by a factor of 4.7 in cultures which had been treated for 16 hr with 0.3 μM AFB_1. At longer post-treatment incubation only little additional recovery of CFA was observed (19). It is likely that residual lesions such as AFB_1-triamino-Py which are apparently irrepairable in 10T1/2 cells are responsible for the incompleteness of the recovery of CFA. Our results concerning the relationship between CFA and AFB_1-DNA adduct concentration strongly suggest that DNA represents the major target for the cytotoxic effect of AFB_1 towards 10T1/2 cells.

Since 10T1/2 cells were found to be remarkably active in the formation of covalent AFB_1-DNA-adducts they might be expected to be highly susceptible to AFB_1-mutagenesis and -transformation without the need for exogenous metabolic activation. However, according to a preliminary report by Mondal et al. (20) no transformation of 10T1/2 cells was observed by AFB_1 in the absence of feeder layers of mouse epithelial liver cells. As shown in Table 2 very low or no transformation was also obtained in our laboratory when confluent 10T1/2 cells at passage 11-12 were treated with 0.2 μM AFB_1 for 4 hr before replating at low cell density for the determination of *in vitro* transformation according to Reznikoff et al. (13).

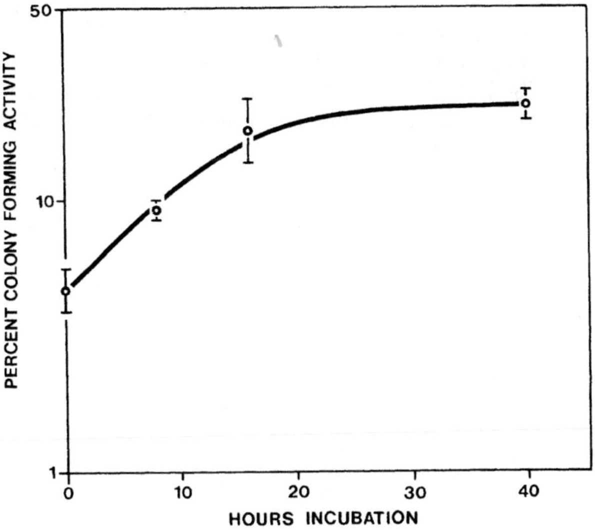

Figure 5
Recovery of colony forming ability of AFB_1 treated 10T1/2 mouse embryo fibroblasts upon holding in confluent state. Confluent cultures of 10T1/2 cells were treated for 16 hr with 0.3 μM AFB_1. After different lengths of incubation in confluency up to maximally 40 hr the cells were replated at low density for the determination of their CFA. The CFA was calculated from the plating efficiencies of AFB_1-treated cells relative to those of sham-treated cells which were kept in confluency for the same lengths of time. The means of two independent experiments are plotted.

TABLE 2 AFB_1-DNA ADDUCT CONCENTRATION, VIABILITY AND TRANSFORMATION OF 10T1/2 MOUSE EMBRYO FIBROBLASTS

AFB_1 conc. [(a)] μM	DNA-adduct conc. μM/Mol DNA-P	% survival of CFA	Transf. frequency [(b)]
0	0	100	0/5880
0.06	1.0	77	---
0.11	2.4	45	0/10500
0.17	3.3	25	---
0.23	4.5	11	1/6460

(a) Treatment of confluent cultures for 4 hr

(b) 200-300 viable cells per T25 Falcon flask; the data is expressed as type II foci per viable cells

Abbreviations used : AFB_1, aflatoxin B_1 ; AFB_1-N^7-Gua, 2,3-dihydro-2-(N^7-guanyl)-3-hydroxyaflatoxin B_1 ; AFB_1-triamino-Py, 2,3-dihydro-2-(N^5-formyl-2',5',6'-triamino-4'oxo-N^5-pyrimidyl)-3-hydroxyaflatoxin B_1 ; AFB_1-dhd, 2,3-dihydro-2,3-dihydroxyaflatoxin B_1 ; AFB_1-DNA, DNA prepared by the reaction of AFB_1 with calf thymus DNA in the presence of rat liver microsomes ; PSC, 0.01 M sodium phosphate - 0.1 M sodium chloride -0.01 M sodium citrate ; HPLC, high-pressure liquid chromatography ; CFA, colony forming ability.

ACKNOWLEDGEMENTS

This work was supported by Grant 3'305.78 of the Swiss National Science Foundation.

REFERENCES

1) Garner,R.,Miller,E.,and Miller,J.:1972,Cancer Res.32,pp.2058-2066.
2) Essigmann,J.,Croy,R.,Nadzan,A.,Busby,W.Jr.,Reinhold,V.,Büchi,G.,and Wogan,G.:1977,Proc.Natl.Acad.Sci.USA 74,pp.1870-1874.
3) Swenson,D.,Lin,J.,Miller,E.,and Miller,J.:1977,Cancer Res.37,pp.172-181.
4) Lin,J.,Miller,J.,and Miller,E.:1977,Cancer Res.37,pp.4430-4438.
5) Croy,R.,Essigmann,J.,Reinhold,V.,and Wogan,G.:1978,Proc.Natl.Acad. Sci.USA 75,pp.1745-1749.
6) Swenson,D.,Miller,J.,and Miller,E.:1975,Cancer Res.35,pp.3811-3823.
7) Neal,G.,and Colley,P.:1979,FEBS Lett.101,pp.382-386.
8) Wang,T.,and Cerutti,P.:1980,Biochemistry 19, in press.
9) D'Andrea,A.,and Hazeltine,W.:1978,Proc.Natl.Acad.Sci.USA 75,pp.4120-4124.
10) Autrup,H.,Essigmann,J.,Croy,R.,Trump,B.,Wogan,G.,and Harris,C.:1979, Cancer Res.39,pp.694-698.
11) Wang,T.,Hottinger,A.,Preisig,R.,and Cerutti,P., unpublished.
12) Lieber,M.,Smith,B.,Szakal,A.,Nelson-Rees,W.,and Todaro,G.:1976,Int. J.Cancer 17,pp.62-70.
13) Reznikoff,C.,Brankow,D.,and Heidelberger,C.:1973,Cancer Res.33, pp.3231-3238.
14) Feldman,G.,Remsen,J.,Shinohara,K.,and Cerutti,P.:1978,Nature 274, pp.796-798.
15) Brown,H.,Jeffrey,A.,and Weinstein,I.B.:1979,Cancer Res.39,pp.1673-1677.
16) Wang,T.,and Cerutti,P.:1979,Cancer Res.39,pp.5165-5170
17) Stich,H.,and Laishes,B.:1975,J.Cancer 16,pp.266-274.
18) Sarasin,A.,Smith,C.,and Hanawalt,P.:1977,Cancer Res.37,pp.1786-1793.
19) Wang,T.,and Cerutti,P.:1980,Cancer. Res.40, in press.
20) Mondal,S., Lillehang,J.,and Heidelberger,C.:1979,Proc.Am.Ass.Cancer Res.20,p.62 (Abstract).

THE ROLE OF DNA REPAIR IN PREVENTING THE CYTOTOXIC AND MUTAGENIC EFFECTS OF CARCINOGENS IN HUMAN CELLS

Veronica M. Maher, Li L. Yang, and J. Justin McCormick
Carcinogenesis Laboratory - Fee Hall
Department of Microbiology and Department of Biochemistry
Michigan State University, East Lansing, MI 48824 USA

ABSTRACT

We compared excision repair-proficient and repair-deficient diploid human fibroblasts, derived from normal individuals or xeroderma pigmentosum patients, for their response to the mutagenic and/or cytotoxic effect of ultraviolet radiation, or reactive derivatives of aromatic amines or polycyclic aromatic hydrocarbons. Each of these carcinogens forms DNA helix-distorting photoproducts or adducts. The results indicated that excision repair processes in these human cells act to prevent cell killing and the induction of mutations; that is, they decrease the potentially cytotoxic or mutagenic effect of such agents. The rate of removal of radioactively-labeled carcinogen adducts from cellular DNA was found to be highly correlated with the rate of increase in percent survival and rate of decrease in the frequency of mutations to 6-thioguanine resistance induced by these carcinogens. The data indicate that the observed mutations are not the result of "error-prone" excision repair but arise, directly or indirectly, from semi-conservative DNA synthesis on templates containing unexcised lesions. The ultimate biological consequences of exposure to these agents, i.e., the probability of a cell not being able to form a colony and the probability of mutations being introduced, appear to be determined by the rate of excision repair and the amount of time available between the introduction of the lesions and the onset of the "critical event responsible for cell killing" or the "critical event responsible for mutations." The data suggest that the latter is semi-conservative DNA synthesis.

1. INTRODUCTION

As part of a study of carcinogenesis, we have been investigating the mechanisms by which mutations are introduced into the DNA of diploid human fibroblasts following exposure to chemical carcinogens or radiation. If we define a mutation as a change in the structure of the genetic material, we can imagine this change arising in at least two ways. The nucleotide sequence could be altered slightly or part of the sequence could be deleted altogether. The former change could result from misreplication (base-pair substitutions, positive frame shifts, etc.), whereas the latter might arise from a failure to replicate a portion

B. Pullman, P. O. P. Ts'o and H. Gelboin (eds.), Carcinogenesis: Fundamental Mechanisms and Environmental Effects, 479–490.

of the DNA (negative frame shifts, deletions of stretches of nucleotides, etc.). Both kinds of mutations can be visualized as resulting from a DNA polymerization process and the faulty polymerization could occur a) during excision repair replication or b) during semiconservative DNA replication.

We carried out a series of experiments to determine which of these two processes are involved in the induction of mutations by ultraviolet (UV 254nm) radiation (1-3) or by reactive derivatives of aromatic amines (4) or polycyclic aromatic hydrocarbons (5,6). The availability of excision repair-deficient xeroderma pigmentosum (XP) cells which could be compared with excision repair-proficient cells derived from normal individuals greatly facilitated such studies. The results indicate that mutations to 8-azaguanine or 6-thioguanine resistance (resulting from the loss of active hypoxanthine(guanine)phosphoribosyltransferase, HPRT, enzyme) are not introduced during excision repair (1,5,6). Rather, they result, directly or indirectly, from semiconservative DNA synthesis on a template containing unexcised lesions.

Our data further suggest that as soon as DNA lesions are introduced by these agents, there is an immediate response by excision repair processes and that the ultimate biological consequences depend upon the rate of excision and the time available for excision before the onset of some "critical event responsible for cell killing" or "critical event responsible for mutation induction." Thus, cells may be protected from the potentially harmful effects of exposure to these agents by having a very rapid rate or an extended length of time before these critical events occur. Conversely, a cell population with a slow rate of excision or with a very short doubling time is predicted to be the one with the highest frequency of mutations for a given dose. In analyzing the role of excision repair in preventing cell death or preventing mutation induction, we have taken both approaches, i.e., we have varied the rate of excision or the length of time between exposure of the cells to the carcinogen and the onset of semiconservative DNA synthesis. The results are discussed below.

2. EFFECT OF VARYING THE RATES OF EXCISION

Figure 1 shows the results of a comparison between three strains of diploid human fibroblasts, each with a cell cycle of approximately the same length, but with a different rate of excision of UV-induced pyrimidine dimers. In these experiments, each strain of cells received approximately the same number of initial DNA lesions by being exposed to the same doses of radiation. The surviving cells were allowed to form colonies or to undergo expression of mutations to 8-azaguanine resistance in situ. The closed circles give the results from cells derived from normal persons (NF) which have a relatively rapid rate of excision; the open circles are from xeroderma pigmentosum (XP) cells which have a much slower rate than normal cells (7); and the open triangles represent data from XP cells which have virtually no capacity for the excision of UV radiation damage (8). This particular strain, XP12BE, does not excise pyrimidine dimers or remove UV endonuclease sensitive sites, and does not carry out detectable amounts of repair replication measured over a period of seven days following irradiation (data not shown). This makes it especially valuable for determining the absolute potential of UV radiation damage for cell killing and for

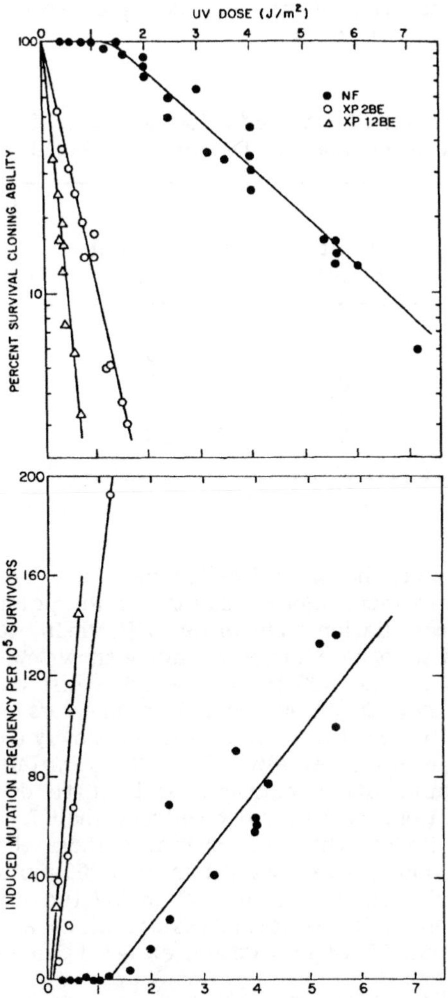

Figure 1. Comparison of the cytotoxic and mutagenic effect of increasing doses of UV radiation in normal fibroblasts (NF) and in XP cells with no detectable excision-repair capacity (XP12BE). Cells were plated into culture dishes at appropriate densities, allowed ~12 h to attach, irradiated, and allowed to develop into colonies. Selection in situ with 10 μM AG was begun after 5-8 days of expression (>3 population doublings). Taken from Maher et al. (1) with permission.

causing mutations in human fibroblasts. For example, the following relationships hold for this strain: (From Figure 1.)

Table 1. Absolute Potential of UV Radiation to Cause Cell Killing and Induce Mutations to Azaguanine Resistance as Defined in XP12BE Cells.

UV Dose (J/m^2)	Percent Survival	Mean Lethal Events	Mutations per 10^5 Survivors
0.1	60	0.5	23
0.3	22	1.5	70
0.6	5	3.0	140
1.5	0.05	7.5	(350)

Figure 1 indicates that the normal cells which have a rapid rate of excision of UV-induced damage are much more resistant to the potentially lethal and/or mutagenic effect of the radiation than either XP strain. Given a dose of 0.6 J/m^2, the XP12BE cells with no excision capacity show 5% survival and ca. 140 mutants per 10^5 survivors. The XP2BE cells which excise damage slowly exhibit a survival of ~25% and ca. 75 mutants per 10^5 survivors as would an XP12BE population irradiated with 0.24 J/m^2. It is as if the slowly excising cells reduced their load of DNA damage to a level 40% of what they initially received. The NF cells with a still more rapid rate of excision exhibit little or no measureable cell killing or mutation induction, as if they had reduced their load to an insignificant number of lesions. Similarly, when we irradiated these three strains with 1.5 J/m^2, the XP2BE cells showed ~4% survival and ca. 200 mutants per 10^5 cells as though they had reduced their dose from 1.5 to 0.7 (i.e. to 40% of what they received). Again, the NF cells exhibited results which an XP12BE population would show if given only ca. 6% of that dose (i.e., ~0.1 J/m^2).

The data in Figure 1 indicate that at equicytotoxic doses, the frequency of mutations induced in all three strains is comparable. Since an essential difference between these three strains is their respective rates of excision repair of UV-induced DNA damage (7), the data are consistent with the hypothesis that the loss of colony-forming ability and the frequency of induced mutations reflect the number of unexcised lesions remaining in their DNA at the time of some "critical event". The results predict that within a specified "critical time", the two strains capable of excision are able to remove a majority of the lesions initially introduced. While one may measure the rates of excision of UV induced pyrimidine dimers in human cells with reasonable accuracy at doses of 10 J/m^2 and above (9), this is not true for the very low doses used in these biological experiments. It was not practical to test this prediction directly in UV-irradiated cultures, but we have recently completed a similar comparative study using a radioactively-labeled chemical carcinogen of high specific activity in the

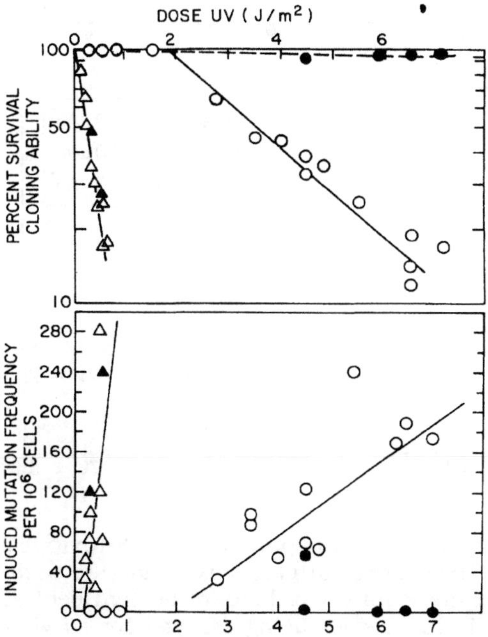

Figure 2. Loss of potentially cytotoxic and mutagenic lesions induced by UV irradiation in NF in comparison to XP12BE cells. Both sets of closed symbols represent cells grown to confluence, irradiated, and then held at confluence for 7 days before being assayed for survival or for frequency of induced mutations to TG resistance (1). Open triangles represent XP12BE cells irradiated in confluence, G_0 resting state, and assayed immediately or irradiated as growing cultures. Open circles indicate NF cultures irradiated in confluence and assayed immediately for survival, or cells from exponentially-growing cultures plated, irradiated, and then assayed for frequency of UV-induced mutations. Taken from Maher et al. (1) with permission.

place of UV. The preliminary evidence confirms that this is what actually does take place during that period (Yang et al., data to be published elsewhere).

3. EFFECT OF VARYING THE LENGTH OF THE CRITICAL TIME

If mutations are caused by semi-conservative DNA replication on a damaged template, and one were to extend the period of time between the introduction of the lesions into the DNA and the beginning of DNA synthesis, it should be possible for a cell which has at least some capacity for excision repair to remove those lesions before they can cause mutations. We tested this

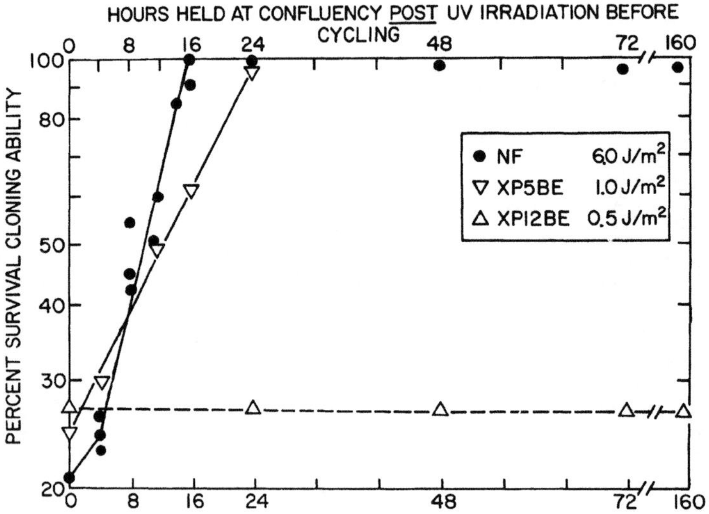

Figure 3. Rate of recovery of cells from potentially cytotoxic lesions induced by UV radiation. A series of confluent cultures was irradiated with the designated doses. Cells were released immediately or held in the confluent state for various lengths of time post-irradiation before being assayed for survival. The cloning efficiency of the untreated cultures (35% for NF, 10% for XP5BE and 7% for XP12BE) did not decrease during the time in confluence. Taken from Maher et al. (1) with permission.

hypothesis by growing the cells to confluence so that they would cease replicating as a result of "contact" or "density inhibition". We measured the rate of excision repair in these populations of nonreplicating cells in the G_0 state and found that they were quite capable of repair replication (9-11). We then irradiated NF cells with a large enough dose of UV light to cause cell killing and to introduce mutations in growing populations, e.g. 5-10 J/m^2, but then, instead of allowing the cells to undergo replication, prevented replication and held them in a nonreplicating state for a period of time (in this case, seven days) before releasing them and assaying them for survival or for mutation induction. Autoradiography studies demonstrated that in cultures maintained in the G_0 state under our conditions (1), less than 0.2% of the cells were labeled with ^{3}H-thymidine during an 8 hour labeling period. As the data in Figure 2 indicate, the normal cells held in this G_0 state completely eliminated the potentially cytotoxic and mutagenic lesions induced by these levels of UV radiation. In contrast, no sign of biological recovery was observed in the excision-minus XP12BE strain treated in a similar manner. This is further evidence that the recovery exhibited by the NF cells did not reflect mere cell turn-over, but rather could be attributed to an "error-free" excision repair process.

Further evidence for this hypothesis came from the experimental data shown in Figure 3. All three strains were grown to confluence and irradiated with appropriate doses so that the cells released immediately would exhibit ca. 20% survival of colony forming ability. Replicate sets were held in confluence for the indicated lengths of time before being released by trypsinization and assayed for survival. The XP5BE cells have an intermediate rate of excision repair of dimers similar to that of XP2BE(4).

Since these XP5BE cells have a slow, but detectable rate of excision repair, it is not surprising that they were able to remove lesions from DNA and exhibit a rate of recovery from low dose which was similar to the rate of recovery of NF cells irradiated with a much higher dose of UV (1). The data in Figure 3 suggest that the slowly-repairing XP5BE strain removed its set of lesions slowly; the normally-repairing strain removed its set of lesions faster, but since it had ~6 times more lesions, the two sets of cells recovered at about the same rate. During the 24 hours held in confluence, both sets of cells were capable of removing sufficient numbers of lesions, so that when they were released from the non-replicating state, they exhibited 100% survival. We also tested the normal cells for rate of recovery from potentially mutagenic lesions and found that if cells in the resting state were irradiated with 9 J/m^2 and then assayed every 6 hours for the number of potentially mutagenic lesions remaining, the frequency of mutations decreased gradually until in cells released 18 hours post-irradiation, only a background level of mutations was observed. (Data not shown.)

4. BIOLOGICAL RECOVERY FROM POTENTIALLY CYTOTOXIC OR MUTAGENIC LESIONS INDUCED BY CHEMICAL CARCINOGENS

In collaboration with Dr. John Scribner of the Pacific Northwest Research Foundation, we conducted similar comparative studies with NF and XP12BE cells exposed to a series of radioactively labeled aromatic amine derivatives. viz., N-acetoxy-4-acetylaminobiphenyl (N-AcO-AABP), N-acetoxy-2-acetylaminofluorene (N-AcO-AAF), N-acetoxy-2-acetylaminophenanthrene (N-AcOAAP), and N-acetoxy-4-acetylaminostilbene (N-AcO-AAS). When the two strains were exposed to similar concentrations of these compounds, the XP12BE cells always proved significantly more sensitive (10, 11). (Figure 4.) It may be well to note that this is not the case if these strains are exposed to methylating agents such as N-methyl-N′-nitro-N-nitrosoguanidine (12).

The data in Figure 4 suggested that DNA lesions introduced by N-AcO-AAP might be removed faster in NF cells than are those induced by N-AcO-AAS. We, therefore, compared the rates of removal of labeled residues from the DNA of NF cells held in confluence. The results are shown in Figure 5. Non-replicating normal cells and XP12BE cells were treated with radioactively-labeled carcinogen and immediately after the treatment one set was assayed for the amount bound to DNA and/or plated at cloning densities to measure survival. Another set was held in the non-replicating state for the time indicated -- up to six or seven days. At the indicated times, cells were released from confluence, the number of residues remaining bound was determined, and/or the corresponding survival of the colony-forming ability after the release

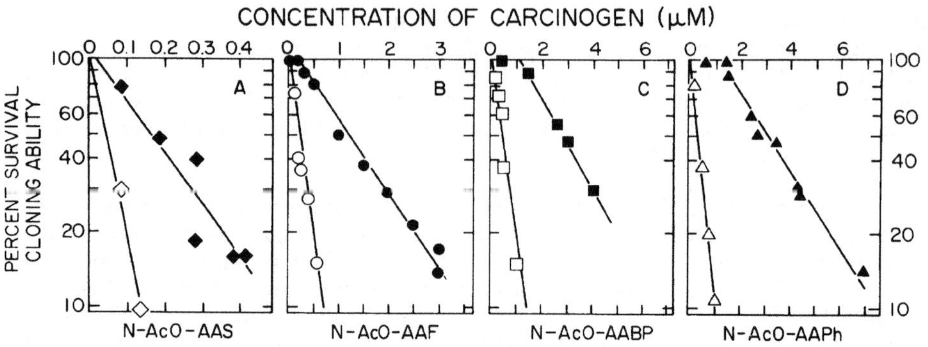

Figure 4. Percent survival of colony-forming ability of NF (closed symbols) or XP12BE (open symbols) as a function of the concentration of carcinogen administered. The cells exposed to N-AcO-AAS were treated in the G_0 state and replated at cloning density. For the other three compounds the cells were plated at cloning density and treated in situ. Cells were in serum-free medium during the 3-hr. exposure time. Some of this data was taken from Heflich et al. (10).

was determined. As can be seen in Figure 5, the rate of recovery of the cells, is correlated with the rate of removal of the bound carcinogen residues. Note that the XP12BE cells, the open symbols, showed little or no removal of the bound material and no recovery, over a period of six or seven days. The data for the XP12BE cells treated with the stilbene compound are not shown here, but recent results indicate that these cells cannot recover from the potentially cytotoxic lesions introduced by this carcinogen. (Heflich et al., data to be published elsewhere.)

Recently completed studies of the kinetics of removal of potentially mutagenic lesions formed in human cells by N-AcO-AAF indicate that as the number of DNA-bound residues is decreased over a period of six or seven days, so also the potentially mutagenic lesions are removed. In contrast, XP12BE cells held for 6 or 7 days showed no recovery from potentially mutagenic lesions (Maher, et al., data to be published elsewhere).

5. KINETICS OF RECOVERY FROM LESIONS INDUCED BY THE 7,8 - DIOL-9,10 - EPOXIDE OF BENZO(A)PYRENE

We have used this approach to investigate the kinetics of removal of DNA adducts formed by the (±)-7β,8α-dihydroxy-9α, 10α-epoxy-7,8,9,10-tetrahydrobenzo(a)pyrene (anti BPDE) and of the rate of biological recovery from the potentially cytotoxic and mutagenic effects (6). The data are shown in Figure 6. A series of cells were grown to confluence and treated with anti BPDE. One set was harvested immediately and assayed for the number of adducts bound to DNA and for survival and mutation frequency. The rest were held for 2, 4, or 8

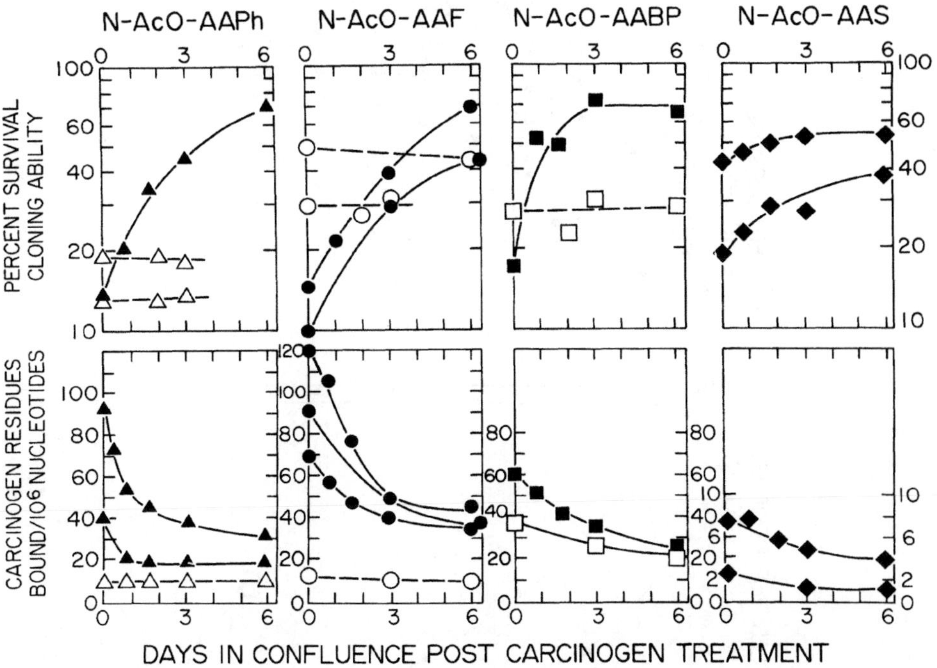

Figure 5. Comparison of the rates of recovery from the potentially lethal effects of these aromatic amine derivatives with the rate of removal of radioactively labeled residues from the DNA of NF (closed symbols) or XP12BE (open symbols). Cells were treated at confluence as described and then assayed after the designated period of time in the G_0 state. Some of this data was taken from Heflich et al. (10).

days before being similarly assayed. The data show that residues were removed by the confluent NF cells over a period of 4 days but then excision virtually ceased. Similarly, the survival increased during the same period and the mutation frequency decreased, indicating that the potentially cytotoxic and mutagenic lesions were being removed with the same kinetics as the total number of adducts excised (6).

HPLC characterization of the DNA adducts indicated one major adduct, viz., the N^2-guanyl derivative. The kinetics of decrease of tritium label in this specific HPLC peak corresponded to the decrease in radioactivity of the total DNA with time and with the kinetics of biological recovery of the cells. Although the existence of unstable or minor adducts cannot be ruled out, these results indicate that the potentially mutagenic and cytotoxic lesion caused by anti BPDE is the N^2-guanine adduct and that excision of this adduct is virtually "error-free".

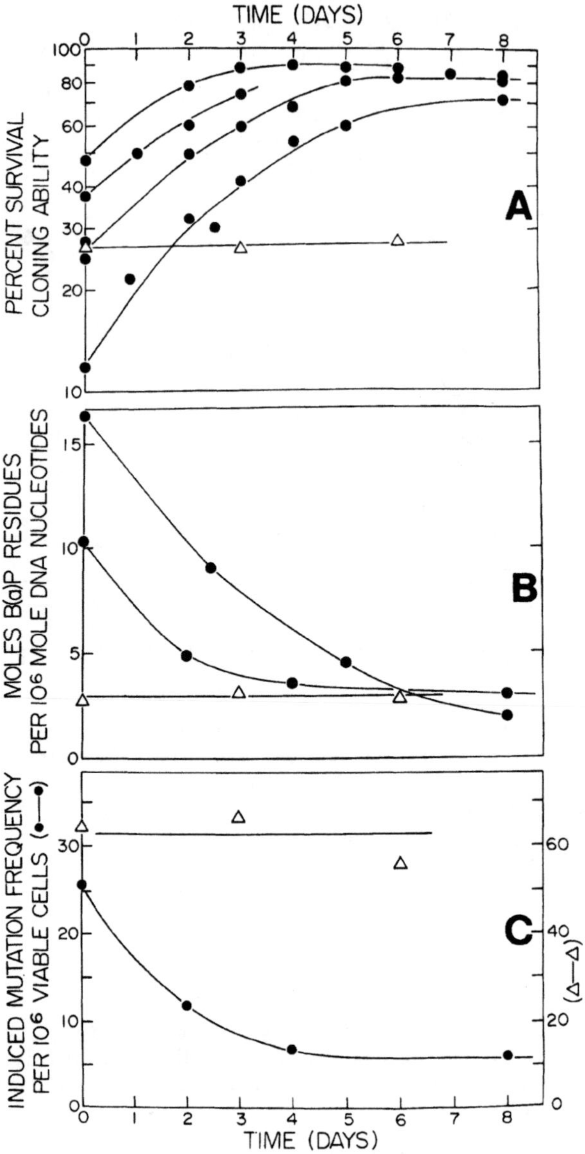

Figure 6. Kinetics of removal of covalently bound adducts (B) and recovery of NF (circles) or XP12BE (triangles) from the potentially cytotoxic (A) or mutagenic (C) effects of anti-BPDE. The cells were treated in the G_0 state, released on the designated days, and assayed for survival of colony-forming ability, for the number of residues bound to DNA, and after a suitable expression period, of induced mutations. Taken from Yang et al. (6) with permission.

6. MUTAGENIC POTENCY OF SPECIFIC DNA HYDROCARBON ADDUCTS

The data shown so far indicate that the frequency of mutations is directly related to the number of lesions in the DNA released from confluence. However, to know whether the mutagenic process recognized specific adducts we treated cells with anti BPDE or with the 4, 5–oxide of BP. We found that it takes a higher concentration of BP 4,5–oxide than anti BPDE to achieve an equal amount of cell killing or mutation induction. We measured the number of residues initially bound to DNA and found it to be identical for both compounds. Thus, for an equal number of bound residues, there was equal cell killing, and an equal frequency of mutations induced by the two compounds. To make certain that this equal mutagenic effect of the two derivatives was not merely the result of different rates of removal of specific DNA adducts, we measured the amount bound to XP cells per lethal and mutagenic event and found these were equal.

In summary, our data indicate that the ability of normal diploid human fibroblasts to carry out excision repair of potentially cytotoxic and/or potentially mutagenic lesions induced in DNA by UV radiation or these classes of chemical carcinogens is what determines the biological consequences of such exposure. The data suggest that the "critical time for mutation induction" is the time between the introduction of the lesions and the beginning of semi-conservative DNA synthesis. The data indicate further that excision repair processes in human fibroblasts are essentially "error free" and that their role is to prevent the mutagenic effects of carcinogen exposure. We find that the mechanism of mutation is correlated with semi-conservative DNA replication on a damaged template containing unexcised adducts or photoproducts and furthermore, that the mechanism does not distinguish between adducts which are structurally related.

ACKNOWLEDGEMENTS

We wish to express our indebtedness to our colleagues, A.E. Aust, R.H. Heflich, B. Konze-Thomas, and J. W. Levinson for their invaluable contributions to the research summarized here. The excellent technical assistance of R. Corner, D. J. Dorney, R. M. Hazard, T. E. Kinney, L. Lommel, A. Mendrala, T. G. O'Callaghan, and T. VanNoord is gratefully acknowledged. Sincere thanks are extended to Dr. John Scribner of the Pacific Northwest Research Foundation for providing us with the labeled aromatic amine derivatives of high specific activity, without which much of this work could not have been accomplished. The labeled anti BPDE was generously provided by the Cancer Research Program of the National Cancer Institute. The research summarized in this report was supported in part by Contract ES-78-4659 from the Department of Energy and by Grants CA 21247 and CA 21253 from the Department of Health and Human Services, National Cancer Institute, NIH, and by a grant from the Elsa U. Pardee Foundation.

REFERENCES

1. Maher, V.M., Dorney, D.J., Mendrala, A.L., Konze-Thomas, B., and McCormick, J.J.: 1979, Mutation Res. 62, pp. 311-323.

2. Maher, V.M., Ouellette, L.M., Curren, R.D., and McCormick, J.J.: 1976, Nature 261, pp. 593-595.

3. Maher, V.M., Ouellette, L.M., Curren, R.D., and McCormick, J.J.: 1976, Biochem. Biophys. Res. Commun., 71, pp. 228-234.

4. Maher, V.M., Curren, R.D., Ouellette, L.M., and McCormick, J.J.: 1976, In Vitro Metabolic Activation in Mutagenesis Testing, Elsevier/North Holland, Amsterdam, pp. 313-336.

5. Maher, V.M., McCormick, J.J., Grover, P.L., and Sims, P.: 1977. Mutation Res. 43, pp 117-138.

6. Yang, L.L., Maher, V.M., and McCormick, J.J.: 1980, Proc. Nat'l. Acad. Sci. U.S.A. in press.

7. Robbins, J.H., Kraemer, K.H., Lutzner, J.A., Festoff, B.W., and Coon, H.G.: 1974, Ann. Intern. Med. 80 pp. 221-248.

8. Petinga, R. A., Andrews, A.D., Tarone, R.E., and Robbins, J.H.: 1977, Biochem. Biophys. Acta 479, pp. 400-410.

9. Konze-Thomas, B., Levinson, J.W., Maher, V.M., and McCormick, J.J.: 1979, Biophys. J. 28, pp. 318-325.

10. Heflich, R.H., Hazard, R.M., Lommel, L., Scribner, J.D., Maher, V.M., and McCormick, J.J.: 1980, Chem.-Biol. Interactions 29, pp. 43-56.

11. Maher, V.M., Heflich, R.H., and McCormick, J.J.: 1980. Proc., Int. Conf. on Carcinogenic and Mutagenic N-Substituted Aryl Compounds, Rockville MD, Nov. 7-9, 1979, in press.

12. Heflich, R.H., Dorney, D.J., Maher, V.M., and McCormick, J.J.: 1977, Biochem. Biophys. Res. Commun. 77, pp. 634-641.

TRANSFORMATION OF DIPLOID HUMAN FIBROBLASTS BY CHEMICAL CARCINOGENS

J. Justin McCormick, K. Charles Silinskas, and Veronica M. Maher
Carcinogenesis Laboratory - Fee Hall
Department of Microbiology and Department of Biochemistry
Michigan State University, East Lansing, MI 48824 U.S.A.

ABSTRACT

Using N-methyl-N'-nitro-N-nitrosoguanidine (MNNG) and following a protocol modeled after that published by Kakunaga (1978, Proc. Natl. Acad. Sci. U.S.A. 75, pp. 1334-1338) we succeeded in causing diploid human fibroblasts (KD cells) to transform into cells which can form foci on the top of confluent monolayers. Cells derived from such foci were isolated, allowed to multiply into large populations, and their progeny assayed for tumorigenicity by injection into athymic mice. Nodules formed in some of the animals. Of these, some later regressed, but histological examination by several pathologists of a large nodule which did not regress revealed a fibrosarcoma. Cells from this fibrosarcoma exhibited a human karyotype. Thus, although the frequency of induction of focus-forming cells in the population is very difficult to quantitate, we have confirmed the results reported earlier by Kakunaga. More recently, we also succeeded in inducing transformation of diploid human fibroblasts with MNNG following a protocol modeled after that of Milo and DiPaolo (1978, Nature 275, 130-132). In this protocol, selection for transformed cells among the progeny of carcinogen-treated cells derived from foreskins was applied by growth in medium containing high concentrations of amino acids followed by growth in soft agar. Cells capable of forming colonies in agar were produced with high frequency and cells derived from such colonies were isolated, propagated, and assayed for tumorigenicity in athymic mice. Again, nodules were observed and a tumor analyzed independently by several pathologists was diagnosed as a fibrosarcoma. The karyotype of the cells from the tumor was human. Experiments were carried out to compare the frequency of acquisition of anchorage independence with that of 6-thioguanine resistance, the genetic marker commonly employed in our mutation assays. Using populations exposed to carcinogens under the same conditions and given similar expression periods, we found the frequency of cells able to grow in agar to be ca. 20-fold higher than that of thioguanine resistant cells. However, the carcinogen-induced frequency of both traits showed a dose-dependent increase. We are presently analyzing these protocols to adapt them for studies of the mechanisms involved in carcinogenesis.

B. Pullman, P. O. P. Ts'o and H. Gelboin (eds.), Carcinogenesis: Fundamental Mechanisms and Environmental Effects, 491–498.

1. INTRODUCTION

For many years it was widely understood that diploid human cells in culture were refractory to malignant transformation induced by chemical carcinogens or radiation. However, a quantitative system for inducing such transformation could be extraordinarily valuable for comparative purposes in studying the mechanisms of tumorigenesis in man and might also prove useful as a short term test for agents hazardous to man. Therefore, in the fall of 1977 we began a concerted effort to accomplish this goal. In March 1978, Kakunaga published a report of his successful transformation of diploid human fibroblasts into tumor-forming cells by exposure to N-methyl-N'-nitro-N-nitrosoguanidine (MNNG) or 4-nitroquinoline-1-oxide. Because of the potential importance of a quantitative assay for such transformation, we directed our subsequent effort to experiments designed to confirm the results of Kakunaga and to determine the usefulness of his assay for studies on the mechanism of human carcinogenesis. Since that time, Milo and DiPaolo also reported (2) the successful transformation of human cells derived from foreskin material using various chemical carcinogens, including MNNG (see below). Still more recently, Borek reported the successful transformation of human cells using ionizing radiation (3).

2. IDENTIFICATION OF TRANSFORMED HUMAN CELLS BY FOCUS FORMATION

In our initial investigation, we took 1×10^7 KD cells (the diploid fibroblasts successfully used by Kakunaga) and treated them with a concentration of MNNG such that only 30% of the cells survived. An initial 3×10^6 untreated cells were similarly examined as a control. The protocol followed, which is shown schematically on the upper half of Figure 1, was essentially that of Kakunaga. The cells were kept in exponential growth by continuously subculturing the flasks at a 1:4 ratio whenever the cells approached confluence. At each subculture, the remaining three flasks were not trypsinized, but rather the cells were allowed to become confluent. To encourage focus formation on the confluent monolayer, these flasks were given medium changes three times a week over the entire length of the experiment (~3 months). Each of the more than 100 flasks was repeatedly monitored for foci (dense areas with cells growing on top of each other).

In the flasks containing cells which had been subcultured 1:4 five or more times (~10 population doublings), small foci appeared on the monolayer. We calculated the frequency of such foci to be from $1 - 4 \times 10^{-6}$ surviving cells originally exposed to MNNG. Several foci were isolated from these flasks using the well technique and transferred to dishes for further study. The result was vigorously growing cells, but no new foci developed on the monolayers in the new flasks. The cells in these latter flasks were tested and shown to have a finite life span. We interpreted these results, i.e., the lack of foci in the new flasks, to be the result of the technique used for the transfer. The well technique allows one to transfer not only transformed cells, but also a large number of non-transformed cells lying below the focus. Therefore, normal cells may have been transferred with the transformed cells and might have been able to overgrow the culture.

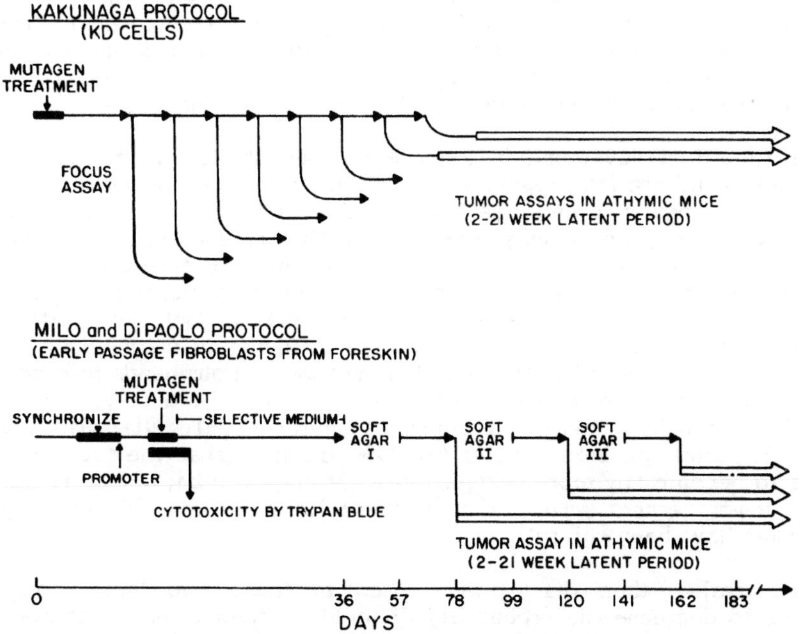

Figure 1. Highly simplified generalization or schematic representation comparing the major features of the protocol successfully used by Kakunaga (*1*) (upper half) or by Milo and DiPaolo (2) (lower half) to detect malignant transformation of diploid human fibroblasts in culture. The purpose of this diagram is not to describe their published techniques but to point out some of the similarities and differences between the two approaches. For example, in both protocols, there is a long expression period involving numerous population doublings before foci or anchorage-independent cells are detected in progeny of carcinogen-treated populations. Furthermore, before cells isolated from foci or agar colonies are assayed for ability to form tumors, they necessarily are allowed another lengthy period of exponential growth involving numerous population doublings in order to obtain sufficient numbers of cells for injection.

At that time we learned that the frequency of transformation estimated by Kakunaga was also very low (1 - 3 x 10^{-7} surviving cells). Since we had been working with only 3 x 10^{6} survivors we devised an expanded protocol using a target population of 3 x 10^{7} so as to have at least 10^{7} survivors after a dose of MNNG sufficient to kill 66% of the population. An additional 10^{7} untreated cells were also carried as a control. Essentially the same protocol was followed. Because the surviving population of 10^{7} cells had to be kept in exponential growth during the 10 to 20 cell generations required for expression and because of the large number of flasks that had to be handled with constant refeeding of the confluent cells in an effort to demonstrate foci formation, the number of T-75 flasks (250 ml, 75 cm^2 surface area) for this single experiment totaled more than 800. This experimental protocol took about 5 months and during the middle period we were dealing with almost all of the 800 flasks simultaneously. This massive effort required three technicians for full-time culture work.

This time foci developed in a large number of flasks. Some exhibited distinctly abnormal growth properties so that they could easily be distinguished from the more regular growth pattern of the background. However, in the majority, the "altered morphology" consisted of the presence of rounded up cells growing on top of the monolayer. Such foci were extremely difficult to distinguish from areas in very dense monolayers where the normal swirl patterns characteristic of human fibroblasts in culture overlap. Only very late in the large experiment did the technicians feel any confidence in distinguishing "real" foci from "interacting swirl patterns". Furthermore, in our hands the technique of DeMars (4), designed to enhance foci growth on top of monolayers of diploid human fibroblasts by repeated feeding of mutagenized cells with medium containing 5% fetal calf serum instead of 15%, did not cause the foci formed in our flasks to become any more recognizable. It may well be, however, that had we used a lower percent serum or a different lot of serum, this technique of DeMars would have been helpful.

Flasks rather than dishes were used in these two large, long-term experiments to decrease the probability of mold contamination. However, this made it more difficult to isolate and successfully transfer candidate foci. Nevertheless, a large series of "typical-looking foci" from the progeny of MNNG-treated cells as well as similar-looking areas from the progeny of untreated cultures were transferred successfully and grown up over a period of many weeks into populations sufficiently large to be injected into athymic mice to assay for tumor formation. However, because of technical problems, the mice were not immediately available, and so these cells had to be stored in liquid nitrogen.

After several months the cells were thawed and $1 - 10 \times 10^6$ cells were injected intradermally into 30 athymic mice. A number of animals injected with progeny of MNNG-treated cells developed small nodules at the site of injection one to four weeks later. These nodules increased in size for a week or two, but when we waited for them to develop further, they regressed. Therefore, no histological examination of these nodules was carried out. However, five months after injection one animal developed a small, rapidly-growing nodule at the site of injection. When this nodule had attained a diameter of > 1 cm and had maintained this size for more than a month without any regression, it was excised. Cells from a small portion of the tumor were returned to culture and histological section made of the rest. The slides were independently examined by two pathologists and both diagnosed the nodule as a fibrosarcoma. The tumor-derived cells returned to culture were found to have a human karyotype. No growths developed in any of the animals injected with the control cells (non-carcinogen treated) carried in parallel to the MNNG-treated cells.

Our conclusion as a result of this study was that we had obtained evidence for transformation of human fibroblasts using Kakunaga's protocol. The low yield of tumors was, of course, disappointing, but the fact that many other workers had failed to obtain tumors with similar protocols and that the tumor was composed of human cells was encouraging. However, these experiments also convinced us that the quantitation of focus formation is difficult and serves at best as a first sign of possible transformation. Although this type of experiment is useful to obtain chemical carcinogen-transformed human fibroblasts and to demonstrate that such transformation can occur, it cannot be used as a quantitative measure of transformation frequency, especially since only an indeterminate fraction of the foci ultimately produce tumors in the mice. A further conclusion was that for a transformation system to be useful as an assay, it is necessary that some type of strong selection pressures be applied to limit the growth of normal cells while allowing the tumor-precursor cells to multiply.

3. SELECTION OF TRANSFORMED CELLS USING ABILITY TO FORM COLONIES IN SOFT AGAR

The report of Milo and DiPaolo (2) was published just as we were finishing the previous experiment. In their protocol (which is shown schematically in the lower half of Figure 1), tumor-precursor cells are selected by growth in medium containing 8X non-essential amino acids (8X AA) beginning shortly after exposure to the carcinogen and continuing for 20 population doublings. Furthermore, in their procedure, tumor-precursor cells are selected by growth in soft agar medium. Cells which grow in soft agar (Agar I) are recovered, grown up to large populations and selected again for growth in agar two more times. Since the selection system results in an increase in frequency of tumorigenic cells in the treated population, only a small number of flasks (30 to 40) are handled during a typical experiment. At each 1:10 passage, most of the population is discarded.

Since this protocol appeared to contain the selection techniques necessary for our purposes, we contacted Milo and sought his advice and counsel in attempting to duplicate the work. We made use of diploid fibroblasts derived from foreskins, synchronized them according to their protocol (2), pretreated them with 1 μg/ml anthralin (a tumor promoter), and then exposed them to MNNG. We modified the protocol of Milo and DiPaolo slightly by assaying the cytotoxicity of MNNG by measuring percent survival of colony-forming ability instead of using the dye exclusion assay they employed. The former method is more accurate, allows one to ascertain whether exposure to increasing concentrations results in a dose-dependent cell killing, and provides a means of comparing one experiment with another as well as one agent with another.

Using initial concentrations of MNNG of 0.8 and 1.2 μg/ml, we obtained survivals of 14% and 4%, respectively. The surviving cells were subcultured in the 8X AA selective medium for 20 population doublings and then 1×10^6 cells from each treated culture and from controls were assayed for ability to form colonies in soft agar by plating 5×10^4 cells per 60 mm diameter petri dishes (2).

The cells from the experimental flasks formed colonies in soft agar (Agar I) with high efficiency ($1 - 3 \times 10^{-2}$). Cells fron non-treated control cultures formed colonies with much lower frequency. Large colonies were isolated from agar after three weeks of growth and propagated into large numbers of cells. A portion of these cells were retested a second time for colony formation in soft agar (Agar II) and the rest used to assay for tumorigenicity by being injected intradermally (>1 x 10^6 cells per injection) into athymic mice. Cells from control cultures were also injected. This cycle was repeated a third time (Agar III). Cells derived from agar colonies showed an even higher cloning efficiency in agar the second time, suggesting an enrichment of the population for cells which have acquired anchorage independence. The latter characteristic has been shown to correlate highly with the ability of human cells to form tumors in athymic mice (5).

As reported above with the cells derived from foci in the Kakunaga protocol, a number of animals injected with cells derived from "Agar I" developed nodules which later regressed. (It should be noted that, in contrast to the published procedure of Milo and DiPaolo, we did not use mice which had been irradiated with 450 rads of ionizing radiation because in our hands this dose resulted in 50% mortality. The residual immunological competence of the athymic mice may account for the lower frequencies of non-regressing tumors observed to date.) However, one nodule, derived from cells taken from Agar II, grew to a size of more than 1 cm in diameter, and did so in a period of two to three weeks. This tumor was analyzed and independently characterized by three pathologists to be a fibrosarcoma and shown to consist of cells with a human karyotype. No mice injected with control cells developed nodules. The animals are still being monitored weekly for tumor development.

4. DETERMINING THE RELATIONSHIP BETWEEN MUTAGENESIS AND INDUCTION OF ANCHORAGE INDEPENDENCE.

Once we had gained first hand experience with the protocol of Milo and DiPaolo (2), we repeated the assay using propane sultone. The dose-dependent response in cytotoxicity and induction of soft agar growth (Agar I) (anchorage independence) are shown in Figure 2. In an accompanying experiment we exposed a second population of the same strain of fibroblasts to propane sultone in an identical manner and at the same concentrations as used for transformation and measured the induction of 6-thioguanine (TG) resistant cells according to our published procedures (6) so that we could compare this response with that of induced growth in soft agar.

The dose response curve for mutation induction in Figure 2 is remarkably similar to that obtained for induction of ability to grow in agar. Note, however, that the frequency of cells able to form colonies in agar is ~20 times higher than that of TG resistant cells. If one assumes for the sake of argument that both phenomena actually represent mutagenic responses, the data can be interpreted in several ways. For example, the gene which controls growth in agar might be 20 times larger than the gene responsible for TG resistance (i.e. coding for the structure or regulation of hypoxanthine(guanine)phosphoribosyltransferase, HPRT), or the gene for transformation may contain mutational "hot spots". The frequency of anchorage-independent colonies may merely reflect a selective advantage present in the original "induced mutants" so that they multiply faster than the non-mutated population during the expression period. It should also be noted that observed mutation frequencies may reflect the stringency of the selection conditions. For example, if one assays a population of mutagenized

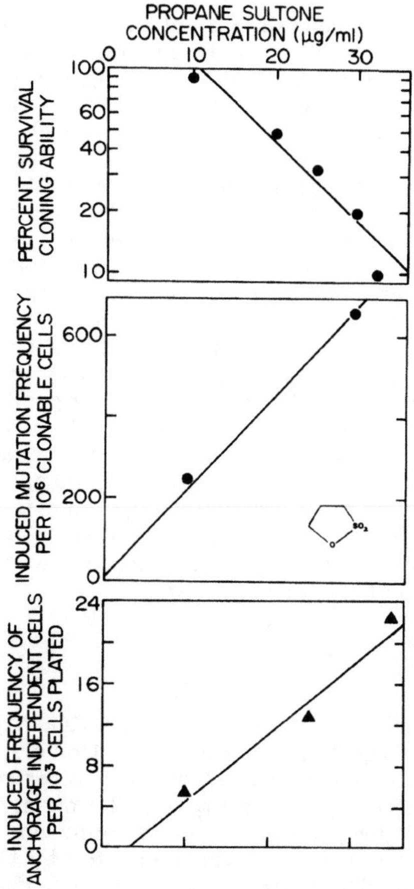

Figure 2. Comparison of the cytotoxicity (upper panel), mutagenicity (middle panel) and ability to cause loss of anchorage dependence (lower panel) in human fibroblasts as a function of the concentration of propane sultone administered. Cells were synchronized by 24 hr of arginine-glutamine starvation (G_1/S block), released and allowed to proceed to S in the presence of anthralin[1] (1 µg/ml). After 10 hr, carcinogen was added at the indicated concentrations. After 14 hr of exposure, the cells were trypsinized, replated at a ratio of 1:2 and a portion plated at cloning density to assay percent survival. The population of cells undergoing phenotypic expression was maintained in exponential growth for ~ 6 population doublings before being assayed for TG-resistant cells and 10 population doublings before being assayed for growth in soft agar (2). A series of control cultures, treated in the identical way but without receiving carcinogen, was carried through each step of this protocol. For details on procedures used to assay for TG resistance, see Maher, et. al. (6). The experimental details of the protocols used to assay for growth in agar will be published elsewhere (Silinskas, et. al., manuscript in preparation).

cells for frequency of 8-azaguanine resistant cells and also for TG resistance, the frequency of the former can be two to five-fold higher (Maher, unpublished data). No simple technique allows the measurement of stringency for soft agar growth. However, the induced frequencies shown in Figure 2 are very similar to those which we previously observed using MNNG (data not shown).

In summary, in spite of the preliminary nature of this work, we consider that we have succeeded in confirming the transformation of diploid human fibroblasts reported earlier by Kakunaga (1) and by Milo and DiPaolo (2). As explained above, we consider their protocols to be qualitative rather than quantitative. This is because the ultimate assay for transformation, i.e. the production of a tumor, necessarily occurs after a period of time greatly removed from the initial carcinogen treatment and only after a very large number of population doublings. Thus, there is, as yet, no simple way to infer the number of events required for transformation. Nevertheless, their published results and our present findings clearly indicate that chemicals can induce diploid human fibroblasts in culture to transform into malignant cells, capable of producing fibroscarcomas in athymic mice. However, extensive and painstaking research to improve the quantitation involved is critically important if one is to develop a reproducible, quantitative system with which to analyze the number and kinds of steps which are involved in the transformation process to gain insight into the nature and interrelationships of the various steps.

ACKNOWLEDGEMENTS

We take this opportunity to thank Dr. Takeo Kakunaga of the National Cancer Institute and Dr. George Milo of Ohio State University for their helpful advice and suggestions on how best to reproduce their experiments. We also wish to thank Rebecca Corner, Suzanne Kateley, Lori Lommel, Lonnie Milam, and John Tower whose extraordinary diligence in carrying out these experiments made this work possible. This research was supported by NIH grant CA 21289 from the National Cancer Institute and by a grant from the College of Osteopathic Medicine, Michigan State University.

REFERENCES

1. Kakunaga, T.: 1978, Proc. Natl. Acad. Sci. U.S.A. 75, pp. 1334-1338.
2. Milo, G.E., Jr., and DiPaolo, J.A.: 1978, Nature 275, pp. 130-132.
3. Borek, C.: 1980, Nature 283, pp. 776-678.
4. DeMars, R., and Jackson, J.L.: 1977, J. Environ. Pathol. Toxicol. 1, pp. 55-77.
5. Freedman, V.H. and Shin, S.I.: 1974, Cell 3, pp. 355-359.
6. Maher, V.M., Dorney, D.J., Mendrala, A.L., Konze-Thomas, B. and McCormick, J.J.: 1979, Mutation Res. 62, pp. 311-323.

MONOCLONAL ANTIBODIES IN THE STUDY OF HUMAN CANCER

Hilary Koprowski
The Wistar Institute of Anatomy and Biology
36th Street at Spruce
Philadelphia, Pennsylvania 19104

Immunization of mice with either intact cells or membrane preparations of cells from either human melanoma or colorectal carcinoma cultures resulted in the production of hybridomas after fusion of mouse splenocytes with mouse myeloma cells. Several species of antibodies secreted by these hybridomas reacted specifically with antigens either present on membranes of homotypic tumors or, as in the case of anticolorectal antibodies, with antigens secreted by the tumor cells into the medium. Antibody-secreting hybridomas were also produced after immunization of mice with cells obtained directly from either gastric carcinoma or colorectal carcinoma removed by surgical procedures. Antimelanoma antibody secreted by three clones of three hybridomas was found to recognize the DR antigen present on melanoma cells; normal and EB-transformed lymphocytes, Raji and HEP 2 cells. This antigen was expressed by cells of 12 out of 17 melanomas maintained in culture and 18 out of 28 melanomas tested within 10 culture days after surgical removal of the tumor. Some of the clones of melanomas did not react with the anti-DR hybridoma antibody. On immunoprecipitation of the DR antigen by the monoclonal antibodies from all the surfaces of melanoma cells labeled with I^{125}, two chains of molecular weight 31,000 (α chain) and 38,000 (β chain) were obtained on polyacrylamide gel electrophoresis. Differences in the modality of the β-subunits derived from cells of various melanomas or from normal lymphocytes obtained from the same patients were observed. Monoclonal antibody secreted by hybridomas other than those reactive with DR antigens reacted with cells of all 17 melanomas kept in culture and with 28 out of 29 melanomas freshly obtained from patients. This antibody did not react with normal human lymphocytes or fibroblasts or with any other human tumor cells except for malignant astrocytes derived from some but not all astrocytomas. In contrast to DR antigens, the antigen(s) recognized by this antibody did not segregate with clones of melanomas. On immunoprecipitation the antigen detected by Nu4B antibody appeared to be composed of two polypeptide chains with molecular weights of 139,600 and 102,200. In contrast to the DR antigen, the molecular weight of this antigen appeared to be similar in all the melanoma

B. Pullman, P. O. P. Ts'o and H. Gelboin (eds.), Carcinogenesis: Fundamental Mechanisms and Environmental Effects, 499–501.

cell lines and one astrocytoma cell line tested. The presence of cross-reacting antigenic determinants on surfaces of melanoma and astrocytoma cells may be explained by the fact that both cell types derived originally from cells migrating from the neural tube during early fetal life. Some of the non-malignant melanocytes derived either from Spitz tumor or giant hairy nevus express this antigen and some do not. One other monoclonal antimelanoma antibody, produced by Soldano Ferrone of the Department of Molecular Immunology at Scripps Clinic and Research Foundation in La Jolla, California, reacted only with cells of human melanoma and precipitated a monomolecular antigen of molecular weight 269,000. Finally, one other antimelanoma antibody, showing specificities somewhat different from the two classes described above, did not precipitate any molecules present on the surface of melanoma cells. All monoclonal antimelanoma antibodies were capable of antibody-dependent cell-mediated cytotoxicity (ADCC) directed specifically against melanoma cells. None of the antimelanoma hybridoma antibody could participate in complement-mediated cytotoxicity (CDC).

Five different classes of immunoglobulins were secreted by anticolorectal hybridomas. Of these, only IgG1 and IgM were found to bind specifically to colorectal carcinoma cells. The IgM antibodies mediated an immunofluorescence reaction with antigens present on colorectal carcinoma cells and also mediated a specific CDC. Antibodies present in the IgG1 fraction bound specifically to colorectal carcinoma cells but did not mediate CDC. One of these antibodies which mediated ADCC in vitro was found to suppress markedly the growth of colorectal carcinoma in nude mice. In these experiments, the antibody was found to bind specifically to the tumor mass. None of the anticolorectal carcinoma antibodies reacted with cells of normal intestinal mucosa or with any other normal or malignant human cells. Three hybridomas secreting antibodies reacting with human lung adenocarcinoma and one with astrocytoma cells were also produced. In these cases, however, in contrast to results obtained with either antimelanoma or anticolorectal carcinoma antibodies, the anti-lung adenocarcinoma and anti-astrocytoma antibodies reacted only with the cells of the tumor used for immunization of mice and not with any other lung cancer or astrocytoma cells.

Up to now, the study of antigens of human cancers was hampered by the unavailability of reagents that could be used to recognize specificities characterizing tumor cells in contrast to normal cells. Use of monoclonal antibodies specific for given tumor cells and nonreactive with other human tumors or normal cells permits not only identification and characterization of human tumor antigens but opens the door to new approaches in the diagnosis and immunotherapy of human malignancies. For instance, inhibition of a reaction between anti-colorectal carcinoma monoclonal antibody and colorectal tumor by antigen(s) secreted by an otherwise undiagnosed colorectal carcinoma may lead to a much more accurate diagnosis than the presently used, nonspecific assay for carcinoembryonic antigen. One can also visualize that a radiolabeled, highly purified monoclonal antibody may be injected into a patient for scanning for presence of tumor tissue which may not be found by any other means. In light of the fact that monoclonal antibody binds specifically to human tumors in situ and not to normal tissue of the affected organ makes it feasible to develop approaches to direct immunotherapy of human cancer. The most effective method may rely on linking antibody with an α-fragment of a toxin such as diphtheria or ricin and have the antibody carry the attached active fragment of the toxin to tumors which would be destroyed by the action of the toxin. The possibility of linking antibody with a chemotactic factor or with an isotope providing a source of internal radiation of the tumor can also be taken into consideration.

OCULO-CUTANEOUS AND INTERNAL NEOPLASMS IN XERODERMA PIGMENTOSUM: IMPLICATIONS FOR THEORIES OF CARCINOGENESIS

Kenneth H. Kraemer, M.D.
Laboratory of Molecular Carcinogenesis,
National Cancer Institute, Bethesda, Maryland 20205

ABSTRACT

Cultured cells from patients with xeroderma pigmentosum have increased susceptibility to ultraviolet radiation-induced killing and mutagenosis. In support of the somatic mutation theory of carcinogenesis, such patients have a markedly increased frequency of cutaneous and ocular neoplasms. Protection of patients from ultraviolet radiation has been shown to prevent these neoplasms. Xeroderma pigmentosum cells also have increased susceptability to killing and mutagenesis by chemical mutagens which may become internalized. There is a substantial increase in neoplasms of the tongue. A few patients have been reported with leukemia, primary brain tumors, and other internal neoplasms.

The somatic mutation theory of carcinogenesis holds that cells damaged by environmental agents so as to produce mutations in their DNA have an increased probability of becoming neoplastic(1). Experimental evidence has shown that cells from patients with xeroderma pigmentosum, a rare genetic disease (frequency about 4 per million) with defective DNA repair, have increased susceptibility to killing and mutagenesis by a given dose of ultraviolet radiation(1-6). In support of the somatic mutation theory, such patients also have a markedly increased rate of cancer of ultraviolet exposed portions of the skin and eyes(2,4-6).

Patients with xeroderma pigmentosum often have acute cutaneous sun sensitivity and ocular photophobia in infancy followed by markedly increased freckling. With continued sun exposure they usually develop multiple pre-malignant lesions such as actinic keratoses. Nearly 100% of the patients develop multiple primary basal cell or squamous cell carcinomas of the skin or eye at an early age. Malignant melanomas have been reported in prevalence from 1% to 50% of several small series. (2,5). This is much greater than the prevalence of cutaneous malignant melanoma in the general population of 1.1 per 10,000(2). Some patients have developed multiple primary cutaneous malignant melanomas.

Cutaneous abnormalities in patients with xeroderma pigmentosum are strikingly limited to sun exposed areas. Shielded portions of the body such as the buttocks and axillae rarely develop cutaneous neoplasms.

B. Pullman, P. O. P. Ts'o and H. Gelboin (eds.), Carcinogenesis: Fundamental Mechanisms and Environmental Effects, 503–507.

Protection from ultraviolet radiation in a few patients has resulted in almost normal skin without neoplasms. For example, twin boys with xeroderma pigmentosum have been shielded from ultraviolet radiation since infancy. As of 1980, they are 20 years old, attending college, and have had no cutaneous neoplasms. Their three older affected siblings were not protected from ultraviolet radiation until later in life. All had developed cutaneous neoplasms before the third decade (5,7).

A little while ago you heard from Dr. Maher that cultured cells from xeroderma pigmentosum patients have increased susceptibility to killing and mutagenesis by some polycyclic aromatic hydrocarbons (8-10). Xeroderma pigmentosum cells have also been reported to have increased sensitivity to other mutagens such as nitroquinoline oxide, acetoaminofluorine or aflatoxin (11-14). Environmental exposure to these chemicals is not limited to the skin and eyes as with ultraviolet radiation but occurs internally as well. The somatic mutation theory would imply that xeroderma pigmentosum patients should have an increased risk of internal carcinogenesis due to these or other internalized mutagens.

As a test of this theory, I have reviewed the literature searching for reports of primary internal neoplasms in xeroderma pigmentosum patients. Table 1 shows these results of this search as well as those patients I have personally observed. Thirteen xeroderma pigmentosum patients with primary non-skin, non-eye neoplasms were acertained. There were 6 reports of xeroderma pigmentosum patients with tumors of the tip of the tongue or gum. All oral cavity neoplasms where the age was mentioned occured in children 13 years old or younger. One patient had two primary squamous cell tumors at the tip of her tongue. This frequency of oral cavity neoplasms appears to be substantially increased above that in the general population. The tip of the tongue however, may be exposed to ultraviolet radiation as well as to chemical carcinogens.

Neoplasms were reported from ultraviolet shielded sites in 7 patients. Two patients were reported to have leukemia, two other patients had primary brain tumors, and there was one report each of testicular sarcoma, breast carcinoma, and a benign tumor, thyroid adenoma. At this stage it is premature to conclude that the tumors outside of the oral cavity have increased prevalence in xeroderma pigmentosum since the total observed population is unknown. It is clear however, that xeroderma pigmentosum patients are susceptable to these internal neoplasms.

Several problems have become apparent in ascertainment of xeroderma pigmentosum patients with internal neoplasms. Patients often die at an early age due to cutaneous neoplasms. If they lived longer they might possibly develop more internal neoplasms. A primary internal neoplasm may be mistaken for the more common metastasis from a cutaneous neoplasm. Internal neoplasms in a patient with many other cutaneous and ocular neoplasms may not be considered as unusual enough to be reported in the literature. I am now attempting to accumulate further data on internal neoplasms in xeroderma pigmentosum and I would appreciate receiving such information.

Table 1 Internal Neoplasms in Xeroderma Pigmentosum+

NEOPLASM	AGE/SEX		REFERENCE
Oral Cavity-			
Squamous Cell Carcinoma, gum above cuspid	9yr	F	15
Squamous Cell Carcinoma, tip of tongue	11yr	M*	16
Squamous Cell Carcinina, tip of tongue (2 primaries)	13yr	F*	16
Squamous Cell Carcinoma, anterior tongue	12yr	M	17
Tongue Tumor - type unspecified	5yr	M§	18
Tongue Tumor - type unspecified	Not stated		19
Leukemia-			
Acute Lymphatic	3yr	M	20
Brain-			
Medulloblastoma	14yr	M	6
Glioblastoma Multiforme	15yr	M	22,23
Testis -			
Sarcoma	12yr	M	24
Breast-			
Carcinoma	38yr	F	24
Thyroid-			
Adenoma	23yr	F	15

+Primary non-skin, non-eye neoplasms. (Modified from reference 5).
*Siblings
§Patient XP2JO (GM2032)

In summary, studies of xeroderma pigmentosum provide strong support for the somatic mutation theory of carcinogenesis. Such patients have markedly increased frequency of neoplasms of the skin and eyes. These neoplasms can be prevented by the shielding from ultraviolet radiation of cells which have been shown to have increased sensitivity to mutagenesis by ultraviolet radiation. Laboratory studies have recently demonstrated an increased susceptibility of xeroderma pigmentosum cells to mutagenesis by chemicals which may be internalized. Prospective epidemiologic studies will be needed to determine if xeroderma pigmentosum patients have a similar increased susceptibility to carcinogenesis by these internalized mutagens.

References

1. Burnet, F.M.: Cancer: somatic-genetic considerations. Adv. Cancer Res. 28: 1-29 (1978).
2. Robbins, J.H., Kraemer, K.H., Lutzner, M.A., Festoff, B.W. and Coon, H.G.: Xeroderma pigmentosum - an inherited disease with sun sensitivity, multiple cutaneous neoplasms, and abnormal DNA repair. Ann. Internal Med. 80: 221-248 (1974).
3. Cleaver, J.E. and Bootsma, D.: Xeroderma pigmentosum: biochemical and genetic characteristics. Ann. Rev. Genet. 9: 18-38 (1975).
4. Kraemer, K.H.: Progressive degenerative disease associated with defective DNA repair: Xeroderma pigmentosum and ataxia telangiectasia. In: Nichols, W.W. and Murphy, D.G. (Ed). <u>DNA Repair Processes</u>. Miami,

Symposium Specialists, 1977, pp 37-71.
5. Kraemer, K.H.: Xeroderma pigmentosum. In: Demis, D.J., Dobson, R.L., and McGuire, J. (Eds). Clinical Dermatology. New York, Harper and Row (In Press, Unit 19-7).
6. Pawsey, S.A., Magnus, I.A., Ramsay, C.A., Benson, P.F. and Gianelli, F.: Clinical, genetic and DNA repair studies on a consecutive series of patients with xeroderma pigmentosum. Quart. J. Med. (New Series). 48: 179-210 (1979).
7. Lynch, J.T., Frichot, B.C., and Lynch, J.F.: Cancer control in xeroderma pigmentosum. Arch. Dermatol. 113: 193-195 (1977).
8. Maher, V.M., McCormick, J.J., Grover, P.L. and Sims, P.: Effect of DNA repair on the cytotoxicity and mutagenicity of polycyclic hydrocarbon derivatives in normal and xeroderma pigmentosum human fibrobasts. Mutat. Res. 43: 117-138 (1977).
9. Regan, J.D., Francis, A.A., Dunn, W.C., Hernandez, O., Yagi, H. and Jerina, D.M.: Repair of DNA damaged by mutagenic metabolites of benzo-[a]pyrene in human cells. Chem. Biol. Interact. 20: 279-287 (1978).
10. McCaw, B.A, Dipple, A., Young, S. and Roberts, J.J.: Excision of hydrocarbon - DNA adducts and consequent cell survival in normal and repair defective human cells. Chem. Biol. Interact. 22: 139-151 (1978).
11. Stich, H.F., San, R.H.C. and Kawazoe, Y.: Increased sensitivity of xeroderma pigmentosum cells to some chemical carcinogens and mutagens. Mutat. Res. 17: 127-137 (1973).
12. Amacher, D. and Liberman, M.W.: Removal of acetoxyaminofluorine from the DNA of control and repair deficient human fibroblasts. Biochem. Biophys. Res. Comm. 74: 285-290 (1977).
13. D'Ambrosio, S.M. and Setlow, R.B.: Defective and enhanced post replication repair in classical and variant xeroderma pigmentosum cells treated with N-Acetoxy -2- Acetyl Amino Fluorine. Cancer Res. 38: 1147-1153 (1978).
14. Sarasin, A.R., Smith, C.A. and Hanawalt, P.C.: Repair of DNA in human cells after treatment with activated aflatoxin B1. Cancer Res. 37: 1786-1793 (1977).
15. Reese, A. and Wilber, L.: The eye manifestations of xeroderma pigmentosum. Am.J. Ophthalmol. 26: 901-911 (1943).
16. Plotnick, H.: Xeroderma pigmentosum and mucocutaneous malignancies in 3 black siblings. Cutis 25: 311-313 (1980).
17. Kraemer, K.H.: Personal observation, 1979.
18. Jenkins, T. In: List of Genetic Variants in the Human Genetic Mutant Cell Repository. 6th Edition (NIH Publication No. 80-2011). Institute for Medical Research. (1979) p. 20.
19. Cockayne, E.A.: Inherited Abnormalities of the Skin and Its Appendages. London: Oxford University Press, (1933) pp. 93-102.
20. Reed, W.B., Sugarman, G.I. and Mathis, R.A.: De Sanctis Cacchione syndrome. A case report with autopsy findings. Arch Dermatol. 113: 1561-1563 (1977).
21. Berlin, C. and Tager, A. Xeroderma pigmentosum, report of 8 cases of mild to moderate type and course: A study of response to various irradiations. Dermatologica 116: 27-42 (1958).
22. Goldstein, N. and Hay-Roe, V.: Prevention of skin cancer with a PABA in alcohol sunscreen in xeroderma pigmentosum. Cutis Jan: 61-64 (1975).

23. Goldstein, N.: Personal communication, 1979.
24. Miller, R.W.: Childhood cancer and congenital defects: A study of U.S. death certificates during the period 1960-1966. Pediat. Res. 3: 389-397 (1969).
25. Van Patter, H.T. and Drummond, J.A.: Malignant melanoma occurring in xeroderma pigmentosum. Cancer 6 942-947 (1953).

THE INDUCTION, EXPRESSION AND MODULATION OF RADIATION INDUCED ONCOGENESIS *IN VITRO* IN DIPLOID HUMAN AND RODENT CELLS

Carmia Borek
Radiological Research Laboratory, Department of Radiology, Department of Pathology, Cancer Center/Institute of Cancer Research, Columbia University, College of Physicians and Surgeons, New York, N.Y. 10032

Techniques developed to identify radiation induced oncogenesis in cultured rodent cells have enabled the qualitative and quantitative assessment of the oncogenic potential of radiation and the identification of factors which modulate these events. Because of the revelance to estimating human cancer risk development of analogous systems using diploid human cells is of great importance. Recently we have shown for the first time that human diploid cells can be transformed *in vitro* by X irradiation into malignant cells which give rise to tumors in animals. The frequency of human cell transformation is lower compared to that observed in animal cells given the same dose.

INTRODUCTION

In recent years public concern has focused on the potential biological hazard of low doses of ionizing radiation, levels to which we may be exposed in the environment through nuclear plants, or from routine medical diagnostic x-rays. Radiation is the most universal carcinogen, though by comparison with chemical carcinogens it is a weak one. Most information on the carcinogenic potential of radiation in man has come from epidemiological data on the high cancer incidence in early radiation workers and from the tragic events in Hiroshima and Nagasaki (1).

Animal studies have contributed much to the knowledge of the carcinogenic potential of radiation (2) but these have their limitations especially at low dose levels when inordinately large numbers of animals are required to obtain reliable data. In addition, the complexity of the physiological mechanisms in the whole animal masque early events in the induction and development of malignancy.

ONCOGENIC TRANSFORMATION OF ANIMAL CELLS *IN VITRO*

For quantitative as well as qualitative studies cell culture

B. Pullman, P. O. P. Ts'o and H. Gelboin (eds.), Carcinogenesis: Fundamental Mechanisms and Environmental Effects, 509–516.

systems offer a powerful tool for evaluating the neoplastic potential of agents found in our environment or used in medical diagnosis and therapy (3).

Our introduction of diploid hamster embryo cell culture systems to investigate radiation induced malignant transformation (4-6) has enabled us (7,8) to assess quantitatively the dose response relationship for transformation over a wide range of doses. We found that doses as low as 0.01 Gy of x-rays and 0.001 Gy of neutrons (1 Gy = 1 gray = 100 rad) can induce transformation, and that while the efficiency of neutrons in inducing transformations was greater than x-rays, so was their effectiveness in cell killing. We were able to demonstrate that splitting x-ray doses of 0.5 or 0.75 Gy into two equal fractions can enhance transformation at low doses by 70% as compared to the same total radiations given as one single dose. In the high dose range a sparing effect of fractionation and lower transformation rates were observed (9,10). A sparing effect for cell survival was observed at all split doses as compared to the single dose. These data suggested that the use of linear interpolation from high to low dose levels may lead to cancer risk estimates that are neither conservative nor prudent, depending on the distribution of dose in time. The data also suggest that there exist different repair mechanisms for survival and transformation.

Studies on the modulation of the neoplastic potential of radiation can also be approached *in vitro*. Thus, we have shown that the protease inhibitor antipain at non toxic doses will *enhance* transformation when present in the cultures during irradiation but will inhibit the initiation and/or expression of malignant transformation when added within minutes after radiation (11). We found that vitamin A analogs inhibit the expression of radiation induced transformation in the hamster cells (12) as it did in mouse cells (12,13). In addition the analog eliminated any promotional effects by the tumor promoter TPA (12). These effects on transformation were not reflected in sister chromatid exchanges (SCE) (12) indicating that the expression phase rather than the initiating phase of transformation is being modulated by the compounds. We further found that while there was no effect on SCE, the membrane enzyme Na/K ATPase was markedly altered by the treatments (12,14). Retinoid depressed this enzyme which is closely associated with sodium transport, TPA enhanced its level while the combination of the 2 agents counteracted one another as in the transformation studies and brought the enzyme level back to control levels.

Animal cell cultures therefore serve well in the qualitative and quantitative evaluation of neoplastic development following radiation and in the study of the nature of the transformed cell (for review see 3). Still, the question exists how do these data pertain to the human situation especially in terms of risk. Available data from epidemiology reflect cancer incidences for radiation induced malignancy at doses over 1 Gy. Information of risk at doses of 0.01 Gy or lower which are of importance in daily exposure are obtained by extrapolation from the high dose levels. While animal cell culture studies on direct induction at low doses have attempted to bridge the gap, to circumvent the necessity of extrapolation for estimating risk the question still exists how these data relate to the human situation, and to the effectiveness of radiation in transforming human cells. Thus, it seems clearly desirable to develop a human cell culture system and to carry out similar experiments as described above in cells of human origin.

ONCOGENIC TRANSFORMATION OF HUMAN CELLS *IN VITRO*

Recently, we have succeeded in transforming human diploid skin cells, the KD cells (15,16) by 4 Gy of x-rays (17) into cells which progressed *in vitro* to malignancy and were able to grow in agar and give rise to tumors when injected into nude mice (17).

We used an early passage of the KD strain. The diploid nature of the cells was ascertained by chromosomal G banding analysis. Cell doubling time was 30-32 hrs. Survival curve analysis indicated that survival fraction following a dose of 4 Gy was close to 12% of the total population (Fig. 1).

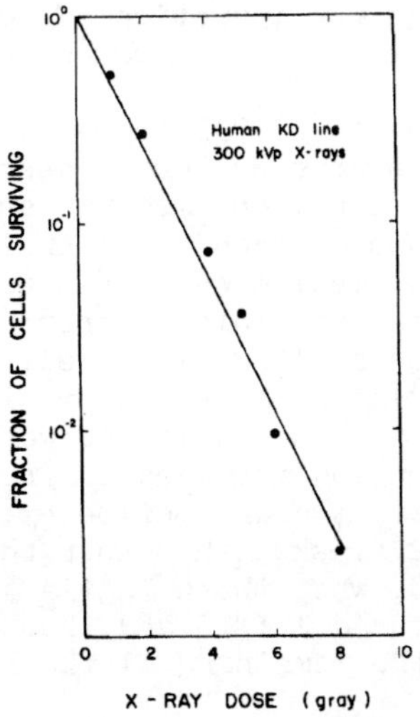

Figure 1. Cell survival of KD cells following irradiation by x-rays. Reproduced from (17).

Cells were routinely maintained in MEM fortified with 10% fetal bovine and 1% human serum (17). The scheme for transformation is presented in Fig. 2 and was as follows:

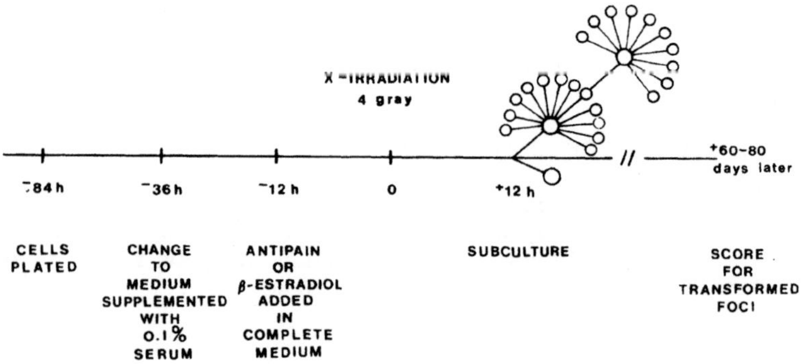

Figure 2. Experimental design for induction of human cell transformation by x-rays. Reproduced from (17).

stock of KD cells were trypsinised (0.25% trypsin), suspended in complete medium and plated in tissue culture flasks (75 cm^2)(Falcon) at 2.6 x 10^5 cells. Two days later, the medium was replaced with MEM containing 0.1% serum for 24 hrs, whereby cell proliferation was greatly reduced. After 24 hrs, medium was again exchanged, this time for complete medium (11% serum) containing either 6 μg ml^{-1} of the protease inhibitor antipain (6) or 1 μg ml^{-1} β-estradiol (Sigma). Experimental cultures were irradiated 12 hrs later with 4 gray of x-rays as previously described (5). All cultures were divided into two 10-12 hrs after irradiation and reseeded in fresh complete MEM without antipain or estradiol. At near confluency, one of the two flasks was subcultured into 12 flasks, the other being left undisturbed. In each experiment, when these 12 flasks reached confluency, one flask was subcultured 1 : 10 and thus defined as 'continuously passaged', and the remaining 11 flasks were left at high density.

The above experimental protocol took advantage of the following a) The cells which were well adapted to conditions of serum deprivation for 24 hrs reached a complete quiescence. b) The addition of complete medium 24 hrs later led to a synchronous wave of DNA synthesis and a treatment of the cells at that point enabled the capturing of cells entering S phase.

Within 2 months of irradiation discrete foci were detected in the irradiated cultures pretreated with antipain of β-estradiol. The foci were composed of cells with subtle irregular orientation. The foci, originally appearing at low frequency (3 foci per experiment) increased in number and size within 3 months and the cells piled up over one another (Fig. 3). By exchanging the medium to low calcium containing medium the foci were altered dramatically while the normal cells degenerated within 24 hrs (17), several foci were isolated. Chromosome G banding indicated a near diploid range of chromosome of 46-49. Saturation density was twofold over that of the normal KD cultures, the transformed but not the normal

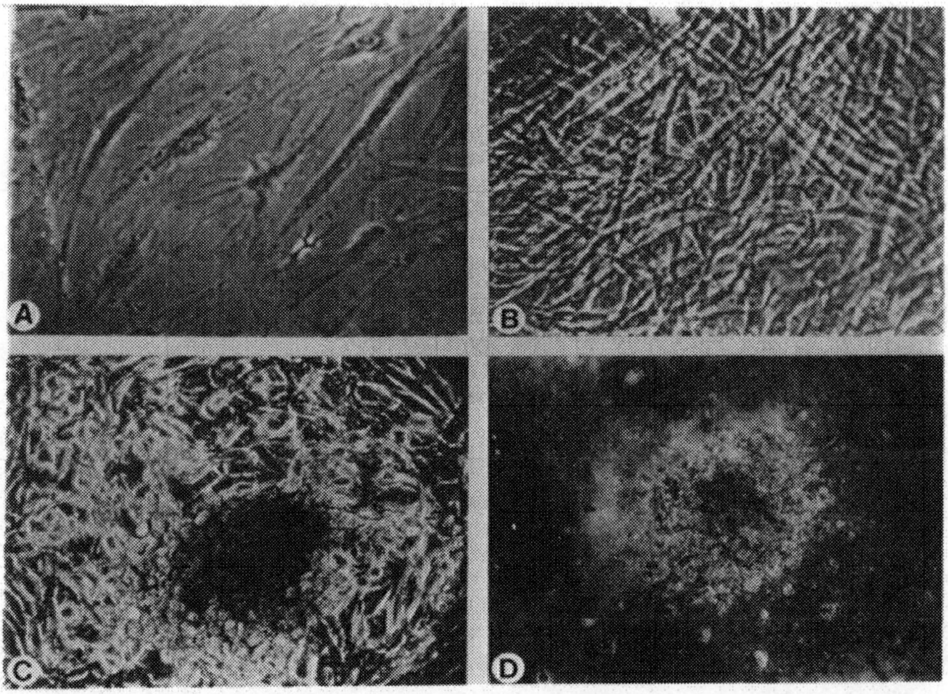

Figure 3. a, Human KD cells in irradiated but untransformed cultures. Phase x58. b, Human KD cells transformed *in vitro* by 4 gray. Note the criss-cross pattern of the cells. Phase x58. c, The same area as in b 24 hrs after incubation in low calcium. Phase x58. d, X-ray-transformed KD cells growing in 0.33% agar. Reproduced from (17).

cells were agglutinable by 25 μg/ml concanavalin A, and were able to form colonies in 0.33% agar. The ultimate proof of malignant transformation lies in the ability of the transformed cells to form tumors in the appropriate host.

While the five isolated transformed cell lines grew in agar, three gave rise to fibrosarcomas when injected intradermally into immunodeficient nude mice. This indicated that growth in semisolid medium such as agar is only suggestive of a neoplastic state of the cells transformed *in vitro* and that injection into animals is imperative for complete assurance of the malignant nature of the cells. No tumors arose upon the injection of the normal KD cells into nude mice.

Cultures which were treated, irradiated but not allowed to replicate more than 4 or 5 times before reaching confluency did not exhibit transformation. This indicated that as in the rodent cells (5,6) replication is required following radiation for the fixation and expression of the transformed state.

In pursuit of the mechanism by which antipain potentiated the x-ray induced transformation we carried out a series of experiments on the effects of antipain on DNA damage and repair in the human skin fibroblasts (18,19). We assayed for DNA damage and repair following treatment with 2 mM antipain by

Inhibition and recovery of DNA synthesis,
Presence of single strand breaks,
Accumulation of single strand breaks during growth in cytosine arabinoside,
Repair replication,
Unscheduled DNA synthesis.

While in parallel experiments these assays readily detected damage and repair following ultraviolet light irradiation they failed to reveal any change in DNA following antipain treatment (18,19). Thus the antipain effectiveness in potentiating radiation induced transformation with rodent cells (11) and in human cells must occur by a mechanism which may not involve interaction of the compound with DNA.

We therefore remain in relative ignorance of the mechanism by which antipain and the equally effective estradiol potentiate radiation induced transformation. Estradiol has been shown to potentiate chemically transformation of human cells (20). Since estradiol possesses some anti-protease activity (11), it is possible that antipain and estradiol act to potentiate transformation via a common mechanism related to this property of protease inhibition in a manner yet to be elucidated.

At the present time the human cell transformation system is not yet quantitative in nature. However, from preliminary experiments it seems clear that the frequency of radiation induced transformation in human cells *in vitro* is much lower than that observed in rodent cells given the same dose of radiation. While in the human cells the frequency per treated cells is approximately 10^{-6} at 4 Gy, a frequency associated with mutational events, rodent cells show a significantly higher incidence of 10^{-4} at that dose level. It is of interest to note from initial observations that the number of doublings required for the expression of the transformed state of the human cells was approximately 10-13, similar to that observed in some rodent cells (5). The longer doubling period of the human cells (30-32 hrs) compared to that of rodent cells (16 hrs) may account in part for the increased length of time required to detect morphological transformation in the human diploid cultures as compared to that in diploid rodent cells. Other observations on the transformed human cells can be stressed. a) Initial loss of contact inhibition is not as striking as that seen in rodent cells. b) In contrast to rodent cells the ability to proliferate in medium with low serum (1%) is not confined to the transformed cells; our normal KD cells as well as the transformed proliferated in medium containing low serum. c) The transformed state is associated with membrane changes and as in rodent cells, agglutinability by plant lectins can be used as a distinguishing probe (21). d) Surface topography in the x-ray transformed human cells is altered but not as dramatically as in the rodent cells. Microvilli found in abundance on rodent cells (22) were increased to a lesser degree in the transformed human cells (C. Borek in preparation).

The ability to transform human cells *in vitro* by x-irradiation opens up a wide range of studies which could aid in the assessment of human risk to radiation induced carcinogenesis. The system could serve in evaluating treatments and conditions which may inhibit the induction and/or expression of radiation induced neoplasm. It could serve in the identification of cocarcinogens or compounds which may serve as promoters. Using cells derived from various individuals it may be possible to identify groups which are high risk due to age or to a particular genetic disposition.

ACKNOWLEDGEMENT

This investigation was supported by grant CA 12536 to the Radiological Research Laboratory/Department of Radiology, and grant CA 13696 to the Cancer Center/Institute of Cancer Research awarded by the NCI, DHEW and by contract DE-AC02-78EV04733 from the Department of Energy.

REFERENCES:

1. Rossi, H.H., and Kellerer, A.M.: 1974, Radiat. Res., 58, pp. 131-140.
2. Upton, A.C., Randolph, M.L., and Conklin, J.W.: 1970, Radiat. Res., 41, pp. 467-491.
3. Borek, C.: 1979, Radiat. Res., 79, pp. 209-232.
4. Borek, C., and Sachs, L.: 1966, Nature (London), 210, pp. 276-278.
5. Borek, C., and Sachs, L.: 1976, Proc. Natl. Acad. Sci. U.S.A., 57, pp. 1522-1527.
6. Borek, C., and Sachs, L.: 1968, Proc. Natl. Acad. Sci. U.S.A., 59, pp. 83-85.
7. Borek, C., and Hall, E.J.: 1973, Nature (London) 243, pp. 450-453.
8. Borek, C., Hall, E.J., and Rossi, H.H.: 1978, Cancer Res. 38, pp. 2997-3055.
9. Borek, C., and Hall, E.J.: 1974, Nature (London) 252, pp. 499-501.
10. Borek, C.: 1979, Br. J. Radiol. 522, pp. 845-847.
11. Borek, C., Miller, R., Pain, C., and Troll, W.: 1979, Proc. Natl. Acad. Sci. U.S.A., 76, pp. 1800-1803.
12. Borek, C., et al.,: Annals of N.Y. Acad. of Sci. In press.
13. Harisiadis, L., Miller, R.C., Hall, E.J., and Borek, C.: 1978, Nature (London) 274, pp. 486-487.
14. Guernsey, D., and Borek, C. (submitted for publication).
15. Day, R.S.: 1974, Cancer Res. 34, pp. 1965-1970.
16. Kakanaga, T.: 1978, Proc. Natl. Acad. Sci. U.S.A. 75, pp. 1334-1338.
17. Borek, C.: 1980, Nature (London) 283, pp. 776-778.
18. Borek, C., and Cleaver, J.E. Radiat. Res. (In press).
19. Borek, C., and Cleaver, J.E. Mutation Res. (In press).
20. Milo, G.E., and DiPaolo, J.A.: 1978, Nature (London) 275, pp. 130-132.
21. Borek, C., and Burger, M.M.: 1973, Exp. Cell Res. 77, pp. 207-215.
22. Borek, C., and Fenoglio, C.M.: 1976, Cancer Res. 36, pp. 1325-1334.

PROGRESS IN CLONING THE TRANSFORMING GENE FROM CHEMICALLY-TRANSFORMED MOUSE CELLS

Ben-Zion Shilo, Chia Ho Shih, Marnin Merrick,
L. C. Padhy and Robert A. Weinberg
Massachusetts Institute of Technology
Center for Cancer Research and Department of Biology
77 Massachusetts Ave., Cambridge, MA 02139

ABSTRACT

DNA was extracted from a variety of chemically-transformed mouse fibroblast lines, and used as a donor for the transfection of NIH3T3 cells. Some of these DNAs were able to induce the appearance of transformed foci on the NIH3T3 monolayer following transfection. These foci grow in soft agar and are tumorigenic in newborn mice. The ability to tag and modify the DNA prior to its transfection, allowed us to initiate the molecular cloning of the transforming gene from one of the chemically transformed lines. The experimental cloning strategy currently underway can be used as a general strategy for other single-copy selectable cellular genes.

B. Pullman, P. O. P. Ts'o and H. Gelboin (eds.), Carcinogenesis: Fundamental Mechanisms and Environmental Effects, 517–526.

INTRODUCTION

The effects of chemical carcinogens on DNA structure are thought to be of majc importance in the generation of transformed and tumorigenic cells. This notion is based on the correlation between the carcinogenic potential and the mutagenic effec of a variety of chemical carcinogens (1-4). The hypothesis of the mutational origi of transformation implies that these are discrete normal genes which can induce transformation following their mutagenesis. Recent developments in DNA mediated ge transfer into eukaryotic cells have allowed us to test this hypothesis. Furthermor we present here a strategy to isolate the transforming gene from an in vitro transformed mouse cell line.

The technique of DNA mediated gene transfer into mammalian cells was pioneered by Graham and van der Eb (5). The DNA is coprecipitated with calcium phosphate and introduced to the recipient cells in this form. The DNA transfer methodology was initially used for the transfer of viral genomes, but has subsequently been used tc transfer single copy cellular genes such as the thymidine kinase gene (6). The DNA transfer technique has several characteristic features: 1. It has a very low efficiency. This is due to the fact that only a small fraction of the cells is actually "competent" to take up and express the transfected DNA. Those cells which are "competent" don't take up more than about a hundred different fragments (7). Therefore, the probability of successful transfer and expression of a single copy mammalian gene is as low as 10^{-5} - 10^{-6}. 2. Only a limited number of cell lines c serve as recipients that will take up and express the transfected DNA efficiently. Since the limiting steps in DNA transfer are not known, the basis for this variabil between different lines is not understood. 3. The use of naked DNA allows the modification of the DNA prior to its transfer, by various DNA restriction endonucle or DNA ligases that can link defined DNA sequences to the transfected DNA.

The DNA transfer technique has been widely used to transfer genomes of transfc viruses in various forms. Among the viral DNAs studied in this way were the DNAs c murine and avian retroviruses (8, 9, 10). The initial transfer experiments depended upon the ability to prepare large amounts of DNA from cultures recently infected with Moloney murine leukemia viruses. The experiments were later extended to study Moloney and Harvey sarcoma viruses. These viruses are replication defective and no virus is released therefore, following transfection. However, the successfully transfected cells develop into foci of transformed cells. The existence of a sarcoma virus genome in those foci could be demonstrated by a subsequent infection with a helper virus. These helper viruses provided the virion proteins for the release of the defective sarcoma viruses (11, 12). The Southern blotting technique was also used to show the presence of the defective sarcoma viruses in these foci (13).

One of the potent DNAs which was able to induce the appearance of foci upon transfection was the DNA of cells containing only one or a few copies of Harvey sarcoma virus per diploid cell DNA complement. This suggested to us that the DNA transfer technique was sensitive enough to identify transforming genes, even if they are present only as a unique copy per genome. We decided to use the DNA transfer technique to analyze the DNAs of chemically-transformed cells.

RESULTS AND DISCUSSION

A. Transfer of the Transformed Properties of Chemically-Transformed Cells to NIH3T3 Cells by DNA Transfection

A series of experiments was conducted to investigate the possibility of oncogenic transformation, when using DNA of chemically transformed cells as a donor in DNA transfection. DNAs were prepared from a variety of mouse fibroblast lines that were transformed by exposure to chemical carcinogens either *in vitro* or *in vivo*. The results of these experiments have been reported recently (14). Briefly, five out of fifteen donor DNAs that were examined were able to induce the appearance of foci at a rate similar to the rate of transfer of single copy cellular genes. The DNAs of the other 10 lines and DNAs from untransformed cell lines or from mouse liver, were unable to induce the appearance of foci, or occasionally yielded only one or two foci. These foci may be spontaneous, or as has been recently suggested (15), may result from activation of genes present in the non-transformed DNA. It should be stressed that the DNAs of the five chemically transformed lines that were positive in the assay, were able to induce the appearance of foci at a frequency that was about two orders of magnitude higher than that of DNAs from non-transformed cells.

If the transforming genes introduced to the NIH3T3 recipients were stably integrated and replicated in the foci, one would expect the DNAs prepared from the foci to serve as potent donors for subsequent transfections. Indeed, we were able to maintain the same transfection efficiency (or in some cases even increase it), in the second (14), third and fourth serial transfections.

To rule out the possibility of transformation by retroviruses in the foci, attempts were made to rescue focus forming viruses, by superinfecting the foci with a helper Moloney murine leukemia virus. We were unable to rescue any focus-forming virus in this way (14). In addition, no detectable production of viral gag protein was produced by the foci. These results rule out the possibility of involvement of C type retroviruses in the transformation of the foci.

The foci were picked by a morphological criteria for transformation, namely the criss-cross growth pattern of refractile cells. However, the donor cells had additional transformed features such as the ability to grow in soft agar and induce tumor appearance in mice. We checked whether these properties have also been transferred to the foci during transfection. Indeed, the cells of the foci were able to form colonies in 0.3% agar in contrast to the non-transformed NIH3T3 cells. In addition, injection of 10^6 cells from the foci to newborn Balb/c mice caused the appearance of tumors within 1-2 weeks. Biochemical evidence will be presented later to show that the tumor cells are indeed descendants of the injected cells.

A few conclusions can be drawn from the results presented so far. First, the ability to transfer the transformed properties of chemically-transformed cells by naked DNA proves that DNA is indeed the ultimate target for chemical carcinogens.

We do not know whether the transforming gene that is being transferred is the actual gene that was modified by the carcinogens, or a gene that was modified later on, as a result of the first mutagenic event. Second, the information for transformation (including morphology, soft agar growth and tumorigenicity) must be encoded in a single DNA fragment smaller than 30 kbp. The DNA is routinely sheared to a size of 30 kbp prior to transfection. Due to the low efficiency of the transfection procedure, it would be statistically impossible for two different unique cellular fragments to be successfully integrated and expressed in the same recipient cell.

We feel that the ability to transform NIH3T3 cells with a single DNA fragment does not allow us to comment on the issue of single versus multi stage transformation by carcinogens. NIH3T3 cells, being an established cell line may already be "semi-transformed", and therefore may require only one gene to express full transformation. The lack of a highly transfectable primary cell line, prevents us from further investigation of this aspect. Finally, the ability to transform NIH3T3 which presumably have the normal counterpart of the transforming gene suggests that the gene is dominant. Since DNA from non-transformed cells was unable to induce the appearance of foci, the effect does not seem to be a gene dosage effect. Rather, only the modified gene from the chemically transformed cells, but not its normal counterpart, can transform the recipient cells.

B. Molecular Cloning Attempts of The Transforming Gene

In order to be able to study the molecular biology of the transforming gene and its normal counterpart, we decided to clone the gene from a line of C3H10T1/2 mouse cells, transformed *in vitro* by 3-methyl cholanthrene, called MCA16 (16). Since we did not have a hybridization probe for the gene, it was impossible to use any of the conventional cloning strategies for unique cellular genes. Instead, a cloning technique that was pioneered in yeast (17) was attempted. The technique consists of the following stages: 1. Identification of a restriction endonuclease that does not cut within the transforming gene. The DNA was cleaved with a variety of restriction enzymes prior to transfection. Complete cleavage was verified by addition of small amounts of marker molecules for cleavage such as adenovirus DNA or pBR322 DNA. Enzymes that cut within the transforming gene are expected to destroy its ability to induce foci following transfection and vice versa. After screening a variety of enzymes, the restriction endonuclease BamHI was identified as an enzyme that does not cut within the gene. 2. The DNA of the MCA16 cells was digested with BamHI. In parallel, the bacterial plasmid pBR322, which has a single BamHI site, was cleaved with the same enzyme. pBR322 was ligated to the MCA16 DNA under conditions in which the plasmid was at a 2 fold molar excess, to make sure that the majority of the MCA16 DNA fragments would be linked to pBR322. 3. The ligated DNA was mixed with a 20 fold excess of sheared NIH3T3 DNA, and transfected onto NIH3T3 cells. The foci that appeared were picked, grown to 5×10^7 cells and DNA was prepared from them. 4. The DNA of the primary foci was digested with the restriction endonuclease XbaI (which does not cleave within pBR322), separated on an agarose gel, transferred to a nitrocellulose filter (18) and hybridized with ^{32}P labelled, nick translated, pBR322 probe. Each one of the foci had multiple bands that hybridized with the labeled pBR, indicating the uptake of pBR322 by the focus (19). This is further evidence for the generation of foci by actual DNA uptake in transfections. Only a few pBR molecules in the primary foci are expected to be covalently linked to the transforming gene. The others have been cotransferred with the transforming gene, in spite of the lack of actual linkage. 5. In order to remain only with the pBR molecules that are actually linked to the gene, the DNA of the primary foci is used as a donor for a second cycle of transfection. In this transfection, due to the vast excess of non-pBR sequences in the DNA of the primary focus, only those pBR sequences that are linked to the transforming gene are expected to be transferred to the cells of the foci. The secondary foci were picked, grown and DNA was prepared from them. 6. The DNA of the secondary foci was analyzed in a similar way to that of the primary foci. Indeed, most secondary foci had only a limited number of pBR molecules (19). Moreover, two secondary foci derived from the same primary focus had common pBR bands, indicating that they have both taken up the same molecule with its flanking cellular sequences. This supports our previous suggestion that a unique DNA fragment is responsible for the generation of the transformed phenotype. When cells of a secondary focus were injected into newborn mice, they caused the appearance of tumors. The DNA extracted from these tumors was cleaved and blotted, and it showed the same pattern of pBR molecules as the DNA of the secondary focus. Thus, the tumors arose from the cells of the secondary focus that were injected into the mouse.

The linkage of the transforming gene to a known sequence such as pBR in the secondary foci, provides the basis for several techniques to be used for the final isolation of the transforming gene. One possibility is to cleave the DNA of a secondary focus with an enzyme that does not cut pBR322 or the transforming gene. Then the cleaved DNA can be circularized by DNA ligase. Since the origin of replication and ampicillin resistance gene of pBR322 should be intact after BamHI digest, the circularized DNA can be used to transform bacterial cells and select for ampicillin resistant colonies. Those colonies should contain a chimeric plasmid consisting of pBR322 and the eukaryotic sequences that were adjacent to it. Unless the restriction enzyme that was used prior to circularization cleaves between pBR322 and the transforming gene, the chimeric plasmid should contain the intact gene. A second technique that may be used to isolate the transforming gene involves the generation of "libraries" in one of the Charon type of lambdaphage vectors (20). A library could be made from the DNA of a secondary focus, and pBR322 used as probe to identify the plaques containing the pBR322 molecules and the sequences attached to them (21).

It is important to emphasize that all cloning techniques depend on a close linkage between pBR322 and the transforming gene. If they are separated by more than 20 Kbp, both methods would not be practical. We are now using a variety of methods to identify secondary foci that have a close linkage between pBR322 and the gene. Once such a focus is identified, it will be used as a source for one or both of the isolation techniques mentioned above.

CONCLUSION

The interaction of chemical carcinogens with DNA is likely not specific with respect to the genes with which adducts are formed. The specificity lies in the fact that only an extremely limited number of cellular genes can be activated to transform the cell. The main basis for this assumption is the similarity that has been found between the rate of mutagenesis of unique cellular genes and transformat both resulting from treatment with chemical carcinogens. The transforming genes are the ones being transferred in the DNA transfer technique discussed here. Thus, the technique offers a new way of studying, and eventually isolating, the unique transforming genes which are the specific targets for carcinogens.

The isolation of the gene can provide the tools for addressing a series of questions at the molecular level. 1. How many sets of targets are there in the cell? In other words, are all of the different lines that are positive as DNA donors in the transfection assay,transferring the same gene to the recipient cells? 2. The isolation of the transformed gene will allow the isolation of its normal counterpart, and the two can be compared at various levels, to study the molecular nature of the change that leads to transformation.

Assuming that the transforming gene that is being cloned is the actual target for the carcinogen, the isolation of its normal counterpart can be used to generate new ways to study chemical carcinogenesis both _in vitro_ and _in vivo_. The cloned, normal gene can be modified in various ways _in vitro_ and then used as a donor in transfections. Thus, various modifications of the normal gene can be assayed by th appearance of foci. In addition, the normal gene may be introduced to the recipien cells in multiple copies, to generate a cell line that would be much more sensitive to carcinogens, because of the increased target size now present in the cell.

ACKNOWLEDGEMENTS

This work was supported by the Rita Allen Foundation of which R.A.W. is a fellow and by the National Instiutes of Health Core Grant #CA14051 to S. Luria. B.S. is a Chaim Weizmann post-doctoral fellow.

REFERENCES

1. McCann, J., Chai, E., Yamasaki, E. and Ames, B.N. (1975) Proc. Natl. Acad. Sci. USA 72, 5135.
2. McCann, J. and Ames, B.N. (1976) Proc. Natl. Acad. Sci. USA 73, 950.
3. Bridges, B.A. (1976) Nature (London) 261, 195.
4. Bouck, N. and diMayorka, G. (1976) Nature (London) 264, 722.
5. Graham, F.L. and van der Eb, A.J. (1973) Virology 52, 456.
6. Wigler, M., Pellicer, A., Silverstein, S. and Axel, R. (1978) Cell 14, 725.
7. Wigler, M., Sweet, R., Sim, G.K., Wold, B., Pellicer, A., Lacy, E., Maniatis, T., Silverstein, S. and Axel, R. (1979) Cell 16, 777.
8. Hill, M., Hillova, J., Dantcher, D., Mariage, R. and Goubin, G. (1974) Cold Spring Harbor Symp. Quant. Biol. 39, 1015.
9. Cooper, G.M. and Temin, H.M. (1974) J. Virol. 14, 1132.
10. Smotkin, D., Gianni, A.M., Rozenblatt, S. and Weinberg, R.A. (1975) Proc. Natl. Acad. Sci. USA 72, 4910.
11. Andersson, P., Goldfarb, M.P. and Weinberg, R.A. (1979) Cell 16, 63.
12. Goldfarb, M.P. and Weinberg, R.A. (1979) J. Virol. 32, 30.
13. Goldfarb, M.P. and Weinberg, R.A., submitted for publication.
14. Shih, C., Shilo, B., Goldfarb, M.P., Dannenberg, A. and Weinberg, R.A. (1979) Proc. Natl. Acad. Sci. USA 76, 5714.
15. Cooper, G.M., Okenquist, S. and Silverman, L. (1980) Nature (London) 284, 418.

16. Reznikoff, C.A., Bertram, J.S., Brankow, D.W. and Heidelberger, C. (1973) Cancer Research 33, 3239.

17. Hicks, J.B., Hinnen, A. and Fink, G.R. (1979) Cold Spring Harbor Symp. Quant. B 43, 1305.

18. Southern, E.M. (1975) J. Mol. Biol. 98, 503.

19. Shilo, B., Shih, C., Merrick, M., Dannenberg, A., Goldfarb, M.P. and Weinberg, 12th Miami Winter Symposium, in press.

20. Maniatis, T., Hardison, R.C., Lacy, E., Laver, J., O'Connel, C., Quon, D., Sim, and Efstratiadis, A. (1978) Cell 15, 687.

21. Benton, W.D. and Davis, R.W. (1977) Science 196, 180.

RELATIONSHIP BETWEEN TRANSFORMATION AND MUTATION IN MAMMALIAN CELLS.

TAKEO KAKUNAGA, KO-YU LO, JOHN LEAVITT*, AND MITSUO IKENAGA**
Laboratory of Molecular Carcinogenesis, National Cancer Institute, Bethesda, MD 20205, U.S.A.,*Division of Virology, Food and Drug Administration, Bethesda, MD 20205, U.S.A. and **Faculty of Medicine, Osaka University, Osaka 530, Japan

ABSTRACT : Two approaches have been taken to clarify the mechanism of cell transformation. First, cell variants which show the different susceptibility to the transformation by various carcinogens were isolated from a subclone of Balb/3T3 cell line and characterized for many aspects. The results obtained suggest that the expression process following fixation process is crucial for the susceptibility of the cells to the transformation by chemical and physical carcinogens and that DNA damages and their repair are involved in the fixation of transformation. On the other hand, the attempts to identify the macromolecules responsible for controlling the expression of transformed phenotype revealed that only a small portion of the cellular proteins were altered in the chemically transformed human cells and it was suggested that at least one alteration of proteins was resulted from the mutation.

INTRODUCTION

The relation of mutagenesis and carcinogenesis has been long argued (1-3) and there have been numerous circumstantial evidences to support their close relationship. However, it is still possible to consider the nonmutational process as a mechanism of chemical and physical carcinogenesis. It is uncertain how many and what kind of steps are involved in the transformation, and the mutagenic process speculated is quite vague. No cellular molecules have been unequivocally identified as factors directly involved in the induction of cell transformation. It seems that the more direct and clearer evidence will be obtained if the cell mutants or variants affected in the process of transformation are isolated, characterized, and utilized, and more directly if the cellular genes and their products controlling the normal and transformed phenotype are identified. These two approaches are not novel as an idea. Particularly, enormous attempts have been made to find the biochemical abnormality or cellular components specific for transformed cells. In the past, however, these biochemical

B. Pullman, P. O. P. Ts'o and H. Gelboin (eds.), Carcinogenesis: Fundamental Mechanisms and Environmental Effects, 527–541.

studies have come to a standstill or nothing due to lack of means to isolate the particular genes from mammalian cells and to determine the biological function of the isolated molecules in the cells. Recent progress in the techniques of molecular and cell biology such as gene engineering, monoclonal antibody, and transfer of macromolecules into mammalian cells appears to have provided us with enough chance to identify the cellular molecules responsible for controlling the expression of normal or transformed phenotype.

On the other hand, isolation and utilization of cell variants are possible only with established cell lines unless the mutated cells are constantly available directly from mutated animals or humans. Cells are available from various patients who are predisposed genetically to a higher incidence of cancer. However, the transformation system using human cells are still far from being a rapid and quantitative assay system(4,5). There have been also reports of animal strains of different susceptibility to polycyclic hydrocarbon-induced carcinogenesis(6-8). The differences in the susceptibility between strains, however, has been ascribed only to the differences in genetic factors that control carcinogen metabolism(6-8). Development of a quantitative system for assay of neoplastic transformation using an established mouse cell line (9,10) has enabled us to isolate cell variants affected in the transformation process.

CELL VARIANTS SHOWING DIFFERENT SENSITIVITY TO TRANSFORMATION

Cell variants of different sensitivity to the transformation were isolated by examining the spontaneously arised variant cells in the mass culture of late passage of Balb/3T3-A31-I clone which was previously isolated in our laboratory from Balb/3T3-A31 (Fig. 1). In order to

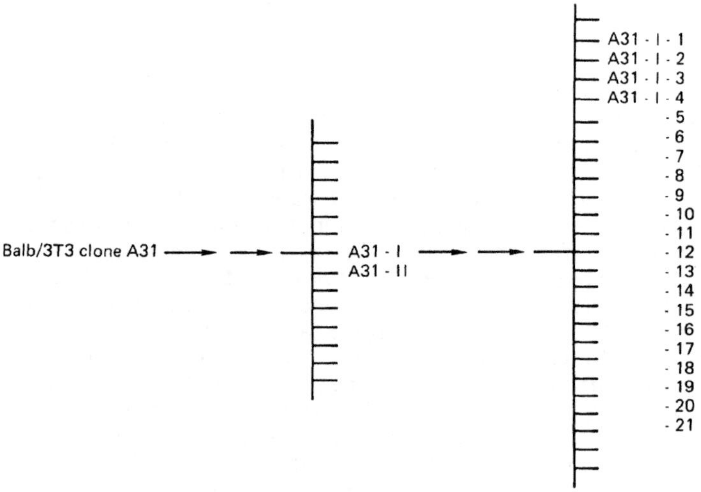

Fig. 1 Outline of pedigree of cell variants

avoid isolating variants having either altered membrane permeability to chemical carcinogens or altered carcinogen-metabolizing activity, ultraviolet light (UV) radiation was used as the transforming agent for screening the susceptibility of many subclones to the transformation. Malignant transformation was assayed by scoring transformed foci as reported previously using A31-714 cells (10). Among many subclones tested, 6 clones were found to show constant susceptibilities to UV-induced transformation through more than 50 cell generations after isolation, and were classified into three groups, highly and intermediately sensitive and resistant to transformation, respectively. The differences in the frequency of transformation induced by UV radiation at dose of 75 erg/mm^2 were approximately 10 fold between highly sensitive and intermediate group and between intermediate and resistant group, and thus 100 fold between highly sensitive and resistant group. Dose-response curves for UV-induced transformation indicate that a highly sensitive variant, A31-I-13, has a significantly higher susceptibility at all UV doses tested and that the slope of curve is almost same between A31-I-13 and intermediately sensitive variant, A31-I-1 (Fig. 2A). Resistant variants produced too few number of transformed foci to draw the dose-response curve in our affordable scale of experiment. On the other hand, the susceptibility to the killing effect of UV-radiation did not significantly differ between variants (Fig. 2B).

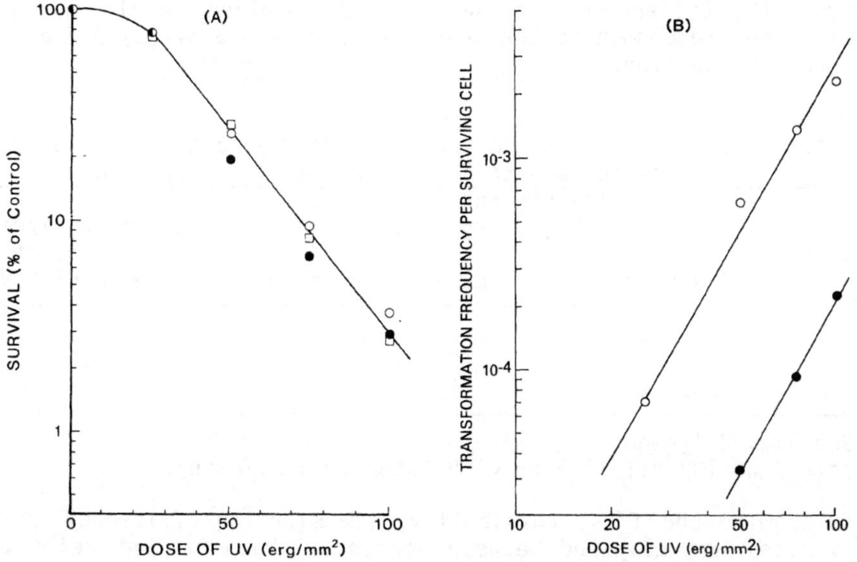

Figure 2. Dose response curves for transformation(A) and cytotoxic effects(B) produced by UV-radiation in highly sensitive variant, I-13 ( ); intermediately sensitive variant, I-1 ( ); and resistant variant, I-15 ( ).
Each value is the average of counts on 14 - 27 dishes.

The differential susceptibility of the variants to transformation was also observed when various chemical carcinogens such as 3,4-benzo-(a)pyrene, 3-methylcholanthrene, 4-nitroquinoline-1-oxide, and N-methyl-N'-nitro-N-nitrosoguanidine were used as the transforming agents(11). The cell variants that are highly or intermediately sensitive, or resistant to UV-induced transformation were also highly, intermediately, or hardly sensitive to the transformation by those chemical carcinogens, respectively.

The transformed cell lines that were isolated from individual transformed foci showed typical transformed phenotype such as higher saturation density, ability to grow in soft agar, and tumorigenicity in nude mice. There has been no significant differences in the transformed phenotypes due to the differences in the original cell variants from which the transformed lines were derived.

The results described above demonstrate that it is possible to isolate cell variants of different sensitivity to transformation, of which differential susceptibility is not due to changes in carcinogen metabolism. As the cause for the difference in the susceptibility to the transformation between variants, the following three possibilities were considered most interesting and were examined experimentally: (1) Differences in the fundamental potential to express the transformed phenotype. (2) Differences in the factors involved in the mutagenic process which are common to the transformation process. (3) Differences in DNA repair function.

Table 1. Susceptibilities of variant cells to UV-induced transformation and mutation and to Ki-MSV-induced transformation

Variants[a]	Frequencies per survival induced by UV (75 erg/mm^2)		No. of foci/dish[b] after infection with Ki-MSV
	Transformation	Ouabain^r-mutation	
I - 13	1.8×10^{-3}	1.8×10^{-5}	36
I - 1	0.6×10^{-4}	2.2×10^{-5}	35
I - 8	0.5×10^{-5}	1.6×10^{-5}	33

a : See Fig. 2 legend.
b : Values at 10^5 dilution of virus stock(0.1 ml/dish).

To examine the first possibility, transforming efficiency of RNA tumor viruses was compared between variants. When variant cells were infected with Kirsten strain of mouse sarcoma virus and Friend leukemia virus as a helper virus at various dilution of virus stock, the number of the morphologically transformed foci formed was almost same between variants at any virus dilution tested (Table 1). This result indicate that even the resistant variant cell has the fundamental potential to be converted into the transformed cells.

Susceptibility to the induction of mutation was compared between variants in order to test the second possibility. We have developed a simultaneous assay system for transformation and mutation using BALB/3T3 subclone and replating procedure, in which system all the cells are exposed to carcinogens and allowed to grow for expression under the quite same conditions. After the expression time, the pooled cells in suspension are divided into three groups for transformation, mutation, and survival assay. It has been widely recognized that cell density at the time of carcinogen-treatment and of expression time has marked effects on the frequency of transformation as well as mutation (9-13). Because different kinds and doses of carcinogen-treatment result in the different surviving fraction of the cells and because the actual surviving fraction varies depending on the cell density, adjustment of initial inoculum size of target cells is not enough to overcome the problem of cell density effects. Introduction of replating procedure in our system solved this problem(14). The results obtained using this simultaneous assy system show that variant cells have the similar susceptibility to the induction of ouabain-resistance mutation by UV-radiation (Table 1). The marked differences in the susceptibility to UV-induced transformation were also observed when the assay system with replating procedure was used (Table 1).

The above results and no difference in the sensitivity to the killing effects by carcinogens (Fig. 2B) between variants appears to be unfavorable for the third possibility, i.e., the differences in DNA repair activity between variants. However, there is a possibility that some factor involved in repair system which does not account for the recovery from the killing effects nor the induction of oua^r-mutation but for the induction of transformation may be different between variants. It has been reported that different alkyl-DNA nucleoside adducts are removed by different repair system at different removal rate, and it has been suggested that some type of repair activity may correlate with organ specificity of carcinogenesis by alkylating agents (see Chapters by Rajewsky and by Montesano in this book). Formation of many different type of BP-DNA nucleoside adducts have been reported in the cells (15 - 21). Although it is not known which type of these DNA modification are correlated with the carcinogenic process, there has been speculation that minor BP adducts may be responsible for induction of mutation or transformation (17,19,20).
Thus, in order to test the third possibility on the cause of differential susceptibility of variants and to know which type of BP-adducts will be most strongly correlated with the induction of transformation, the amount and the type of BP-DNA nucleoside adducts formed were extensively examined in the variant cells. Because the amount and the type of adducts formed and its removal rate from the cells are dependent on the dose of BP exposed to the cells, analysis of binding were carried out under the same condition as used for transformation experiments so that the results on chemicals binding could be relevantly interpreted in relation to the induction of transformation. DNA isolated from variant cells exposed to [6-^{3}H]BP was subjected to enzymatic hydrolysis and then

nucleoside separated through Sephadex LH20 were analyzed by high pressure liquid chromatography (HPLC). HPLC profiles were compared with the profiles of authentic markers. All [^{3}H]BP-treated cellular samples revealed only one radioactive peak in HPLC profiles regardless of the time of BP-treatment and the variant cell lines (Fig. 3). Its elution position coincided with the second peak of deoxyguanosine adducts, which corresponds to 10-trans-7R-BPDE I(anti)-dGua. Other adducts such as deoxycytidine or deoxyadenosine adducts and BPDE II(syn)-dGua that have been found in other cells or in the in vitro DNA-BPDE reaction mixture were not detected. If these minor adducts be formed in the variant cells, they represent less than 0.1 % of the total adducts formed. This implies that these minor adducts can not be the major cause of transformation unless they are thousand fold more efficient in inducing transformation than deoxyguanosine adduct, and that formation of BP-deoxyguanosine adduct itself seems to be enough to initiate the transformation process induced by BP in variant cells if the covalent binding of carcinogen to DNA is a necessary step in chemical carcinogenesis. Removal of BP-dGua adduct was very slow in all variant cells tested (Fig. 4A). There was no difference in the removal rate of adduct during post-incubation between variants.

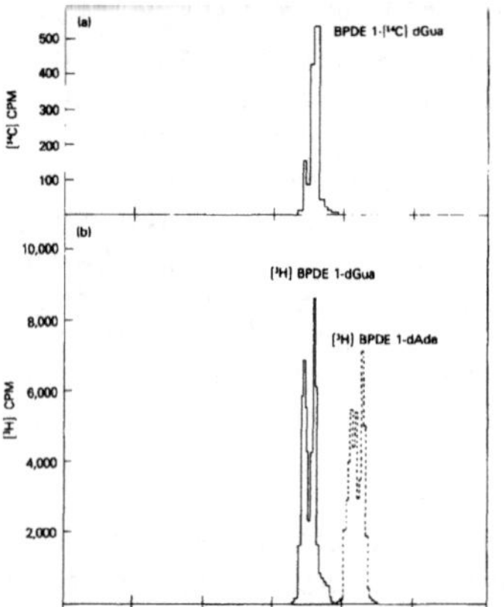

Figure 3. HPLC profiles of DNA adducts formed in Balb/3T3 clone I-13(c), I-1(d) and I-8(e) cells incubated with [^{3}H]BP.
(a) & (b): in vitro marker.

These results on variant cells can be interpreted in two alternative ways, i.e.; (1) mutation or covalent binding of carcinogen to DNA may be nothing to do with the induction of transformation; or (2) transformation process may include not only the formation of DNA-carcinogen adduct or mutational step but also another step which is specifically required for expression of the transformed phenotype after the fixation of transformation, and variant cells may be different in this particular process involved in the expression. Although it is necessary to proceed more experiments to conclude either possibility, at present we are favorable for the second possibility based on the findings and assumptions described in the following section.

MUTATIONAL AND NON-MUTATIONL STEPS IN THE CELL TRANSFORMATION

One of the findings obtained in the past that suggested the crucial role of DNA damage and its repair system in the transformation induction is the loss of the abilities of the cells to be transformed while held at nongrowing condition after carcinogen-treatment (12,13,22,23). This phenomenon (termed as "confluent holding recovery", CHR) which is similar to liquid holding recovery observed for mutagen-induced cell killing and mutagenic effects in microorganisms (24-29) was first reported by Borek and Sachs on radiation-induced morphological transformation of hamster embryonic cells (22) and extensively examined by us on the chemically induced malignant transformation of subclone of Balb/3T3, A31-714 cells (12,13,23).

When A31-714 cells were maintained at confluent and nongrowing condition after 4NQO-treatment, the ability of the cells to be transformed was decreased with the time of post-incubation (Fig. 4B). On the other hand, the ability to be mutated was decreased very slowly (Fig. 4B). Treatment of A31-714 cells with 4NQO resulted in the formation of 4NQO-purine adducts in DNA (23). These adducts were removed from cells during post 4NQO-treatment incubation. When the amounts of DNA adducts were plotted as a function of time of post 4NQO-treatment incubation, the removal rate of adducts was parallel to the decrease in the abilities of the cells to be transformed (Fig. 4A & B). Because the 4NQO-purine adducts were not removed from the cells derived from xeroderma pigmentosum(XP) patient (complementation group A, defective in excision repair), the removal of 4NQO adducts under the experimental conditions used were considered to be ascribed to the excision repair system of the cells. On the other hand, when A31-714 cells were allowed to divide once immediately after 4NQO-treatment, they retained their ability to produce transformed foci even after being kept in the nongrowing condition for 5 days thereafter (12). Similar results were obtained when MCA was used as the transforming agent (13). These results with A31-714 cells were explained assuming that carcinogen-induced DNA damage was converted into stable form as the first step of transformation (fixation) through cell division and that the damage was excised from cells under nongrowing condition before being converted into stable and replicable form.

As shown previously, all variant cells did not remove BP adduct efficiently in contrast to the results with A31-714. Thus, it was examined whether variant cells lose the ability to produce transformed foci while held at nongrowing condition after BP-treatment or UV-radiation. The ability of variants to be mutated was not lost at all and the ability to produce transformed foci decreased only slightly while held at nongrowing condition(unpublished and Fig.4). It has also been found that human diploid fibroblasts lose the ability to be mutated when UV-radiated cells were held under nongrowing condition before being released and allowed to undergo suitable expression period (our unpublished data and see Chapter by Maher in this book). Thus, there is an

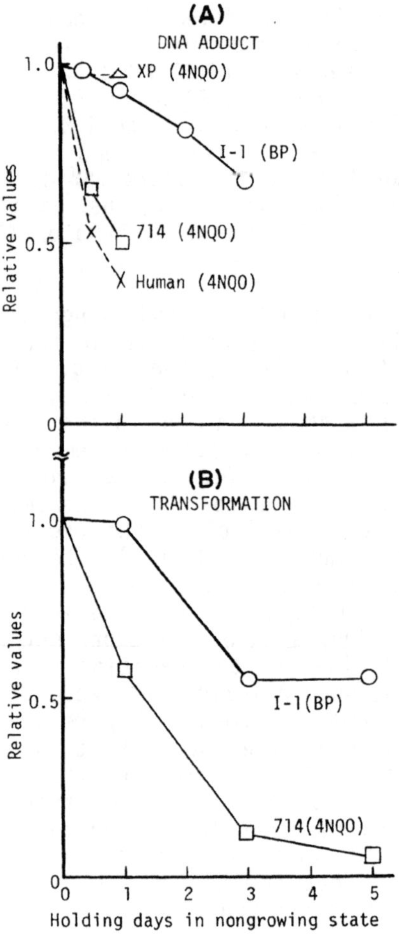

Figure 4. Relative decrease in the amounts of carcinogen-DNA adducts(A) and the abilities of the cells to be transformed(B) incubation in the different cells. Partially from Int. J. Cancer, 14, 736 (1974) and from Cancer Res., 37, 3672 (1977).

apparent parallelism between the removal rate of carcinogen-DNA adduct and the decrease in the fixation of transformation (Fig. 4A & B). These results are consistent with the hypothesis that mutagenic events, i.e., DNA damage and subsequent alteration of DNA structure, are the main cause in the initiation of transformation.

Then, two questions arise. First, why some variant cells are resistant to the transformation in spite of their poor ability to remove carcinogen adduct? Our present hypothesis is that the efficiency of expression of the already fixed information of transformation is very low in the resistant clone. Such a expression mechanism unique for transformation may be related to the target of promoters.

Another question is whether the resistance of human diploid cells to transformation is exclusively ascribed to their high exision repair activity or not. Even in our most successful results of transformation experiments, the transformation occured at very low frequency and only after long-term cultivation (4). If the strong resistancy of human diploid cells to transformation is solely due to their high excision repair (error free ?) activity, the excision repair deficient cells like XP cells should be extraordinarily sensitive to the transformation. However, the data in our (unpublished) and other (30,31) laboratories indicate that XP cells are not as sensitive as mouse, rat and hamster cells to chemical or UV transformation (this does not imply that XP cells are not more sensitive to the transformation than normal human diploid cells).

While we were attempting to develop a rapid and quantitative assay system for chemical transformation of human diploid cells, we observed a transient increase in the colony formation in soft agar of 4NQO-treated fibroblasts (5).

However, the cells which grew in soft agar did not overcome the "crisis" and did not become tumorigenic. The partial, abortive or transient expression of the transformed phenotype was also observed by several other groups (31-34, and personal communication by Broocks and by Ono). It is well known that most of human diploid cells transformed by tumor viruses do not become permanent cell line nor tumorigenic in spite of their clear expression of some transformed phenotypes in culture (35-37). These findings indicate that partiallly transformed human cells have to undergo additional alterations to become tumorigenic. The additional changes, probably steps to overcome senescence, seem to occur at very low frequency in human cells of diploid origin compared to mouse, rat and hamster cells of diploid origin.

From these results described above and many other circumstantial evidences, a hypothesis is proposed for the process of cell transformation by chemical and physical carcinogens (Fig. 5 and 6).

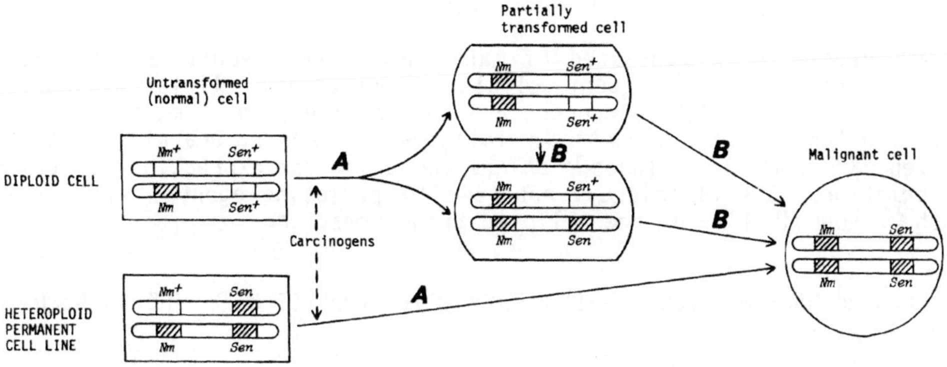

Figure 5. schema of hypothetic process of malignant cell transformation by chemical or physical carcinogens.
Nm : Genes responsible for normal morphology and pattern of cell growth.
Sen : Genes regulating limited life-span or senescence.
Function of Nm and Sen is not directly related.

Conversion of normal diploid cells into malignant cells is a multistaged, at least two stages, development. First stage will include the loss of normal regulation of cell growth and movement which are manifested as the changes in cell morphology, growth pattern, cell to cell arrangement, and requirement for growth factors. The actual number and the nature of late step represented as process B in Fig. 5 are unknown. However, the most important property which cells acquire through process B is the escape from senescence or the loss of program for limited life-span. Process B may correspond to the so-called "progression" phenomenon (38,39).

On the other hand, heteroploid and permanent but still untransformed untransformed cell line had already acquired the properties which the cells of diploid origin has to acquire through process B. Thus, the malignant transformation of established cell line is basically one step progress. Human diploid cells seem to be resistant to B conversion as well as A conversion.

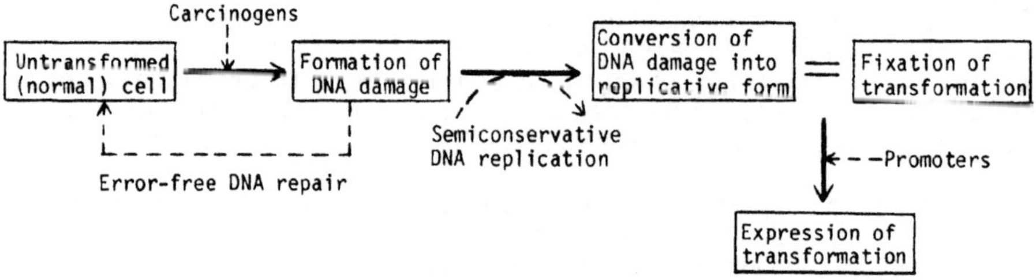

Figure 6. Details of process A in Fig. 5.

Process A is consisted of many many chemical-cellular components interactions, and subsequent cellular responses and cellular metabolisms (Fig. 6). Step "X" in process A is not entirely the same to the so-called "expression" process in mutagenesis such as ouabain^r-mutation. Step "X" probably includes a unique mechanism for expression of transformation. Its efficiency, enhanced by promoting agents, will affect the susceptibility of the cells to transformation.

CELLULAR MACROMOLECULES RESPONSIBLE FOR EXPRESSION OF THE TRANSFORMED PHENOTYPE

As described in introduction, it is obviously important to identify the cellular genes responsible for expression of the transformed phenotype and to define the transformed phenotype at molecular level. One approach to this purpose we have been taking is to isolate and characterize the cellular DNA sequence which has the potency to induce or reverse the transformed phenotype when introduced into the cells. The second approach is to identify the macromolecules, changes of which are associated with the transformation by chemical or physical carcinogens. Combination of chemically transformed human cells and their parental untransformed human diploid fibroblast cells seems to provide one of the best cell systems available for these approaches.

When the [^{35}S]methionine-labeled, Triton-soluble cellular proteins were compared between the 4NQO-transformed and the untransformed human fibroblast by two-dimensional gel electrophoresis, there were 20 to 24 differences in polypeptides out of greater than 1000 electrophoretically distinguishable species (Fig 7). Some changes like appearance of new polypeptides a2 and a3 (Fig 7D) were common to the transformed lines tested and others were not. Only one polypeptide out of these changes

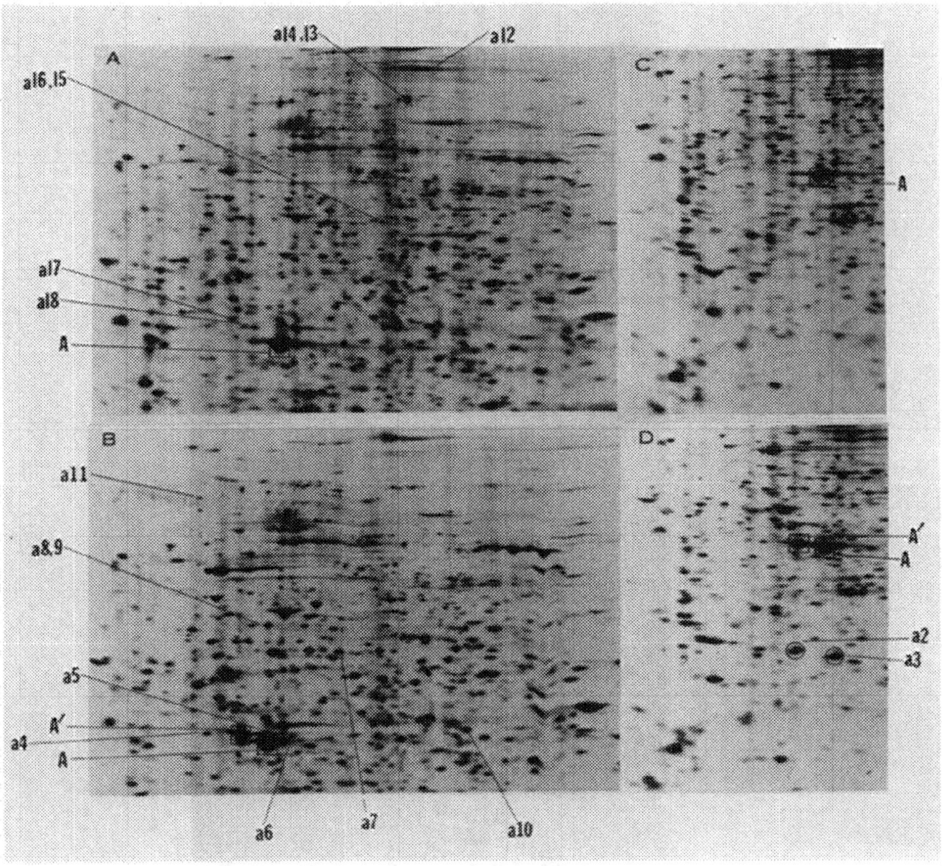

Figure 7. Autoradiographs of two-dimensional gels containing [^{35}S]-methionine polypeptides separated by isoelectrofocusing in the first dimension. Polypeptides from ; A , Normal human fibroblasts, KD, in a 9% acrylamide slab gel; B, a 4NQO-transformed KD, HuT-14, in a 9% acrylamide slab gel; C, KD in a 12.5% acrylamide slab gel; D. HuT-14 in a 12.5% acrylamide slab gel. Polypeptide A is β- and γ-actin. Polypeptides A', and a2 through a18, are polypeptide species present in only one of the two cell types. From J. Biol. Chem. 255, 1650 (1980).

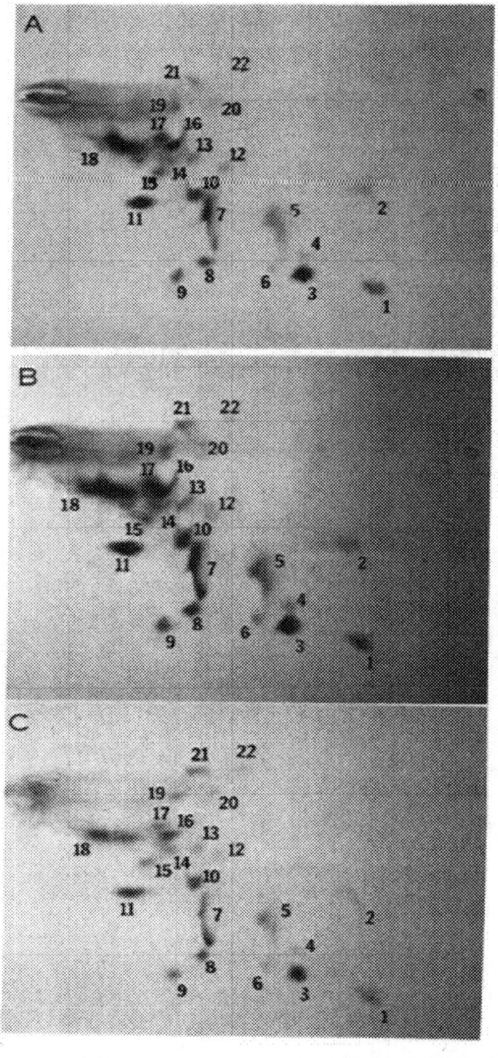

Figure 8. Tryptic [^{35}S]methionine peptide patterns of isolated β-, γ-actin and A' polypeptides. A, β-and γ-actin of KD cells; B, β- and γ-actin of HuT-14 cells; C, A' of HuT-14 cells. Chromatography in the first dimension is from right to left. + and - identify the relative charge of the peptides in the second electrophoretic dimension. The numbered spots indicate the positions of common tryptic peptides in each pattern. From J. Biol. Chem., 255, 1650 (1980).

in the transformed cells had been previously known as LETS (large external tumor specific) protein which disappeared in the transformed human cells. In order to identify these new, or disappeared polypeptides in the transformed cells, it is now being attempted to isolated monoclonal antibody against these polypeptides and to clone DNA sequences coding these polypeptides. Although most of the changes in the polypeptides are supposed to be resulted from the altered transcriptional, post-transcriptional, or post-translational process, it seems that at least one of the changes in the polypeptides observed in a transformed human cell line is ascribed to a mutation based on the following results.

A new polypeptide, called A' (Fig 7B & D), with the molecular weight 44,000 and pI 5.2, was observed in one of 4NQO-transformed human fibroblast cells. The A' polypeptide was identified as a variant form of actin because it immunoprecipitates with antibodies specific for actin and tryptic peptide analysis of [^{35}S] methionine-labeled A' polypeptide andβ- and γ-actin polypeptides showed that patterns of their 22 resolved peptide species were virtually identical (Fig 8). The tryptic peptide patterns generated by [^{14}C] phenylalanine-labeled A' polypeptide exhibited two new negatively charged tryptic peptide species which were not found in the pattern produced by [^{14}C] phenylalanine-labeled β- and γ-actin (42).

A variant actin, A' polypeptide, was found in the products of <u>in vitro</u> translation of mRNA extracted from HuT-14 using reticulocyte lysate or wheat germ extract. Partial amino acid sequencing of A' polypeptide indicates that A' polypeptide is a variant form of -actin (K. Weber et al. personal communication). Cloning of genomic A' polypeptide DNA sequence into plasmid and phage is now in progress.

ACKNOWLEDGMENT

We thank Ms. J.D.Crow, Ms. C.Augl and Mr. A.H.Leavitt for their invaluable contributions to these studies.

REFERENCES

1. Boveri, T.: Zug Frage der entstehung maligner Tumoren. Fischer, Jena (1914)
2. Bauer, K.H.: Mutationstheorie der Geschwulst-Entehung. Ubergang von Korperzellen in Geschwulstzellen durch Gen-Anderung, Springer-Verlag, Berlin (1928)
3. Boveri, T.: The origin of malignant tumors. Williams & Wilkins Co., Baltimore (1929)
4. Kakunaga, T.: Proc. Natl. Acad. Sci., U.S.A., 75, 1334 (1978)
5. Kakunaga, T.: In Short Term Test for Chemical Carcinogens, H.F. Stich and R.H.C. San (eds), Springer-Verlag, New York, in press.

6. Nebert, D.W., Benedict, W.F., and Kouri, R.E.: In Chemical Carcinogenesis, P.O.P. Ts'o and J.A. DiPaolo (eds), Marcel-Dekker, New York, pp. 271 (1974)
7. Kouri, R.E.: In Polynuclear aromatic hydrocarbons chemistry, metabolism and carcinogenesis, R.I. Freudenthal and P.W. Jones (eds), Raven Press, New York, pp. 139 (1976)
8. Nebert, D.W., Atlas,S.A., Guenthner, T.M. and Kouri, R.: In Polycyclic hydrocarbons and cancer, H.V. Gelboin and P.O.P. Ts'o (eds), Academic Press, New York, pp. 345 (1978)
9. Reznikoff, C.A., Bertram, J.S., Brankow, D.W. and Heidelberger, C.: Cancer Res., 33, 3239 (1973)
10. Kakunaga, T.: Int. J. Cancer, 12, 463 (1973)
11. Kakunaga, T.: In Polycyclic Hydrocarbons and Cancer, H.V. Gelboin and P.O.P. Ts'o (eds), Academic Press, New York, pp. 293 (1978)
12. Kakunaga, T.: Int. J. Cancer, 14, 736 (1974)
13. Kakunaga, T.: Cancer Res., 35, 1637 (1975)
14. Kakunaga, T.: In Radiation Research, S. Okada, M. Imamura, T. Terashima, and H. Yamaguchi (eds), JARR, Tokyo, pp. 589 (1979).
15. Grover, P.L., Hewer, A., Pal, K. and Sims, P.: Int. J. Cancer, 18, 1, (1976)
16. Jeffrey, A.M., Jennette, K.W., Blobstein, S.H., Weinstein, I.B., Beland, F.A., Harvey, R.G., Kasai, H., Miura, I. and Nakanishi, K.: J. Am. Chem. Soc., 98, 5714 (1976)
17. Shinohara, K. and Cerutti, P.A.: Proc. Natl. Acad. Sci., U.S.A., 74, 979 (1977)
18. Baird, W.M. and Diamond, L.: Biochem. Biophys. Res. Commun., 77, 162 (1977)
19. Ivanovic, V., Geacintov, N.E., Yamasaki, H. and Weinstein, I.B.: Biochemistry, 17, 1597 (1978)
20. Cerutti, P.A., Sessions, F., Hariharam, P.V. and Lusky, A.: Cancer Res., 38, 2118 (1978)
21. Brown, H.S., Jeffrey, A.M. and Weinstein, I.B.: Cancer Res., 39, 1673 (1979)
22. Borek, C. and Sachs, L.: Proc. Natl. Acad. Sci., U.S.A., 59, 83 (1968)
23. Ikenaga, M. and Kakunaga, T.: Cancer Res., 37, 3672 (1977)
24. Hollander, A. and Curetis, J.T.: Proc. Soc. Exptl. Biol. Med., 33, 61 (1935)
25. Castellani, A., Jagger, J. and Setlow, R.B.: Science, 143, 1170 (1970)
26. Parry, J.M. and Parry, E.M.: Genetic Res.,19, 1 (1972)
27. Little, J.B.: Radiation Res., 56, 320 (1973)
28. Hahn, G.M.: Radiation Res., 64, 533 (1975)
29. Ikenaga, M., Ichikawa-Ryo, H. and Kondo, S.: J. Mol. Biol., 92, 341 (1975)
30. Igel, H.J., Freeman, A.E., Spiewak, J.E. and Kleinfeld, K.L.: In Vitro, 11, 117 (1975)
31. Shimada, H., Shibata, H. and Yoshikawa, M.: Nature, 264, 547 (1976)
32. Demars, R. and Jackson, J.L.: J. Env. Path. Toxicol., 1, 55 (1977)

33. Freeman, A.E., Lake, R.S., Igel, H.J., Gernard, L., Pezzutti, M.R. Malone, J.M., Mark, C. and Benedict, W.F.: Proc. Natl. Acad. Sci., U.S.A., 74, 2451 (1977)
34. Freedman, V.H. and Shin, S.: J. Natl. Cancer Inst., 58, 1873 (1977)
35. Graham, F.L., Smiley, J., Russell, W.C. and Nairn, R.: J. gen. Virol., 36, 59 (1977)
36. Shevliaghy, V.J. Karazas, N.V. and Amchenkova, A.M.: Neoplasma, 24, 375 (1977)
37. Takemoto, K.K., Bond, S.B., Haase, A.T. and Ting, R.C.: J. Viol., 25, 326 (1978)
38. Kakunaga, T. and Kamahora, J.: Symposia Cell Chem., 20, 135 (1969)
39. Barrett, J.C. and Ts'o, P.O.P.: Proc. Natl. Acad. Sci., U.S.A., 75, 376 (1978)
40. Leavitt, J. and Kakunaga, T.: J. Biol. Chem., 255, 1650 (1980)

MEMBRANE AND OTHER BIOCHEMICAL EFFECTS OF THE PHORBOL ESTERS AND THEIR RELEVANCE TO TUMOR PROMOTION

I. Bernard Weinstein, R. Alan Mufson, Lih-Syng Lee, Paul B. Fisher, Jeffrey Laskin, Ann D. Horowitz and Vesna Ivanovic
Division of Environmental Sciences and Cancer Center/ Institute of Cancer Research, Columbia University, New York, New York 10032, U.S.A.

ABSTRACT

The pleiotropic effects of TPA and related phorbol esters on a variety of cell cultures provide important clues to the process of tumor promotion and the multistep nature of carcinogenesis. These effects can be divided into three categories: 1) mimicry of transformation in normal cells, and enhancement of transformation by chemical carcinogens or oncogenic viruses; 2) modulation (inhibition or induction) of differentiation; and 3) membrane and receptor effects. Recent evidence suggests that TPA acts by binding to specific high affinity cell surface membrane receptors and that this then leads to rapid alterations in the composition of membrane phospholipids. Presumably, these changes in the lipid matrix of cell membranes produce signals or mediators which lead to the subsequent cytoplasmic and nuclear effects of TPA. Thus, whereas the critical target in the action of initiating carcinogens appears to be cellular DNA, the critical target of the phorbol ester tumor promoters appears to be cell membranes. As a unifying concept of two-stage carcinogenesis, we postulate that during the initiation phase in carcinogenesis the covalent binding of carcinogens to DNA induces a host response somewhat analogous to that of the SOS response in bacteria. However, in mammalian cells this response results in abberations in the commitment of the target cells. This may involve ordered events, for example, gene transpositions, rather than random point mutations. Tumor promoters, via their effects on growth, gene expression and differentiation, enhance the selective outgrowth of initiated cells and induce them to express their newly acquired but previously dormant committed state. Thus, initiation and promotion parallel events during normal development and differentiation, but during carcinogenesis the new cell populations are aberrant in terms of specialized functions and growth control.

INTRODUCTION

Most naturally occurring cancers probably result from a complex interaction between endogenous (host) factors and exogenous (environ-

B. Pullman, P. O. P. Ts'o and H. Gelboin (eds.), Carcinogenesis: Fundamental Mechanisms and Environmental Effects, 543–563.

mental) factors. In addition, the carcinogenic process frequently proceeds via several discrete steps and over a time span that may occupy a considerable fraction of the lifespan of the individual. For the past few years, our laboratory has been engaged in studying cell culture systems that display interactions between chemical carcinogens, tumor promoters and viral agents in terms of their effects on cell phenotype, differentiation and malignant transformation. Cell culture systems appear to be a valid model of carcinogenesis in the intact animal since, as in the intact animal, the malignant transformation of cells in culture often proceeds by a multistep process (1-3). This paper will review the salient features of these cell culture studies and speculate about their possible significance with respect to multistage carcinogenesis in the intact animal.

The pioneering and elegant studies of Isaac Berenblum and his colleagues led to the definition of two distinct processes during carcinogenesis on mouse skin, initiation and promotion (4). These studies, as well as subsequent related studies by others (5,6,7), represent a major landmark in experimental pathology and carcinogenesis research. Tumor promoters can be defined as compounds which lack significant carcinogenic activity when tested alone but markedly enhance the yield of tumors when applied after a low dose of an initiating carcinogen, for example benzo(a)pyrene (BP). The most potent tumor promoter on mouse skin is 12-O-tetradecanoyl-phorbol-13-acetate (TPA), and related phorbol diesters (5,6,7,8). Elsewhere, we have contrasted the properties of initiating agents with those of the phorbol ester tumor promoters (9,10,11). The major difference is that whereas initiating carcinogens usually generate electrophiles that bind covalently to cellular DNA (and are, therefore, mutagenic) this is not the case for tumor promoters.

A major tribute to the studies by Berenblum and other investigators of two stage mouse skin carcinogenesis is that their findings have served as a paradigm for more recent studies on the multistage aspects of carcinogenesis in several other tissues and species. Evidence that hepatocellular cancer, bladder cancer, colon cancer, and breast cancer also proceed via processes analogous to initiation and promotion has been reviewed elsewhere (8,12). In addition, there are several recent studies indicating that the tumor promoting effect of the phorbol esters is not confined to mouse skin (Table 1). Thus, these and related compounds, and the mechanisms by which they act, may play a role in various types of neoplasia.

The cellular targets for the action of the phorbol ester tumor promoters are not known with certainty, but recent studies in cell culture systems have provided important clues. The effects seen in cell culture can be classified into three categories (all of which conveniently begin with the letter "m"): mimicry of transformation, modulation of differentiation, and membrane effects (For recent reviews on cell culture effects of the phorbol esters see 8-12,14).

MIMICRY OF TRANSFORMATION AND SARC GENES

TPA induces several properties in normal cells that mimic those often seen in transformed cells. This mimicry includes changes in cell morphology, growth properties, cell surface properties and induction of plasminogen activator (PA), ornithine decarboxylase and prostaglandin synthesis (8-12, 14, see also Table 4). Recent findings that add to this growing list include TPA induced decreases in growth requirements for serum (15) and Ca^{2+} (16), TPA induced loss of actin cables (17), and TPA induced loss of metabolic cooperation (18,19). All of these properties are frequently associated with cell transformation. TPA can enhance the stable transformation of fibroblast cultures previously exposed to a chemical carcinogen, uv or x-irradiation (20,21,22,23), an adenovirus (15,24), or Epstein-Barr virus (25). As described below, TPA can also induce the irreversible acquisition of anchorage-independent growth in "partially" transformed cells (2,3,28). These stable or irreversible effects of TPA indicate that "initiated" cells have a qualitatively different response to TPA than normal cells.

When TPA was applied to Avian Sarcoma Virus (ASV) transformed chick embryo fibroblasts (CEF), which already synthesize a high level of PA, there was a further increase in PA synthesis (9). We refer to this phenomenon as "enhancement". Studies with CEF transformed by a temperature-sensitive mutant of ASV showed that enhancement of TPA induced PA synthesis required continuous expression of the sarc gene of ASV (9). This finding has been confirmed and extended by other investigators (26,27). These results suggested that TPA might act by enhancing expression of the endogenous sarc gene present in normal vertebrate cells and/or the expression of the integrated ASV sarc gene in cells transformed by ASV. The products of the endogenous and ASV sarc genes have recently been identified as 60k phosphoproteins which have protein kinase activity (29). However, recent studies (30) done in collaboration with Ray Erikson have indicated that TPA does not cause a significant increase in the amount or activity of this protein kinase in either normal or ASV transformed CEF (Table 2). Further studies are required to determine if TPA induces a different kinase or interacts with the sarc gene product through other mechanisms.

MODULATION OF DIFFERENTIATION

Since it is likely that carcinogenesis involves major disturbances in differentiation, it is of interest that TPA is a highly potent inhibitor or inducer of differentiation in a variety of cell systems (Table 3). The examples include a variety of programs of differentiation and cells from such diverse species as avian, rodent, human and even echinoderm. It is possible that the ability of TPA to either induce or inhibit differentiation depends on the nature of membrane constituents of the target cell. Reciprocal effects of the same agent on differentiation, depending on the target cells, have been seen with other agents including glucocorticoid hormones, cyclic AMP and BUdR.

Certain macrocyclic plant diterpenes inhibit the growth of a transplantable mouse lymphoma (31). We wonder whether this is due to an inductive effect on differentiation and whether this approach can be further exploited in cancer therapy since certain neoplasms retain the capacity to respond to inducers of differentiation. It might be possible to derive analogs that are potent inducers of terminal differentiation for certain neoplasms yet lack tumor promoting activity.

A number of years ago, Berenblum (4) postulated that tumor promoters act by inducing disturbances in differentiation. The results in cell culture systems provide direct evidence that the phorbol esters can act as potent modifiers of terminal differentiation (Table 3). We have previously emphasized that this effect may be an important clue to their ability to act as tumor promoters on mouse skin (9-11). The basal cells in the adult epidermis are continually dividing, yet the tissue is in a state of balanced growth. Presumably, this is because of asymmetric division of stem cells. One daughter cell remains a stem cell and the other daughter cell is committed to keratinize and terminally differentiate, thus, irreversibly losing its growth potential. If an "initiated" stem cell were restrained to this mode of division, it could not increase its proportion in the stem cell pool. If, however, the stem cell division mode was interrupted by the action of a promoting agent, the initiated cell could undergo exponential division thus yielding a clone of similar cells. Since TPA can also induce phenotypic changes in cells that mimic those of transformed cells, the microenvironment of a clone of such cells might itself enhance their further outgrowth and development into a tumor. In addition, clonal expansion of the population of initiated cells would provide a larger population from which variants that have undergone progression to later stages of neoplasia might emerge.

Recently, we have examined the profile of proteins synthesized by normal mouse epidermis and epidermis exposed to TPA, utilizing the technique of 2-dimensional gel electrophoresis (32). We found that within 3-24 hours after a single exposure to TPA, the synthesis of specific proteins was inhibited, whereas the synthesis of other proteins was induced (Figure 1). Some of the proteins that were diminished appear to be higher molecular weight keratins (peaks 8-10, Fig. 1). The significance of the proteins that are induced by TPA (for example, peaks 43,51 & 74, Fig. 1) remains to be determined. This approach should prove useful for identifying the protein(s) involved in the process of tumor promotion in mouse skin.

MEMBRANE EFFECTS AND PHOSPHOLIPID METABOLISM

Early studies on the effects of TPA in cell culture suggested that the cell surface membrane may be the initial and major target of TPA action (5,34). More recent studies have reinforced this hypothesis. Table 4 is a list of effects of TPA on cell surfaces and membranes. Several of these effects occur within minutes after exposing cell cultures to TPA and are not blocked by inhibitors of protein or RNA synthesis, suggesting that they result from a direct action of TPA on cell membranes. This is true for the enhancement of 2-deoxyglucose uptake, altered membrane "fluidity", altered cell adhesion, the induction of phospholipid turnover and the inhibition of EGF receptor binding.

One of the earliest effects of TPA is a change in membrane phospholipid metabolism. TPA rapidly induces an increase in the incorporation of P^{32} or choline into membrane phospholipids (35,36,37). In addition, TPA induces the release from membrane phospholipids of arachidonic acid and this is associated with increased prostaglandin synthesis (38,39,40). This effect is also not blocked by actinomycin D, although it is blocked by inhibitors of protein synthesis (39). Since trans retinoic acid and related retinoids inhibit the promotion phase of skin carcinogenesis (41), it is of interest that we have found that retinoids also inhibit TPA induced arachidonic acid release in cell culture (39).

More recently, we have studied the release of water soluble choline metabolites from C3H10T½ cells prelabelled for 24 hrs with (^{3}H) choline (42). Within 5 minutes of exposure to TPA, the release of (^{3}H) choline metabolites was enhanced two-fold and by 60-120 minutes the release was 4-5 times that of vehicle controls (Figure 2). Choline metabolite release was concentration dependent between 10 and 100ng TPA/ml. Phorbol 12,13-didecanoate (PDD) was also active but 4αPDD, which is not a tumor promoter, was inactive. The radioactivity released by TPA was derived from phospholipids since changes in the acid soluble pool of choline metabolites were insufficient to account for the amount of material released. The released material was identified by chromatography as choline and phosphoryl choline. Neither cycloheximide (4-40 μg/ml) nor cordycepin (4-40 μg/ml) blocked the TPA induced release. The release was, however, temperature sensitive and did not occur at 4°C. TPA did not induce the release of (^{3}H) inositol from prelabelled cells. The calcium ionophore A23187 did not enhance (^{3}H) choline release from prelabelled cells, but both TPA and A23187 enhanced (^{3}H) arachidonic acid and (^{3}H) prostaglandin release. We believe that TPA induced choline release is due to activation of an endogenous phospholipase C or D. This TPA effect is similar to that of cholinergic agonists which also activate degradation of phospholipids, although in the latter case the target is phosphatidylinositol rather than phosphatidylcholine (43). It appears that TPA induced choline release precedes TPA induced arachidonic acid release and TPA induced incorporation of choline into membrane phospholipids. An effect of TPA on CDP-choline transferase, an enzyme involved in phosphatidylcholine synthesis, has been described by Mueller et al. (this Symposium). A hypothetical scheme for the effects of TPA on phospholipid metabolism is given in Figure 3. As discussed below, the initial interaction between TPA and cells appears to involve specific binding to cell surface receptors. Presumably, this binding triggers activation of a phospholipase C or D enzyme as well as the other events shown in Figure 3. It is not known how these membrane effects might induce signals or second messengers that mediate the subsequent cytoplasmic and nuclear events. These later responses, most of which require de novo RNA and protein synthesis, include induction of PA and ODC synthesis, stimulation of DNA synthesis, altered cell surface glycoproteins and effects on differentiation. An increase in intracellular Ca^{2+} concentration, enhanced protein kinase activity, or release of mediators remain to be explored as potential candidates for these effects. Mediators of the responses of cells to naturally occurring polypeptide growth factors, like epidermal growth factor (EGF), are also poorly understood at the present time. Thus results obtained on the mechanism of action of TPA may provide clues to the action of other growth factors.

MEMBRANE RECEPTORS FOR EGF AND TPA; DOWN-REGULATION

TPA and EGF have similar but not identical effects on cells in culture (44-46). Because of the possibility that TPA might exert some of its actions via the EGF receptor, we examined the effects of TPA on the binding of ^{125}I-EGF to its cell surface receptors (45,47). Kinetic studies indicated that TPA causes almost an immediate inhibition of EGF binding and that it also causes a loss of previously bound EGF from cell surface receptors. As with most of the other cell culture effects of TPA, the effect on EGF receptors occurred with concentrations of TPA in the range of 10^{-8} to 10^{-10}M. When a series of phorbol compounds, or chemically related macrocyclic plant diterpenes were tested, the results correlated quite well with the known potencies of these compounds as tumor promoters on mouse skin . An interesting exception is the compound mezerein which, although it is equipotent to TPA with respect to several cell culture effects (9,11,48), and induction of ODC in mouse skin (49), is much weaker than TPA as a tumor promoter on mouse skin (49). The reasons for this discrepancy are not apparent at the present time. Studies with a variety of inhibitors (47,50), suggest that the TPA inhibition of EGF-receptor binding does not require RNA or protein synthesis, energy metabolism, or cytoskeletal changes. Thus the effect appears to be mediated directly at the level of the plasma membrane. We have presented evidence that the TPA inhibition of EGF binding is not due to increased degradation of EGF or increased internalization of the EGF-receptor complex (47). As a matter of fact, EGF degradation by cells was less in the presence of TPA than in its absence, apparently because in the presence of TPA previously bound EGF is displaced into the medium (47,50).

Although our initial studies on the EGF receptor were motivated by the possibility that TPA might exert its action by occupying the EGF receptor, more detailed studies indicate that the effect of TPA on EGF receptors is an indirect one, perhaps related to TPA induced changes in the lipid microenvironment of the EGF receptors (47,50). This is suggested by the effects of temperature, pH and subcellular studies on EGF-receptor binding (47,50). This interpretation is consistent with results obtained by other investigators (51,52).

When HeLa cell cultures were exposed to TPA (33 ng/ml) in serum containing medium for 0 to 24 hours, and at various times assayed for EGF binding, there was a progressive decrease in the inhibitory effect of TPA on EGF binding. By 24 hours the EGF binding capacity had returned to 85% of that of parallel cultures not exposed to TPA (Fig. 4). This was not due to exhaustion from the medium or degradation of TPA since the addition of fresh TPA at 24 hours also failed to produce significant inhibition of EGF binding. On the other hand, if the cells were treated with TPA (33 ng/ml) for 60 minutes, rinsed once with serum free medium, and incubated in serum free medium for various periods of time, then the ability of the cells to bind ^{125}I-EGF remained significantly depressed for up to 9 hours. Even under these conditions, however, there was gradual recovery so that 86% of the EGF binding capacity was restored at 24 hours after the preincubation with TPA (50).

Tables 5 and 6 indicate the effects of prolonged exposure of growing cultures of HeLa cells to EGF, or to EGF plus TPA, on the level of EGF receptors. When cells were grown in 13 ng/ml EGF there was an initial depression of ^{125}I-EGF binding, but by day 3-4 the binding was partially restored when compared to cultures not exposed to EGF (Table 5). Throughout the four day period the direct addition of TPA to the EGF binding assay resulted in marked inhibition of EGF binding (Table 5). Thus prolonged exposure of cells to EGF does not make them resistant to the inhibitory effect of TPA (50).

In a separate experiment, the cultures were exposed over a four day period to both EGF (13 ng/ml) and TPA (33 ng/ml) (Table 6). Although the initial inhibition by EGF plus TPA was greater than with EGF alone (compare day 0 in Tables 4 and 5), by day 3 EGF binding in the EGF plus TPA exposed cultures was restored to a level approaching that in cultures not exposed to EGF or TPA (Table 5). In contrast to the results obtained with cultures exposed to only EGF, by day 2, cultures exposed to both EGF and TPA had EGF receptor activity which could not be depressed even when fresh TPA was added directly to the binding assay (compare Tables 5 and 6). The latter result also indicates that with continuous exposure to TPA HeLa cell cultures escape from TPA inhibition of EGF receptor binding. Our earlier studies are consistent with this conclusion (47). The data in Tables 5 and 6 indicate that this escape is not affected by the presence of EGF in the medium. These results also provide evidence that the effects of TPA on EGF receptors are not simply a consequence of TPA binding to EGF receptors in place of EGF (47,50).

The finding that with prolonged exposure to TPA cells escape and become refractory to TPA-induced inhibition of EGF-receptor binding is consistent with previous evidence that following prolonged exposure cells can escape from other effects of TPA. The other examples include escape from TPA inhibition of melanogenesis (53) and adipocyte differentiation (54), TPA induced morphologic changes in 10T½ cell cultures (C. Heidelberger et al. this Symposium), and TPA induced hyperplasia in hamster epidermis (C. Barrett et al. this Symposium). We are currently studying whether this reflects down-regulation of the TPA receptors.

This phenomenon could play an important role in tissue specific and dose-scheduling effects of TPA.

The ability of TPA to inhibit the binding of EGF to its receptors to "uncouple" the β adrenergic receptor and to cause a decrease in acetylcholine receptors in chick myoblasts (Table 4), suggest that the pleiotropic effects of TPA on cell growth, function and differentiation may relate, at least in part, to its effects on membrane receptors, thus altering the response of cells to extracellular signals. Possible affects on receptors involved in cell-cell contacts and TPA inhibition of metabolic cooperation (18,19), could also impair cell-cell recognition in TPA exposed tissues. It appears likely that TPA exerts such effects by binding to highly specific cell surface receptors (see below), and that this then leads to multiple changes in membrane structure and function. It is perhaps for this reason that certain membrane perturbing agents like the bee venom polypeptide melittin (55), and the Staph. aureus polypeptide delta hemolysin (56), can induce several effects similar to those of TPA.

We have previously postulated that TPA may usurp the function of a cell surface receptor whose normal function is to mediate the action of a yet to be identified growth regulator or hormone (9,45). Consistent with this hypothesis are: (i) the low concentrations at which TPA acts in cell culture (approximately 10^{-8} to 10^{-10}); (ii) the remarkable similarity in structural requirements seen when a variety of phorbol esters and related macrocyclic diterpenes are tested in diverse systems; (iii) the highly pleiotropic and reversible effects of these compounds; and (iv) indirect evidence that the putative TPA receptor, like several other receptors, displays down regulation (see above). The postulated endogenous factor that normally utilizes the receptor system usurped by TPA could play an important role in the control of stem cell proliferation and differentiation in various normal tissues. The fact that tumor cells can make polypeptide growth factors (57,58) suggests that changes in the function of growth factor receptors may play an important role not only in the carcinogenic process but also in maintenance of the transformed state.

In accordance with the above speculations, Blumberg's laboratory (59), has recently employed ^{3}H-phorbol dibutyrate (PDBu) to obtain direct evidence that the membranes of fibroblast cultures and mouse epidermal cells do contain high affinity saturable receptors for PDBu, TPA and related tumor promoting phorbol esters. Our laboratory has confirmed and extended this finding and developed a simple assay for these receptors using intact monolayers of rat embryo fibroblasts (60). A typical binding curve is shown in Figure 5. Although EGF inhibits the binding of ^{125}I-EGF to its cell surface receptors, EGF does not inhibit the binding of ^{3}H-PDBu to its receptors (59,60). These results provide further evidence that the receptor for the phorbol esters is distinct from that for EGF. Scatchard plots indicate that the rat embryo cells contain about 2×10^{5} high affinity PDBu receptors with a K_D of about 8nM (60). The receptor assay for phorbol esters should prove extremely valuable in further studies on the biochemical mechanisms of action of these agents and in the search for the putative endogenous substance that normally utilizes this class of receptors.

MULTISTEP ASPECTS OF VIRAL TRANSFORMATION AND THE EFFECTS OF TUMOR PROMOTERS

We have recently demonstrated that certain clones of rat embryo (RE) cells transformed by the H5ts125 mutant of type 5 adenovirus initially exhibit a low or negligible capacity to grow in agar; with repeated serial passage they spontaneously acquired this ability (2). Growth of these early passage (-E) clones in agar medium containing as little as 1 to 3 ng/ml of TPA induced agar growth (2). In addition, TPA further enhanced anchorage-independent growth in late passage (-L) clones which had progressed to a point where they grow in agar without TPA. We have also found that EGF and melittin also enhance anchorage-independent growth of H5ts125 transformed RE cells. Detailed studies with a H5ts125 transformed clone, E11, indicated that: (a) growth in TPA results in agar colonies which are larger and more diffuse; (b) the ability of a series of diterpene esters to enhance E11 growth in agar correlated with their biological activity on mouse skin; and (c) acquisition of anchorage-independent growth induced by TPA appears to be irreversible (2). Nucleic acid hybridization studies and Southern blot analyses indicated that these progressive changes are not the result of alterations in the state of integration of the viral DNA in the host genome (61). A possible explanation for TPA induction and enhancement of growth in agar of the adenovirus transformants, as well as spontaneous progression of transformed cells with continued serial passage, is that the transformants are genetically unstable and generate variants at a high frequencey. Chromosomal alterations or changes in the expression of host or viral genes (at the level of transcription and RNA processing or protein synthesis), either occurring spontaneously or as a consequence of exposure to TPA, could lead to the development of variants with decreased anchorage-dependence (2). This system should, therefore, provide a useful model for analyzing the complex phenomena involved in tumor progression (2,28).

Subsequent to our studies demonstrating that TPA enhances the transformation of cells infected with adenovirus, other investigators found that TPA enhances EB virus transformation of human lymphocytes (25). These examples of tumor promoter-viral interactions should be kept in mind in the search for the etiology of certain human cancers. It is possible, for example, that whether or not EB virus causes infectious mononucleosis, Burkitt's lymphoma or nasopharyngeal cancer depends upon tissue specific factors with functions similar to those of the known tumor promoters.

MOLECULAR MECHANISM OF INITIATION AND A UNIFIED THEORY OF INITIATION AND PROMOTION

In the final analysis, the mechanism by which tumor promoters enhance the conversion of initiated cells to tumor cells must take into account the mechanism of action of initiating carcinogens. Although current evidence suggests that covalent binding of carcinogens like BP or AAF to cellular DNA is the critical event in initiation, the subsequent biochemical events that lead to establishment of the initiated cell are not known. Several possible molecular mechanisms are listed in Table 7 and have been discussed in detail elsewhere (62).

There has been a tendency to think of the initiating event in chemical carcinogenesis as a simple random-point mutation resulting from errors in replicating the damaged DNA. However, certain aspects of the carcinogenic process, particularly the apparently high efficiency of initiation (23), and the long latent period required for expression, are not consistent with this simple mechanism. In bacteria, physical and chemical agents that damage DNA, including chemical carcinogens, induce a highly pleiotropic response called "SOS functions" (62,63), which includes induction of an error-prone DNA synthesis mechanism. Our laboratory has recently found that mutagenesis by benzo(a)pyrene diol epoxide (BPDE 1) in E. coli is mediated via this mechanism (63). It is not known whether similar responses to DNA damage occur in eukaryotic cells, and if so, what the components of this response might be.

Recent studies in both prokaryotic and eukaryotic cells indicate that the linear arrangement of coding sequences in DNA may be more complex and also more plastic than previously envisioned. A simple "flip-flop" inversion of a specific phage gene controls its expression in Salmonella (64). Brack and Tonegawa (65) have found that synthesis of a specific mouse immunoglobin is associated with somatic rearrangements of immunoglobin genes coding for the variable and constant regions. The mating type in yeast is controlled by gene transposition via a mechanism referred to as the "cassette model" (66). There is recent evidence that antigenic variation in trypanosomes is also due to gene transposition (67). It is possible, therefore, that genome rearrangements will be found to underlie other aspects of development and differentiation in eukaryotic systems, an hypothesis proposed by McClintock a number of years ago (68).

Rather complex and delicate biochemical mechanisms must underlie the above-described specific rearrangements in DNA sequence. It would not be surprising if chemical modification of the DNA by carcinogens disrupted these mechanisms. One might further speculate that the "SOS" program of response to DNA damage in mammalian cells includes the induction of transposition of specific DNA sequences. In this regard, it is of interest that UV light can induce switches in mating type in yeast. By scrambling an otherwise orderly process of genome rearrangements, carcinogens could produce distortions in cell commitment and thus initiate the carcinogenic process. Phrased in other terms, this theory postulates that during normal development, the establishment of specific populations of stem cells involves gene transpositions. By damaging DNA, initiating carcinogens induce aberrant forms of gene transposition, thus establishing aberrant stem cells. The subsequent role of tumor promoters might be to enhance the outgrowth of these cells (as discussed above), as well as "switch on" their aberrant programs of differentiation, just as normal growth factors might induce normal stem cells to express their specialized functions. Presumably, the phorbol ester tumor promoters accomplish this by binding to and usurping the function of receptors normally occupied by endogenous factors that control stem cell replication and differentiation. Following repeated exposure of initiated cells to TPA, a neoplastic population might eventually emerge which grows autonomously in the absence of TPA, perhaps due to further changes in genome structure.

Obviously, the above model is highly speculative. Fortunately, recent exciting advances in DNA sequence techniques and in methods for the transfer and cloning of mammalian genes should make it possible to directly test this hypothesis.

ACKNOWLEDGEMENTS:

The authors are indebted to Patricia Vickman and Evelyn Emeric for valuable assistance in the preparation of this manuscript and to Janet Bozzone, Ester Okin and James Chi for their valuable technical assistance. We also thank Drs. Ray Erickson, Dean Engelhardt and Harry Ginsberg for their valuable collaboration in various aspects of these studies.

This research was supported by National Cancer Institute Grant CA-26056 and CA-21111.

REFERENCES

1. *Barrett, J.C. and Ts'o, P.O.P.: 1978, Proc. Natl. Acad. Sci. USA, 75, pp. 3761-3765.*
2. *Fisher, P.B., Bozzone, J.H. and Weinstein, I.B.: 1979: Cell 18, pp. 695-705.*
3. *Fisher, P.B. and Weinstein, I.B.: In: Molecular and Cellular Aspects of Carcinogen Screening Tests, Intl. Agency for Res. on Cancer, in press.*
4. *Berenblum, I.: 1975, In: Cancer, (F.F. Becker, ed.) pp. 323-344, Plenum Press, N.Y.*
5. *Van Duuren, B.L.: 1969: Prog. Exp. Tumor Res., 11, pp. 31-68.*
6. *Boutwell R.K.: 1974, CRC Crit. Rev. Toxicol. pp. 419-443.*
7. *Hecker, E.: 1978: In: Mechanisms of Tumor Promotion and Cocarcinogenesis, (T.J. Slaga, A. Sivak and R.K. Boutwell, eds.), pp. 11-48, Raven Press, New York.*
8. *Slaga, T.J., Sivak, A. and Boutwell, R.K., eds.: 1978: In: Mechanisms of Tumor Promotion and Cocarcinogenesis, Vol. 2, Raven Press, New York*

9. Weinstein, I.B., Wigler, M. and Pietropaolo, C.: 1977, In: Origins of Human Cancer, (H.H. Hiatt, J.D. Watson and J.A. Winston, eds.), Cold Spring Harbor, New York.
10. Weinstein, I.B., Lee, L.S., Fisher, P.B., Mufson, A. and Yamasaki, H.: 1979, In: Environmental Carcinogenesis (P. Emmelot and E. Kriek, eds.), Elsevier North Holland Biomedical Press, Amsterdam pp. 265-285.
11. Weinstein, I.B., Lee, L.S., Fisher, P.B., Mufson, A. and Yamasaki, H.: 1979, Jrnl. of Supramolecular Struct. 12, pp. 195-208.
12. Weinstein, I.B.: 1980, In: Systematics of Mammary Cell Transformatio (C. McGrath, ed.), Academic Press, in press.
13. Goerttler, K., Loehrke, H., Schweizer, J. and Hesse, B.: 1980, Cancer Res. 40, pp. 155-161.
14. Diamond, L., O'Brien, T.G. and Rovera, G.: 1978, Life, Sci. 23, pp. 1979-1988.
15. Fisher, P.B., Goldstein, N.I. and Weinstein, I.B.: 1979, Cancer Res. 39, 3051-3057.
16. Fisher, P.B. and Weinstein, I.B.: 1980, Cancer Lett. 10, pp. 7-17.
17. Rifkin, D.B., Crowe, R.M. and Pollack, R.: 1979, Cell 18, pp. 361-368.
18. Murray, A.W. and Fitzgerald, D.J.: Biochem. Biophys. Res. Commun. 91, pp. 395-401.
19. Yotti, L.P., Chang, C.C. and Trosko, J.E.: 1979, Science 206, pp. 1089-1091.
20. Mondal, S., Brankow, D.W. and Heidelberger, C.: 1976, Cancer Res. 36, 2254-2260.
21. Mondal, S. and Heidelberger, C.: 1976, Nature 260, pp. 710-711.
22. Kennedy, A., Mondal, S., Heidelberger, C. and Little, J.B.: 1978, Cancer Res. 38, 439-443.
23. Kennedy, A.R., Murphy, G. and Little, J.B.: 1979, Cancer Res. in press.
24. Fisher, P.B., Weinstein, I.B., Eisenberg, D. and Ginsberg, H.S.: 1978, Proc. Natl. Acad. Sci. USA, 75, pp. 2311-2314.
25. Yamamoto, N. and Zur Hausen, H.: 1979, Nature 280, pp. 244-245.
26. Bissell, M.J., Hatie, C. and Calvin, M.: 1979, Proc. Natl. Acad. Sci. USA, 76, pp. 348-352.
27. Goldfarb, R.H. and Quigley, J.P.: 1978, Cancer Res. 38, pp. 4601-4609.
28. Colburn, N.H., Former, B.F., Nelson, K.A., Yuspa, S.H.: 1979, Nature 281, pp. 589-591.
29. Collett, M., Erikson, E. and Erikson, R.L.: 1979, J. Virol., 29, pp. 770-781.
30. Laskin, J.D., Pietropaolo, C. and Erikson, R.L.: 1980, Proc. Am. Assoc. Cancer Res. (Abstracts), 21, pp. 116.
31. Kupchan, S.M., Uchids, I., Branfman, A.R., Dailey, R.G. and Fei, B.Y.: 1976, Science 191, pp. 571-572
32. Laskin, J.D., Mufson, R.A., Piccinini, L., Engelhardt, D.L. and Weinstein, I.B.: submitted for publication.
33. O'Farrell, P.H.: 1975, Jl. Biol. Chem. 250, pp. 4007-4021.
34. Sivak, A. and Van Duuren, B.L.: 1967, Science 157, pp. 1443-1444.

35. Kinzel, V., Kreibich, G., Hecker, E. and Suss, R.: 1979, Cancer Res. 39, pp. 2743-2750.
36. Suss, R., Kreibich, G. and Kinzel, V.: 1972, Europ. J. Cancer 8, pp. 299-304.
37. Wertz, P.W. and Mueller, G.C.: 1978, Cancer Res. 28, pp. 2900-2904.
38. Levine, L. and Hassid, A.: 1977, Biochem. Biophys. Res. Commun. 79, pp. 477-483.
39. Mufson, R.A., DeFeo, D. and Weinstein, I.B.: 1979, Molecular Pharmacol. 16, pp. 569-578.
40. Yamasaki, H., Mufson, R.A. and Weinstein, I.B.: 1979, Biochem. Biophys. Res. Comm. 89, pp. 1018-1025.
41. Verma, A.K., Shapas, B.G., Rice, H.M. and Boutwell, R.K.: 1979, Cancer Res. 39, pp. 419-425.
42. Mufson, R.A. and Weinstein, I.B.: 1980, Proc. Am. Assoc. Cancer Res. (abstracts) 21, pp. 117.
43. Salmon, D.M. and Honeyman, T.W.: 1980, Nature 284, pp. 344-345.
44. Lee, L.S. and Weinstein, I.B.: 1978, Nature 274, pp. 696-697.
45. Lee, L.S. and Weinstein, I.B.: 1978, Science 202, pp. 313-315.
46. Dicker, P. and Rozengurt, E.: 1978, Nature 276, pp. 723-726.
47. Lee, L.S. and Weinstein, I.B.: 1979, Proc. Natl. Acad. Sci. USA, 76, pp. 5168-5172.
48. Kensler, T.W. and Mueller, G.C.: 1978, Cancer Res. 38, pp. 771-775.
49. Mufson, R.A., Fischer, S.M., Verma, A.K., Gleason, G.L., Slaga, T.J. and Boutwell, R.K.: 1979, Cancer Res. 39, pp. 4791-4795.
50. Lee, L.S. and Weinstein, I.B.: submitted for publication.
51. Brown, K.D., Dicker, P. and Rozengurt, E.: 1979, Biochem. Biophys. Res. Commun. 86, pp. 1037-1043.
52. Shoyab, M., DeLarco, J.E. and Todaro, G.J.: 1979, Nature 279, pp. 387-391.
53. Mufson, R.A., Fisher, P.B. and Weinstein, I.B.: 1979, Cancer Res. 39, pp. 3915-3919.
54. Diamond, L., O'Brien, T.G. and Rovera, G.: 1977, Nature 269, pp. 247-248.
55. Mufson, R.A., Laskin, J.D., Fisher, P.B. and Weinstein, I.B.: 1979, Nature 280, pp. 72-74.
56. Umezawa, K., Weinstein, I.B. and Shaw, W.W.: 1980, Biochem. Biophys. Res. Commun. 94, pp. 625-629.
57. Todaro, G. and DeLarco, J.G.: 1978, Cancer Res. 38, pp. 4147-4154.
58. Fisher, P.B., Lee, L.S. and Weinstein, I.B.: 1980, Biochem, Biophys. Res. Commun. 93, pp. 1160-1166.
59. Delclos, K.B., Nagle, D.S. and Blumberg, P.D.: 1980, Cell 19, pp. 1025-1032.
60. Horowitz, A. and Weinstein, I.B., unpublished studies.
61. Fisher, P.B., Dorsch-Hasler, K., Weinstein, I.B. and Ginsberg, H.S.: 1979, Nature 281, pp. 591-594.
62. Weinstein, I.B., Yamasaki, H., Wigler, M., Lee, L.S., Fisher, P.B., Jeffrey, A. and Grunberger, D.: 1979, In: Carcinogens and Mechanisms of Action, (A.C. Griffin and C.R. Shaw, eds.), pp. 399-418, Raven Press, New York.
63. Ivanovic, V. and Weinstein, I.B.: 1980, Cancer Res., in press.

64. Kamp, D., Kahmann, R., Zipser, D., Broker, T.R. and Chow, L.T.: 1977, Nature 271, pp. 577-580.
65. Brack, C. and Tonegawa, S.: 1977, Proc. Natl. Acad. Sci. USA, 74, pp. 5652-5656.
66. Herskowitz, I., et al.: 1980, In: The Molecular Genetics of Development, (W. Loomis and T. Leighton, eds.), Academic Press, New York, in press.
67. Hoeijmakers, J.H.J., et al.: 1980, Nature 284, pp. 78.
68. Finchman, J.R.S. and Sastry, G.R.K.: 1974, Ann. Rev. Genet. 8, pp. 15-50.
69. Putney, J.W., Weiss, S.J., Van der Walle, C.M. and Haddas, R.A.: 1980, Nature 284, pp. 345-347.

Figure Legends

Figure 1: Effects of TPA on the protein profile of mouse epidermis. TPA or the acetone solvent were applied to a local region on the skin of female CD-1 mice and 24 hours later the exposed skin was removed and incubated *in vitro* for 1 hour with ^{35}S-methionine. Epidermal proteins were then solubilized, chromatographed in two dimensions (IEF and SDS), and autoradiographed, essentially as described by O'Farrell *et al*., (33). (A) Acetone solvent. (B) TPA treated epidermis.

Figure 2: Effect of TPA on release of ^{3}H-choline from prelabelled cells. 10T½ cells were prelabelled with ^{3}H-choline. At time 0, TPA or the DMSO solvent were then added and the release of ^{3}H-choline from the cells into the media was determined. For additional details see ref. 42.

Figure 3: Schematic diagram of effects of TPA and related compounds on phosphatidylcholine turnover. We postulate that the binding of TPA to specific cell membrane receptors activates phospholipase (PLase) C and/or D, resulting in the conversion of phosphatidylcholine (P'TDYL Choline) to diacylglycerol plus choline. Arachidonic acid (AA) is then released by diacylglycerol lipase; prostaglandins (PG's) and other AA metabolites are also formed. AA also may be released by the direct action of PLase A_2 on P'TDYL Choline. The calcium ionophore A23187 and melittin may induce AA release via the latter mechanism. P'TDYL Choline may be resynthesized via CDP choline, as shown. Presumably, during these biochemical transformations, a transmembrane signal to the cytoplasm and/or nucleus is generated, i.e. increased Ca^{2+} uptake or redistribution, activation of a protein kinase, or some other mediator. Phosphatidic acid, serving as a Ca^{2+} ionophore (69), or some other P'TDYL Choline metabolite may stimulate this process.

Figure 4: Escape of HeLa cells from TPA inhibition of EGF binding during incubation in the presence of TPA. Hela cells were incubated at 37°C in serum-containing growth medium for 0 to 24 hours in the presence of TPA (33 ng/ml) or DMSO (0.003%). At the indicated time points, the growth media were removed the cultures were rinsed and assayed for ^{125}I-EGF binding in the absence of TPA at 37°C for 60 minutes as previously described (47). The binding of ^{125}I-EGF to the cells exposed to TPA is expressed as percent of ^{125}I-EGF bound to the cells exposed to DMSO.

Figure 5: Inhibition of specific binding of ^{3}H-PDBu to a rat embryo fibroblast cell line (FRE-8D) by increasing concentrations of TPA. ^{3}H-PDBu binding was not inhibited by phorbol or EGF. (A. Horowitz and I.B. Weinstein, unpublished studies).

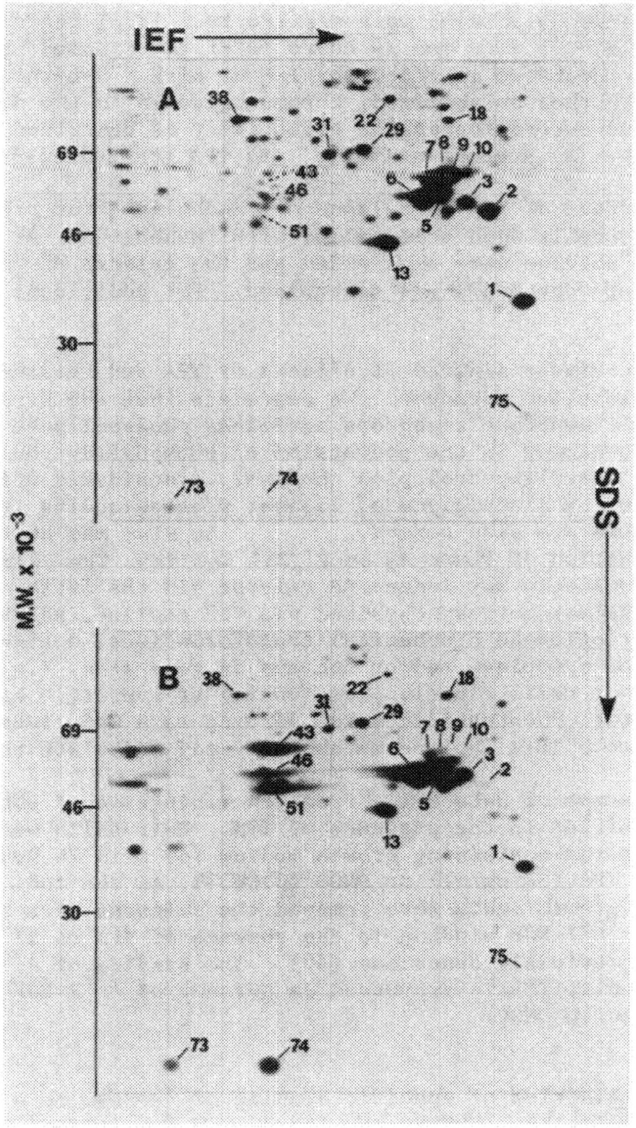
IEF
SDS
M.W. x 10^-3
A
B
69
46
30
38
31
22
29
18
7 8 9 10
43
46
6
3
2
51
5
13
1
75
73
74

Figure 1

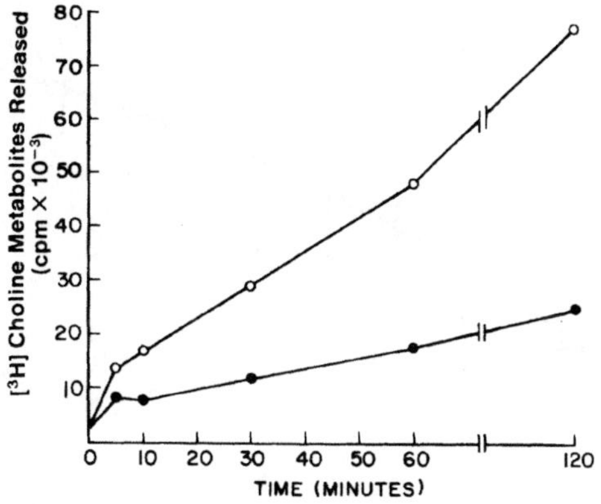

Figure 2

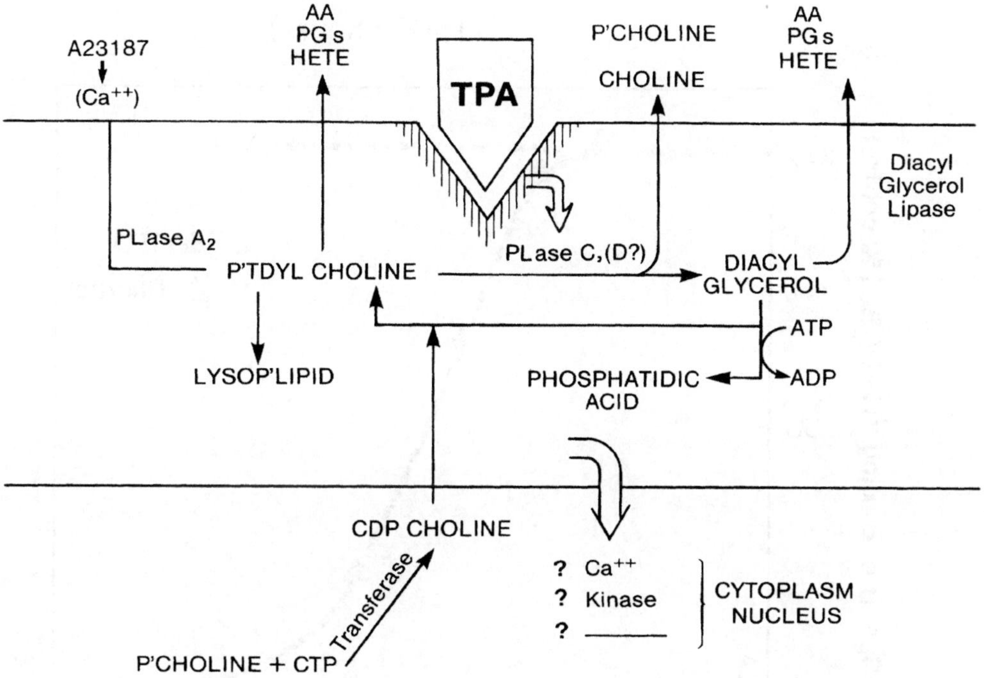

Figure 3

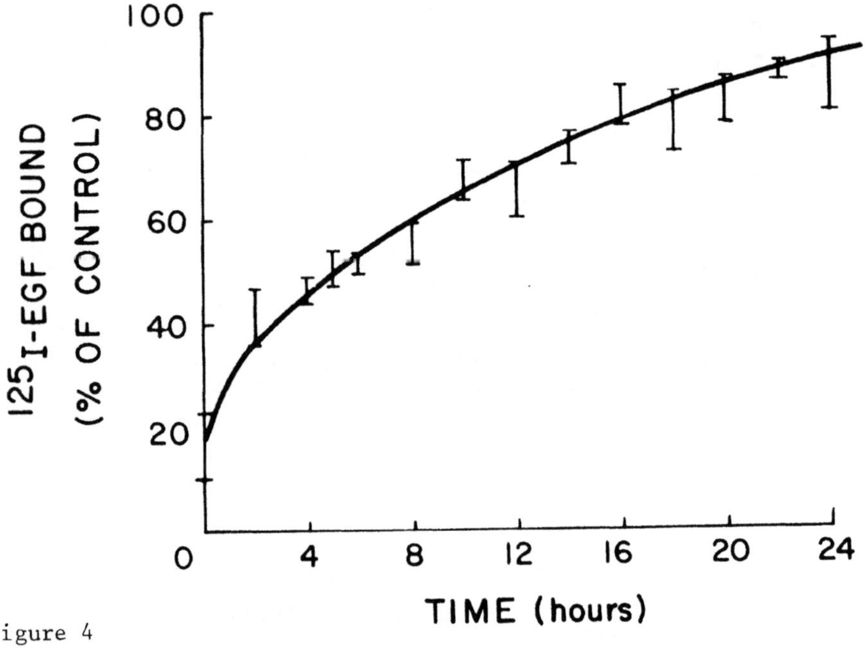

Figure 4

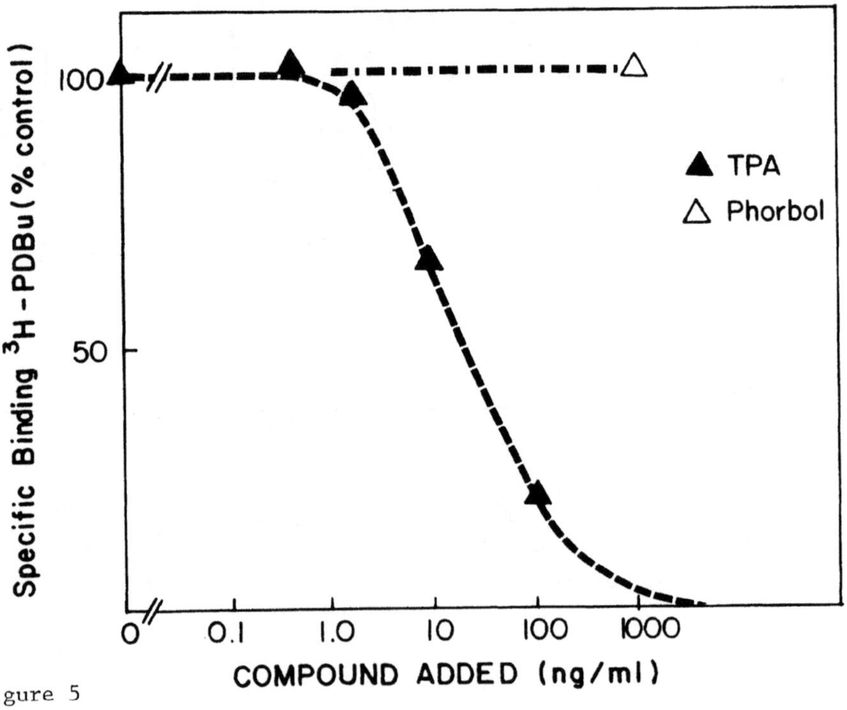

Figure 5

Table 1: Examples in which Phorbol Esters Enhance Carcinogenesis in Tissues other than Mouse Skin.

Type of Tumor	Species	Agents
Ovary, Intestine	Mouse	DMBA or Urethane (Diaplacental) + TPA (Postnatal)
Forestomach	Mouse	DMBA + TPA
Stomach	Rat	MNNG + Croton Oil
Melanoma	Hamster	DMBA + TPA
Esophagous[a]	Human	? + Diterpene from Croton flavens

[a]Suggestive Evidence. For specific refs. see 11, 13.

Table 2: Effect of Exposure of Cell Cultures to TPA on *In Vitro* $pp60^{src}$ Kinase Activity.

Cells	Growth Temperature	TPA	Kinase Activity
CEF	36°	-	3
CEF	36°	+	9
CEF-tsASV	36°	-	65
CEF-tsASV	36°	+	34
CEF-tsASV	40°	-	48
CEF-tsASV	40°	+	78

*fmol ^{32}P/mg protein with CR-TBR antiserum, 1 hour after addition of solvent or TPA (30 ng/ml) to cell cultures. For additional details see ref. 30. For methods see ref. 29.

Table 3: TPA Modulation of Differentiation*

Cell System:	Type of Differentiation:
Examples of Inhibition of Differentiation	
Chick Embryo Fibroblasts	Myogenesis
Chick Embryo Chondroblasts	Chondrogenesis
Chick Embryo Dorsal Root Ganglion	Neurite
Murine Erythroleukemia	Erythroid
Murine 3T3 Cell Line	Adipocytes
Murine Neuroblastoma	Neurite
Murine Melanoma	Melanogenesis

Table 3: TPA Modulation of Differentiation* Cont'd.

Cell System:	Type of Differentiation:
Hamster Epidermal Cultures	Keratinocytes
Mouse Epidermal Cultures	Keratinocytes
Rat Mammary Carcinoma	Dome Formation
Sea Urchin	Embryogenesis
Examples of Induction of Differentiation	
Murine Rauscher Virus Erythroleukemia	Erythroid
Human Myeloid Leukemia Cell Line	Macrophage and Granulocyte
Murine Myeloid Leukemia	Macrophage and Granulocyte
Human Melanoma Cell Line	Melanogenesis

*For refs. see 8-12,14.

Table 4: Effects of TPA on Cell Surfaces and Membranes in Cell Culture

Altered Na/K ATPase
Increased Uptake 2-DG, ^{32}P, ^{86}Rb
Increased Membrane Lipid "Fluidity"
Increased Phospholipid Turnover
Increased Release Arachidonic Acid, Prostaglandins
Altered Morphology and Cell-Cell Orientation
Altered Cell Adhesion
Increased Pinocytosis
Altered Fucose-Glycopeptides
Decreased LETS Protein
"Uncoupling" of β-Adrenergic Receptors
Inhibition of Binding of EGF to Receptors
Decrease in Acetylcholine Receptors
Synergistic Interaction with Growth Factors
Inhibition of Metabolic Cooperation

For refs. see 8-12, 14, and text.

Table 5: Effect of Growth of HeLa Cultures in the Presence of EGF on ^{125}I-EGF Binding.

Growth Condition	Assay Condition	^{125}I-EGF Bound (cpm/10^6 cells) on Day: 0	1	2	3	4
PBS	DMSO	1361	1000	728	524	405
PBS	TPA	189	72	140	109	64
EGF	DMSO	286	39	270	413	349
EGF	TPA	0	0	43	65	58

Table 5: Cont'd.

Cell cultures were grown and binding assays were performed as described in ref. 47. The concentrations of the indicated reagents were: TPA, 33 ng/ml; DMSO, 0.003%; and EGF, 13 ng/ml. Binding assays contained 14,891 cpm ^{125}I-EGF (specific activity 30 μCi/μg). The binding assays on day 0 were performed after the cell cultures were exposed to either PBS or EGF for 15 minutes at 37°.

Table 6: Effects of Growth of HeLa Cultures in the Presence of TPA plus EGF on ^{125}I-EGF Binding.

Growth Condition	Assay Condition	^{125}I-EGF Bound (cpm/10^6 cells) on Day: 0	1	2	3	4
DMSO	DMSO	608	633	529	365	348
DMSO	TPA	103	104	51	80	64
EGF + TPA	DMSO	15	56	283	365	297
EGF + TPA	TPA	0	48	266	360	297

Cell cultures were grown and binding assays were performed as described in ref. 47. The concentrations of the indicated reagents were: DMSO, 0.003%; EGF, 13 ng/ml; and TPA, 33 ng/ml. Binding assays contained 5,977 cpm of ^{125}I-EGF (specific activity 67 μCi/ mg). Binding assays on day 0 were performed after the cell cultures were exposed to either DMSO or EGF + TPA for 15 minutes at 37°C.

Table 7: Possible Molecular Mechanisms of Initiation of the Carcinogenic Process.

A. With permanent changes in DNA sequence:
   1. Random point mutations
      a. Direct: base substitution, frame shift, deletion-in-structural or regulatory gene.
      b. Indirect: induction of "SOS-type" error-prone DNA synthesis.
   2. Ordered gene rearrangements: transposition, amplification, deletion, integration of exogenous sequences, etc.

B. Without permanent changes in DNA sequence:
   altered chromatin structure, altered feedback loops, DNA methylation, etc.

THE CHEMISTRY OF POLYCYCLIC AROMATIC HYDROCARBON-DNA ADDUCTS

A.M. Jeffrey, T. Kinoshita, R.M. Santella, D. Grunberger, L. Katz and I. Bernard Weinstein. Division of Environmental Sciences and Cancer Center/Institute of Cancer Research, Columbia University, New York, New York 10032, U.S.A.

Abstract

There is considerable evidence which suggests that modification of DNA by chemical carcinogens is the critical first event in tumor induction. Various methods available for investigation of the chemistry, conformation and consequences of these reactions are discussed using benzo(a)pyrene and 7,12-dimethylbenzanthracene as examples. Metabolic activation of these hydrocarbons occurs via 'Bay region' dihydrodiol epoxide derivatives. The amino group of guanine is the major site of attack in subcellular systems and in cell cultures of human and rodent tissues. We have found that 7-methylbenzo(a)pyrene is metabolized to the 7,8-dihydrodiol and binding to DNA apparently occurs via a 7,8-dihydrodiol 9,10-epoxide. Preliminary evidence on the structure of DMBA-DNA adduct is also presented.

Benzo(a)pyrene-DNA Adducts, Methods and Structures

In the previous paper, Dr. Jerina gave an introduction to the role of benzo(a)pyrene dihydrodiol epoxide (BPDE) in the metabolism and carcinogenicity of benzo(a)pyrene (BP). We shall review the way in which these and related compounds bind covalently to DNA and the consequences of these reactions. The studies done by our research group at Columbia University have included invaluable collaborations with other research groups including those led by Dr. R. Harvey at the University of Chicago, Dr. K. Nakanishi at Columbia University, Drs. C. Harris and H. Autrup at the National Cancer Institute, and Dr. N. Geacintov at New York University.

Papers which we have published give the detailed experimental evidence for the structure of the major DNA adduct formed in human and rodent cell systems exposed to BP (1-5) (Figure 1). Evidence for this structure was gained through what may now be considered generally applicable methods for polycyclic aromatic hydrocarbon-DNA adducts. To obtain preliminary evidence for the types of adducts formed, cells are exposed to the parent hydrocarbon for about 24 hours, and the DNA

B. Pullman, P. O. P. Ts'o and H. Gelboin (eds.), Carcinogenesis: Fundamental Mechanisms and Environmental Effects, 565–579.

extracted and extensively purified to remove all noncovalently bound material. Analysis of the fluorescence spectrum of this DNA at liquid nitrogen temperatures, at which fluorescent quenching is substantially reduced (6), may indicate the types of chromophores present. The level of detection of, for example, BP-modified DNA is in the order of one adduct per 10^5 to 10^6 bases. These experiments, together with those described by Borgen et al (7), in which they studied the binding to DNA by microsomal oxidations of a number of metabolites of BP, give a good indication of the types of adducts formed, and therefore, the class of compound which may be the ultimate carcinogen. Sims (10) proposed that a dihydrodiol epoxide is the ultimate carcinogen. Defined chemical synthesis of the ultimate carcinogens from a number of polycyclic aromatic hydrocarbons has met with general success. 'Bay region' dihydrodiol epoxides of BP (8) and benz(a)anthracene (9) have been prepared, although attempts to prepare the 'Bay region' dihydrodiol epoxide of 7,12-dimethylbenz(a)anthracene (DMBA) have so far failed.

The next step has been to react these compounds with homopolymers and DNA in vitro. After careful extraction of the non-covalently bound derivatives, the nucleic acids can be digested with a combination of both endonucleases and exonucleases, the latter hydrolyzing the modified DNA from both the 5' and 3' ends, to give nucleoside adducts. These are separated from unmodified nucleosides by Sephadex LH-20 column chromatography. The adducts can then be further resolved by high performance liquid chromatography (HPLC). The similarity between the adducts formed with the homopolymers using synthetic hydrocarbon derivatives and those formed with DNA as a result of metabolic activation of the radiochemically labelled hydrocarbon in the intact cell may then be compared by HPLC. The adducts are detected by their fluorescence, uv absorption or radioactivity. However, several adducts are usually produced in the reactions of dihydrodiol epoxides with the homopolymers and the comparison with metabolic products must be undertaken with procedures that give high resolution. To ensure chromatographic identity, we have compared the adducts from the synthetic and in vivo reactions by co-elution from the HPLC, derivitization to their peracetates and then reanalysis by HPLC. In this way, we were able to confirm the identity of products formed from reactions of BPDE with polydeoxyguanosine in vitro and those formed in DNA as a result of metabolic activation of BP by human bronchial segments (1). A clear demonstration of the importance of this technique comes from our work on deoxyadenosine-BPDE adducts, where two of the adducts could not be separated on a μBondapak C_{18} column, unless they were converted to their peracetates (11). Results reported at this meeting by Professor Sims indicate that certain benz(a)anthracene-DNA adducts formed in vivo could not be separated by HPLC until the peracetates were prepared. Using the techniques described above, we have shown that the same BPDE-dG adduct is the major DNA adduct formed when human bronchial segments (1), human colon (2), or human esophageal samples (5) are exposed to BP (see Figure 1).

The next important step has been to unambiguously determine the structures and stereochemistry of these adducts. In some instances

(12), where they can be produced in large amounts, it has been possible to directly examine the structures of the DNA adducts. In other cases, we have had to adopt an indirect approach in which we determined the structure of ribose adducts, which we were able to prepare in larger amounts than their deoxyribose analogs, and then correlate the structures of the former with the latter by a variety of techniques (11). Nuclear magnetic resonance requires relatively large amounts of material (in the range of 0.2-1mg of adducts to obtain useful spectra). From these spectra, we were able to determine not only the involvement of the exocyclic amino group of guanine in the reaction with BPDE, but also that the attack took place at the 10-position and trans to the original oxide (13). Similar detailed information has been obtained with a number of other adducts (11,12,39). Additional structural information can be gained from mass spectral analysis of these adducts for which much less material is needed than for nmr analysis. We have used two types of mass spectrometry: field desorption (FD), which is ideally suited for this type of work since it generally gives very intense molecular ions and little fragmentation and may be used with relatively involatile compounds; and electron impact spectra, from which, when the peracetate derivatives of the guanine and adenine adducts were prepared, we have been able to obtain weak molecular ions, but more importantly, ions resulting from fragmentation between the exocyclic amino group, to which the hydrocarbon moiety was attached, and the bases. Care must be taken, however, to confirm the identity of such ions by their high resolution mass spectra (11), since in some instance ions of the same nominal mass can be obtained by fragmentations in which the nucleoside rather than the hydrocarbon moiety is detected.

A very important aspect of the structures of these DNA adducts is their absolute stereochemistries since the conformation of these adducts in DNA may be highly dependent upon interactions of these chiral centers. The circular dichroism (CD) spectra of the adducts are generally very complex and result mainly from interactions between the hydrocarbon and the base moieties. Therefore, we have had to use indirect approaches to resolve their absolute stereochemistries. One approach has been to first resolve by HPLC the diastereoisomeric esters of the dihydrodiol precursor of the dihydrodiol epoxide and then determination of the absolute stereochemistries of the former. The resolved dihydrodiols were then converted to their respective dihydrodiol epoxides, which form diastereoisomeric adducts with DNA because of the chirality of the deoxyribose moiety. These were then resolved by HPLC and correlated with the chromatographic properties of the DNA adducts formed when cells were exposed to the parent hydrocarbon. The importance of the absolute stereochemistries of these adducts is seen in several ways. For example, as discussed by Drs. Jerina and Yang at this conference, the metabolism of polycyclic aromatic hydrocarbons to both dihydrodiols and dihydrodiol epoxides is highly stereoselective. In addition, when racemic BPDE is reacted with native DNA, the ratio of reactivity of the 7R and 7S enantiomers with dG is about 19:1 (14,38). However, when we investigated the dA-BPDE adducts formed in plasmid DNA the 7S adduct predominated (11). Thus, the combination of stereoselective metabolism and stereoselective reaction with DNA can have important consequences in the overall carcinogenicity of a compound.

Conformations of DNA Adducts

In addition to knowledge of the absolute stereochemistry of the adduct, it is of interest to know the orientation of the adduct within a DNA helix and its possible effects on the conformation of the DNA. Samples of native calf thymus DNA in which about 1% of the bases were modified by BPDE showed little or no evidence of denaturation when examined by susceptibility to digestion by the single strand-specific nuclease S_1, formaldehyde unwinding or heat denaturation (15). Therefore, in contrast to AAF modification (15), BPDE modification of DNA causes little disruption of the helix. These results, as well as model building studies suggested that the BPDE residue lies in the minor groove, which is the site of the exocyclic amino group of guanine, with the BPDE residue pointing towards the 5' terminus of the strand to which it is attached. Experimental evidence for this model comes from electric dichroism experiments done in collaboration with Dr. Geacintov et al. (16). In this procedure, the modified DNA samples are oriented in an electric field and the difference in absorption of the sample parallel and perpendicular to the field is observed. From these observations, it was possible to calculate that the maximum angle at which the pyrene chromophore could reside, with respect to the long axis of the DNA, was 35°. This is, within experimental error, in good agreement with the value of about $25-30^{o}$ obtained from model building. Studies employing fluorescence quenching techniques also provided evidence that the covalently bound BPDE residue is not intercalated into the DNA helix but is on the exterior of the molecule, in contrast to the situation when BP forms a simple physical complex with DNA (16). A computer-generated stereoscopic display of a model consistent with all of the above data is given in Figure 2.

The reasons for the preferential binding of 7R-BPDE to native DNA are not obvious from the above model. However, if the 7S enantiomer were to adopt an analogous conformation, for which no experimental evidence is available since it has not yet been possible to prepare this adduct in a pure state in DNA, the BPDE residue would have to point in the opposite direction (towards the 3' terminus) in the DNA helix to avoid serious steric hindrance. We are currently studying the possibility that during the reaction of BPDE with native DNA, there is formation of a reaction intermediate in which the BPDE residue is physically bound to the DNA, perhaps by intercalation, and that during formation of the covalent complex there is rearrangement so that the BPDE residue finally lies in the minor groove. If this is the case, then it is possible that the 7R enantiomer of BPDE has a more favorable conformation than the 7S enantiomer for the formation of this intermediate complex. In any case, it is clear that stereoselective reaction of BPDE enantiomers with native DNA is strongly influenced by the conformation of the nucleic acid, since the preferential reaction with the 7R enantiomer is diminished or lost when the substrate is denatured DNA, RNA, homopolymers or mononucleosides (17,18,38).

The model described above for BPDE-DNA adducts is quite different than that proposed (15) for the conformation of the major adduct formed as a result of the binding of N-acetylaminofluorene (AAF) to DNA. In this model, a stereoscopic view of which is shown in Figure 3, the binding of AAF to the C-8 position of guanine involves problems of steric hindrance. To relieve this hindrance, the guanine rotates out from its normal base pairing position with cytosine, the sugar-base bond rotates from the preferential anti conformation to syn, and the AAF residue is inserted into the helix perpendicular to the long axis such that it occupies the former position of the displaced guanine residue. The model also predicts that the dC residue on the opposite strand is also displaced to accomodate the AAF residue (Fig. 3). Evidence for this has been presented by Sage et al. (41), utilizing antibody to single-stranded cytosine. We refer to this model as "base displacement" (15,20). A similar model termed "insertion denaturation" has been proposed by Fuchs and Daune (19). The order of reactions proposed is speculative although there is good evidence for the final conformation (for review see 15). The driving forces for these reactions presumably come from the high steric hinderance of the original conformation, the dynamic nature of the DNA structure, and the energy gained by the stacking of the AAF with adjacent bases.

In vivo a major fraction of the modified dG residues in DNA, following the administration of AAF, contain the deacetylated residue (AF) bound to the C-8 position. Recent evidence indicates that the AF-dG adduct produces less distortion of the DNA helix than does the AAF-dG adduct, presumably because the lack of the acetyl group causes less steric hindrance at the C-8 position (21).Another minor adduct in which AAF is linked to the N^2 of dG (39) also produces less distortion of the DNA helix than AAF C-8-dG adduct (42); presumably it resides in the minor groove of the helix similar to the BPDE-dG adduct.

The above results and interpretations apply largely to native DNA and in our model building efforts were made to accomodate either the BPDE or AAF residues with minimum distortions of the B conformation of DNA. Recent X-ray crystallographic studies of the hexanucleotide $d(G-C)_3$ indicate that it has an unusual conformation termed "Z DNA" in which the helix is left-handed, rather than right-handed as in B DNA, and the dG residues are in the syn rather than the anti conformation (23). Fiber diffraction data on poly d(G-C) and on poly (A^4s-T), (24), CD data on poly d(G-C) in high Mg^{2+} (25) or ethanol solutions (26), and nmr data on poly d(G-C) (27) are also consistent with the Z conformation. The fact that in Z DNA the C-8 position of dG is on the surface of the molecule and is already in the syn conformation and, therefore, much less sterically hindered than in B DNA, suggested the possibility that the Z conformation of poly d(G-C) might be favored when N-acetoxy-AAF (N-AcO-AAF) is reacted with poly d(G-C). In recent studies done in collaboration with Dr. A. Rich of MIT we have examined the conformational changes associated with N-AcO-AAF binding to poly d(G-C), utilizing CD spectrometry. The CD spectrum of poly d(G-C) in 1 mM phosphate and 55% ethanol is shown in Figure 4. In agreement with previous studies, the high ethanol spectrum is an inversion of the low ethanol B form spectrum. A sample of poly d(G-C) in which about 28% of the residues contained the C-8 AAF dG adduct was prepared and similarly analyzed. Even in the absence of ethanol this sample had a CD spectrum characteristic of the Z form of DNA. Thus, it appears that AAF modification favors the transition of poly d(G-C) from the B to Z form. Presumably, this conformation differs from that of the base displacement model since in Z DNA base pairing can still occur between the modified guanine and the cytosine residue on the opposite strand, and in Z DNA the AAF residue could reside outside the helix because the C-8 position of guanosine is accessible to modification without base displacement. Further studies on the precise conformation of this structure are in progress.

Although, it is not known whether Z form DNA occurs in vivo, there are several naturally occurring DNAs containing stretches of alternating purine-pyrimidine sequences with the potential to form Z DNA (23). It is possible that such segments, when modified by AAF or other carcinogens, would adopt the Z conformation. Other regions of DNA without these alternating sequences, however, may exist in a conformation more in agreement with the base displacement model, when modified by AAF. The possible significance of Z DNA to the action of AAF and other chemical carcinogens remains to be determined.

Studies with 7-Methylbenzo(a)pyrene

It would, of course, be desirable to know the factors which determine the relative carcinogenic potency of a series of polycyclic aromatic hydrocarbons. Current attempts based on calculations (28-29) of various molecular parameters of the parent hydrocarbon or its metabolites, while helpful, will probably fail in the absence of further experimental data. Thus, it is already apparent that one must consider subtle factors, ignored in such calculations, for example, those relating to: structure-, regio-, or stereoselective metabolism, stereoselective binding of enantiomeric diol epoxides to DNA, differential rates of excision from DNA; differences in functional effects of the modified DNA; etc. Thus, although the 'Bay region' theory is of interest, it does not adequately explain why benzo(e)pyrene or benz(a)anthracene, which have 'Bay regions', are weak or non-carcinogenic, or why the different isomers and enantiomers of BPDE differ markedly in carcinogenicity, mutagenicity and reactivity with DNA.

To begin to investigate some of these parameters, we have recently examined methylated benzo(a)pyrenes. Most of these compounds fall within the range of carcinogenicity which one might predict. Thus, all are quite active with the exception of 8-, 9-, and 10-MeBP (30). One might assume, therefore, that substituants in the 'Bay region' block the formation of dihydrodiol epoxides. However, this would lead to the prediction that 7-MeBP would be non-carcinogenic, which is not the case (30). An alternative pathway for the metabolic activation of 7-MeBP may exist in which it is oxidized at the 9,10-positions to the oxide which rearranges to the corresponding 9-phenol of 7-MeBP. The 9-phenol of BP is much less carcinogenic than BP (31), although it does bind to DNA through further oxidation at the 4,5-positions (32). This mechanism does not explain, however, the carcinogenicity of 4,5,7-trimethylBP or the lack of carcinogenicity of 8-MeBP (46). To investigate these questions, we have studied the metabolism and DNA binding of 7-MeBP using primary hamster embryo cells, mouse embryo 10T½ cells and microsomal preparations.

Primary hamster embryo cells were exposed to either BP or 7-MeBP, the DNAs isolated and their fluorescence spectra measured in ethylene glycol-water mixtures at 77K. We found that the spectra were very similar to each other qualitatively and to those previously reported for DNA isolated from BP treated cells (6), with emission maxima at 380 and 400 nm. However, the intensity of the spectrum of DNA from 7-MeBP treated cells was about one third that from BP treated cells. Fluorescence spectra of DNA from cells exposed to 10-methyl- or 7,10-dimethyl-BP were the same as those of DNA from control cultures, which is consistent with the marginal carcinogenic activity of these compounds. Studies using radiochemically labelled 7-MeBP also demonstrated covalent binding of the hydrocarbon to DNA in mouse 10T½ cells, although the extent of binding was only about 20% that of BP under the same conditions. An explanation for these observed effects, which is contrary to the general assumption that substitution of a double bond would block epoxidation at such positions, is that 7-MeBP can be oxidized to give a 7,8-dihydrodiol. We compared, therefore, the microsomal oxidation of 7-MeBP with BP. The results are shown in Table 1.

TABLE 1:

HPLC RETENTION TIMES (MIN.) OF METABOLITES OF BP AND 7-METHYLBP

	9,10-diol	4,5-diol	7,8-diol	hydroxy methyl	quinones	phenols	hydrocarbons
BP	17	30	35	-	40-48	53-58	72
7-MeBP	22	39	42	53	56-62	70-74	80

Microsomal incubations were extracted (44) with ethylacetate, the solvent removed and the samples separated by HPLC (μBondapak C-18 column, at 50°C, 30-85% methanol in water linear gradient over 100 min.).

BP showed the expected pattern of metabolites with large quantities of 9,10-dihydrodiol and smaller amounts of the 4,5- and 7,8-dihydrodiols, quinones and phenols. 7-MeBP gave as major oxidation products 7-hydroxy-MeBP and 7-MeBP 9,10-dihydrodiol, plus smaller amounts of 7-MeBP 4,5-dihydrodiol. A minor component, which was quite strongly fluorescent, was identified as 7-MeBP 7,8-dihydrodiol. The compound had absorption maxima at 255, 282, 293, 332, 348, and 368nm, and major fluorescence emission maxima at 400 and 422nm with a shoulder at 435nm. This compound, upon treatment with acid, gave a product with a fluorescence spectrum that closely resembled 8-hydroxyBP. The formation of this 7,8-dihydrodiol is analogous to the recently reported formation of 8-methylbenz(a)anthracene 8,9-dihydrodiol from 8-methylbenz(a)anthracene (43). It appears, therefore, that oxidations of substituted double bonds, presumably via epoxide intermediates, may be a quite general phenomenon.

Since BP 7,8-dihydrodiol is formed by the cytochrome P_{450} complex with a high degree of optical purity, the 7R enantiomer predominating, and since this has important consequences on its further metabolism, we investigated the stereochemistry of the 7-MeBP 7,8-dihydrodiol. To resolve the enantiomers the dihydrodiol metabolite was converted to its menthoxyacetate derivatives (33) to yield diastereoisomeric esters which were separated by HPLC. We found that, in contrast to BP, the metabolism of 7-MeBP results in racemic 7,8-dihydrodiol.

Although we do not have direct proof for the structure of the adducts formed in the DNA of cells treated with 7-MeBP, the fluorescence spectrum of the modified DNA and our identification of 7-MeBP 7,8-dihydrodiol as a metabolite strongly suggest that the adducts are formed from a 'Bay region' epoxide. Thus great care must be taken in attempting to predict the metabolism and carcinogenicity of substituted aromatic hydrocarbons.

Studies with 7,12-Dimethylbenz(a)anthracene

Another carcinogen which we have investigated in terms of DNA adduct formation is 7,12-dimethylbenz(a)anthracene (DMBA) (40). There are two major reasons for our interest in this substituted polycyclic aromatic hydrocarbon. First, significant differences in carcinogenicity occur as a result of methyl substitution in the benz(a)anthracene (BA) nucleus (34). The two most critical sites on which methyl substitution enhances carcinogenicity of BA are the 7 and 12 positions. Second, there is the interesting phenomenon, which has not yet been explained on a molecular basis, of the high sensitivity of 50 day-old female Sprague-Dawley rats to induction of mammary tumors by DMBA (35).

DMBA is highly sensitive to both chemical and photochemical decomposition. For these reasons, it has been more difficult to study than BP. The adducts themselves are also light sensitive (36). Several investigators have so far failed to isolate and characterize dihydrodiol epoxides in the 'Bay region' although some progress is being made (45).

We have used, therefore, indirect approaches to obtain information on the structure of DMBA-DNA adducts. Previous experiments (37) have shown that DNA extracted from cells which were exposed to DMBA, had a characteristic fluorescence emission spectrum with a maximum at 450nm. Dimethylanthracene has a similar emission spectrum. The differences in wavelengths and fine structure between the spectrum of the hydrocarbon and that of the modified DNA may be accounted for by the fact that in the latter case the dimethylanthracene moiety is bound to DNA. Direct evidence for the involvement of DMBA 3,4-dihydrodiol has come from more recent experiments in which we have compared the DNA adducts isolated from human bronchus which had been exposed to ^{3}H-DMBA with those obtained from DNA incubated with DMBA 3,4-dihydrodiol in the presence of liver microsomal preparations or m-chloroperbenzoic acid. Chromatographic identity was established for the major adducts obtained from all three systems, by HPLC of the nucleoside adducts as well as their peracetates. The ultraviolet spectra of the adduct prepared from the dihydrodiol, showed a strong absorption at 266nm with weaker absorptions at 367, 398, and 408nm, and a broad fluorescence emission at 440nm. These data are consistent with further oxidation of the double bond at the 1,2-positions of 3,4-dihydrodiol DMBA, again suggesting that the adduct was derived from a 'Bay region' epoxide.

The major in vivo adduct showed a strong and complex CD spectrum. This indicates that, since only one major adduct was isolated from the m-chloroperbenzoic acid oxidation of racemic DMBA 3,4-dihydrodiol, the reaction of DMBA dihydrodiol epoxides with DNA is similar to that of the racemic BPDE, in that there is preferential covalent binding of one of the enantiomers with DNA. The field desorption mass spectrum of this adduct was very informative. The molecular ion of the adduct indicated that a dihydrodiol epoxide of DMBA had indeed reacted covalently with a guanine residue in DNA. The mass spectrum was however, unusual in that considerable fragmentation was seen including losses of 18 (H_2O), 28 (CO) and 117 (deoxyribose moiety). Analysis of this data supports the partial structure shown in Figure 6.

Thus, the adduct is derived from trans-3,4-dihydroxy-1,2-epoxy-1,2,3,4-tetrahydroDMBA. One enantiomer preferentially binds to DNA although the absolute stereochemistry of the compound remains to be determined. The modified base in DNA is guanine, however, the site on the base to which the dihydrodiol epoxide is attached has not been established with certainty. The N-7 position can, however, be excluded based on the stability of the adduct. It seems most likely that, by analogy with the BPDE-dG adduct, DMBA is linked to the 2-amino group of guanine. Further studies are in progress to completely characterize this compound. Its structure is of considerable interest because we have identified it in both human and rodent systems.

Conclusion

Although BP is perhaps the most extensively studied of the polycyclic aromatic hydrocarbons, we still lack an understanding of how this or any other carcinogen initiates the program of events leading to cancer. Many factors could limit the ability of a compound to express this potential. These may include exposure and absorption by the animal, tissue distribution, metabolism to the ultimate carcinogenic species, stability of the ultimate carcinogen to metabolic or chemical decomposition prior to its reaction with DNA, the ability to react with

DNA and the possibility of reaction with competing nucleophiles. Subsequently, the ability of the target cell to repair the DNA damage in an error free way, or the likelihood that the modified DNA will be impaired in its template activity or will induce error-prone repair or other host responses, may be critical. Subsequent exposure of the host to tumor promoters or other cofactors and the state of proliferation or differentiation of the target cells are also known to be important variables in terms of the likelihood of tumor formation. Much progress has been made in defining some of these variables. This complexity suggests that it is unlikely that any single biochemical or molecular calculation can be developed to accurately predict whether or not a compound is carcinogenic, or its relative potency as a carcinogen, in the absence of consideration of other factors.

Hopefully, future studies on the actual cellular mechanism(s) by which the covalent modification of DNA (and other cellular macromolecules) by chemical carcinogens initiate the carcinogenic process will provide further insight into the factors that determine carcinogenic activity and potency as well as tissue and species specificity.

Acknowledgements:

We should like to thank Ronald and Samuel Teichman for their important contributions in generating the computer graphic models of DNA-carcinogen adducts and to Patricia Vickman for her careful typing of the manuscript. This work was supported in part by NCI Grant CA 21111.

References:

1. *Jeffrey et al.: 1977, Nature 269, pp. 348-350.*
2. *Autrup, H., Harris, C.C., Trump, B.F. and Jeffrey, A.M.:1978, Cancer Res., 38, pp. 3689-3696.*
3. *Brown, H.S., Jeffrey, A.M. and Weinstein, I.B.: 1979, Cancer Res., 39, pp. 1673-1677.*
4. *Autrup et al.: 1980, Int. J. Cancer, 25, pp. 293-300.*
5. *Harris, C.C. et. al.: 1979, Cancer Res., 39, pp. 4401-4406.*
6. *Ivanovic, V., Geacintov, N.E. and Weinstein, I.B.: 1976, Biochem. Biophys. Res. Commun. 70, pp. 1172-1179.*
7. *Borgen, A., Darvey, H., Castagnoli, N., Crocker, T.T., Rasmussen, R.E., and Wang, I.Y.: 1973, J. Med. Chem. 16, pp. 502-505.*
8. *McCaustland, D.J. et al.: 1976, In: Carcinogenesis 1, (Freudenthal, R. and Jones, P.W. eds.), Raven Press, New York, pp. 349-429.*
9. *Harvey, R.G., and Sukumaran, K.B.: 1977, Tet. Lett., pp. 2387-2390.*
10. *Sims, P., Grover, P.L., Swaisland, A., Pal, K., and Hewer, A.: 1974, Nature 252, 326-328.*
11. *Jeffrey, A.M., Grzeskowiak, K, Weinstein, I.B., Nakanishi, K., Roller, P. and Harvey, R.G.: 1979, Science 206, pp. 1309-1311.*
12. *Croy, R.G., Essigmann, J.M., Reinhold, V.N. and Wogan, G. N.: 1978, Proc. Natl. Acad. Sci., USA, 75, pp. 1745-1749.*
13. *Weinstein et al.: 1976, Science 193, pp. 592-595.*
14. *Prusik, T., Geacintov, N.E., Tobiasz, C., Ivanovic, V. and Weinstein,*

I.B.: 1979, Photochem. and Photobiol. 29, 223-232.
15. Grunberger, D. and Weinstein, I.B.: 1979, In: Chemical Carcinogens and DNA, (P.L. Grover, ed.), CRC Press, Boca Raton, Fla. pp. 59-93.
16. Geacintov, N.E., Gagliano, A., Ivanovic, V. and Weinstein, I.B.: 1978, Biochemistry, 17, pp. 5256-5262.
17. Jennette, K.W. et al.: 1977, Biochemistry 16, pp. 932-938.
18. Osborne, M.R., Beland, F.A., Harvey, R.G. and Brookes, P.: 1976, Intl. J. Cancer 18, pp. 362-368.
19. Fuchs, R. and Daune, M.: 1972, Biochemistry 11, pp. 2656-2666.
20. Nelson, J.H., Grunberger, D., Cantor, C.R. and Weinstein, I.B.: 1971, J. Mol. Biol. 62, pp. 331-346.
21. Santella, R.M., Kriek, E. and Grunberger, D.: 1980, Proc. Am. Assoc. Cancer Res., 21, pp. 70.
22. Fuchs, R. and Daune, M.: 1973, FEBS Lett. 34, pp. 295-298.
23. Wang, A.H., Quigley, G.J., Kolpak, F.J., Cranford, J.L., Van Boom, J.A., Van der Macel, G., and Rich, A.: 1979, Nature 282, pp.680-686.
24. Arnott, S., Chandrasekaron, R., Birdsoll, D.L., Leslie, A.G.W. and Ratliff, R.L.: 1980, Nature 283, 743-745.
25. Pohl, F.M. and Jovin, T.M.: 1972, J. Mol. Biol. 67, pp. 374-396.
26. Pohl., F.M.: 1976, Nature 260, pp. 365-366.
27. Patel, D.J., Canuel, L.L., and Pohl, F.M.: 1979, Proc. Natl. Acad. Sci., USA 74, pp. 2508-2511.
28. Jerina, D.M. et al.: 1976, In: In Vitro Metabolic Activation in Mutagenesis Testing, (J.Fouts, J.R. Bend and R.M. Philipot, eds.), Elsevier North Holland Press, Amsterdam, pp. 159-177.
29. Shipman, L.L.: 1978, In: Carcinogenesis, (Jones P.W. and R. Freudenthal, eds.), Raven Press, N.Y., 3, pp. 139-144.
30. Harvey, R.G. and Dunne, F.B.: 1978, Nature 273, pp. 566-568.
31. Slaga, T.J. et al.: 1978, In: Carcinogenesis, (P.W. Jones and R. Freudenthal, eds.), Raven Press, New York Volume 3, pp. 371-382.
32. Jerstrom, B. et al.: 1978, Cancer Res. 38, pp. 2600-2607.
33. Nakanishi, K. et al.: 1977, J. Am. Chem. Soc. 99, pp. 258-260.
34. Newman, M.S.: 1976: In: Carcinogenesis, (P.W. Jones and R. Freudenthal, eds.), Raven Press, New York, 1, pp. 203-207.
35. Huggins, C., Grand, L.C. and Brillantes, F.P.: 1961, Nature 189, pp. 204-207.
36. Baird, W.M. and Dipple, A.: 1977, Intl. J. Cancer 20, pp. 427-431.
37. Ivanovic, V., Geacintov, N., Jeffrey, A.M., Fu, P.P., Harvey, R.G., and Weinstein, I.B.: 1978, Cancer Lett. 4, pp. 131-140.
38. Meehan, T. and Straub, K.: 1979, Nature 277, pp. 410-412.
39. Westra, J.G., Kriek, E., and Hittenhausen, H.: 1976, Chem. Biol. Interactions 15, pp. 146-164.
40. Jeffrey, A.M., Weinstein, I.B. and Harvey, R.G.: 1979, Proc. Am. Assoc. Cancer Res. 20, p. 181.
41. Sage, E., Spodheim-Maurizot, M., Rio, P., Leng, M. and Fuchs, R.P.P.: 1979, FEBS Lett. 108, pp. 66-68.
42. Yamasaki, H., Pulkrabek, P., Grunberger, D. and Weinstein, I.B.: 1977, Cancer Res. 37, pp. 3756-3760.
43. Yang, S.K., Chou, M.W., Weims, H.B. and Fu, P.P.: 1979, Biochem. Biophys. Res. Commun. 90, pp. 1136-1141.

44. *Pietropaolo, C. and Weinstein, I.B.: 1975, Cancer Res. 35, pp. 2191-2198.*
45. *Cooper, S. et al.: 1980, Chem.-Biol. Interactions 29, pp. 357-367.*
46. *Schurch, O. and Winsterstein, A.: 1935, Ztschr. physiol. Chem. 236, pp. 79-91.*

Figure 1: Structure of (I) BPDE-deoxyguanosine and (II) BPDE-deoxyadenosine adducts. See refs. 1 and 11.

Figure 2: A stereoscopic view of the adduct formed between BPDE and B-DNA is shown. In this model the conformation of the guanine and the coplananity of the N^2 amino group and the base to which the BPDE is attached are retained. The 7- and 8-hydroxyl groups and the 9- and 10-hydrogens of BPDE are quasiequatorial. The angle between the plane of the pyrene and the axis of the DNA is 28°. This results in some slight steric hindrance, based on van der Waals radii which is presumably relieved by slight distortions of the helix. (To view this image in stereo two lenses of about 20 cm focal length should be mounted about 14 cm above the images and 6.5 cm apart. Viewers of this type are sold by Taylor Merchant Corp., 24 West 45th Street, New York, New York, 10036 under the name of Stereoptican 707).

Figure 3: The "base displacement" model of AAF-DNA adducts. The guanine to which the AAF is attached has been rotated out of the helix and the AAF moiety is inserted into the helix and stacked with the bases above and below. AC-designates acetyl group of AAF. The cytosine (marked C) residue on the opposite strand would overlap with the AAF residue, therefore it has been removed and the 3' and 4' carbon atoms of the corresponding deoxyribose in the DNA back bone are indicated. In reality, this C probably rotates out from the helix to accomodate the AAF; the exact conformation is not known, although there is evidence that this region of DNA is "single-stranded". See refs. 15,19,20,22,41.

Figure 4: Circular dichroism spectra of poly d(G-C) and poly d(G-C) modified with N-acetoxy-AAF. Poly d(G-C) in 1 mM phosphate buffer (———), poly d(G-C) in 60% ethanol (-----), and poly d(G-C) modified with AAF to an extent of 28% (———). Sample concentrations were about $1x10^{-4}M$.

Figure 5: 7-Methylbenzo(a)pyrene 7,8-dihydrodiol. Only the 7,8,9,10-ring is depicted.

Figure 6: The partial structure of the major adduct formed in vivo between DMBA and DNA. Evidence supports a 3,4-dihydrodiol intermediate which undergoes epoxidation at the 1,2 double bond. This epoxide reacts with guanine bases in DNA, probably at the N^2 amino group (-NH) to give an adduct linked through the 1 or 2 positions. The other position is substituted with a hydroxyl group derived from the original oxirane ring.

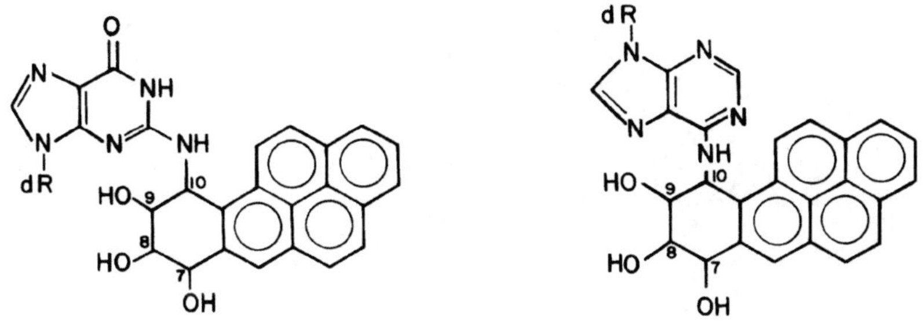

Figure 1.

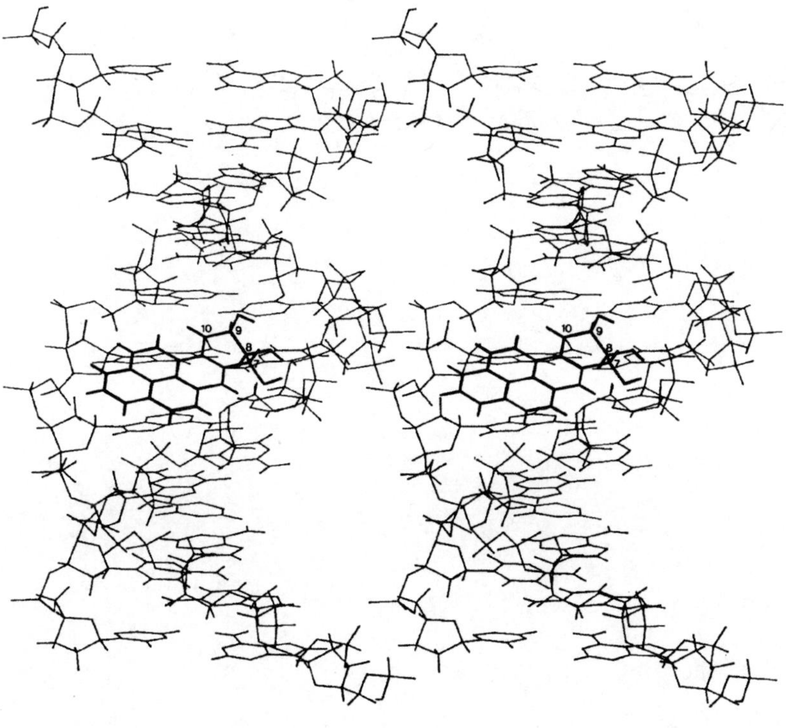

Figure 2.

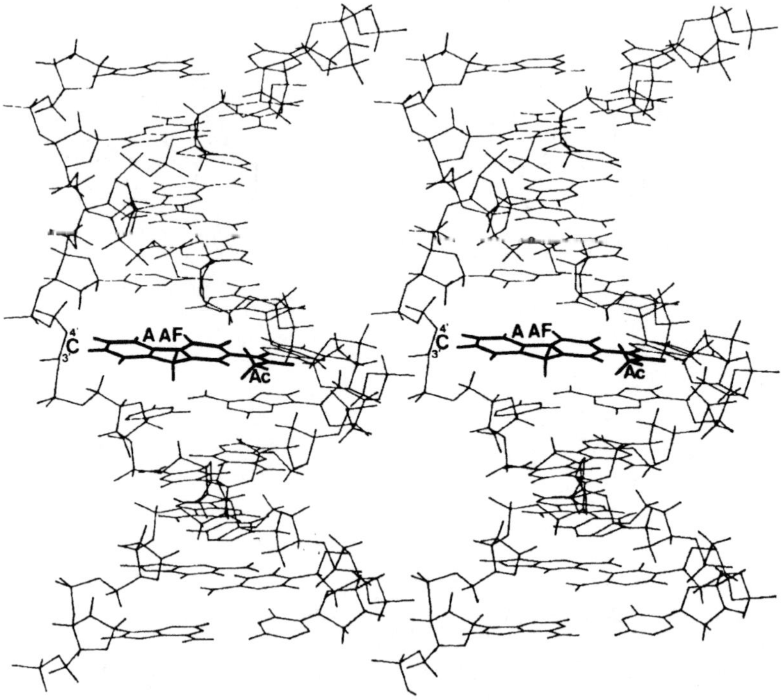

Figure 3.

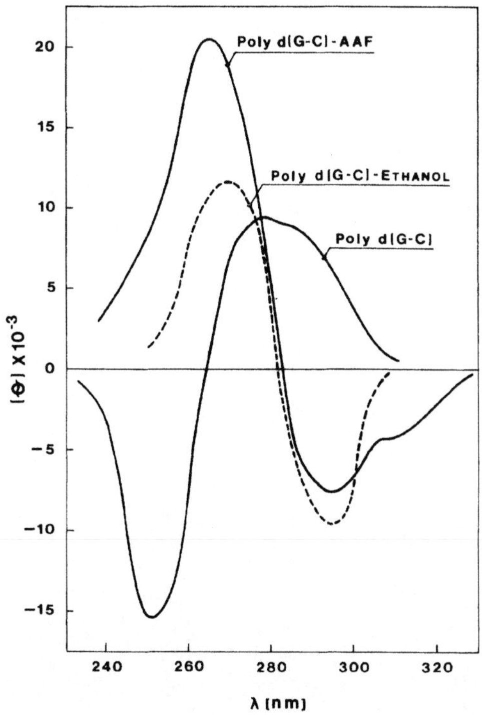

Figure 4.

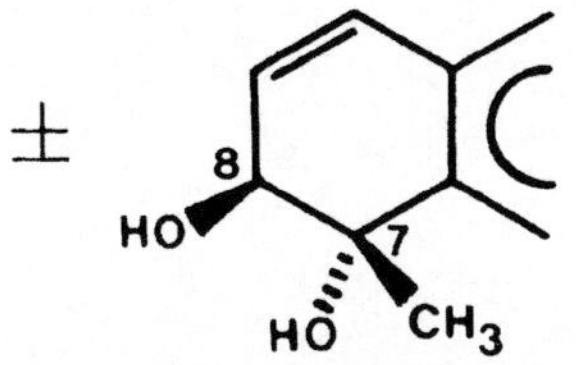

Figure 5.

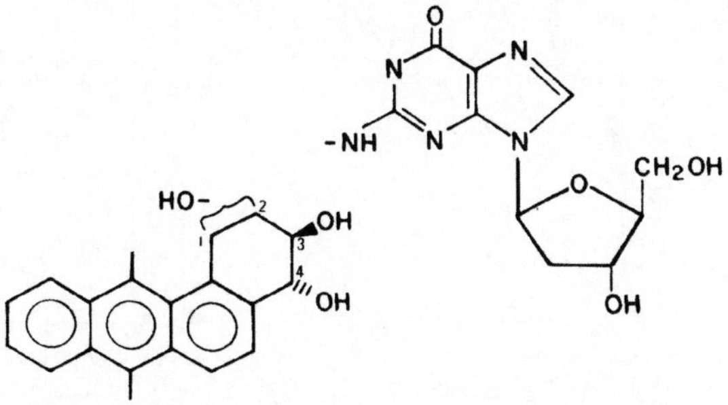

Figure 6.

INDEX OF SUBJECTS

GPSR Compliance

The European Union's (EU) General Product Safety Regulation (GPSR) is a set of rules that requires consumer products to be safe and our obligations to ensure this.

If you have any concerns about our products, you can contact us on ProductSafety@springernature.com

In case Publisher is established outside the EU, the EU authorized representative is:

Springer Nature Customer Service Center GmbH
Europaplatz 3
69115 Heidelberg, Germany

Batch number: 10815974

Printed by Printforce, the Netherlands